Fundamentals of Engineering Tribology with Applications

Tribology is the study of the principles of friction, wear and lubrication of machine elements. As a branch of mechanical engineering and materials science, tribology deals with the design of fluid containment systems like seals and gaskets, and lubrication of surfaces in relative motion. The study of tribology helps in better understanding of design and maintenance of machine elements such as bearings, gears, cam-followers, hard disk drives, seals, pumps, compressors, etc. In order to calculate the load support required by bearings, there is a need to study the structure and nature of fluid flow.

This book discusses the theories and applications of hydrodynamic thrust bearings, gas (air) lubricated bearings and elasto-hydrodynamic lubrication in detail. Explaining the concepts of friction including coefficient of friction, friction instability and stick-slip motion, the book also clarifies the delusion that harder and cleaner surfaces produce better results in wear. The recent developments including Online Condition Monitoring (an integration of moisture sensor, wear debris and oil quality sensors) and Multigrid Technique are also presented in the book. In addition, it provides design problems and their real-life applications for cams, followers, gears and bearings. MATLAB programs, frequently asked and multiple choice questions are interspersed throughout the book for better understanding.

Harish Hirani is Professor, Department of Mechanical Engineering, Indian Institute of Technology, Delhi. He has more than 16 years of teaching and research experience and has published nearly 45 papers in national and international journals. He received his Ph.D. from Indian Institute of Technology, Delhi. He was a visiting researcher at Massachusetts Institute of Technology, USA (2003–2004). He has been teaching courses on lubrication, machine element design, graphics science, introduction to engineering visualization and MR fluid at both undergraduate and postgraduate levels. His areas of interest are semi-active magneto rheological bearing, rolling element bearing, hydrodynamic bearing and seals.

Fundamentals of Engineering Tribology with Applications

Harish Hirani

CAMBRIDGE
UNIVERSITY PRESS

4843/24, 2nd Floor, Ansari Road, Daryaganj, Delhi 110002, India

Cambridge University Press is part of the University of Cambridge.

It furthers the University's mission by disseminating knowledge in the pursuit of education, learning and research at the highest international levels of excellence.

www.cambridge.org
Information on this title: www.cambridge.org/9781107063877

© Harish Hirani 2016

This publication is in copyright. Subject to statutory exception and to the provisions of relevant collective licensing agreements, no reproduction of any part may take place without the written permission of Cambridge University Press.

First published 2016

Printed in India by Thomson Press India Ltd., New Delhi 110001

A catalogue record for this publication is available from the British Library

Library of Congress Cataloging-in-Publication Data
Hirani, Harish.
 Fundamentals of engineering tribology with applications / Harish Hirani.
    pages cm
 Includes bibliographical references and index.
 Summary: "Presents explanation on the theories and applications of hydrodynamic thrust bearing, gas (air) lubricated bearing and elasto-hydrodynamic lubrication"-- Provided by publisher.
 ISBN 978-1-107-06387-7 (hardback)
1. Tribology. I. Title.
 TJ1075.H57 2015
 621.8'9--dc23
 2015012814

ISBN 978-1-107-06387-7 Hardback

Cambridge University Press has no responsibility for the persistence or accuracy of URLs for external or third-party internet websites referred to in this publication, and does not guarantee that any content on such websites is, or will remain, accurate or appropriate.

*To my mother, Sitadevi
who sacrificed her sleep and happiness to make me perfect in my work.
She was the best friend that I ever had.*

Contents

List of Figures — xii
List of Tables — xxi
Preface — xxiii
Acknowledgments — xxiv
Nomenclature — xxv

1. **Introduction** — 1
   1.1 Defining Tribology — 1
      1.1.1 What is tribology? — 1
      1.1.2 Need of tribology as a separate subject — 1
      1.1.3 History of Tribology — 3
   1.2 Tribology in Design — 4
      1.2.1 Mechanical design of seal and gasket — 4
      1.2.2 Tribological design of seals and gasket — 9
   1.3 Tribology in Industry (Maintenance) — 10
      1.3.1 Example: seal — 10
      1.3.2 Example: cam — 11
      1.3.3 Example: journal bearings — 11
      1.3.4 Example: magnetic bearing — 12
      1.3.5 Example: multi-row roller bearing — 12
      1.3.6 Example: gear — 13
   1.4 Defining Lubrication — 13
      1.4.1 Examples — 13
      1.4.2 Applications of lubricant — 14
      1.4.3 What is expected from a lubricant? — 15
   1.5 Basic Modes of Lubrication — 17
      1.5.1 Thick and thin lubrications — 17
      1.5.2 Lubrication mechanisms — 17
   1.6 Properties of Lubricants — 18
   1.7 Types of Lubricants — 19
      1.7.1 Solid lubricants — 20
      1.7.2 Semi-solid lubricant — 26
      1.7.3 Liquid lubricants — 29
      1.7.4 Gaseous lubricants — 34

viii Contents

1.8	Lubricant Additives	34
	1.8.1 Need of Additives	34
	1.8.2 Types of additives	35
	1.8.3 Interference between additives	38
1.9	Lubrication Selection	38
1.10	Defining Bearing Terminology	39
	1.10.1 Comparison between sliding and rolling contact bearings	40
	1.10.2 Rolling contact bearings	40
	1.10.3 Sliding contact bearings	41
Frequently Asked Questions		43
Multiple Choice Questions		47
Answers		50
References		51

2. Friction, Wear and Boundary Lubrication — 52

2.1	Friction	52
	2.1.1 Classification of friction	53
	2.1.2 Laws of friction	54
	2.1.3 Causes of dry friction	56
2.2	Theories of Dry Friction	57
2.3	Friction Measurement	65
2.4	Stick–Slip Motion and Friction Instabilities	68
2.5	Wear	72
	2.5.1 Classification of wear	74
	2.5.2 Factors affecting wear	98
2.6	Theories of Wear	99
2.7	Approaches to Friction Control and Wear Prevention	102
2.8	Boundary Lubrication	103
Frequently Asked Questions		110
Multiple Choice Questions		115
Answers		118
References		118

3. Lubrication of Bearings — 120

3.1	Mechanics of Fluid Flow	121
	3.1.1 Theory of hydrodynamic lubrication	122
	3.1.2 Lubricant Viscosity	124
	3.1.3 Mechanism of pressure development in lubricant film	132
3.2	Reynolds' Equation and its Limitations	134
3.3	Idealized Bearings	138
	3.3.1 Infinitely long plane fixed sliders	138
	3.3.2 Infinitely long plane pivoted sliders	143
	3.3.3 Infinitely long journal bearings	144
	3.3.4 Infinitely short journal bearings	151
3.4	Journal Bearings	153
	3.4.1 Locating journal position	156
	3.4.2 Lubricant supply in bearing	156
	3.4.3 Design of journal bearings	158

	Frequently Asked Questions		178
	Multiple Choice Questions		180
	Answers		183
	References		183
	Program Listing in MATLAB for Problem 3		184

4. Hydrodynamic Thrust Bearing — 187
- 4.1 Introduction — 187
- 4.2 Pressure Distribution — 188
  - 4.2.1 Fixed pad thrust bearing — 198
  - 4.2.2 Tilting pad thrust bearing — 200
- 4.3 Load — 201
- 4.4 Centre of Pressure — 201
- 4.5 Friction — 202
- Frequently Asked Questions — 203
- Multiple Choice Questions — 207
- Answers — 208
- References — 208

5. Hydrostatic and Squeeze Film Lubrication — 209
- 5.1 Hydrostatic Lubrication — 209
  - 5.1.1 Basic concept — 209
  - 5.1.2 Advantages and limitations — 210
  - 5.1.3 Viscous flow through rectangular slot — 210
  - 5.1.4 Types and configurations — 213
  - 5.1.5 Circular step thrust bearing — 215
  - 5.1.6 Rectangular thrust bearing — 219
  - 5.1.7 Hydrostatic journal bearing — 222
  - 5.1.8 Energy losses — 226
- 5.2 Squeeze Film Lubrication — 227
  - 5.2.1 Basic concept — 228
  - 5.2.2 Squeeze action between circular flat plates — 229
  - 5.2.3 Squeeze action between rectangular plates — 231
  - 5.2.4 Squeeze action under variable and alternating loads — 232
  - 5.2.5 Application to journal bearings — 234
- 5.3 Engine Bearing Lubrication — 237
  - 5.3.1 Oil flow — 240
  - 5.3.2 Power loss — 241
  - 5.3.3 Temperature rise — 241
  - 5.3.4 Design procedure — 241
  - 5.3.5 Case studies — 243
- Frequently Asked Questions — 248
- Multiple Choice Questions — 256
- Answers — 259
- References — 259

6. Elasto–Hydrodynamic Lubrication — 261
- 6.1 Principles and Applications — 261
- 6.2 Hertz Theory — 263

6.3	Pressure–Viscosity Term in Reynolds' Equation	269
6.4	Ertel–Grubin Equation	274
6.5	Numerical Method for Determining Oil Film Thickness in Elasto–Hydrodynamic Lubrication	278
6.6	Rolling Element Bearings	281
6.7	EHL of Gear–Teeth Contact	284
	Frequently Asked Questions	288
	References	290
	Program Listing in MATLAB for Figure 6.3.2–6.3.4	290
	Program Listing in MATLAB for Figure 6.5.2–6.5.3	293

7. Gas (Air) Lubricated Bearings — 297

7.1	Introduction	297
7.2	Merits, Demerits and Applications	300
7.3	Aerodynamic Bearings	301
	7.3.1 Pad bearings	301
	7.3.2 Cylindrical bearings	306
	7.3.3 Magnetic recording discs with flying head	307
7.4	Aerostatic Bearings	309
	7.4.1 Flow through restrictors	311
	7.4.2 Radial aerostatic bearings	313
	7.4.3 Thrust aerostatic bearings	316
	Frequently Asked Questions	319
	Multiple Choice Questions	321
	Answers	323
	References	323
	Program Listing in MATLAB for Figure 7.3.7	324

8. Mixed Lubrication — 326

8.1	Introduction	326
8.2	Surface Topography	330
8.3	Characterization of Surface	337
8.4	Boundary Lubrication	338
8.5	Effect of Surface Topography on Mixed Lubrication	350
8.6	Asperity Temperatures in Mixed Film Lubrication	350
8.7	Tribological Performance of Bearing Operating in Mixed Lubrication Regime	351
	Frequently Asked Questions	357
	Multiple Choice Questions	358
	Answers	360
	References	360
	Program Listing in MATLAB for Figure 8.39	361

9. Tribological Aspects of Rolling Motion — 371

9.1	Rolling Element Bearings	371
	9.1.1 Bearing terminology	371
	9.1.2 Classification of rolling bearings	374
	9.1.3 Load capacity	376
	9.1.4 Standardization	382

	9.1.5 Tribology of rolling bearings	383
	9.1.6 Case study: failure analysis of four row cylindrical roller bearing	386
9.2	The Mechanics of Tyre–Road Interactions	392
Frequently Asked Questions		394
Multiple Choice Questions		396
Answers		398
References		398

10. Tribological Aspects of Gears — 400
10.1 Spur Gears — 401
10.2 Friction and Wear of Spur Gears — 401
10.3 Contact Stresses — 407
10.4 Lubrication of Spur Gears — 410
10.5 Surface Failures — 412
10.6 Offline Monitoring of Gears — 412
    10.6.1 Offline condition monitoring – a case study — 415
10.7 Online Monitoring of Gears — 418
    10.7.1 Online condition monitoring – a case study — 419
Frequently Asked Questions — 421
Multiple Choice Questions — 425
Answers — 426
References — 427

Index — 429

List of Figures

1.1.2.1	Solid mechanics	2
1.1.2.2	Fluid mechanics and chemistry	2
1.1.2.3	Material science	3
1.2.1	Oil seal	5
1.2.2	(a) Circular 4 bolt gasket, (b) Non-circular multi-bolt gasket, (c) Circular multi-bolt gasket and (d) Gasket assembly	6
1.2.3	Assembled views of different gasket joints	6
1.2.4	Cylinder head gasket	7
1.2.5	Gasket assembly	9
1.3.1	Carbon graphite seal	11
1.3.2	Pitting of cam surface	11
1.3.3	(a) Abrasive wear, (b) Rubbing wear	12
1.3.4	Wear scar due to edge loading	12
1.3.5	Failure of large size roller bearing	12
1.3.6	Pit on gear surface	13
1.4.1.1	Simple lock and key	13
1.4.1.2	(a) Window lifting mechanism, (b) Pendulum clock	14
1.4.2	Newtonian and non-Newtonian fluids	15
1.4.3.1	Lubricant between cylinder liner and rings	15
1.4.3.2	Hydrodynamic pressure profile in radial journal bearing	16
1.4.3.3	Lubrication in bone joint	16
1.5.1	(a) Thick lubrication, (b) Thin lubrication	17
1.5.2	Comparison between average and root mean square roughness	18
1.5.3	Transition in lubrication regime in the presence of foreign particles or wear debris	18
1.7	Molecular state of lubricants	19
1.7.1.1	Eccentric cam	20
1.7.1.2	Crystal structure of molybdenum disulphide [Lansdown, 1999]	22
1.7.1.3	Carbon transfer layer on stainless steel	23
1.7.1.4	Carbon graphite seal	23
1.7.1.5	Structure of graphite	24
1.7.1.6	Perfect and distorted structure of graphite	24
1.7.1.7	Graphite seal submerged in water	25
1.7.1.8	Graphite seal in water environment	25

1.7.2.1	Semi-solid lubricant	26
1.7.2.2	Irregularities filled by grease	28
1.7.2.3	Cone arrangement to measure consistency	29
1.7.3.1	Stribeck curve	30
1.7.3.2	Chemical forms of mineral oils	31
1.7.3.3	VI improvers	32
1.7.3.4	VI improvers in action	32
1.7.3.5	Viscosity model of multi-grade oil (at high temperature)	33
1.8.1.1	High contact pressure in gears	35
1.8.2.2	Detergent additive action	36
1.8.2.3	Detergent additives in action	36
1.8.2.4	Formation of air bubbles in lubricant	37
1.8.2.5	Rust prevention additives	37
1.9.1	Lubricant selection	38
1.10.1	Radial and thrust bearing	39
1.10.2	Sliding contact bearing	39
1.10.3	Roller contact bearing	40
1.10.2.1	Types of rolling contact bearing	41
1.10.3.1	Hydrodynamic lubrication (a) Journal at rest, (b) Journal starts to rotate, (c) Journal at full speed	42
1.10.3.2	Pressure distribution in hydrodynamic bearing	42
1.10.3.3	Pressure distribution in hydrodynamic bearing	42
2.1.1	Coefficient of friction for various metals	53
2.1.2	Adhesive friction among various materials	53
2.1.3	Difference between the static and kinetic friction may initiate 'stick–slip' [Stachowiak, 2006]	54
2.1.4	Amontons' work	55
2.1.5	Hysteresis loss	56
2.1.6	Adhesion	57
2.1.7	Abrasion (deformation)	57
2.2.1	Coulomb friction model	57
2.2.2	Cold welding in steel and indium	58
2.2.3	Carbon graphite and stainless steel	58
2.2.4	Adhesion theory	58
2.2.5	Friction coefficients for various material pairs	59
2.2.6	Deformation theory	60
2.2.7	Two contacting surfaces	61
2.2.8	Surface contamination	64
2.2.9	Variation in Coefficient of friction due to surface contamination	64
2.2.10	Sliding friction vs. time	65
2.3.1	Inclined plane tribometer	65
2.3.2	Sled tribometer	66
2.3.3	Sliding friction measurement (pin on disk tribometer)	66
2.3.4	Measurement using curved contact surfaces [Blau, 2008]	67
2.4.1	Force required to overcome static friction	68

2.4.2	Stick–slip motion	68
2.4.3	Friction performance of MR (Magneto Rheological) brake	69
2.4.4	Stick–slip	69
2.4.5	Variation of vibration parameters	69
2.4.7	Stiction case	70
2.4.8	Negative and gradient case	70
2.4.9	Positive damping	71
2.4.10	Negative damping	71
2.4.11	Friction instability	72
2.5.1	(a) Zero wear of helical gear	73
2.5.1	(b) Measurable wear of helical gear	73
2.5.2	Formation of pit	73
2.5.3	Worn out rollers	74
2.5.4	Abrasion marks on bearing bore	74
2.5.5	Effect of clearance on load	74
2.5.1.1	Abrasive wear	75
2.5.1.2	2-body abrasion	76
2.5.1.3	3-body abrasion	76
2.5.1.4	Effect of microstructure on 3-body abrasive wear	76
2.5.1.5	M R Particles	77
2.5.6	Adhesive wear	78
2.5.7	Location of shear plane affect wear rate	78
2.5.8	Scoring	79
2.5.9	Contaminant layers on metal surface	79
2.5.10	Steps leading to adhesive wear	80
2.5.11	Wear transition	80
2.5.12	Archard's wear model	81
2.5.13	Pin-on-disk arrangement	83
2.5.14	Mild wear	84
2.5.15	Debris in severe wear	85
2.5.16	Seizure of rolling elements	85
2.5.17	Wear–mechanism map [Lim, 1998]	86
2.5.18	Jaw coupling	86
2.5.19	Passivation of corrosion	87
2.5.20	Continuous corrosion	87
2.5.21	Impingement angle vs. wear rate	88
2.5.22	Pneumatic transportation	88
2.5.23	Helicopter engine	89
2.5.24	Modified engine	89
2.5.25	Fatigue wear	89
2.5.26	Fatigue wear during sliding [Rigney, 1979]	90
2.5.27	Fatigue wear [Kimura, 1983]	90
2.5.28	Cracking	91
2.5.29	Fretting wear	91
2.5.30	Process of fretting wear	92
2.5.31	Cam wear	92

2.5.32	Sketch to illustrate pitting	93
2.5.33	Normal load vs. cam angle	93
2.5.34	Pressure angle vs. cam angle	93
2.5.35	Cam–follower interface at various angular positions	94
2.5.36	Cam–follower mechanism	95
2.5.37	Effect of tangential force	95
2.6.1	Archard's wear model	101
2.8.1	Boundary lubricants 'oiliness additives' [Stachowiak, 2006]	103
2.8.2	Long chain boundary additives	104
2.8.3	Number of layers vs. friction coefficient	104
2.8.4	Physical adsorption of solid additives on boundary surface	104
2.8.5	Temperature vs. μ [Bowden and Tabor, 1950]	105
2.8.6	Effect of temperature on adsorption [Stachowiak, 2006]	105
2.8.7	Temperature gap	106
2.8.8	Combination of physisorption and chemisorption of effective lubrication	106
2.8.9	Comparative study among (1) dry, (2) boundary and (3) hydrodynamic lubrication mechanisms	107
2.8.10	Online monitoring of spur gears	107
2.8.11	Online sensor instrument	108
3.1	Various concepts to separate two solid surfaces	120
3.1.1	(a) Converging wedge shape geometry	121
3.1.1	(b) Squeeze lubrication	122
3.1.1	(c) Hydrostatic lubrication	122
3.1.1.1	Hydrodynamic lubrication	123
3.1.1.2	Shearing of lubricant	123
3.1.1.3	(a) Sketch of Tower's test setup	123
3.1.1.3	(b) Sketch of Tower's test setup with hole	123
3.1.1.4	Journal bearing	124
3.1.1.5	(a) Convergent wedge of liquid	124
3.1.1.5	(b) Lubricant shearing	124
3.1.2.1	Viscosity comparison	125
3.1.2.4	Variation in oil viscosity with temperature	126
3.1.2.5	Stribeck diagram shows viscosity index of few commonly used lubricating oils	127
3.1.2.6	Viscosity index	128
3.1.2.2.1	Viscosity–shear rate relationship	130
3.1.2.2.2	Shear thinning effect of multi-grade oils	130
3.1.3.1	Lubrication between parallel plates	132
3.1.3.2	Lubrication between inclined plates	133
3.1.3.3	Positive pressure gradient at exit and negative pressure gradient at entrance	133
3.1.3.4	Formation of converging wedge	133
3.2.1	Unit volume of hydrodynamic film	134
3.2.2	Fluid element subjected to pressure and viscous forces	134
3.2.3	Pressure profile between inclined plates	136
3.3.1	Slider bearing	138
3.3.2	A long fixed slider	139

3.3.3	Oil wedge in a long hydrodynamic slider bearing	139
3.3.4	Values of dimensionless load plotted against aspect ratio (m)	141
3.3.5	Values of coefficient of friction plotted against aspect ratio (m)	143
3.3.6	Analysis of pivoted slider bearing	144
3.3.7	Infinitely long journal bearing	144
3.3.8	Coordinate system and force components in a journal bearing	145
3.3.9	Pressure distribution for full Sommerfeld solution in a journal bearing	147
3.3.10	Pressure distribution using half-Sommerfeld boundary condition	149
3.3.11	Pressure distribution in a journal bearing using Reynolds' boundary conditions	150
3.3.12	Infinitely short journal bearing	151
3.3.13	Boundary conditions for evaluating pressure distribution	151
3.3.14	Load ratio vs. Eccentricity ratio	153
3.4.1	Long journal bearings	154
3.4.2	Acrylic journal bearing with temperature measuring copper rivets	154
3.4.3	Journal bearing with cavitation	154
3.4.4	Recirculation of lubricant indicating side leakage and cavitation	154
3.4.5	Locations of journal in bearing	156
3.4.6	Hole and groove arrangement in bearing	157
3.4.7	Partial oil groove	157
3.4.8	Journal rotating inside a bearing	158
3.4.9	Viscosity–temperature chart	162
3.4.10	Temperature rise variable [Raimondi and Boyd, 1958] Chart 6	162
3.4.11	Minimum film thickness parameter [Raimondi and Boyd, 1958] Chart 1	164
3.4.12	Position of minimum film thickness [Raimondi and Boyd, 1958] Chart 2	164
3.4.13	Coefficient of friction variable [Raimondi and Boyd, 1958] Chart 3	165
3.4.14	Flow variable [Raimondi and Boyd, 1958] Chart 4	165
3.4.15	Flow ratio [Raimondi and Boyd, 1958] Chart 5	166
3.4.16	Maximum film pressure ratio [Raimondi and Boyd, 1958] Chart 7	167
3.4.17	Position of maximum film pressure [Raimondi and Boyd, 1958] Chart 8	167
3.4.18	Terminating position of film [Raimondi and Boyd, 1958] Chart 9	168
3.4.19	Long static bearing	173
3.4.20	Short static bearing	173
4.1	Thrust bearing supporting vertical shaft	187
4.2	Discretization of surface	189
4.3	Thrust pad of 10 × 100 mm dimensions	194
4.4	Pressure profile at different nodes of the bearing	194
4.5	Thrust pad of 100 × 10 mm dimensions	196
4.6	Non-dimensional pressure profile at different nodes in X- and Z- directions	197
4.7	Geometry of sector pad bearing	197
4.8	Sector pad mesh definition for finite difference scheme	198
4.9	Pivoted slider bearing	201
4.10	Thrust Pad Bearing for Q3.	203
4.11	Pivoted Thrust Pad Bearing for Q4.	203
4.12	Oil flow in a Thrust Pad Bearing for Q5.	204
4.13	Geometry of the sector shaped pad	204

4.14	Relaxation factor vs number of iterations	205
4.15	Geometry of Rayleigh step bearing for Q10.	206
4.16	Geometry of a thrust bearing for Q11.	206
4.17	Fixed pad thrust bearing for Q 5.	207
4.18	Geometry of taper slider bearing for Q6.	208
5.1.1	Hydrostatic lubrication	209
5.1.3.1	Small element of oil film is extruded by pressure difference acting upon it	211
5.1.3.2	Parabolic velocity profile across slot	212
5.1.4.1	Types of hydrostatic bearing	213
5.1.4.2	Pressurized oil in recess directly by pump or through restrictor	214
5.1.4.3	Compensation devices	214
5.1.4.4	Annular thrust bearings	215
5.1.5.1	Flat plate thrust bearing without a compensating element	216
5.1.5.2	Pressure distribution	217
5.1.6.1	Geometry of rectangular hydrostatic pad bearing	219
5.1.6.2	Pressure distribution of rectangular hydrostatic pad bearing	221
5.1.7.1	Hydrostatic lift	222
5.1.7.2	Geometry of multi-recess hydrostatic journal bearing	223
5.1.7.3	Four recess hydrostatic journal bearing	224
5.1.7.4	A loaded hydrostatic journal bearing	224
5.2	Squeeze film engine bearings	228
5.2.1	Squeeze lubrication	228
5.2.2.1	Representation between squeeze film between disk and flat plate	229
5.2.3.1	Squeeze lubrication under variable load	231
5.2.4.1	Representation for calculating oil–cushion effect in a slipper type of bearing	233
5.2.4.2	Load as a function of angles (time)	233
5.2.4.3	Variation in gas pressure as a function of crank angle (time)	233
5.2.5.1	Journal bearing with squeeze film action	234
5.3.1	Schematic of engine bearing	238
5.3.2	Computational scheme for bearing analysis using pressure model	243
5.4.1	Magellan telescope (http://www.gmto.org)	249
5.4.2	Mile High stadium in Denver	249
5.4.3	Schematic representation of hydrostatic lift	250
5.4.4	Schematic representation of multi recess system with orifice/capillary compensation	251
5.4.5	Hydrostatic bearing with moment producing load	252
5.4.6	Hydrostatic bearing with multiple compensating elements where one pump supplies lubricant to several bearings	252
5.4.7	Comparison of stiffness factor for constant flow rate, orifice and capillary compensated bearings	253
5.4.8	Capillary compensating device	253
5.4.9	Orifice compensating device	254
5.4.10	Constant flow valve compensating device	254
5.4.11	Adjustable length capillary compensating device	254
6.1.1	Effect of load on film thickness in elasto–hydrodynamic lubrication	261
6.1.2	Brass and Acrylic bearings	262

6.2.1	Contact between two solid spheres	263
6.2.2	Contact between two solid cylinders	263
6.2.3	Elastic deformation of two contacting solid spheres	263
6.2.4	Elastic deformation between two solid spheres	264
6.2.5	Elastic deformation of Spheres	266
6.2.6	Schematic of elastic deformation	267
6.2.7	Thrust bearing with races and balls	269
6.3.1	Journal bearing	270
6.3.2	The effect of pressure–viscosity relation	271
6.3.3	The effect of pressure–viscosity relation for thicker oils	272
6.3.4	The effect of pressure–viscosity relation on increasing relative velocity	273
6.4.1	Fluid film thickness	274
6.4.2	Film thickness and pressure distribution in EHL	277
6.5.1	Meshing in finite difference method	278
6.5.2	Deflection curve in elasto–hydrodynamic lubrication	280
6.6.1	Four main components of rolling element bearing	281
6.6.2	Rolling element arrangement in the bearing	281
6.7.1	Gear pair showing contact between two pairs of teeth	285
6.7.2	Involute profile of gear/pinion tooth (phi A and phi B, inv phi A and inv phi B, addendum circle)	285
6.7.3	Gear and pinion in mesh	286
6.7.4	Cylindrical contact analogy for gear and pinion tooth interaction	286
7.1.1	General load clearance characteristics of air bearing [Neale, 2001]	298
7.1.2	Cylindrical aerodynamic bearing	299
7.1.3	Cylindrical aerodynamic with pressurised air supply arrangement	299
7.1.4	Annular pocketed aerostatic bearing	299
7.3.1	Tilting pad journal bearing	301
7.3.2	Infinitely long plane slider	302
7.3.3	Discretization of pad surface	304
7.3.4	Magnetic recording device	307
7.3.5	Pad bearing arrangement for hard disc drive	308
7.3.6	Long plane slider	308
7.3.7	Non-dimensional pressure along non-dimensional length	309
7.4.1	Various types of restrictors [Neale, 2001]	310
7.4.2	Radial aerostatic bearing	313
7.4.3	Pressure distribution	316
7.4.4	Circular thrust gas bearing	316
8.1	Model of mixed lubrication	327
8.2	Surface roughness profile of bearing (corresponding to data at Serial No. 17)	329
8.3	Surface roughness produced by common production methods [ANSI B46.1-1985]	330
8.4	Stylus method of surface roughness measurement	331
8.5	Standard method to estimate roughness parameters	331
8.6	Discretization of surface to find $R_{average}$ or R_a	331
8.7	Surface roughness profile of bearing	332
8.8	Mean peak spacing parameter Sm [ISO 4287/1]	333

8.9	Surface profiles having same R_a and S_m values	333
8.10	Different profiles with their kurtosis values [ASME B46.1-1995]	334
8.11	Auto-covariance function	335
8.12	Auto-correlation function	336
8.13	Comparison of bearing surfaces based on R_q values	337
8.14	Inverse surfaces with identical R_a value	338
8.15	Comparison of bearing surfaces	338
8.16	Formation of low shear strength layer at the asperity interface	339
8.17	Physisorption in boundary lubrication	339
8.18	Effect of temperature on friction coefficient in boundary lubrication [Bowden and Tabor, 1950]	340
8.19	Effect of temperature on boundary layer	340
8.20	Chemisorption	341
8.21	Effect of load variations on the wear rate	341
8.22	Effect of temperature variation on the wear rate	342
8.23	Photograph of experimental set up for online condition monitoring of gear wear.	342
8.24	Photograph of online condition monitoring unit	342
8.25	Photograph of the failed sugar mill bearing [Muzakkir et al., 2011]	345
8.26	Journal bearing test rig [Muzakkir et al, 2011]	346
8.27	Drawing of the test bearing [Muzakkir et al., 2013]	347
8.28	Coefficient of friction of test bearings 1, 2, 3 [Muzakkir et al, 2011]	348
8.29	Coefficient of friction of test bearings 7 and 8 [Muzakkir et al, 2011]	348
8.30	Coefficient of friction of test bearings 5 and 6 [Muzakkir et al, 2011]	349
8.31	Photograph of bearing 1 showing excessive wear after the test [Muzakkir et al, 2011]	349
8.32	Effects of the solid/fluid interaction on asperity thermal [Zhai and Chang, 2001]	351
8.33	(a) Circularity	352
8.33	(b) Cylindricity	352
8.34	(a) Test bearing	352
8.34	(b) Test bearing drawing	352
8.35	Actual surface profile of bearing 3 with respect to journal surface [Muzakkir et al., 2014]	353
8.36	Surface profile of bearing No. 3 after test [Muzakkir et al., 2014]	354
8.37	(a) Actual location of maximum wear zone located towards the left of minimum film thickness zone [Muzakkir et al., 2014], (b) Actual location of maximum wear zone located towards the right of minimum film thickness zone [Muzakkir et al., 2014]	355
8.38	Profile of the worn out bearing 1 [Muzakkir et al., 2014]	356
8.39	Cumulative wear with respect to journal rotations for different bearing profiles [Muzakkir et al., 2014]	356
9.1	Four main components of rolling element bearings	372
9.2	Ball bearings	373
9.3	Cylindrical roller bearings	373
9.4	Taper roller bearing	374
9.5	Spherical roller	374
9.6	Single row spherical roller bearing	374
9.7	Arrangements of rolling element bearings	375
9.8	Rolling element arrangement in the bearing	376

9.9	Load transfer under combined loading	380
9.10	Stages in pitting failure in rolling motion	381
9.11	Different outer diameter of bearings with common shaft diameter	382
9.12	Bearing standard series	383
9.13	Failure of four-row roller bearing	387
9.14	Spalling failure of outer ring of pin-type roller bearing	387
9.15	Mild abrasive wear of rollers	388
9.16	Fretting corrosion of outer ring	388
9.17	Beach marks indicating fatigue failure of the outer ring	389
9.18	Load (N) distribution in roller bearing	390
9.19	Material handling holes in bearing outer ring	391
9.20	Fracture near hole	392
9.21	Excessive deformation of tyre	393
10.0	Various types of gears	400
10.1.1	Contact between asperities of gear and pinion	401
10.2.1	Mixed lubrication between gear and pinion surface	402
10.2.2	SEM analysis at gear tooth	403
10.2.3	Relative life with various film thickness / surface roughness ratio	403
10.2.4	Gear tooth profile	405
10.2.5	Meshing points on the gear	405
10.2.6	Destructive wear	406
10.3.1	Initial pitting	408
10.3.2	Destructive pitting	409
10.3.3	Spalling failure	409
10.4.1	White colour contamination layer on gear and pinion surfaces	410
10.4.2	Non-uniform distribution of operating temperature	411
10.6.1	Flowchart of oil analysis techniques included in this study	413
10.6.2	Direct Reading Ferrograph (http://www.tricocorp.com/product/direct-reading-ferrograph)	414
10.6.3	Ferrogram analyser	414
10.6.4	Separation of ferrous particles from oil on ferrogram/slide	415
10.6.1.1	SEM result of the wear debris extracted from sample 1	415
10.6.1.2	SEM result sample 2 (largest particle size is 6.61 μm)	416
10.6.1.3	SEM result sample 3	416
10.6.1.4	SEM result sample 4 (largest particle size is 28.82 μm)	417
10.6.1.5	EDX of debris from sample 4	417
10.7.0	Flowchart stating oil analysis techniques	418
10.7.1	Spur gear experimental test rig	419
10.7.2	Online oil analysis sensor suite	420
10.7.3	Fresh oil and used oil	420
10.7.4	Fe concentration in ppm for 9000 wear cycles (speed = 300 rpm, Torque = 15 N-m)	420
10.7.5	Results oil sensor suit at 500 rpm, 50 N-m torque after 90 minutes of operation	421

List of Tables

1.7.1	Classification of greases based on thickeners	27
1.7.2	Comparative chart	28
1.7.3	Role of base oil	28
1.7.4	National Lubricating Grease Institute (NLGI) grease classification	29
1.8.1	Importance of lubricant additives	35
1.8.2	Lubricant additives	35
1.9.1	Lubricant selection	38
2.1.1	Coefficient of friction (under dry condition) between two similar metals	52
2.1.2	μ for wood–on–wood reported in various articles	54
2.2.1	Cone angle Vs Coefficient of Friction due to deformation	60
2.2.2	Ratio of shear stresses Vs coefficient of friction	63
2.3.1	Reproducibility of pin-on-disk friction data [Blau, 2008]	67
2.5.1	Data related to friction coefficient and wear constant	82
2.5.2	Experimental data	83
2.5.3	Values of wear constant	84
2.5.4	Contact stress	94
2.5.5	Stresses accounting tangential force	96
2.5.6	Material strength data for rolling with 9% sliding	97
2.5.7	Material data for pure rolling case	97
2.5.8	Cam life for various materials	98
2.8.1	Percentage of boundary additive vs. friction coefficient	105
2.8.1	Fe concentration	108
2.8.2	Changing operating condition changes the dynamics of B.L.L	109
2.8.3	Particle size vs. time at N = 500 rpm	109
2.8.4	Particle size vs. time at N = 2000 rpm	109
2.8.5	Fe concentration vs. load	110
3.1.1.1	Typical operating viscosity ranges	125
3.1.2.1	ISO viscosity grades	125
3.1.2.3	Variation of viscosity with temperature for commonly used engine oil	127
3.1.2.4	Shear stability of multi-grade oils	131
3.1.2.5	Representative values of viscosity—pressure index Z [Roelands, 1966]	132
3.1.2.6	Viscosity variation with pressure	132

3.4.3.1	Dimensionless parameters in Raimondi and Boyd method	160
3.4.3.2	Performance characteristics of full journal bearing [Raimondi and Boyd, 1958]	169
3.4.3.3	Performance characteristics of 180° arc journal bearing [Raimondi and Boyd, 1958]	170
3.4.3.4	Performance characteristics of 120° arc journal bearing, [Raimondi and Boyd, 1958]	171
3.4.3.5	Performance characteristics of 60° arc journal bearing, [Raimondi and Boyd, 1958]	172
3.4.3.6	Design table for journal bearing	175
3.4.3.7	Comparison of results for example 3	177
3.4.3.8	Comparison of results for example 4	177
4.1	Values of $\bar{x}$ at different nodes	193
4.2	Pressure values at different iterations	195
5.3.4.1	Guidance on dangerous levels of film thickness Booker [Booker, 1979]	243
5.3.5.1	Force component for VEB connecting rod [Booker, 1971]	244
5.3.4.2	Force component for main bearing Paranjpe and Goenka. [Paranjpe and Goenka, 1990]	246
5.3.5.3	Values of different parameters	248
5.4.1	Ranking of compensating elements [Rippel, 1963]	256
6.1.1	Torque, friction coefficient, temperature and maximum pressure values for different combination of speed and load for acrylic and brass bearing	262
7.1	Viscosity of gases [Booser, 1984]	297
8.1	Surface roughness and height parameters (Experimental values)	328
8.2	Effect of additive concentration on friction coefficient in boundary lubrication	340
8.3	Experimental results showing concentration of wear particles (no load condition)	343
8.4	Experimental results of Fe concentration after oil replacement	344
8.5	Experimental results under loaded condition	344
8.6	Theoretical load carrying capacity of the sugar mill bearings	346
8.7	Diameter and radial clearance of fabricated bearings	347
8.8	Wear of test bearing measured in terms of weight loss	349
8.9	Measurement data of the test bearings	353
8.10	Experimental results (load 7500 N, speed 10 rpm)	353
8.11	Parameters used in mathematical model	355
9.1	Failure probability vs. factor a_1	378
9.2	Rolling element bearing vs. load	379
9.3	Misalignment capabilities of rolling element bearing	380
9.4	Rolling element bearing catalogue	381
9.5	Comparative study of coefficient of friction	384
9.6	Lubrication factor f_L	385
9.7	Friction factors for seals [Brandlein et al., 1999]	385
9.7	Load carrying capacity of pin type and brass type cage	386
10.6.1.1	Direct reading ferrography Results	417
10.7.1	Specification of gearbox	419

Preface

This introductory yet comprehensive book provides information on tribology and its related systems. Tribology is study of interacting surfaces in relative motion. It encompasses the field of friction, lubrication, wear, and related surface phenomena. An understanding of tribological principles and their appropriate application is an essential requirement for optimum design, operation and maintenance of tribo-systems. Knowledge of tribology helps improve service life, safety and reliability of interacting machine components and it yields substantial economic benefits.

This book targets senior undergraduate and postgraduate students of mechanical engineering, production engineering and metallurgical engineering. It can also serve as a reference for researchers and designers. For this purpose the book includes the study of solid mechanics which is required to estimate the contact stresses developed at asperity interactions; study of fluid mechanics which is required to establish the lubricant behavior and film formation between various geometric shapes of sliding surfaces; study of material science which focuses on atomic and micro-scale mechanisms that cause solid surface degradation or alteration occurring during relative motion; and study of chemistry that describes the reactivity between lubricants, additives and solid surfaces. It provides sufficiently strong scientific background to tribological concepts, with their applications to real problems. It is the result of the author's background in tribology research and consultancy, experience in teaching tribology courses to practicing engineering professionals, graduate and senior undergraduate students and conducting industrial courses over a period of about 16 years. A prominent feature of the book is the description of the basic scientific principles detailing the inter-disciplinary approach and the corresponding computer based solutions. Presenting all requisite concepts with clarity and conciseness, this book aims to fill the wide gap between the scientific basis of the solution strategies in dealing with tribological problems and prevalent experience based industrial practices.

Real life case studies have been meticulously selected and incorporated at the end of each chapter to make the reader aware of the scientific approach in solving tribological problems in contrast to the prevalent trial and error methods based on industrial experience that lacks scientificity. A selected list of references at the end of each chapter may be used by the reader for further in-depth study. Frequently asked questions and multiple choice questions have been included at the end of the chapters so that the book may be readily adopted for university and industrial courses. This book is intended for adoption as a text for a single semester course in tribology at senior undergraduate and graduate levels. It is hoped that this book will motivate students to explore tribology in space-systems and nano-systems. This book will also be useful for practicing engineers and researchers in tribology.

Acknowledgments

The author is thankful to his students, industry professionals and teachers of engineering colleges whose participation in the professional courses conducted by him motivated him to write the present book on tribology. The author is also thankful to his Ph.D. students S. M. Muzakkir, Chiranjit Sarkar and K. P. Lijesh in providing assistance during the course of writing of this book.

The author is thankful to the Quality Improvement Program, under the aegis of the AICTE of Government of India, for providing the necessary financial support.

The author is obliged to IIT Delhi for providing the opportunity to write this book by way of sabbatical leave.

The author is grateful to his wife Meera and mother-in-law Kokilam for timely assistance to complete the book. The author is appreciative of his children Netra and Daksha for bearing him during the course of writing this book.

Nomenclature

a	Radius of asperity
a_g	Axial length of groove
A	Area of contact, Bearing support area
A_o	Cross section area of orifice
b	Width of slab, radius of contact patch
B	Length of the slider
c	Radial clearance
C	Damping Coefficient, Dynamic load rating
C_0	Static load rating
C_1, C_2	Integral constants
C_d	Discharge coefficient, Diametric clearance
d_c	Diameter of capillary tube
d_e	Pitch diameter
d_h	Diameter of the hole
d_{p1}	Pitch circle diameter of pinion
d_{p2}	Pitch circle diameter of gear
e	Bearing eccentricity
E	Young modulus
E_1	Young's modulus of cam
E_2	Young's modulus of follower
F	Force, Total friction of the pad
$F(t)$	Harmonic force
F_a	Axial load, Force required to tear the cold junction
F_d	Force required to cause deformation of junction
F_r	Load capacity, Radial load
g	Switch function
g_s, g_0	Correction factors
G	Modulus of rigidity/shear
$G_{(\omega)}$	Power spectral density function (PSDF)
h	Separation distance, Thickness of asperity, Minimum film thickness
H	Power loss, Kinematic viscosity of lubricant at 40°C having the same kinematic viscosity at 100 °C as the oil whose viscosity index is to be calculated

h	Thickness of slab
h_{min}	Minimum film thickness, m
h_{min}	Minimum film thickness
h_o	Film thickness in Hertzian contact region
h_s	Film thickness outside Hertzian contact region
H_d	Hardness of soft material
H_p	Pumping loss
H_{total}	Total power loss
H_v	Viscous loss
k_e	Dimensionless ellipticity parameter
k_m	Constant in Rowe's wear equation
K	Stiffness, Wear constant, Non-dimensional Knudsen number, Shear stability parameter
K_1	Dimensionless constant, indicating probability of removing a wear particle
K_e	Empirical constant
K_r	Reference strength
l	Length of asperity
l_c	Length of capillary tube
L	kinematic viscosity at 40 °C of an oil with VI = 100 having the same kinematic viscosity at 100 °C as the oil whose viscosity index is to be calculated
L_d	Sliding distance of travel
m	Aspect ratio, Module
$m_{a,R}$	Axial flow from station R
$m_{bearing}$	Mass flow through an annulus
$m_{c,R \to S}$	Circumferential flow from station R to S
$m_{c,S \to R}$	Circumferential flow from station S to R
$m_{restrictor}$	Mass flow through restrictor
M	Mass, Number of nodes in Z-direction
n	Number of asperities, velocity exponent, Polytropic gas expansion exponent
n_1	Atomic layers of inter planer spacing
N	Journal rpm, Normal reaction, Number of nodes in X-direction
N_s	Journal speed in revolutions per second
p	Pressure
p_0	Fluid pressure estimated by short bearing approximation
p_∞	Fluid pressure estimated by long bearing approximation
p_{avg}	Average fluid pressure
$p_{i,j}$	Nodal pressure
p_m	Maximum pressure
P	Pressure of the gas
P_a	Atmospheric pressure
P_{av}	Average fluid pressure
P_i	Instantaneous point of contact between gear and pinion teeth
P_r	Recess pressure, Restrictor pressure
P_s	Supply pressure, Flow rate under supply pressure
P_{shear}	Power loss due to shearing
$P_{squeeze}$	Power loss due to squeeze action

P_w	Probability of wear particle formation
q	Reduced pressure
Q	Lubricant flow rate, Volume swept by all penetrated asperities, Correction factor for slip
Q_{con}	Flow rate of the continuum Poiseuille flow
Q_H	Hydrodynamic flow
Q_L	Side leakage
Q_p	Feed pressure flow
Q_p	Poiseuille flow
r_g	Pitch circle radius of gear
r_p	Pitch circle radius of pinion
R	Universal gas constant
$R_{(\tau)}$	Auto-covariance function (ACVF)
R^*	Effective radii of curvature
R_1, R_2	Radii of meniscus in mutually perpendicular planes, m
R_2	Radius of sphere 2
R_J	Journal radius
R_{ku}	Kurtosis
R_p	Maximum profile peak height
$R_{rms,\ bearing}$	RMS surface roughness of bearing
$R_{rms,\ journal}$	RMS surface roughness of journal
R_{sk}	Skewness parameter
R_z	Maximum Height of Profile
$R_{\Delta z}$	Root mean square slope
$s_{(\tau)}$	Structure function (SF)
S	Sommerfeld number
S_m	Mean peak spacing
T	Friction torque, Absolute temperature
$T_{variable}$	Temperature Variable
u	Linear sliding velocity of lubricant layers, m/s
U	Relative velocity
V	Volume of the gas
V_e	Erosive wear
w	Wear rate, Flow velocity in Z-direction
w_i	Weighting factor
W	Load, Load on journal, Applied load
W_{max}	Maximum load on rolling element
W_{pad}	Load capacity of pad bearing
W_x	Resultant load capacity in X-direction
W_y	Resultant load capacity in Y-direction
W_ψ	Load on the rolling element
Y	kinematic viscosity, at 100 °C, of the oil whose viscosity index is to be
Z	Number of balls, Pressure–viscosity index, No of atoms removed per atomic encounter, Number of rollers in roller bearing or number of balls in ball bearing
Z_p	Maximum Profile Peak Height
Z_v	Maximum Profile Valley Depth

$\dot{\varnothing}$	Rate of change of attitude angle
τ_ι	Interface shear stress
τ_ψ	Shear strength
α	Pressure angle, Pressure-viscosity coefficient, Angle between the axis and plane of curvature of the rolling element
α_c	Fluid compressibility due to presence of air bubbles/gas
β	Fractional surface film defect
β_c	Bulk modulus
γ	Shear rate, Surface tension
δ_1, δ_2	Spherical contact deformation
δA	Elemental area
δF	Elemental force
Δp	Pressure due to capillarity, Pa
δ_r	Radial shift of outer ring
Δt	Time of approach
ΔV	Volume swept by penetrated asperity
δW	Load on asperity
$\dot{\varepsilon}$	Rate of change of eccentricity ratio
ε	Eccentricity ratio
ζ	Damping Factor, material constant
η	Dynamic viscosity, Pa.s
η_0	Viscosity at atmospheric zero pressure
θ	Friction angle
θ_I	Constant for material 1
θ_II	Constant for material 2
λ	Material constant, Molecular mean free path
μ	Coefficient of friction
μ_d	Coefficient of friction due to deformation
μ_k	Kinetic coefficient of friction
μ_s	Static coefficient of friction
Λ	Specific film thickness, Bearing compressibility number
λ	Thermal conductivity
ρ	Density
$\rho_{(\tau)}$	Autocorrelation function (ACF)
ρ_c	Density of lubricant at cavitation pressure
σ	Composite surface roughness
σ_1	First principal stress
σ_2	Second principal stress
σ_c	Contact stress
σ_f	Friction stress
σ_x	Stress in x direction
σ_{xn}	Normal stress
σ_{xt}	Tangential stress
τ	Shear Stress, MPa
τ_{bli}	Shear strength of bond lubricant interface

τ_{EHLi}	Shear strength caused by elasto–hydrodynamic mechanism
τ_{mi}	Shear strength of metal interface
τ_{xy}	Shear Stress in XY plane
υ_1	Poisson's ratio of cam
υ_2	Poisson's ratio of follower
ϕ	Shear Strain rate, Position of minimum film thickness, Attitude angle
ω	Journal angular velocity
$\bar{\omega}$	Average angular velocity between the journal and bearing relative to load line
ω_b	Angular velocity of the connecting rod bearing
ω_j	Angular velocity of the journal
ω_n	Natural Frequency

Chapter 1

Introduction

1.1 Defining Tribology

1.1.1 What is tribology?

Tribology, derived from the Greek word 'Tribos', is a science that deals with friction, lubrication and wear in all contacting pairs. Tribological knowledge helps in reducing the requirement of maintenance and improves reliability of interacting machine components. Essence of tribology at design stage yields substantial economic benefits.

1.1.2 Need of tribology as a separate subject

Friction, wear and lubrication have been taught in science and engineering classes, but at a rudimentary level. It means empirically derived trends (friction force is proportional to loading force, static friction is greater than kinetic friction, viscous friction in a fluid is proportional to the normal contact force, etc.,) are often the only predictive tools available. These approaches have the drawbacks of being predictive only over a limited range of parameters. Since the underlying physical mechanisms are not well understood, often one does not even know the important parameters or the range over which the observed trends are valid. This poor predictive power has given birth to the field of tribology, being pursued in many scientific quarters as a complete subject.

Most tribological phenomena are inherently complicated and interconnected, making it necessary to understand the concepts of tribology in detail. For example, calculations of contact stresses and surface temperature during sliding require understanding of 'solid mechanics'. Similarly, study of lubricant film formed between different geometric shapes of interacting surfaces demands knowledge of fluid mechanics. Mechanical wear at atomic and micro scales involves thorough understanding of material science. Formation of boundary layer on the solid surface demands information about chemistry. In short, integration of knowledge from multifaceted disciplines (solid mechanics, fluid mechanics, material science, chemistry, etc.,) is essential, and therefore, a separate subject is required.

Solid Mechanics

Solid mechanics governs the response of solid material to applied force, as shown in Fig. 1.1.2.1.

- Based on storage of energy and loss modulus, materials can be categorized as elastic, visco–elastic and plastic materials. Behaviour of these materials with fluid at the interface of contacting solids affects the performance of the system.

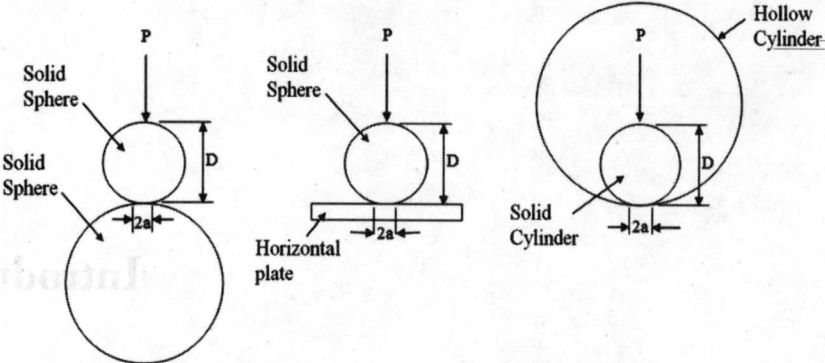

Fig. 1.1.2.1   Solid mechanics

- Surface roughness and real area of contact between surfaces play very important role. Neither zero roughness nor high roughness is desirable. Real area of contact may be 10–50 percent of apparent area. So to achieve high performance from contacting surfaces, account of roughness and real area of contact is essential.
- Behaviour modelling of thin layer coatings having different elastic properties than the substrate is required. The layer may slip or stick at the substrate interface.
- Heat source and heat conduction equations are required to estimate temperature distribution.
- In addition, theories related to crack nucleation, crack propagation and delamination are required.

Fluid Mechanics

A fluid (gas or liquid) is often used to separate two contacting surface as shown in Fig. 1.1.2.2. Following theories/relations are required to estimate tribological behaviour of incorporated fluid:

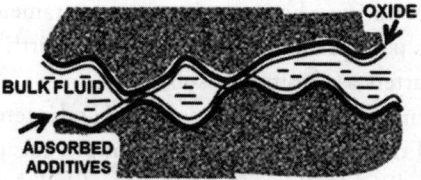

Fig. 1.1.2.2   Fluid mechanics and chemistry

- Hydrodynamic, aerodynamic, hydrostatic, and aerostatic theories of fluid film lubrication.
- Theories related to convective heat transfer.
- Rheological behaviour of liquid to semi-solids.
- Boundary, mixed and elasto–hydrodynamic lubrication mechanisms.
- Study of viscosity thinning and thickening effects.
- Mathematical modelling of thin lubricant film.

Material Science

This science is required to estimate the behaviour of materials in contact as shown in Fig. 1.1.2.3.

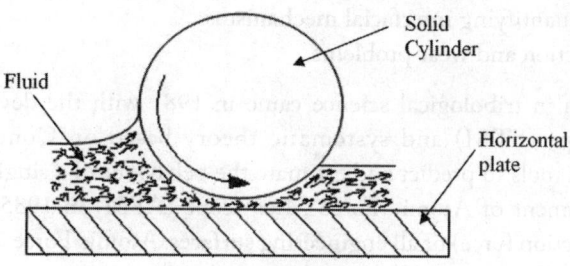

Fig. 1.1.2.3 Material science

Following aspects are important.
- Surface hardening/treatment.
- Development of high/low temperature coatings to provide non-sticking surfaces in bearings, gears, and other tribo–pairs. To deal with formation of adhesive junctions at high temperatures, coatings of ceramics, glass, and metals can be engineered to withstand 75 °C to nearly 200 °C.
- Manufacturing processes to apply nanometre to micrometer thick coating on various materials (material compatibility). Often intermediate coatings are used for better adhesive performance.
- Modelling of thin coatings and linings (thick coatings).

Chemistry

Knowledge of chemistry is required for
- Synthesis of additives: antiwear additives and extreme pressure additives
- Understanding compatibility of lubricants with process fluids and contacting surfaces
- Estimating shelf life of lubricant and its additives
- Predicting performance of lubricant layer as a function of temperature, sliding, etc.
- Optimizing concentration of lubricant additives: covalent, metallic and Van der Waal bonds

As tribology requires understanding of so many subjects, often it is scary to study tribology. In the present book, a step by step approach is followed to learn the subject. The empirical formulae, fundamentals and numerical solutions have been described to study this course.

1.1.3 History of Tribology

In September 1964, a conference on 'Lubrication in Iron and Steel Works in Cardiff (UK)' took place, realizing considerable losses due to lack of knowledge related friction and wear of machine components. After this realization the UK Minister of State for Science formed a committee to investigate into the education, research and the need of industry related to lubrication. The committee after deliberations concluded that only lubrication engineering could not provide complete solution to deal with friction and wear of machine components. An interdisciplinary approach embracing solid and fluid mechanics, chemistry, and material science is essential. Because there was no word for such new concept, a new name 'Tribology' was coined in 1966.

After 1966, the word Tribology has been used for:

1. Basic mechanisms governing interfacial behaviour

2. Basic theories quantifying interfacial mechanisms
3. Solutions to friction and wear problems

Major breakthrough in tribological science came in 1981 with the development of Scanning Tunnelling Microscope (STM) and systematic theory based on Contact Mechanics. Such developments provided tools to predict and estimate the behaviour of a single asperity contact.

Subsequent development of Atomic Force Microscope (AFM) in 1985 allowed measurement (surface topography, friction force) of all engineering surfaces. Atomic Force Microscope can be used to study friction, wear and lubrication at the nanometer level.

The development of tip-based microscopes (STM and AFM) and simulation software to imitate tip–surface interactions and corresponding elastic/plastic deformation, has allowed scientific investigations for interacting surfaces. Tribology is enriched with new findings from such investigations.

1.2 Tribology in Design

Tribology is required for sustainability of the systems having relatively moving machine elements such as gears, cam–follower, bearings, seals, valves, etc. Decrease in the system performance due to aging depends on the rate of wear. Therefore, it is very important to understand the change in system with time and frame suitable maintenance schemes for that system. In addition, if the technology incorporated in the system is expected to be replaced with a newer technology, then the design of the system may be economized by incorporating the tribological aspects alone. Under these conditions the design may be based on the estimated average wear rate for the component to obtain a finite life.

To differentiate mechanical and tribology designs, the following examples of gasket and seal are considered.

1.2.1 Mechanical design of seal and gasket

Seal

The seals are required for closing the unintentional openings or gaps between two or more jointed members for producing a joint that can prevent leakage and withhold fluid under pressure by providing a physical barrier. Basic functions of seal are prevention of fluid leakage between two relatively moving machine components and prevention of entry of foreign particles, like dust or abrasive material into the operating medium. To fulfil these functions, a seal unit requires the following:

- Sealing medium such as an elastomer (i.e., nitrile, silicon and fluoro rubber). The usage of elastomer is advantageous to tolerate, to some extent, misalignment of the shaft and the vibrations.
- Loading mechanism such as spring to deal with seal wear
- Casing that contains seal and restricts its degrees of freedom

The size of seals varies from few millimeters (for micro-bearings) to few thousand millimeters (for canal locks) because of their wide range of applications, e.g. in motor industry, household appliances, power and pump industries, offshore applications, oil refining, automotive, aerospace, construction, agriculture, hydraulics and pneumatics industries. The construction of a typical commercial oil seal unit is shown in Fig.1.2.1. This seal unit consists of an elastomeric seal, circumferential spring, called garter spring, and metallic (carbon steel, aluminium or brass) casing. A press fit between casing and seal is recommended to avoid rotation of the seal material in the casing. To reduce the extent of press-fit casing-bore and seal outer diameters must be machined to the required tolerances and surface roughness. For proper functioning of the oil seals, the shafts should have a highly polished surface free from scratches and tool marks.

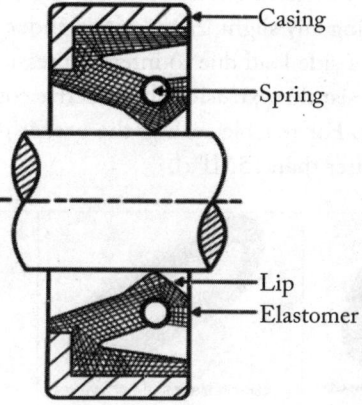

Fig. 1.2.1 Oil seal

Any improper selection or design of the seal may result in failure in the form of leakages, pressure losses, contamination or heat generation that causes reduced efficiencies, energy losses, degraded performance and possibly environmental pollution. The failure costs are generally high as it includes down time costs in addition to seal cost and disassembly & installation costs. As per mechanical design, the diameter of the sealing lip must be slightly less than the shaft diameter so that the seal is elastically deformed while being mounted on the shaft. The elastic deformation and spring action creates contact pressure between the sealing lip (small portion of the elastomer which makes mechanical contact over the surface of the rotating shaft) and the rotating shaft. Too large magnitude of the contact pressure causes excessive friction, resulting in high temperature and rapid wear of the sealing lip. On the other hand, excessive leakage happens due to too little contact pressure. An estimation of the correct contact pressure between the seal lip and the shaft is essential to reduce friction, increase seal life and getting good sealing.

Gasket

If the members of the joint do not have any relative motion between them, then the sealing is achieved by the static seals known as gasket (Fig. 1.2.2). A sealing between jointed members using gasket is achieved by the compression (Fig. 1.2.3) between them. This compression causes gasket material to flow into the imperfections on the gasket seating surfaces for ensuring complete contact between the gasket and its seating surfaces thereby preventing the leakage of the pressurized fluid. Like seal, the basic function of gasket is to prevent leakage from or into the system having relatively stationary components. To be more specific, it can be said that:

- A gasket prevents leakage in a relatively stationary joint (velocities of both the components are the same).
- A seal prevents leakage in a relatively moving joint (velocity of one of the component is greater than velocity of other component).

As there is no relative motion, there is no need of tribology and gaskets can be designed based on mechanical design guidelines. From mechanical design point of view, a gasket is required to fill the space between two mating surfaces and should be able to sustain compressive loading. Due to absence of relative motion between mating parts, 'less-than-perfect' mating surfaces are allowed and surface irregularities are filled by gasket material. In other words gasket material must be able to deform and tightly fill the space including any slight irregularities under compressive loading. The gasket is also many a times subjected to a side load due to internal pressure that tends to extrude it through the flange clearance space. To resist this extrusion the effective compression pressure must be greater than the internal fluid pressure. For reliable sealing the gasket material must be able to withstand high compressive pressure (greater than 15MPa).

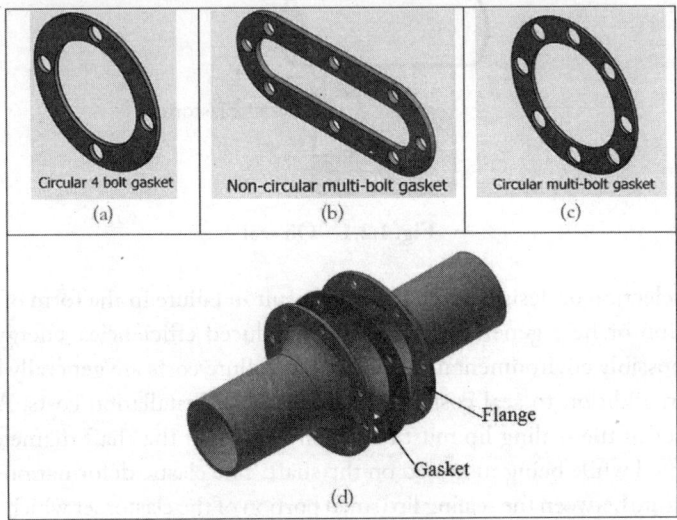

Fig. 1.2.2 (a) Circular 4 bolt gasket, (b) Non-circular multi-bolt gasket, (c) Circular multi-bolt gasket and (d) Gasket assembly

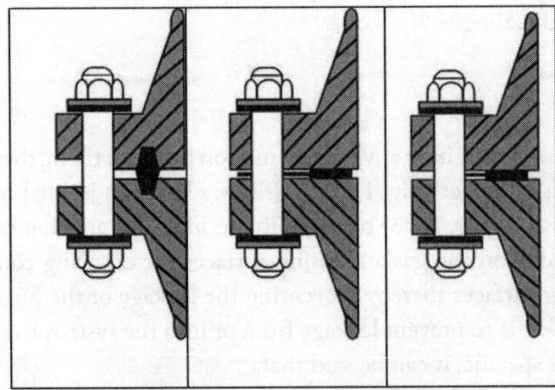

Fig. 1.2.3 Assembled views of different gasket joints

Gaskets are commonly produced by cutting from sheet materials, such as gasket paper, rubber, silicone, metal, cork, felt, neoprene, nitrile rubber, fibreglass, polytetrafluoroethylene (otherwise known as PTFE or teflon) or a plastic polymer (such as polychlorotrifluoroethylene). An example of a gasket used to seal cylinder block and the cylinder head is shown in Fig. 1.2.4.

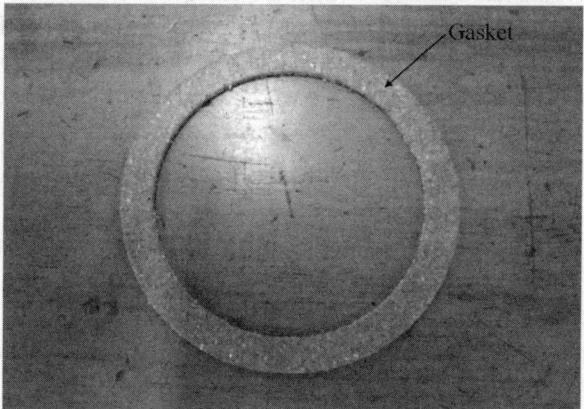

Fig. 1.2.4 Cylinder head gasket

The design of a gasket requires careful consideration of the following aspects:

i. Since the gasket material needs to fill surface irregularities, it must be resilient enough to flow into and fill any irregularities in the surfaces being sealed.
ii. It must remain rigid enough to resist extrusion into the clearance gap between the surfaces under the full system pressure being sealed.
iii. Since a resilient flow is required for gasket closure sufficient closure loading and consequent compressive stresses is required. This requires evaluation of the internal pressure causing leakage problems, the gasket contact pressure produced by the bolt forces, and the gasket materials, which are selected to withstand the operating conditions.
iv. Since the performance of the gasket is degraded due to stress relaxation sufficient contact pressure is required to be maintained to store elastic strain energy to resist relaxation effects. Greater the stored elastic strain energy of the seal greater will be the margin available to resist any relaxation effects during use.
v. Sufficient stresses are required to cause flow of gasket into the imperfections in the seating surfaces. The load required for this deformation is dependent on the gasket material.
vi. The selection of the gasket material must be such that it will withstand the operating pressures and temperature. A material with a low relaxation is preferable as can be used with lower initial compression pressure (otherwise provides higher factor of safety at the same pressure).
vii. Surface finish of gasket seats dictates the thickness and compressibility necessary in the gasket material for providing a physical barrier in the clearance gap.
viii. Flange faces must be parallel and sufficiently rigid to resist distortion on being tightened down and under hydrostatic end loads. Distortion under working loads, often called flange rotation, can appreciably affect the working conditions of the gasket.
ix. The mechanical design of gasket should aim to distribute the load evenly over the whole area of the gasket rather than have a few points of high loading with reduced stress at mid points between the bolts. A more satisfactory arrangement is achieved by employing a large number of smaller diameter bolts rather than fewer bolts of larger diameter. The preload on the bolt must be large enough to achieve minimum seating pressure.

The static sealing of the joint is achieved by proper design of the flanges, selection of gasket type and its material along with its proper installation. The type of flange for a particular application may be selected from ASME standard flanges (which are categorized based on the pressure rating) as given in ASME B16.5. The ASME Unfired Pressure Vessel Code Section VIII, Division I defines the different types of gaskets and their materials to enable selection by the designer as per the operating conditions. The material for the gasket must have good chemical resistance, heat resistance and compressive strength. There are three categories of gasket materials:

1. Non-metallic (elastomers, compressed non-asbestos, PTFE, flexible graphite, mica slabs etc) as given in ASME B16.21
2. Semi-metallic (metallic gaskets containing filler materials like PTFE, flexible graphite, mica, ceramics etc. as given in ASME B16.20 and to be used with ASME standard flanges (ASME B16.5), and
3. Metallic, mostly available in ring type with oval, octagonal and other cross sectional shapes.

The first step in the design of a gasket joint, the selection of the gasket type and its material is made that suits the operating conditions (internal fluid pressure, operating temperature, environment, etc.). The yield value (minimum pressure required to maintain a leak proof gasket joint in the absence of the internal fluid pressure) of the gasket, denoted by 'y', is read from table [3]. Similarly, the ratio between the resultant contact pressure and internal fluid pressure, denoted by 'm', is also read from table [3] corresponding to the given gasket type. This ratio must not be less than a critical value for maintaining leak proof joint because the internal fluid pressure reduces the gasket contact pressure. Since the compression of the gasket is not uniform over its entire compression area due to distortion caused by bolt tightening and fluid pressure, only a narrow band on the outer edge is considered for gasket pressure estimation. It is known as effective gasket yielding width and is denoted by 'b'. The equations for determining this width is available in [1]. The determination of the flange type, bolt size, number of bolts & their arrangement and bolt preload are then determined.

The force in the bolt due to initial tightening is given by:

$$F_g = A_g \cdot q \qquad \ldots (1.1)$$

where F_g is force in the bolt, A_g is the effective gasket area and q is the pressure on the gasket due to bolt tightening.

With the consideration of internal fluid pressure, a force is produced where A_i is the area subjected to internal fluid pressure and p is the internal fluid pressure. The total force on the gasket is thus equal to $A_g \, m \, p$. The total bolt force is given by

$$F_b = p(A_i + A_g \cdot m) \qquad \ldots (1.2)$$

The force on gasket must then be equal or greater than F_b.

On equating equation 1.1 and 1.2, we obtain

$$A_g \cdot q = p(A_i + A_g \cdot m) \qquad \ldots (1.3)$$

Example: Figure 1.2.5 depicts a joint proposed to be fitted with a gasket for preventing leakage of pressurized fluid. The internal fluid is non-corrosive and is at a pressure 'p' of 12×10^6 N/m² (gauge). 24 bolts of 1/2"-12 UNF series are required to keep the cover in place. Select a suitable gasket and bolt loading for ensuring a leak proof joint.

Solution: The figure 1.2.5 depicts a possible gasket arrangement.

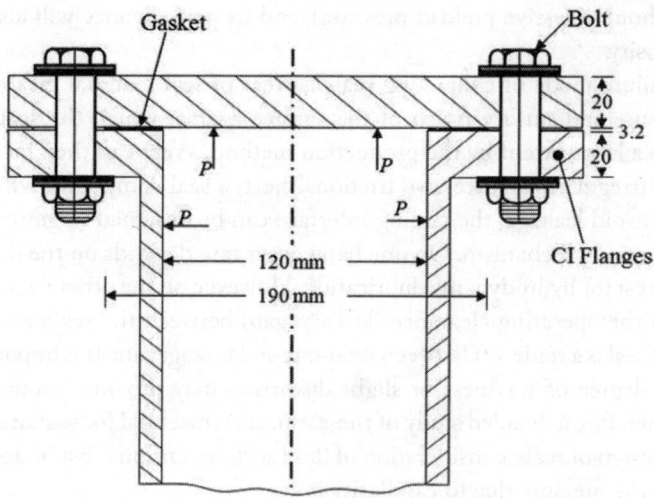

Fig. 1.2.5 Gasket assembly

The fluid is non-corrosive therefore aluminum may be selected as the gasket material. For the aluminum corrugated gasket, the geometry given in figure 1.2.5 may be used. The effective gasket width 'b' is taken to be 3.2mm. The gasket factor 'm' is 2.5 and yield value for the gasket material 'y' is 20×10^6 N/m² (Table 2-5.1, pp 356, Reference [3]).

The area subjected to internal pressure, $A_i = \pi \cdot 0.120 \cdot 0.0032 = 0.0012$ m²

The effective gasket area, $A_g = \frac{\pi}{4}\left(0.190^2 - 0.120^2\right) = 0.017$ m²

Using equation 1.3 with m=2.5, $q = \dfrac{12 \times 10^6 \left(0.0012 + 0.0170 \times 2.5\right)}{0.0170} = 30.85 \times 10^6 \ N/m^2$

Since the gasket pressure is more than the yield value for the gasket material (20×10^6 N/m²), hence this pressure causes the yielding of the gasket that fills the surface asperities and produces a leak proof joint.

The bolt force required for the joint is $F_b = p\left(A_i + A_g \cdot m\right) = 0.5$ MN.

1.2.2 Tribological design of seals and gasket

Since a gasket requires sealing of stationary members, its design procedure does not involve any tribological considerations. However a seal requires consideration of tribological aspects for its proper design. The seals may be classified into contact seals and clearance (non-contacting) seals. Contact seals bears against its mating surface under positive pressure whereas clearance seals operates with positive clearance (i.e. no rubbing contact). For both kinds of seals, knowledge of tribology is essential.

The majority of the seal types are contact seals, operating with rubbing contact. Friction and wear can be held to a minimum by ensuring that the contact surface is adequately lubricated. In designing a seal, the thickness of oil film between seal faces play an important part in determining seal performance and seal life. If thickness of oil film too small it can be bridged by surface irregularities, producing high friction and rapid seal wear. If too thick, the meniscus will break down, producing a high leakage rate. In practice, the seal will not be perfect under dynamic conditions (i.e. it will not be zero–leakage without excessive preload pressure), and its performance will also depend on load, speed and fluid viscosity.

The prevailing failure mode of contacting seals is wear of seal material. Seal wear is dependent on the contact pressure and surface finish of the surface against which the seal rubs, this in turn being determined to a large extent by the production method. Wear can then be aggravated by lack of lubrication, shaft irregularities, excessive frictional heat, a seal compound which is too soft, etc. To reduce wear and avoid leakage, the sealing interface can be designed targeting almost nil to full hydrodynamic lubrication mechanism. On one hand, wear rate depends on the designed lubrication regime and is the lowest for hydrodynamic lubrication. However, on the other hand the rate of leakage strongly depends on the operating clearance (leakage gap) between the sealing surfaces. Therefore, tribological design of seal is a trade-off between wear rate and leakage rate. It is important to understand that even a modest degree of waviness or slight distortion may provide favourable lubrication or excessive leakage. Therefore a detailed study of these aspects is essential for sustainable seal operation.

For tribological design of seals, consideration of fluid surface tension must be accounted. The usual formula to calculate the pressure due to capillarity is

$$\Delta p = \gamma \left[\frac{1}{R_1} + \frac{1}{R_2} \right] \qquad (1.4)$$

where γ = surface tension, R_1 and R_2 = radii of the meniscus in mutually perpendicular planes.

In case of parallel plane surfaces, R_2 can be taken as infinity and R_1 as approximately h/2 where h is the separation of the surfaces. From equation (1.4) it can be said that higher surface tension is desirable to reduce leakage rate. For example, the contact angles of oil against synthetic rubber and steel under industrial conditions are found to be high, so that the sealed oil does not spread along the steel shaft and leakage rate is negligible if pressure difference across the sealing surfaces is lesser than Δp predicted from Eq. 1.4.

The surface tension is a function of temperature; therefore, increase in temperature will increase the leakage rate. In addition, wear rate also increases with increase in temperature; this means tribological design of seal must incorporate thermal analysis.

If the fluid pressure across the seal is over and above the required pressure to overcome the surface tension (Eq. 1.4), an estimate of the volume of the leakage may be made using the following formula:

$$\text{flow/ unit width} = -\frac{1}{12} h^3 \frac{dp}{dx} \qquad (1.5)$$

where, η = viscosity,
h = separation distance between the surfaces, and
$\frac{dp}{dx}$ = pressure gradient

1.3 Tribology in Industry (Maintenance)

Let us consider few failed machine components, failure of which could have been avoided using tribological knowledge.

1.3.1 Example: seal

As shown in Fig. 1.3.1, carbon graphite seal is employed to avoid leakage of steam from rotary joints of paper industry. Failure of this component occurs due to adhesive wear. Adhesive wear causes uneven surface that leads to reduction in mechanical contact area. For the same imposed load, reduction in mechanical contacts increases the level of stress and hence chances of failure.

Fig. 1.3.1   Carbon graphite seal

1.3.2   Example: cam

Figure 1.3.2 shows pitting wear on the cam surface. Cams are used to transmit rotary motion in reciprocating motion. These components are subjected to jerks in sliding, which lead to form some pits on the cam surface. Creation of pits on cam surface increases noise pollution and reduces mechanical performance. Understanding the mechanism of pit formation helps to estimate the life of component and find methods to reduce such pitting failures.

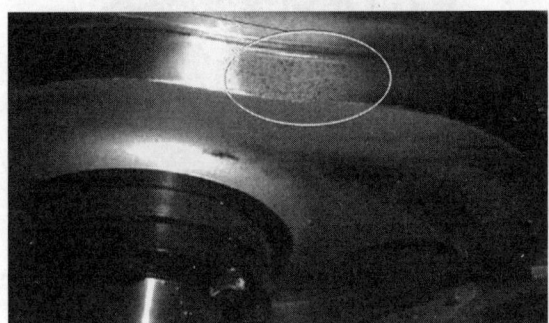

Fig. 1.3.2   Pitting of cam surface

1.3.3   Example: journal bearings

The Figs. 1.3.3 (a) and 1.3.3(b) are examples of journal bearings. There is the photograph of centrally grooved engine journal bearing on the left hand side. It appears that bearing is worn out due to foreign particles. On the right hand side there is a photograph of an aluminium bearing subjected to heavy load that causes shaft surface to run over bearing inner surface. In these examples of journal bearing, wear increases the clearance between shaft and bearing, and leads reduction in bearing load capacity. Often such failures occur in mixed lubrication regime due to relatively low speed. Learning tribology cultivates an understanding that at low speeds the main purpose of oil is the lubrication, and high viscosity oil will be preferred to low viscosity oil; while at high speeds the major purpose of oil is to act as a coolant and low viscosity lubricants are preferred to carry away frictional heat of operation. Here lubrication is a secondary consideration.

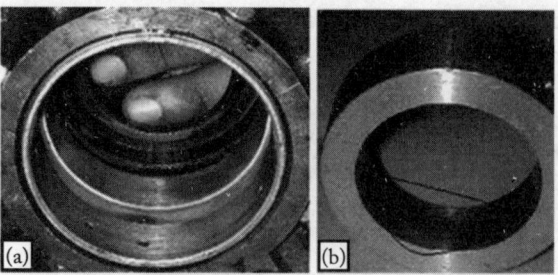

Fig. 1.3.3   (a) Abrasive wear, (b) Rubbing wear

1.3.4   Example: magnetic bearing

Magnetic bearings are known for non-contact levitation. Figure 1.3.4 shows a repulsive type permanent magnetic bearing. Due to improper design and external noise factors, bearing failed within three hours of operation at relative speed of 115 rpm.

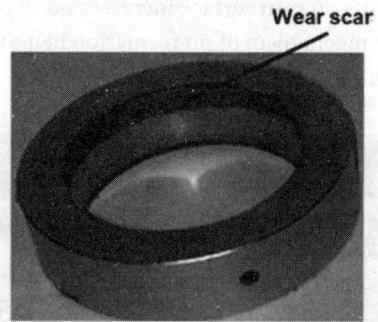

Fig. 1.3.4   Wear scar due to edge loading

1.3.5   Example: multi-row roller bearing

Cracking of outer ring is shown in Fig. 1.3.5. Here cracking means deep cracks which breaks outer ring in number of pieces. Such a failure occurs due to faulty manufacturing and wrong assembly of roller bearing. Understanding tribology may help estimating contact stresses due to misalignment of shaft and improper mounting of bearing surfaces, and approximating a reduction in service life.

Fig. 1.3.5   Failure of large size roller bearing

1.3.6 Example: gear

A pit on the surface of gear tooth is shown in Fig. 1.3.6. A pit generally occurs due to excessive contact stress. Understanding the effect of contact stress helps in developing an equation for estimation of perspective gear life.

Fig. 1.3.6   Pit on gear surface

Study of fluid film bearings, rolling element bearings, seals, gears, cams and brakes are some of the applications in which tribology is required. Basic knowledge gained by tribology course is very useful for industries, related to power, steel, cement, oil, etc. Practicing such knowledge in problems ranging from household appliances to large size ships earns great economic benefits. Therefore, tribology course is often named as: 'Industrial Tribology' and 'Applied Tribology'.

1.4   Defining Lubrication

It is known since ages that liquids and greases reduce the friction between sliding surfaces, by filling the surface cavities and making the surfaces smoother. Action of liquid or grease is known as lubrication. In other words, lubrication is a process by which the friction and wear rates in a moving contact are reduced by using suitable lubricant. Almost every assembly consisting of relatively moving components requires lubrication.

Lubricant is a substance introduced between relatively moving parts to reduce friction (μ = 0.1–0.0001) and wear rate. The lubricant decreases adhesion component of friction to a greater extent compared to abrasion component of friction.

1.4.1   Examples

1.4.1.1   *A standard lock*

On turning the key, the bolt slides into a notch on the door frame. Force required for turning key and move bolt will be reduced by lubrication.

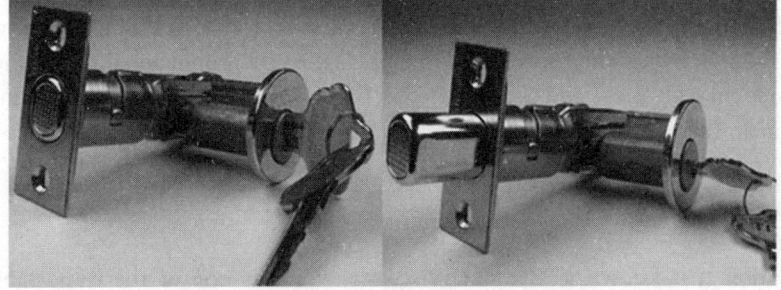

Fig. 1.4.1.1   Simple lock and key

1.4.1.2 *Window lifting mechanism*

Windows in most cars use linkages to lift the window glass [Fig. 1.4.1.2 (a)]. A small electric motor is attached to a worm gear to lift the window. For smooth functioning lubrication is used.

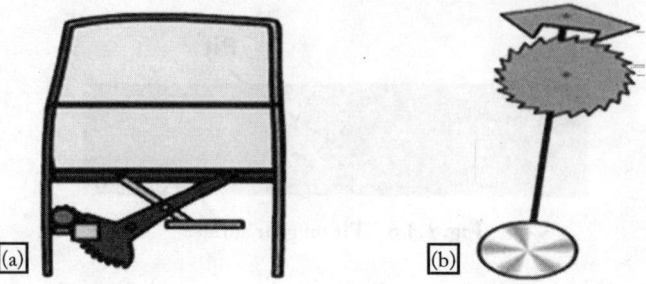

Fig. 1.4.1.2 (a) Window lifting mechanism, (b) Pendulum clock

Sometimes the choice of lubricant type depends on the properties of system. For example, in watches or instruments [Fig. 1.4.1.2(b)], any lubricant type could meet the load and speed requirements, but because of need for low friction, it is normal to use a very low viscosity oil. However, for open gears, wire ropes, or chains the major problem is to prevent the lubricant from thrown off the moving parts, and it is necessary to use a thick bituminous oil or grease having special adhesive properties.

Advantages of lubrication, in addition to reducing friction and wear rate, are:

- Reducing instant failures
- Reducing fatigue failure (a lubricant reduces the force required in tangential direction so reduces the fatigue failure)
- Reducing surface failures
- Reducing stress concentration

1.4.2 Applications of lubricant

1. Transmission parts
2. Bearings
3. Cams and followers
4. Seal faces
5. Any situation involving metal to metal contact

Lubricants are often classified as 'Newtonian' and 'non-Newtonian' fluids. This classification is on basis of relation between shear stress and shear strain rate (Fig. 1.4.2). In Fig. 1.4.2, line a-a illustrates a Newtonian fluid, in which the shear strain rate (ϕ) is directly proportional to the shear stress (τ). Remaining three curves b-b, c-c and d-d depicted in Fig. 1.4.2 show non-Newtonian fluid behaviour.

For Newtonian fluid, shear stress is given by Eq. 1.6.

$$\tau = \eta \phi = \eta \frac{\partial u}{\partial y} \tag{1.6}$$

where, $\eta = \tan \theta$

In this relation η is known as dynamic viscosity, which is one of the important lubrication parameters. Method of replenishing lubricant decides overall performance of the system.

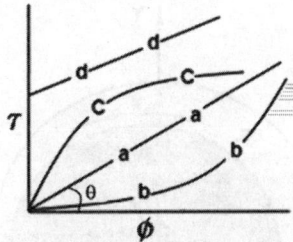

Fig. 1.4.2 Newtonian and non-Newtonian fluids

1.4.3 What is expected from a lubricant?

Required lubricant properties are specific to applications. The main objective of the lubricant is to reduce metal-to-metal contact and thereby minimize friction and wear. A lubricant is also required to perform other functions like cooling, cleaning and suspending, protection and transfer of power. We expect some requirements from the lubricant which can be explained by considering a few examples:

1.4.3.1 *Lubricant between cylinder liner and rings*

The requirements for such type of lubrication (shown in Fig. 1.4.3.1) are as follows. Lubricant

- must form a film to separate the surfaces and reduce the friction between metal to metal contacts in order to improve the efficiency of the system
- needs to adhere to the surfaces (attachment of thin lubricant layer on the surfaces)
- must neutralize the corrosive products of combustion
- must withstand high temperature inside the cylinder.

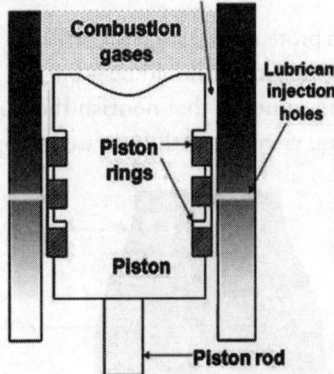

Fig. 1.4.3.1 Lubricant between cylinder liner and rings

1.4.3.2 *Lubrication in journal bearings*

The requirements for lubrication in journal bearing are as follows. Lubricant should

- support shaft and loads on it
- avoid contact stresses
- have ability to dampen vibrations

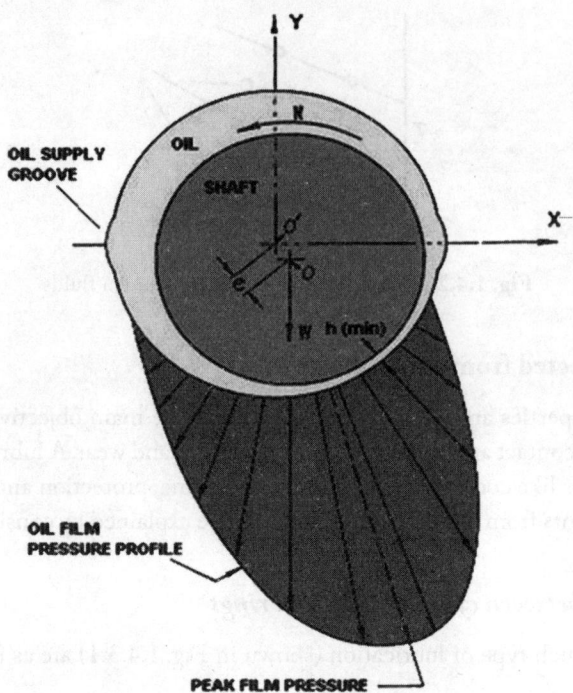

Fig. 1.4.3.2 Hydrodynamic pressure profile in radial journal bearing

1.4.3.3 *Lubrication in bone joints*

The requirements for lubrication in bone joints are:

- The lubricant should contain proteins that stick to cartilage layer resulting in smooth sliding.
- Coefficient of friction of the lubricant should be ~ 0.01.
- The lubricant should contain minerals that nourish the cartilage cells.
- The lubricant should increase viscosity with increase in applied pressure.

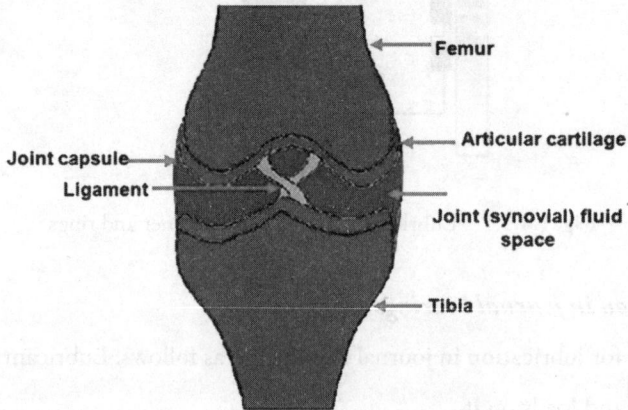

Fig. 1.4.3.3 Lubrication in bone joint

1.5 Basic Modes of Lubrication

Lubricant thickness between two solid surfaces must be thick to avoid wear, but thin enough to minimize lubricant shearing. As thin lubrication and thick lubrication terms are subjective, there is a need to differentiate these terms.

1.5.1 Thick and thin lubrications

- When the interacting surfaces are completely separated and the influence from the surface asperities is negligible, lubrication is termed as thick lubrication. Thick lubrication is governed by Reynolds theory.
- Thin lubrication is far more complex. When the lubricating film is unable to completely separate the sliding surfaces and the surface asperities or roughness on engineering surfaces begin to interfere, the lubrication under such condition is termed as thin lubrication. It requires scientific study at nanometre to micrometre level. In such a situation, the total applied load is partly carried by mechanical contact and partly by fluid between sliding surfaces. Analysis of thin lubrication is very complex as it involves colliding surface asperities, mechanical deformation, lubricant flow transport and chemical reaction.

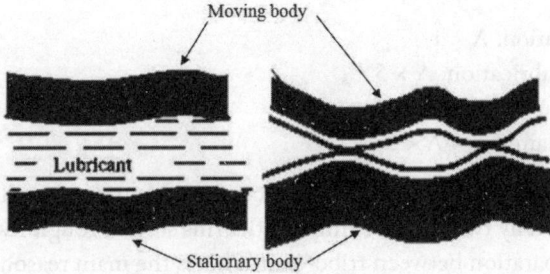

Fig. 1.5.1 (a) Thick lubrication, (b) Thin lubrication

1.5.2 Lubrication mechanisms

Lubrication mechanisms are often classified as 'hydrodynamic', 'elasto–hydrodynamic', 'mixed' and 'boundary lubrication'. As thin and thick lubrications (Fig. 1.5.1) are related to surface roughness of contacting surfaces, it is necessary to understand the lubrication mechanisms relative to the surface texture of tribo–pair. A dimensionless film parameter Λ (often referred as 'specific film thickness') is used to classify the aforementioned four lubricant regimes.

$$\Lambda = \frac{h_{\min}}{\sqrt{R_{rms,a}^2 + R_{rms,b}^2}} \tag{1.7}$$

$R_{rms,a}$ is the root mean square (rms) surface roughness of surface a, and $R_{rms,b}$ is the rms surface roughness of surface b. Interestingly, here the rms value is used, while in industry arithmetic avg. roughness is often preferred. To clarify this, let us examine Fig. 1.5.2. From tribological point of view, a surface without any asperity but with a number of valleys (that retains lubricants and provides a room for debris collection) is always preferred. Measurement of average roughness imposes a linear penalty on all points whether a point is too close to nominal line or too far. However, the rms roughness parameter uses square term. If there are three points: A one unit, B two units and C three units away from nominal line, RMS roughness parameter put penalty of one, four and nine on points A, B, C, respectively. Therefore, RMS value is a better roughness parameter compared to average roughness.

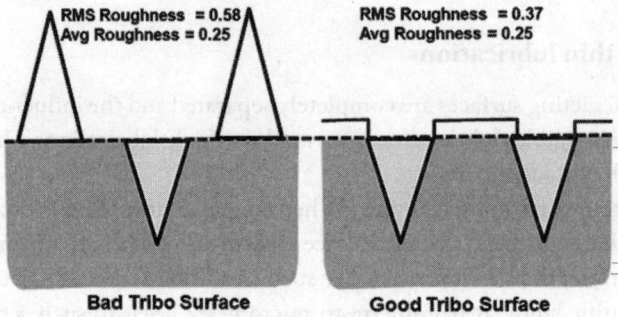

Fig. 1.5.2 Comparison between average and root mean square roughness

Based on the value of dimensionless film parameter (Λ) in Eq. 1.7, lubrication mechanisms are classified as follows:

- Boundary lubrication: $\Lambda < 1$
- Hydrodynamic lubrication: $\Lambda > 5$
- Mixed lubrication: $1 < \Lambda < 3$
- Elasto–hydrodynamic: $3 < \Lambda < 5$

Film parameter depends on film thickness and composite surface roughness of tribo–pair. Peak surface roughness is generally two to three times of the rms surface roughness. Therefore, $\Lambda > 1$ does not indicate the clear separation between tribo–pair. This is the main reason to keep film parameter lesser than three but greater than one to identify the mixed lubrication mechanism. To avoid any wear and minimize friction a complete separation, between asperities of two relative moving surfaces is essential. This requires film parameter more than three. Often foreign particles or wear debris changes the hydrodynamic/elasto–hydrodynamic lubrication to mixed or boundary lubrication mechanism, as shown in Fig. 1.5.3.

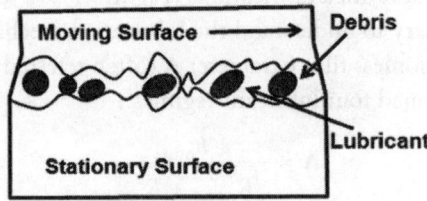

Fig. 1.5.3 Transition in lubrication regime in the presence of foreign particles or wear debris

1.6 Properties of Lubricants

Selection of a lubricant for any particular tribo–application (continuous/ periodic/ intermittent working), depends on the lubricant properties. Lubricant must be able to reduce friction between tribo–pair and minimize the wear and surface damage. In addition, a lubricant should eliminate the possibilities of thermal heating (causing thermal deformation and thermal stress). For a liquid lubricant viscosity and its resistance to tolerate temperature play very important role. Therefore, the following properties of a liquid lubricant must be considered before the selection of lubricant.

(a) Viscosity (resistance to flow) at 40 °C and at 100 °C.
(b) Total Acid Number (*TAN:* the lower the better) or Total Base Number (*TBN:* the higher the better).
(c) Ash content (0.005–0.25%)
(d) Flash point(minimum temperature at which vapourization of lubricant starts)
(e) Pour point(minimum temperature at which a lubricant continuous to flow)
(f) Resistance to foam

In practice, a lubricant has to fulfil many other requirements, such as: thermal stability, chemical stability, compatibility, thermal and heat conductivity, heat capacity, and toxicity.

Knowledge of lubricant properties and their effect on lubrication mechanisms always helps in selecting appropriate lubricant. Therefore, it is worthwhile to study that in tribology course.

1.7 Types of Lubricants

A lubricant is a substance that reduces friction and wear at the interface of two materials by reducing the shear strength of interface. Based on the shear strength of lubricant or molecular state (Fig. 1.7), lubricants are classified in four categories.

- Solid lubricants
- Semi-solid lubricants
- Liquid lubricants
- Gaseous lubricants

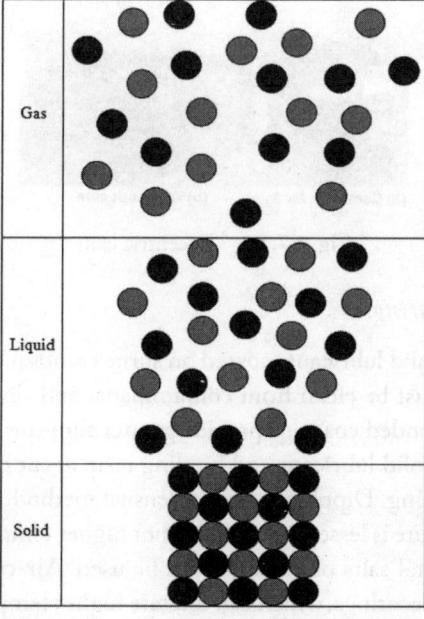

Fig. 1.7 Molecular state of lubricants

1.7.1 Solid lubricants

A solid lubricant is basically any solid material having the following two properties:

- Ability to sustain the applied load without measurable deformation or material loss. In other words, rate of wear must be acceptably low.
- Coefficient of friction must be reasonably low. For this, shear strength of formed interface must be lower than shear strength of contacting solids.

Advantages of solid lubricants are:

1. Solid lubricants are more effective than liquid lubricants at high loads/pressures.
2. Solid lubricants have high resistance to deterioration in storage.
3. They are highly stable in extreme temperature conditions.
4. In absence of pump, tubing, etc., solid lubricants permit equipment to be lighter and economic.
5. Solid lubricants exhibit superior cleanliness.

Disadvantages of solid lubricants are:

1. A broken solid film cannot repair on its own and shorten the useful life of the lubricants.
2. Poor thermal conductivities of solid lubricants may increase the operating temperature.
3. Solid lubricants have greater coefficient of friction than liquid lubricants.

One way to apply solid lubricant is powder coating. We can use powder form of solid lubricant and rub against the tribo–surface. A lubricant layer shall form on the tribo–surface but may not last longer. To explain this, two eccentric cams are shown in Fig. 1.7.1.1. Left hand side cam (Fig. 1.7.1.1 (a)) is rubbed with MoS_2 powder. After three hours of operation, MoS_2 coating gets detached from the cam surface and the worn out cam is shown in Fig. 1.7.1.1(b). In short, if the self-healing mechanism is missing, then operating life of solid lubricated component will be very short. Therefore, carrier fluids are used to coat MoS_2 under operating conditions and replenish MoS_2 wherever required.

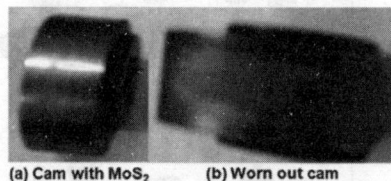

(a) Cam with MoS_2 (b) Worn out cam

Fig. 1.7.1.1   Eccentric cam

Solid lubricants as bonded coating:

To increase the durability of solid lubricants coated on surfaces, often binders along with lubricating pigments are used. Surface must be clean from contaminants and slightly rough to provide greater area to adhere to lubricant. Bonded coatings provide greater film thickness; so they increase service life. Coatings consisting of a solid lubricant and binding resin agent can be applied to the substrate by spraying, dipping, or brushing. Dipping is less expensive method. Resins, binder agents, remain effective if operating temperature is lesser than 250 °C. For higher temperature application (> 300 °C), inorganic binders, such as metal salts or ceramics can be used. Air-cured coatings are temperature sensitive, therefore heat-cured coatings, which can tolerate higher temperature, are used for inorganic binders. Applications of bonded coating of solid lubricants are:

(a) Cylindrical bushes (plain bearings)
(b) Separator (cage of rolling bearing)
(c) Electrical brushes (additive to carbon–graphite)

Solid lubricants in use are self-lubricating composites. These composites are classified as:

1. Polymer
2. Metal–solid
3. Carbon and graphite
4. Ceramic and cermet

1. Polymers

Polymers are a form of solid lubricants and are suitable to bear lighter (lighter compared with other solid lubricants and heavier compared with fluid lubricants) loads. With recent research advances in polymers, these make the largest group of solid lubricants. In polymer sub-class of solid lubricants, PTFE, nylon and synthetic polymers are common solid lubricants. Polytetrafluoroethylene is a polymer produced from ethylene in which all the hydrogen atoms have been replaced by fluorine atoms. Teflon is trade name of PTFE, given by Du Pont. It is used in very light load applications. Poor adhesion of PTFE to other materials is responsible for very low coefficient of friction ($\mu < 0.1$). Some of the strengths of PTFE are:

- High chemical stability. Therefore, it can be used in variety of applications.
- Great chemical inertness because of carbon fluorine bonds. Low tendency to make adhesive bonds.
- Very low surface energy. Low adhesive friction.
- Nontoxic. Therefore useful in pharmaceutical and food industries.

Some of the PTFE weaknesses are:

- Too soft and high wear rate; therefore, restricted to low to moderate load but low relatively velocity
- Poor thermal conductivity, poor creep resistance and high thermal expansion; therefore restricted to low temperature applications (temperature limit<250 °C)
- Vacuum is detrimental to performance

Most of the disadvantages of PTFE can be overcome by using fillers (glass, carbon) and impregnating it with metal (bronze, lead) structures. With a suitable rigid (metal) backing, PTFE can bear extremely high loads (100 MPa or more) with a friction coefficient of 0.1 or less and virtual freedom from stick–slip sliding. The wear rate of polymer composites depends on the surface roughness of mating metallic faces. Initially polymer composite gets worn out and transfers its material as a film on the mating surface. In addition, there is possibility of polishing the mating surface, which means surface roughness of mating metallic surface is reduced with progress in run-in time. In other words, as the run-in period is completed, the wear rate is reduced due to polymer film transfer or by polishing action between the sliding surfaces.

Nylon is similar to PTFE but slightly harder (Specific wear rate 10^{-6}–10^{-5} mm^3/ min). There are the following two main limitations of polymers, which must be considered during selecting polymers as lubricant:

- Low heat transfer rate from polymers inhibits heat dissipation, which causes operating temperature to rise and premature failure due to softening.
- Low thermal conductivity of polymers permits the permissible speed, which is much lower than that permitted by a polymer against a metal surface. Therefore, polymers should not be used as bulk material for both the tribo–surfaces; at least one bulk surface must be metallic surface.

2. Metal–solid lubricants

Molybdenum disulphide (Fig. 1.7.1.2) is one of the most commonly used metal–solid lubricants. Some of the strengths of MoS_2 are:

- High compressive strength greater than 700 MPa
- Low friction
- Lubricant properties even at high temperature; molybdenum disulphide starts to oxidize at temperature greater than 350 °C in oxygen-environment and at 450 °C in air, but the main oxidation product is molybdic oxide, which is a fairly high temperature lubricant itself.
- Very good lubricant in space; in high vacuum the disulphide is said to be stable even at 1000 °C and it evaporates very slowly, so that it has been widely used in space.

Some of the weaknesses of MoS_2 as a lubricant are:

- Moisture detrimental to performance; in moist environment MoS_2 loses its lubrication properties.
- Stable film thickness in the range of 0–15 μm. Thicker films apparently do not last as long; may be because it becomes easier for wear particles consisting of the MoS_2 resin material to come off in lose form. Therefore, there is a need to provide a replenishing source of MoS_2.

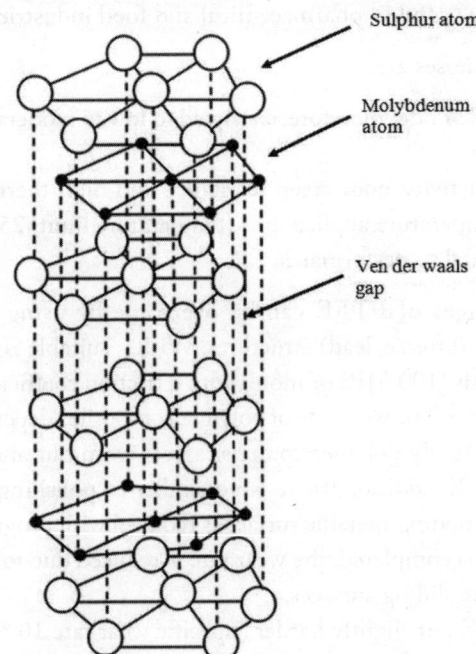

Fig. 1.7.1.2 Crystal structure of molybdenum disulphide [Lansdown, 1999]

The metal–solids lubricants containing lamellar solids rely on film transfer to achieve low friction. But continuous transfer of film may reduce the life of component; therefore often No Lamellar Solids (NLS) are added to lamellar solids. To achieve these objectives often holes are drilled in machine parts and those holes are packed with solid lubricants.

3. Carbon and graphite

The carbon has low tensile strength and ductility due to which it cannot be used in shape of any element. However, it is extensively used in powder form as a solid lubricant. Their high thermal and oxidation stability at temperatures of 500–600 °C enables the use of this solid lubricant at high temperatures and high sliding speeds.

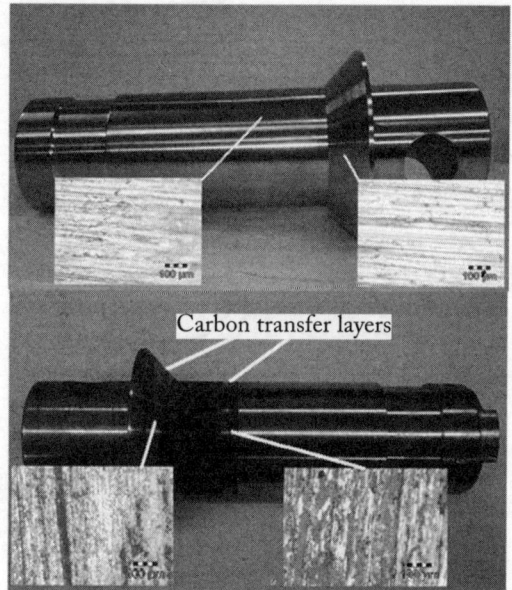

Fig. 1.7.1.3   Carbon transfer layer on stainless steel

Fig. 1.7.1.4   Carbon graphite seal

Carbon–graphite seals (Fig. 1.7.1.4) are the most common example of Carbon and Graphite solid lubricant group. These seals transfer layers of graphite on mating surface (Fig. 1.7.1.3) and provide low friction, but tight seal.

	Strengths of graphite	Weaknesses of graphite
1	Moderate loads (< 275 MPa).	Corrosion.
2	Low friction.	Vacuum detrimental to performance.
3	High temperature stability.	

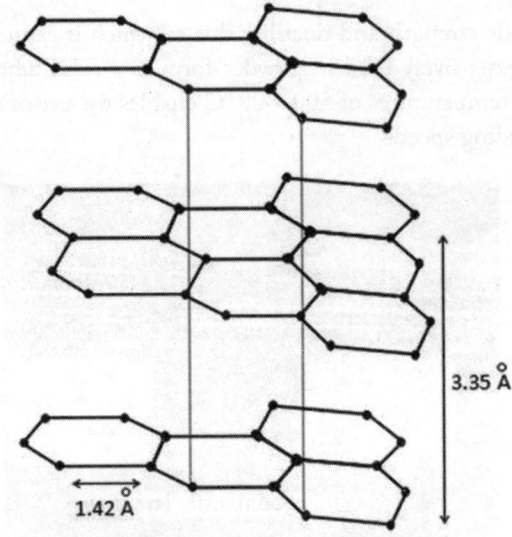

Fig. 1.7.1.5 Structure of graphite

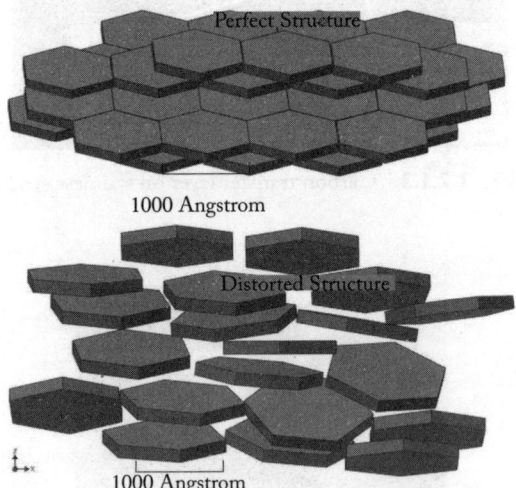

Fig. 1.7.1.6 Perfect and distorted structure of graphite

Figure 1.7.1.5 indicates that graphite is a lamellar solid. Mechanical distortion of graphite is shown in Fig. 1.7.1.6, which limits its usage to moderate pressure (< 275 MPa). It is interesting to note that presence of water helps graphite in lubrication, while the presence of water is detrimental to MoS_2. On the other hand vacuum is detrimental to graphite, but favourable for MoS_2.

To observe the performance of graphite, an experiment on graphite seal submerged in water, as shown in Fig. 1.7.1.7, was performed. Experiment results indicated excessive wear rate of mechanical seal (at 100–300 rpm). Second experiment on same seal but much lesser water, as shown in Fig. 1.7.1.8, was performed. The results pointed out much lesser wear but almost same friction performance.

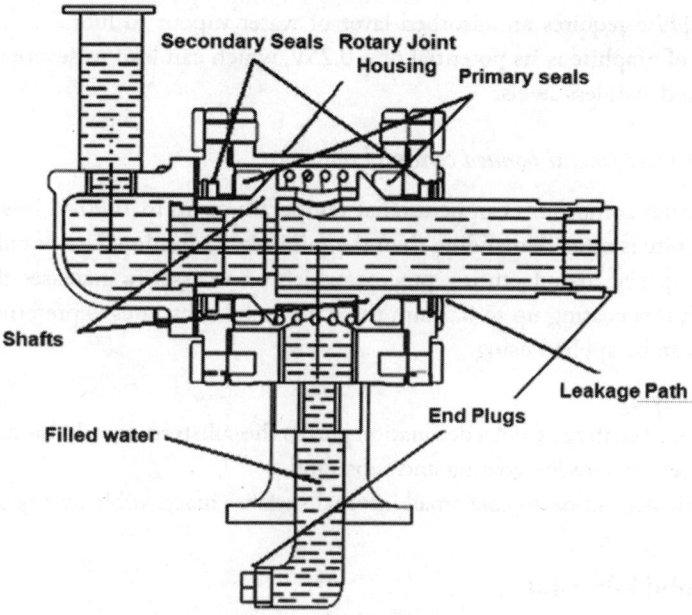

Fig. 1.7.1.7 Graphite seal submerged in water

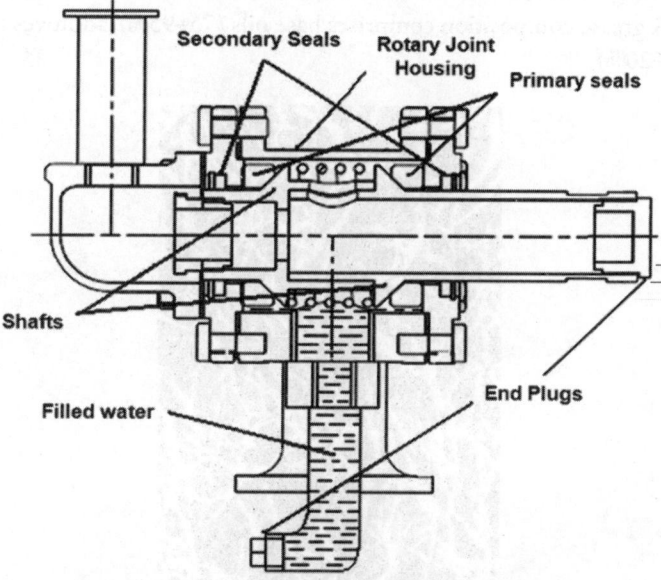

Fig. 1.7.1.8 Graphite seal in water environment

It can be said that graphite relies on adsorbed moisture or vapours to achieve low friction behaviour. At temperature lesser than 100 °C, possibility of adsorbed moisture or vapours is reduced; therefore graphite may not be effective lubricant. Lubrication performance of graphite increases with increase in temperature, but beyond 500°C the possibility of corrosion also increases. During the World War II, aircraft flew at higher altitude and electric motor brushes failed. Research into this problem revealed that graphite requires an adsorbed layer of water vapour to lubricate effectively. One of major restriction of graphite is its potential of + 0.25V, which can lead to severe galvanic corrosion of copper alloys and stainless steels.

4. Ceramic and cermet (metal bonded ceramic) coatings

Ceramics and cermet composites can be used at temperatures up to 1000 °C in applications where minimizing wear rate is more demanding than the minimizing friction. Use of bulk ceramic/cermet material requires special manufacturing process, which unnecessarily increases the manufacturing cost. To deal with this coating up to 0.5 mm thick on metal substrates is preferred. The coating of ceramic/ cermet can be applied using:

- Plasma spraying
- Impingement coatings from a detonation gun to the substrate metal; this method provides better adhesion between coating and substrate.
- Electrolytic deposition to coat small internal surfaces inaccessible by any other technique

1.7.2 Semi-solid lubricant

Grease

In layman's language, grease is a black or yellow sticky mass used in the bearings for lubrication purpose. Lubricating greases are low viscosity oils thickened by means of finely dispersed solids called thickeners. A grease composition comprises base oils (75–95%), additives (0–5%) and minute thickener fibres (5–20%).

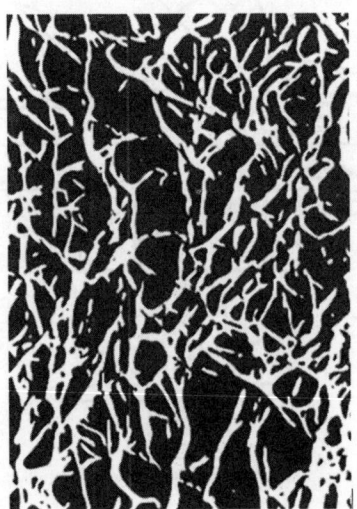

Fig. 1.7.2.1 Semi-solid lubricant

Base Oil: The viscosity of oil (naphthenic, paraffinic, PAO's, esters, silicones, glycols, etc.,) is selected based on the requirement of low/ high operating temperature. Lower viscosity base oil is selected to synthesize low temperature greases, while higher viscosity base oil is chosen to synthesize high temperature greases.

Additives: Chemicals are added to grease in order to enhance its performance and compatibility and to deal with environment. Solid lubricants, such as graphite, MoS_2, EP (extreme pressure) additives, enhance the performance and application of grease. Addition of graphite (graphite grease), PTFE (teflon grease) and MoS_2 (MOLY grease) reduces friction by increasing the adhesive film strength to avoid surface to surface contact. In other words, these additives provide a cushioning effect and keep metal surfaces apart from each other. Under very large load or relatively high operating temperature, extreme pressure additives (soluble compounds of sulphur, chloride, phosphorous) are mixed with oil. These additives chemically react with sliding metal surfaces to form films which are insoluble in the lubricant. Most widely used EP additives are:

- Tricresyl phosphate (TCP)
- Dibenzyldisulphide
- Zinc dialkyldithiophosphate (ZDDP).

Thickener: Organic (soap-based or non-soap based) and inorganic (non-soap based) thickeners are used to make greases. The fibre structure (Fig. 1.7.10) provided by the metal soap determines the mechanical stability and physical properties of the finished grease. The most commonly used economic grease is lime (calcium)-based grease (max. temperature 55–80 °C). Soda (sodium)-based grease (max. temperature 90–120 °C) is preferred over lime-based grease in rolling bearings.

Complex grease: Complex grease is similar to regular grease except that the thickener contains two dissimilar fatty acids, one of which is the complexing agent. This type of mixed soap structure has special properties that enable the grease to be heated to a higher temperature without losing its structure or oil separation from the thickener. This maximum temperature is referred to as the dropping point. The dropping point is critical because it is the point at which the grease reverts back to a liquid (the oil separates from the thickener). In other words to enhance performance characteristics of grease a complex agent is added to the soap thickener to convert it to a soap salt complex thickener. The greases are then referred to as 'complexes' and can be operated at relatively high temperature. A classification of greases based on simple soap, complex soap and non-soap thickeners is listed in Table 1.7.1.

Table 1.7.1 Classification of greases based on thickeners

Simple Soap			Complex Soap		Non Soap	
Lithium	Calcium	Aluminium	Lithium	Calcium	Silica Gel	Clay
Barium	Sodium	Strontium	Aluminium	Barium	Alumina	PTFE
Mixed Soap			Titanium		Polyurea	

Comparison among various types of greases has been provided in Table 1.7.2. The role of base oil to decide the operating temperature is given in Table 1.7.3.

Table 1.7.2 Comparative chart

Properties	Regular grease			Complex grease				Clay	
	Ca	Li	Na	Al	Ca	Ba	Li	Polyurea	Bentone
Dropping points	80	175	170	260	260	260	260	250	–
Max Temp	65	125	125	150	150	150	160	150	150
Low Temp	Fair	Good	Poor	Good	Fair	Poor	Good	Good	Good
Water Resist.	Exc	Good	Poor	Exc	Exc	Exc	Exc	Exc	Fair
Oxi. Res	Poor	Good	Good	Exc	Exc	Poor	Good	Exc	Good

Table 1.7.3 Role of base oil

Types of base fluid	Min. temp	Max. temp	Cost
1. Lithium + mineral oil	– 40° C	150°C	Medium
2. Lithium + esters	– 75° C	120°C	High
3. Lithium + silicones	– 55° C	205°C	High

Advantages of grease:
- Remains at application point and adhere to surface; therefore less frequent application needed; this characteristic makes grease very useful for inclined/vertical shafts.
- Seals out contaminants and less expensive seals needed.
- Water resistant and reduce oil vapour problems.
- Prolongs the life of worn parts by filing irregularities as shown in Fig. 1.7.2.2.
- Provides better mechanical lubrication cushion for extreme conditions such as shock loading, reversing operations, low speeds and high loads.
- Reduces noise and vibration.

Fig. 1.7.2.2 Irregularities filled by grease

Disadvantages of grease:
- Because of semi-solid nature of grease, it does not perform the cooling; so, poor dissipation of heat.
- Once dust or dirt enters the grease, it cannot be easily removed and would act as deterrent in performance.
- No filtration; so contaminants/ wear–debris cannot be separated.

Grease characteristics:

One of the important characteristics of grease is consistency, which is defined as the degree of hardness. Consistency is assessed by measuring the distance in tenths of mm to which a standard metal cone penetrates the grease (Fig. 1.7.2.3) under a standard load; the result is known as the penetration. A widely used classification of grease is that from the American National Lubricating Grease Institute

(NLGI). Table 1.7.4 lists the NLGI grades of greases. To find the consistency of grease, the following method can be used:

1. Grease surface (maintained at 25 °C) is smoothed out to make it uniform.
2. Cone release mechanism (Fig. 1.7.2.3) is activated and cone is allowed to sink for 5 seconds. Based on the depth of penetration (as listed in Table 1.7.4) consistency (in terms of NLGI grade) is defined. For example if depth of penetration of cone ranges in 17.5–20.5 mm, then consistency of grease is 4.

Table 1.7.4 National Lubricating Grease Institute (NLGI) grease classification

NLGI Grade	Penetration @ 25°C (1/10th mm)
000	445 – 475
00	400 – 430
0	355 – 385
1	310 – 340
2	265 – 295
3	220 – 250
4	175 – 205
5	130 – 160
6	85 – 115

Fig. 1.7.2.3 Cone arrangement to measure consistency

1.7.3 Liquid lubricants

The most common liquid lubricants are oil. SAE (Society of Automotive Engineers) and API (American Petroleum Institute) standards are used to classify oils. In SAE viscosity grade the letter 'W' is incorporated to indicate winter oils, i.e. 5W30, 10W30, 15W40. In general, the lower the first number, the better the oil performance in cold conditions. Conversely, the higher the second number, the better the oil protection at higher temperatures.

To specify lubricating oils for gasoline or diesel automotive applications, the API designation is used. In this gasoline engines oils are designated with 'S', while a 'C' designation is used for diesel engines.

Classification of liquid lubricants:

One way of classifying liquid lubricants is on the basis of their sources such as:
  i. Vegetable oils.
  ii. Animal fats.
  iii. Mineral oils.

(i) Vegetable (castor, rapeseed) oils:

- Less stable (rapid oxidation) than mineral oils at high temperature
- Contain more natural boundary lubricants than mineral oils; suitable for slow speed tribo–pairs operating at relatively low temperature

(ii) Animal fats:

These oils are extracted from animals and fishes. They are composed of fatty acids and alcohols. They are often called fixed oils because they do not volatilize unless they decompose, they are slow to oxidation and possess extreme pressure properties. Due to these properties, animal fats are added to mineral oils to improve film formation under high load and high temperature conditions. One of major problem of this class of lubricants is the limited availability.

(iii) Mineral oils:

Mineral oils are extracted from crude oils. Mineral oil consists of hydrocarbons (composed of 83–87% carbon and 11–14% hydrogen by weight) with approximately 30 carbon atoms in each molecule (composed of straight and cyclic carbon chains bonded together). Also it contains sulphur, oxygen and nitrogen. Based on the sulphur content, these oils are classified as Pennsylvanian oil (< 0.25%), Middle east (~1%), Venezuelan (~2%) and Mexican (~5%). Sulphur percentage ranging 0.1%–1.0% is preferred. This means Mexican and Venezuelan are least preferred.

Figure 1.7.3.1 indicates 'zone 1', 'zone 2' and 'zone 3' based on product of viscosity, speed and inverse of apparent pressure. Mineral oils are suitable for 'zone 3', while grease and mineral oil with fatty acids may be used for 'zone 2' and solid lubricants for 'zone 1'.

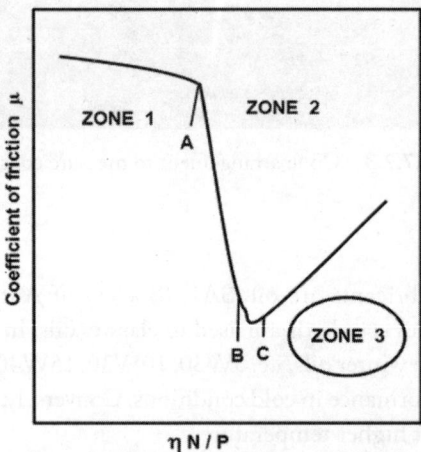

Fig. 1.7.3.1 Stribeck curve

Classification of mineral oils on the basis of their chemical form:

Mineral oils (as shown in Fig. 1.7.3.2) are classified as paraffins, naphthene and aromatic. For lubrication, paraffins are preferable choice compared to napthenes or aromatics.

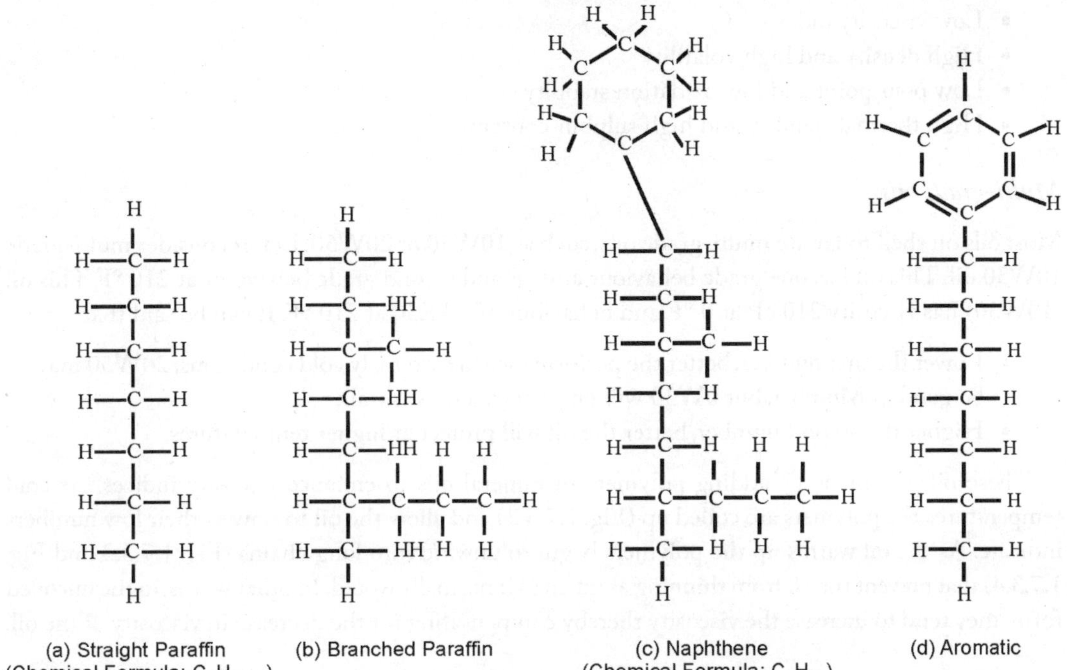

Fig. 1.7.3.2 Chemical forms of mineral oils

(a) Straight Paraffin (Chemical Formula: C_nH_{2n+2})
(b) Branched Paraffin
(c) Naphthene (Chemical Formula: C_nH_{2n})
(d) Aromatic

(i) Paraffinic oils:

These oils have good natural resistance to oxidation. But on oxidation they form acids; which means, when burnt, they leave a hard carbonaceous deposit. Some of the characteristics of paraffinic oils are:

- Good thermal stability
- Low volatility
- High viscosity index (VI = 90–115)
- High flash point
- Pour point higher than naphthenic or aromatic

(ii) Naphthenic oils:

Some of the characteristics of naphthenic oils are:

- Lower VI (15–75)
- Less resistant to oxidation
- Lower flash points than paraffinic
- Lower pour point than paraffinic; therefore good for low temperature applications
- When burnt soft deposits are formed, therefore abrasive wear is lower
- Oxidation leads to undesirable sludge type deposits

(iii) Aromatic oils:

The aromatic oils have very low viscosity index due to which they have very limited use. It is preferred to extract aromatic oil components during refining process of mineral oil. Some of the characteristics of the aromatic oils are:
- Low viscosity index (<50)
- High density and high volatility
- Low pour point and low oxidation stability
- High thermal stability and high sulphur contents

Multi-grade oils:

Most oils on shelf today are multi-grade oils, such as 10W30 or 20W50. Let us consider multi-grade 10W30 oil. This oil has one grade behaviour at 0 °F and second grade behaviour at 210 °F. This oil (10W30) has viscosity 210 cP at 0 °F and behaviour of SAE30 at 210 °F. It can be said that

- Lower the first number, better the performance in extremely cold conditions; 20W50 may be good in Mumbai, but 0W30 will be preferred in Kashmir.
- Higher the second number, better the oil will protect at higher temperatures.

These oils are made by adding polymers in mineral oils to enhance viscosity indices. At cold temperatures the polymers are coiled up (Fig. 1.7.3.3) and allow the oil to flow as their low numbers indicate. As the oil warms up the polymers begin to unwind into long chains (Fig. 1.7.3.3 and Fig. 1.7.3.4) that prevent the oil from thinning as much as it normally would. In other words, in the uncoiled form, they tend to increase the viscosity thereby compensating for the decrease in viscosity of the oil.

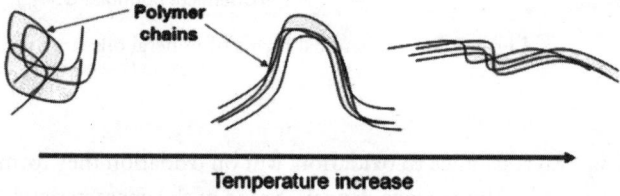

Fig. 1.7.3.3 VI improvers

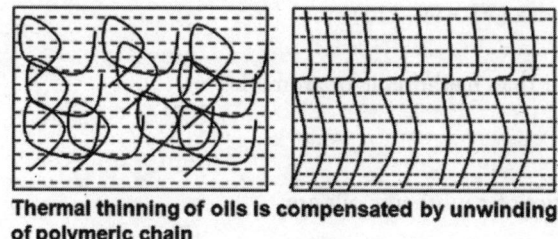

Fig. 1.7.3.4 VI improvers in action

Effectiveness of multi-grade oils is affected by the shear rate, the rate at which the oil has to pass through confined spaces. At high shear rate, viscosity of multi-grade oil may be little or no different from that of base oil as shown in Fig. 1.7.3.5.

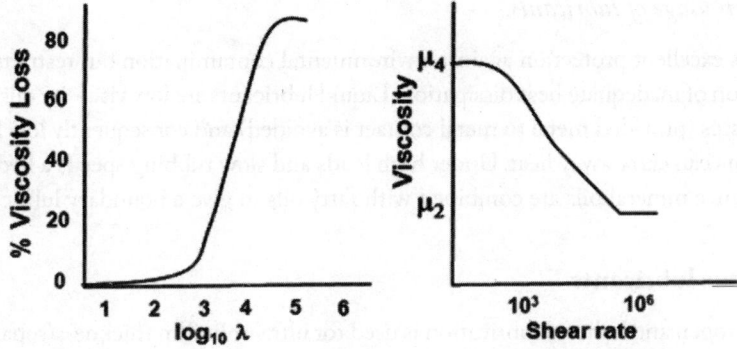

Fig. 1.7.3.5 Viscosity model of multi-grade oil (at high temperature)

Synthetic oils:

Synthetic oils are used in extreme operating conditions where conventional oils cannot be used. Conditions such as ambient temperature lesser than -120 °F, rotational speed more than 60000 rpm, and temperature higher than 500 °F cannot be handled by conventional oil. These harsh conditions require usage of synthetic oils that are engineered specifically in uniformly shaped molecules with shorter carbon chains to resist heat and stress. Almost every jet engine in the world uses synthetic lubricants. Synthetic oils are costlier compared to mineral oil; therefore, synthetic oils are selected where mineral oils are inadequate. Some of the advantages of synthetic oils are:

- Viscosity does not vary as much with temperature as in mineral oil.
- Rate of oxidation is much slower.

Common synthetic oils are:
(i) Polyglycols (polyalkylene glycol):
    These oils can absorb water and have viscosity index of 200. These oils have been used as brake fluids and are described in *Engineering Tribology*, by G. Stachowiak. Polyglycols offer distinct advantages as lubricants for systems operating at high temperatures such as furnace conveyor belts, where the polyglycol burns without leaving any carbonaceous deposit. It is also used in textile industry.
(ii) Esters:
    Esters are better than mineral oils in reducing friction, resisting oxidation, prolonging draining period and volatility. Their cost is only a little more than mineral oils.
(iii) Silicon:
    ☺ VI = 300. Good for space application; chemically inert
    ☹ Poor boundary lubricant; low solubility; high production cost
(iv) Perfluoro polyalkyl ether: good oxidation and thermal stability; VI= 200; in vacuum can be used for thin film lubrication
(v) Perfluoropolyethers:
    High oxidation (320 °C) and thermal (370 °C) stability; has low surface tension; chemically inert

Few remarks on usage of lubricants:

Grease provides excellent protection against environmental contamination but restricts to a speed of 2 m/s for the reason of inadequate heat dissipation. Liquid lubricants are low viscosity oils which have low fluid friction losses (provided metal to metal contact is avoided) and consequently low heat generation. Liquid lubricants can carry away heat. Under high loads and slow rubbing speeds a hydrodynamic film cannot form, hence mineral oils are combined with fatty oils to give a boundary lubrication layer.

1.7.4 Gaseous lubricants

Gas (i.e., air, nitrogen and helium) lubrication is used for ultra-thin film thickness(separation) between tribo–pairs.

Advantages:

- Temperature range (-200–2000 °C); no vapourization, cavitation, solidification and decomposition
- Very low viscosity (0.001 times of the thinnest mineral oil); therefore, ultra-low friction; possible high speed
- Cleanliness
- No seal requirement for lubrication

Disadvantages:

- Very low load capacity; low damping; ultra-low film thickness
- Smooth surfaces and very low clearance (to maximize load capacity and minimize flow rate); needs a specialist designers and manufacturer (close tolerance)
- Less forgiving of errors in estimating loads or of deviations from specifications during manufacture and installation.

1.8 Lubricant Additives

1.8.1 Need of Additives

In applications where hydrodynamic lubrication films are formed there is no need of additives. But to counteract high speed, high temperature, high load, etc., additives are required. Practically all lubricants contain additives to enhance existing properties, or to impart new properties. Additives need to be soluble or uniformly dispersed throughout the carrier media (liquid base such as mineral oil, synthetic fluid, etc., or grease). Some of the applications of additives are:

- Gears are subjected to very high contact pressure (Fig. 1.8.1.1), and experience metal to metal contact at gear teeth. Lubricants with extreme pressure (EP) additives are required.
- I.C. engine parts are subjected to high temperatures. Additives to delay oxidation are required.
- Detergent and dispersant additives to remove combustion and breakdown products of the oil from the surfaces.
- Corrosion inhibitors to prevent corrosion caused by combustion and oxidation products.

- Refrigeration system lubricants encounter the low temperatures (below 0 °C), the additives are needed to lower the pour points.

Fig. 1.8.1.1 High contact pressure in gears

Importance of lubricant additives is indicated in Table 1.8.1. Engine oil (base oil + detergent and dispersant additives + corrosion inhibitors + antiwear additives) reduces wear to mild wear regime.

Table 1.8.1 Importance of lubricant additives

Surface 1	Surface 2	Lubricant	Wear Constant (k_1)
52100 steel	52100 steel	None	1.0×10^{-3}
52100 steel	52100 steel	Paraffinic oil	3.2×10^{-7}
52100 steel	52100 steel	Paraffinic oil + Additive	3.3×10^{-8}
52100 steel	52100 steel	Engine Oil	2.0×10^{-10}

1.8.2 Types of additives

Table 1.8.2 Lubricant additives

Action	Element
Detergent of dispersant additive	Barium (Ba), Calcium (Ca), Magnesium (Mg), Sodium (Na)
Extreme-pressure additive	Boron (B)
Anti-wear additive	Copper (Cu), Lead (Pb), Phosphorus (P), Zinc (Zn)
Friction modifier	Molybdenum (Mo)
Anti-foaming additive	Silicon (Si)
Anti-oxidant additive	Zinc (Zn)

(1) Detergent and dispersant additive:

Detergents and dispersants additives are blended into lubricants to remove and neutralize harmful products.

Detergents:

Detergent additives are soaps of high molecular weight, soluble in oil (functional) group. Detergents use metallic basis, such as barium, calcium, magnesium and sodium. Detergents form a protective layer on the metal surfaces to reduce the amount of acidic materials produced and prevent deposition of sludge and varnish. The protective ability of detergent is measured by its total base number or its reserve alkalinity.

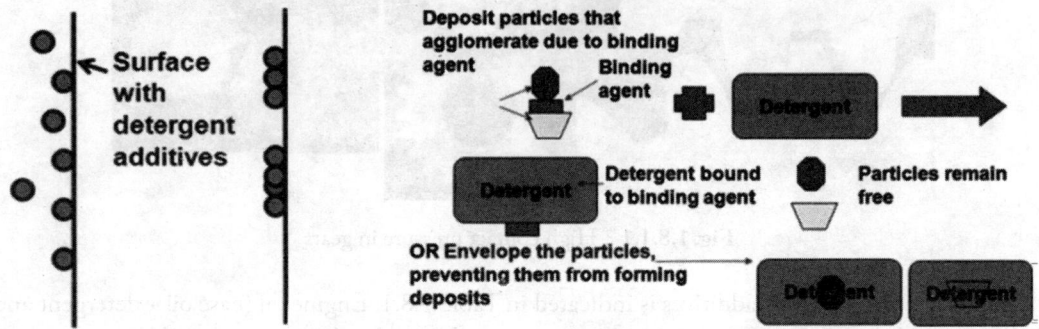

Fig. 1.8.2.2 Detergent additive action   **Fig. 1.8.2.3** Detergent additives in action

Dispersants:

Dispersants have a large hydrocarbon tail and a polar group head. Tail section serves as a solubilizer in the base oil, while polar (functional) group attracts particulate contaminants in the lubricant. Purpose of dispersant additives is to suspend or disperse harmful products (i.e., dirt, water, fuel, process material, and lube degradation products such as sludge, varnish, oxidation products) within the lubricant.

(2) Anti-wear additive:

Anti-wear additives typically contain zinc and phosphorus compounds. These additives prevent metal to metal contact under lighter to moderate loads. With increase in load, anti-wear additive may be ineffective and EP additives are required in heavy load applications, such as gearboxes.

Anti-wear additives are chemically active. They form a coat of a protective layer on the metal surface by chemical decomposition and absorption. The Zinc dialkyldithiophosphate (ZDDP) is probably the most widely used in formulated engine oils as anti-wear additives. Stearic acid is also used as anti-wear additive. Molybdenum disulfide and graphite additives are a special form of anti-wear additives known as anti-seize agents. Once these additives get depleted oil needs to be changed.

(3) Anti-foaming agents (foam inhibitors):

Lubricant foams (Fig. 1.8.2.4) due to agitation and aeration that occur during operation. Foaming interfere with flow rate and heat transfer; and increase oxidation. The anti-foam additives (usually long chain silicon polymers are used in small quantities of about 0.05%–0.5% by weight) lower the surface tension between the air and liquid to the point where bubbles collapse. It is interesting to note that detergent and dispersant additives tend to promote foam formation and always require anti-foaming additives in same oil to reduce the formation of foam.

Fig. 1.8.2.4 Formation of air bubbles in lubricant

(4) Anti-oxidant additive (oxidation inhibitors):

Under high temperature and pressure oxidation of lubricating oil occurs. Products of oxidation are sticky in nature (gummy) and if deposited on material (i.e. cadmium, copper and lead alloys) surface increases the chances of corroding those materials. This result in increased power loss due to increased viscous drag and difficulties in pumping the lubricant also increases. Therefore it is recommended to replace oil if TAN > 3. TAN stands for Total Acid Number, and it is the amount of potassium hydroxide in milligrams that is needed to neutralize the acids in one gram of oil.

Antioxidant additives are often classified into corrosion and rust inhibitors.

Corrosion inhibitors: Used for non-ferrous metals (i.e. copper, aluminium, tin and cadmium) used in bearings and seals. These additives protect surfaces against any corrosive agents (sulphur, phosphorus, chlorine and oxidation products) present in oil.

Rust inhibitors: Needed for ferrous metals particularly to trap (Fig. 1.8.2.5) oxygen dissolved in oil and water. These polar additives are adsorbed strongly upon metal surface and neutralize acids formed by oxidation. Sulphonate, phenate and amines are few examples of rust inhibitors additives.

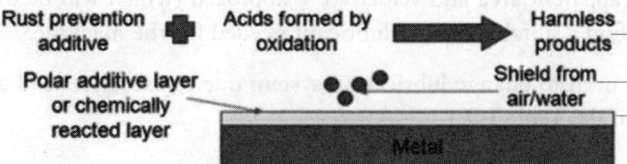

Fig. 1.8.2.5 Rust prevention additives

(5) Pour point depressants:

Pour point is the lowest temperature at which the lubricant will flow. Waxy crystals are formed at a temperature lower than pour point; therefore additives like methacrylate polymers, polyalkylphenol esters are required to encapsulate waxy crystals to stop their growth. In this way pour point depressants effectively reduce the pour point and lubricants operate at lower temperatures.

1.8.3 Interference between additives

Additives often interfere with each other; for example, detergents and rust inhibitors can significantly suppress lubricating action of ZDDP. Corrosion inhibitors contaminated with ammonia lead to extensive plant damage. Dispersants accelerate the oxidation of oil and anti-oxidants must be included when these additives are used.

1.9 Lubrication Selection

A selected lubricant must match the design, operation and environmental requirements. Controllable friction and wear are two essential operational requirements from a lubricant. Sometimes additional requirements are cooling, dealing with contamination and corrosion problems. Following step by step procedure can be followed to choose appropriate lubricant:

Step 1: To reduce the initial cost we must avoid re-lubrication (motor–pump unit, pipe lines, a number of pipe fittings) requirement. Therefore it is necessary to analyze the requirement. If there is no need of continuous supply of lubricant, such as in watches, clocks, door locks, sewing machines, etc., lubricant based on load and speed can be selected as shown in Fig. 1.9.1. Environment and sealing requirements are additional factors which affect lubricant selection. Apparent area, material conductivity and friction coefficient decide the operating temperature.

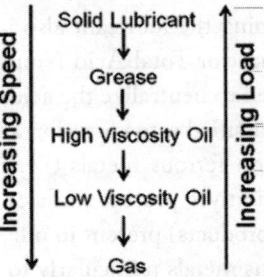

Fig. 1.9.1 Lubricant selection

To account load, apparent area and velocity PV approach (which will be discussed in following chapters) is used to find a suitable type of lubricant needed for the machine.

Step 2: Is there any need to change lubrication system due to excessive load and/or speed, heat or debris? If yes, refer to the Table 1.9.1.

Table 1.9.1 Lubricant selection

Too much speed	Lesser viscous oil, Oil circulation system with greater oil flow rate, gas lubrication.
High operating temperature	High VI oil with anti-oxidant additives, greater oil-flow rate, Solid lubricants.
Too much debris	Circulation system with filtration.
Requirement of long life	Oil/grease with additives, provision for re-lubrication.
Too much load	More viscous oil, grease, EP additives, solid lubricants

Step 3: If lubricant selection is to be made for complete assembly (i.e., I.C. Engine), then it will be preferable to use same lubricant for all tribo–pairs of that assembly.

(a) Single oil reservoir and circulation system can be used which would prove economic, reliable, easily storable and there is lesser chances of wrong usage of lubricant.

(b) Due to self-compensating behaviour of oil viscosity, slightly higher value of viscosity can be selected.

1.10 Defining Bearing Terminology

Bearing is a load support that permits relative motion between two parts, such as the shaft and the housing, with minimum friction. Based on the direction of load support, bearings are classified as radial and thrust (axial) bearings as shown in Fig. 1.10.1. A radial bearing supports the load, P, that is perpendicular to the axis of the shaft. A thrust bearing supports the load, W, that acts along the axis of the shaft.

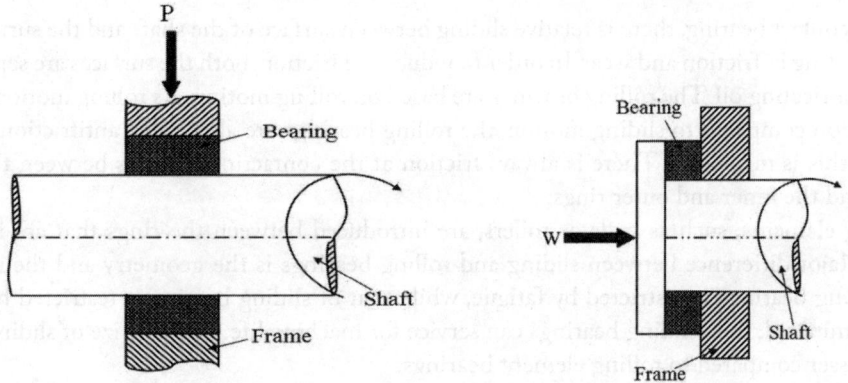

Fig. 1.10.1 Radial and thrust bearing

Bearings may also be classified based on the relative (sliding and rolling) motion between bearing surfaces into sliding contact bearings and rolling contact bearings, as shown in Figs. 1.10.2 and 1.10.3.

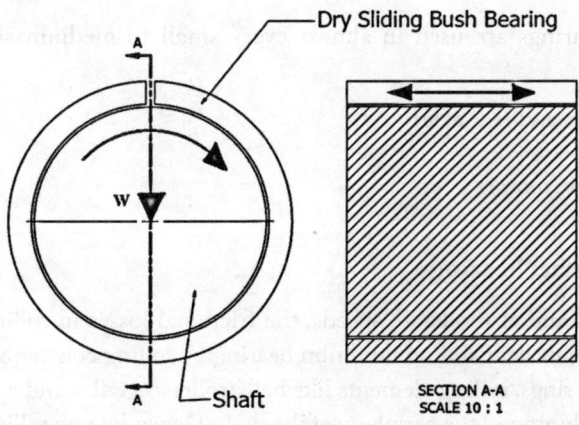

Fig. 1.10.2 Sliding contact bearing

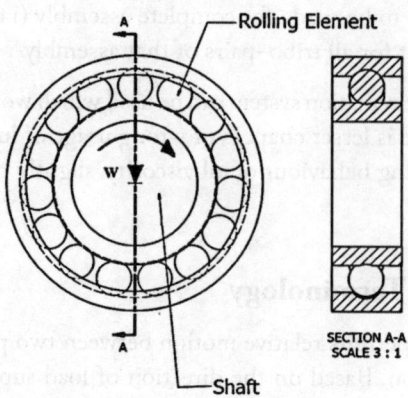

Fig. 1.10.3 Roller contact bearing

1.10.1 Comparison between sliding and rolling contact bearings

In sliding contact bearing, there is relative sliding between surface of the shaft and the surface of the sleeve, resulting in friction and wear. In order to reduce the friction, both the surfaces are separated by a film of lubricating oil. The rolling bearings are based on rolling motion. As rolling motion provides lesser friction compared to sliding motion, the rolling bearings are also called antifriction bearings. However, this is misnomer. There is always friction at the contacting surfaces between the rolling element and the inner and outer rings.

Rolling elements, such as balls or rollers, are introduced between the rings that are in relative motion. Major difference between sliding and rolling bearings is the geometry and their life. The life of rolling bearings is restricted by fatigue, while that of sliding bearing is restricted by wear. If wear is eliminated, then sliding bearings can service for machine life. Further size of sliding bearing is much lesser compared to rolling element bearings.

Few applications of sliding contact bearings are:

    i. crankshaft and connecting rod bearings in petrol and diesel engines
   ii. centrifugal pumps
  iii. large size electric motors and ships
  iv. steam and gas turbines

Rolling contact bearings are used in almost every small to medium size machines. Typical applications are:

    i. automobile front and rear axles
   ii. gear boxes
  iii. small size electric motors

1.10.2 Rolling contact bearings

For starting conditions and at moderate speeds, the frictional losses in rolling contact bearing are lower than that of an equivalent sliding fluid film bearing. A rolling contact bearing consists of four parts—inner and outer rings, rolling elements like balls/rollers/needles and a retainer that holds the rolling elements at evenly around the periphery of the shaft. Depending on rolling elements, the rolling bearings are classified as ball bearings, cylindrical roller bearings, taper roller bearings and needle bearings. The types of rolling contact bearing, which are frequently used, are shown in Fig. 1.10.2.1 The most frequently used rolling bearing is deep groove ball bearing. In this type of bearing, the radius of the ball is slightly less than the radii of curvature of the grooves in the rings. Kinematically this gives a point of contact between the balls and the rings. When larger load carrying capacity is required in a given space, the point of contact in ball bearing is replaced by the line contact of rolled bearing. Cylindrical roller bearing consists of relatively short rollers that are positioned and guided by the cage.

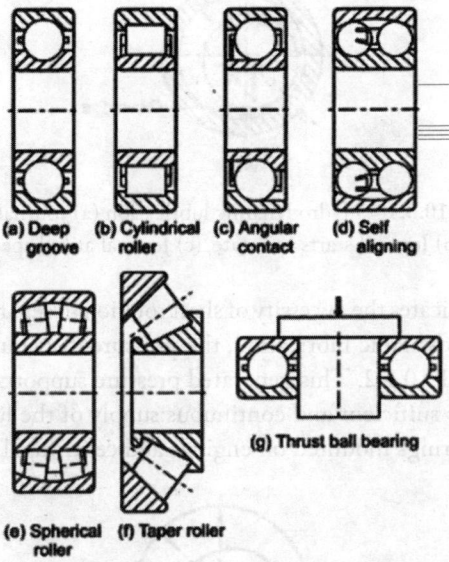

Fig. 1.10.2.1 Types of rolling contact bearing

To deal with misalignment problems, there are two types of self-aligning rolling contact bearings, viz. self-aligning ball bearing and spherical roller bearing. In these cases, the assembly of the shaft, the inner ring and the balls/roller with cage can freely roll and adjust itself to the angular misalignment of the shaft. Both types of self-aligning bearing permit minor angular misalignment of the shaft relative to the housing and are used in agricultural machinery, ventilators, and railway axle–boxes.

The taper roller bearing consists of rolling element in the form off-rustum of a cone. They are arranged in such a way that the axes of individual rolling elements intersect in a common apex point on the axis of the bearing. In kinematics analysis, this is the essential requirement for pure rolling motion between conical surfaces. A taper roller bearing subjected to pure radial load induces thrust component and vice versa. Therefore, taper roller bearings are always used in pair to balance the thrust component. Taper roller bearings provide better rigidity and can be easily assembled and disassembled due to separable parts.

1.10.3 Sliding contact bearings

Idle form of sliding bearings is those operating in hydrodynamic lubrication regime so that maintenance free infinite bearing can be achieved. Hydrodynamic lubrication (shown in Fig. 1.10.3.1) is defined as a system of lubrication in which the load supporting fluid film is created by the shape and relative motion of the sliding surfaces.

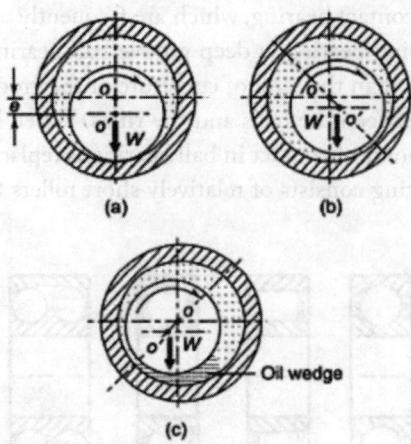

Fig. 1.10.3.1 Hydrodynamic lubrication (a) Journal at rest,
(b) Journal starts to rotate, (c) Journal at full speed

Figure 1.10.3.1 clearly indicates the necessity of shaft rotation to separate bearing from shaft surface. In full film regime of hydrodynamic lubrication, the pressure distribution around the periphery of the journal is shown in Fig. 1.10.3.2. This generated pressure supports the external load, W. In this case, the only requirement is sufficient and continuous supply of the lubricant. Such hydrodynamic lubrication is realized in bearings mounted on engines and centrifugal pumps.

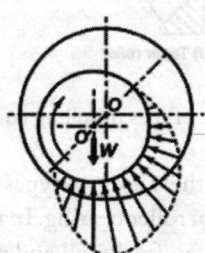

Fig. 1.10.3.2 Pressure distribution in hydrodynamic bearing

From Fig. 1.10.3.2 it is clear that pressure develops only in half of the bearing. Based on this understanding, two types of hydrodynamic journal bearings can be considered namely, full journal bearing and partial bearing. The construction of full and partial bearings is illustrated in Fig. 1.10.3.3.

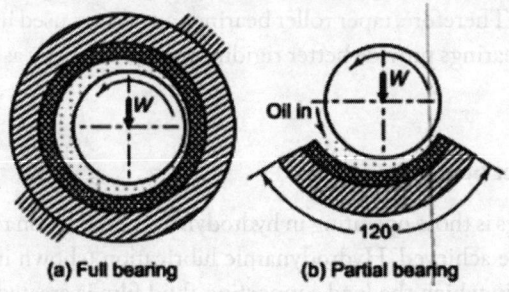

Fig. 1.10.3.3 Pressure distribution in hydrodynamic bearing

Frequently Asked Questions

Q.1. How is tribology useful in the conservation of energy?

Ans. The knowledge of tribology is useful in reducing the unnecessary friction and wear between two rubbings surfaces (tribo–pair). Using tribology appropriate lubricant and lubrication mechanism can be adopted to minimize the friction and eliminate wear which would reduce the wastage of energy and enhance working life of tribo–pair.

Q.2. How do tip-based microscopes like Atomic Force Microscope (ATM) and Scanning Tunnelling Microscope (STM) help in the study of tribology?

Ans. Atomic force microscopy (AFM) and scanning force microscopy (SFM) are high-resolution type of microscopes, having resolution of the order of fractions of a nanometre, more than 1000 times better than the optical diffraction limit. Such high resolution microscopes help in understanding the interaction of asperities between the two rubbing surfaces and the behaviour of lubrication on the surface, and therefore the study of tribology.

Q.3. What is/are the various science subjects whose knowledge is required for tribology?

Ans. Tribology requires the knowledge of multifaceted disciplines like solid mechanics, fluid mechanics, material science, chemistry, etc. For example calculations of contact stresses and surface temperature during sliding require understanding of 'solid mechanics'. Similarly study of lubricant film formed between different geometric shapes of interacting surfaces demand knowledge of fluid mechanics. Mechanical wear at atomic and micro scales involve thorough understanding of material science. Formation of boundary layer on the solid surface demands information on chemistry. In short, integration of knowledge from multifaceted disciplines (solid mechanics, fluid mechanics, material science, chemistry, etc.) is essential and therefore a separate subject is required.

Q.4. What is a better method for quantifying surface roughness: average roughness or root mean square roughness and why?

Ans. Surface roughness is vertical deviations from nominal surface/line. Often surface roughness is quantified as average and root mean square roughness. Root mean square method is a better measure of quantifying surface roughness as it involves the integration of deviations from the nominal surface.

Q.5. How can one utilize knowledge in tribology to make a mechanical system much more efficient?

Ans. The tribology knowledge can be utilized by lubricating all the joints and moving/ rubbing pairs in the mechanical system which would not only reduce friction, wear, corrosion, etc., but would also make the system much more efficient and reliable by reducing mechanical wear. Due to reduced friction, wear and corrosion, the longevity of the system would improve. The tribology knowledge would also help in identifying the right kind of lubricant and lubrication mechanism for the system.

Q.6. Can knowledge on tribology be utilized in the initial design of a component / system / product to improve its efficiency?

Ans. Tribology knowledge can be very useful during the initial design of the component. The geometry, surface finish and material selection of the component can be done as per the environment and conditions in which the component would be operating.

Q.7. Can efficiency of a system be increased to 100% with the use of tribology knowledge?

Ans. It is difficult to achieve 100% efficiency for any system but the use of tribology knowledge would definitely help in improving the efficiency level.

Q.8. Is the use of tribology only confined to the use of mechanical systems involving two rubbing surfaces?
Ans. Tribology is not only confined to two rubbing surfaces in a mechanical system but also covers a broad area of lubricants like magneto rheological lubricants whose viscosity can be varied as per the load requirements. It also covers bio-systems (all joints), electric system (pin–socket joint), hard-disk drive, etc.

Q.9. Out of all the lubrication mechanisms, which is the best method and why?
Ans. Different lubrication mechanisms are used as per different requirements and applications and it would be difficult to categorize as one of them as best. There are different types of lubrication mechanisms like fluid film lubrication, elasto–hydrodynamic lubrication, boundary lubrication; and each one has different usage. Elasto–hydrodynamic lubrication provides minimum friction and zero wear, therefore it is preferable compared to other lubrications, but operating regime of EHL is very narrow.

Q.10. As per the tribology, which are the two best surfaces that produce minimum friction levels?
Ans. In general the surfaces that possess very high surface finish and are hard enough to withstand wear are the best surfaces to minimize friction levels.

Q.11. Is the scope of tribology only limited to reduce friction levels between two mating surfaces?
Ans. No, main aim of tribology is to eliminate wear by controlling friction between tribo–surfaces. Reduction in wear enhances the service life and provides higher returns compared to returns obtained by reducing friction.

Q.12. What are additives and how can additives like friction modifiers enhance the quality of lubricant?
Ans. Basic lubricating oils, having appropriate viscosity, may be useful for idle tribological conditions. In practice, very high load, very low relative speed, environment humidity, surrounding temperature, etc., affects the performance of basic oils. Therefore, to make basic oil more useful for practical situations additives are mixed with basic oils. In other words, additives are soluble compounds that enhance the characteristics of lubricant. Typically lubricants contain 90–95% base oil (most often petroleum fractions, called mineral oils) and 5–10% additives. Friction modifiers are additives that reduce friction by increasing the adhesive film strength to avoid surface to surface contact. In other words these additives provide a cushioning effect and keep metal surfaces apart from each other.

Q.13. What is the effect of adding additives to the lubricants?
Ans. In practice, liquid lubricants are subjected to a number of odd situations like formation of foam, oxidation, thermal thinning, squeezing out from tribo–surface, sludge formation, etc. To reduce the effect of those odd situations on lubrication, additives are used. Anti-foam additive reduce the surface tension of lubricant and allow better lubrication. Oxidation resistant additive reduce the chances of liquid lubrication. Boundary additives reduce the effect of liquid squeezing out. Similarly, polymer additives reduce the effect of temperature on the liquid viscosity.

Q.14. What is extreme pressure lubrication and what are the environmental hazards associated with it?
Ans. Extreme pressure term for additive is a misnomer. 'Extreme pressure' additive acts at high temperature and reacts chemically with metal to form a low shear interface film on metals. For example EP additives containing chlorine, sulphur, and phosphorus react with metal surfaces at relatively high temperature and form chloride, sulphide and phosphate lubricant layers respectively. Extreme pressure additives are usually used in high contact pressure applications such as gearboxes. Major difficulty with extreme pressure lubricants is their carcinogenic nature

and environmental pollutant, therefore extreme pressure additives should be avoided as far as possible.

Q.15. What is the difference between absorption and adsorption?

Ans. Adsorption implies adhesion and absorption implies diffusion. In adsorption a gas or liquid solute accumulates on the surface of a solid and forms molecular or atomic films (the adsorbate). This process is different from absorption, in which a substance diffuses into a liquid or solid to form a solution. Absorption is a bulk phenomenon in which the particles of gas or liquid get uniformly distributed throughout the body of the solid whereas adsorption is the phenomenon of higher particle concentration of gas or liquid on the surface so that film formation occurs. It can be said that absorption is a 'bulk phenomenon', but adsorption is a 'surface phenomenon'.

Q.16. Why does the viscosity of a lubricant decrease with increase in temperature?

Ans. Lubricant may be gaseous or liquid. In liquid, with increase in temperature molecular separation increases and their resistance to flow decreases, therefore decrease in viscosity with increase in temperature occurs. But in the case of gaseous lubricants, the molecular activities increase with increasing temperature. This means viscosity of gaseous lubricant increases with increase in temperature.

Q.17. Is there any liquid lubricant that shows negligible changes in viscosity with variation in temperature?

Ans. There is no liquid lubricant that shows negligible change in viscosity with variation in temperature. Some synthetic lubricants, such as silicone which has very high VI = 300, have relatively weaker viscosity–temperature relation.

Q.18. Is there any standard parameter which denotes the viscosity variation of liquid lubricant with respect to temperature?

Ans. Viscosity of all liquid lubricant decreases as the temperature increases, but rate of decrease varies considerably. The sensitivity of viscosity thinning on increase in temperature is expressed by VI = Viscosity Index. In other words VI parameter may be used to denote the viscosity–temperature relation.

Q.19. What are solid lubricants?

Ans. Solid lubricants are self-lubricating composites, having relatively low shear strength.
These lubricants are able to transfer a layer on the tribo–surfaces and as a result reduce the friction coefficient. The most common solid lubricants are polymers [polytetrafluoroethylene (PTFE), nylon], metal–solid [molybdenum disulfide], carbon and graphite, ceramics, etc.

Q.20. Is it necessary for solid lubricants to possess lesser hardness as compared to the surfaces being lubricated?

Ans. Yes it is necessary for the solid lubricants to possess lesser hardness as compared to the surfaces being lubricated in order to reduce the wear of the surfaces. Use of harder lubricants will damage the surfaces and would increase the surface roughness.

Q.21. How does a solid lubricant help in reducing friction and what are their specific advantages as compared to liquid lubricants?

Ans. A solid lubricant is basically any solid material which can be placed between tribo pair and which will shear more easily under a given load than the tribo–materials themselves. In general solid lubricants help filling the surface irregularities in order to reduce the surface roughness and thus reducing friction. Solid lubricants are effective at high loads, high temperature and high pressure as compared to liquid lubricants.

Q.22. What are the characteristics of good lubricants? What is lubricity?

Ans. A good lubricant shall possess the following characteristics:
- High boiling point
- Low freezing point
- High viscosity index
- Thermal stability
- Corrosion prevention
- High resistance to oxidation

The property of lubricant of reducing friction is known as lubricity.

Q.23. What are the major factors which affect the selection of lubricants?

Ans. Load and speed are two major factors which affect selection of lubricants. Requirement of fluid sealing, material conductivity, friction coefficient, and surrounding environment are additional factors which affect lubricant selection.

Q.24. Can the oil used for lubricating a diesel engine be used in a gasoline engine?

Ans. It is not advisable to use the lubricating oil suitable for diesel engine for lubricating gasoline engine because diesel-engine lubricants are more viscous as compared to lubricants used in gasoline engine. Use of too thick oil in a gasoline engine will make it run hotter. Higher viscosity oils impose start-up problems and are difficult to pump the lubricant. This may lead to lubricant starvation of tribo–surfaces.

Q.25. Why foam inhibitors are used in lubricants and what purpose do they serve?

Ans. Lubricant foam formation occurs due to aeration. Foaming interfere with flow rate and heat transfer, and increase oxidation. The foam inhibitors, usually long chain silicon polymers, reduce the surface tension between the air and liquid to the point where bubbles collapse thereby reducing the foaming effect. These additives are used in small quantities of about 0.05% to 0.5% by weight.

Q.26. Different additives are added in the lubricants to enhance the specific properties of the lubricants. Does addition of multiple additives create interference with the desired characteristics for which they are added?

Ans. Yes, addition of multiple additives does create interference. For example, dispersants accelerate the oxidation of oil and anti-oxidants must be included when these additives are used. Hence it is very important to have complete detail of the additives, i.e., about their effects and interferences to existing properties of the lubricant.

Q.27. If a lubrication mechanism has to be designed for a complete mechanical system, then should a single lubricant be used for the complete system or different lubricants shall be preferred for different sub systems.

Ans. It is better to use a single lubricant for the complete mechanical system because single oil reservoir and circulation system can be used for the complete system which would prove economical, reliable and easily storable. Use of single reservoir system also reduces the chances of wrong usage of lubricant in different sub-systems. Although keeping in mind the functional and design requirements, if a particular subsystem requires the use of different lubricants, then more emphasis must be paid to the choose solid lubricants or inert coating for critical tribo–surface in addition to single lubricant for whole machine.

Q.28. Are there any general guidelines or thumb rules for lubrication selection?

Ans. There are some guidelines which can definitely help in identifying the general characteristics of lubricants for a tribological system such as:

- Too much speed: lesser viscous oil, oil circulation system with greater oil flow rate, gas lubrication
- High operating temperature: high viscosity index oil with anti-oxidant additives, greater oil flow rate, solid lubricants
- Too much debris: circulation system with filtration
- Requirements of long life: oil/grease with additives, provision for re-lubrication
- Too much load: More viscous oil, grease, EP additives, and solid lubricants

Q.29. What is journal bearing and what are the common components of journal bearing?

Ans. Journal or plain bearings consist of a journal which rotates freely in a supporting sleeve. Journal is the part of the shaft in contact with the bearing which slides over the bearing surface. There are no rolling elements in these bearings. The common components of journal bearing are:
- Housing
- Journal or shaft
- Bearing liner
- Oil inlet
- Drain

Q.30. Why seals are increasing friction even when there is no contact with rolling element?

Ans. Seals, in rolling element bearings, remain in mechanical contact with inner and outer rings. Either inner or outer ring needs to relatively rotate; therefore, seals slide against rotating rings. Such sliding motion between ring and seal causes additional friction force in rolling element bearings.

Multiple Choice Questions

Q.1. In which year the word Tribology was coined to represent an interdisciplinary approach that was required to contain the significant losses arising out of lack of knowledge in friction and wear.
  (a) 1960 (b) 1966
  (c) 1964 (d) 1970

Q.2. Out of the following disciplines which one is not considered for an interdisciplinary approach in tribology?
  (a) Solid and fluid mechanics (b) Chemistry
  (c) Material science (d) Industrial engineering

Q.3. The meaning of the Greek word 'Tribos' from which the word Tribology is coined is
  (a) Rubbing (b) Movement
  (c) Fluid (d) Heat

Q.4. Which one of the following is not the purpose of Tribology?
  (a) Improve service life (b) Increase safety and reliability
  (c) Reduce fatigue (d) Increase heat generation

Q.5. Asperities are basically
  (a) Sharp tips on surface (b) Edge of a surface
  (c) Corner of a surface (d) Hole in a surface

Q.6. Which one is not a standard method for quantifying surface roughness?
(a) Root mean square roughness
(b) Average roughness
(c) Geometric Dimensions and Tolerance (GD&T)
(d) Rating method on any arbitrary scale

Q.7. Which one of the following statements is true?
(a) Wear rate increases with increasing load.
(b) Wear rate decreases with increasing temperature.
(c) Wear rate decreases with increasing speed.
(d) Wear rate is independent of load/ temperature.

Q.8. The purpose of lubricant filter system is
(a) To remove the debris from the lubricant.
(b) To enhance the viscosity of lubricant.
(c) Reduce the temperature of lubricant.
(d) Reduce the quantity of lubricant.

Q.9. Which one of the following parameter is not included by Stribeck curve?
(a) Viscosity of the lubricant.   (b) Speed of the surfaces
(c) Load at the interface   (d) Surface roughness

Q.10. The purpose of lubrication is
(a) To reduce friction.   (b) To reduce wear.
(c) Transfer heat produced.   (d) All of above.

Q.11. Which of the following is not a function of lubricant in IC engine?
(a) Form a film to separate the surfaces.
(b) Adhere to surface.
(c) Withstand high temperature inside the cylinder.
(d) Reduce the size of the asperities and improve the surface finish.

Q.12. Synovial fluid is a lubricant that is found in
(a) Human bone joints.   (b) Journal bearings.
(c) IC engine.   (d) None of the above.

Q.13. Which one of them is a correct combination?
1. Boundary lubrication   (i) Dimensionless film thickness <1
2. Hydrodynamic lubrication   (ii) Dimensionless film thickness lies between 1 and 3.
3. Mixed lubrication   (iii) Dimensionless film thickness lies between 3 and 5.
4. Elasto–hydrodynamic   (iv) Dimensionless film thickness is greater than 5.
   lubrication

(a) 1-(i), 2-(iv), 3-(ii), 4-(iii)   (b) 1-(iv), 3-(iii), 2-(i), 4-(ii)
(c) 2-(i), 3-(iv), 4-(iii), 1-(ii)   (d) 3-(iv), 2-(iii), 1-(i), 4-(ii)

Q.14. As the temperature is increased, the coefficient of friction
(a) Increases   (b) Decreases
(c) Remains unchanged   (d) May increase or decrease

Q.15. Which of the following is a desirable property of boundary lubricant?
(a) Dissolvability in lubricating oils.
(b) Reactivity with metals in lubricating oils.
(c) Low shear strength and high melting point.
(d) All of above.

Q.16. The major disadvantage with extreme pressure lubricants is
(a) Carcinogenic nature of the lubricant.
(b) Low melting point.
(c) It is ineffective.
(d) All of above.

Q.17. In hydrodynamic lubrication the major source of friction is
(a) Shearing of lubricant film.
(b) Viscosity of oil lubricant.
(c) Both (a) and (b).
(d) None of the above.

Q.18. Which of the following statements is true about viscosity?
(a) Dynamic viscosity is the ratio of shear stress to the resultant shear rate.
(b) Kinematic viscosity is equal to dynamic viscosity divided by density.
(c) The CGS unit of dynamic viscosity is centipoise and CGS unit of kinematic viscosity is centistokes.
(d) All of above

Q.19. The lubricant film thickness in a radial journal bearing operating in hydrodynamic lubrication regime is dependent upon
(a) Applied load and relative velocity
(b) Viscosity
(c) Both (a) and (b)
(d) None of these

Q.20. Viscosity of multi-grade oils
(a) Reduces with temperature
(b) Increases with temperature
(c) Is less sensitive to temperature
(d) None of the above

Q.21. Viscosity Index denotes
(a) Relationship between the dynamic and kinematic viscosity
(b) Relationship between viscosity and temperature
(c) Relationship between viscosities of different lubricants
(d) Relationship between viscosity and pressure

Q.22. Which of the following is true in modelling viscosity temperature relationship?
(a) Dynamic viscosity is expressed by Vogel relation.
(b) Kinematic viscosity is expressed by Walther's relation.
(c) Both (a) and (b).
(d) None of the above

Q.23. Viscosity index of a mineral oil can be improved by
(a) Removing aromatics (low VI) during refining stage
(b) Blending with high viscous oil
(c) Using polymeric additives
(d) All of the above.

Q.24. Which one is the common system for oil classification?
(a) SAE (Society of Automobile Engineers)
(b) API (American Petroleum Institute)
(c) ISO (International Organization for Standardization)
(d) All of the above

Q.25. Barus relation, shows the relationship between
(a) Lubricant viscosity and temperature.
(b) Lubricant viscosity and pressure.
(c) Dynamic viscosity and kinematic viscosity.
(d) Lubricant temperature and lubricant pressure.

Q.26. Which of the following is not an advantage/benefit of solid lubricant?
 (a) More effective at high loads
 (b) Resistance to deterioration
 (c) Good heat dissipation
 (d) Highly stable in extreme temperature and environment

Q.27. Out of the following which is not an example of solid lubricant?
 (a) Carbon and graphite
 (b) Molybdenum sulphide
 (c) PTFE (Polytetrafluoroethylene) / Nylon
 (d) Mineral oils and multi-grade oils.

Q.28. Which of the following is/are the constituents of grease?
 (a) Base oil
 (b) Additive
 (c) Thickness fibre
 (d) All of above

Q.29. Which of the following is not the advantage of grease?
 (a) Remains at application point and adhere to the surface
 (b) Less frequent application needed
 (c) Good for inclined / vertical shaft
 (d) Good dissipation of heat

Q.30. The common friction modifiers used in grease is/are
 (a) Tricresyl phosphate (TCP)
 (b) Dibenzyl disulphite
 (c) Zinc dialkyldithiophosphate (ZDDP)
 (d) Molybdenum Disulphide

Q.31. Synthetic oils are used in aerospace applications because
 (a) They can withstand very high range of temperature from -120 °F to 500 °F.
 (b) Very high shaft rpm of the order of 60,000 rpm.
 (c) They have shorter carbon chains which are more resistant to heat and stress.
 (d) All of the above.

Q.32. Identify the incorrect statement about the additives.
 (a) The purpose of dispersant is to suspend harmful products like dirt and sludge.
 (b) Anti-wear additives typically contain zinc and phosphorus compounds.
 (c) Anti-foaming agents tend to lower the surface tension between air and liquid to the point where bubbles collapse.
 (d) Pour point additives increase the pour point of the lubricants.

Q.33. Apart from reducing friction and wear, the secondary purpose(s) of lubricants is/are
 (a) Heat dissipation
 (b) Reducing corrosion
 (c) Both (a) and (b)
 (d) None of these

Answers

Q.1. (b)	**Q.2.** (d)	**Q.3.** (a)	**Q.4.** (d)	**Q.5.** (a)	**Q.6.** (d)
Q.7. (a)	**Q.8.** (a)	**Q.9.** (d)	**Q.10.** (d)	**Q.11.** (d)	**Q.12.** (a)
Q.13. (a)	**Q.14.** (d)	**Q.15.** (d)	**Q.16.** (a)	**Q.17.** (c)	**Q.18.** (d)

Q.19. (c) Q.20. (c) Q.21. (b) Q.22. (c) Q.23. (d) Q.24. (d)
Q.25. (b) Q.26. (c) Q.27. (d) Q.28. (d) Q.29. (d) Q.30. (d)
Q.31. (d) Q.32. (d) Q.33. (c)

References

ASME Boiler and Pressure Vessel Code 2007, VIII Division I, Rules for Construction of Pressure Vessels, ASME.

Lansdown, "Molybdenum Disulphide lubrication", Elsevier; 1999.

Rossheim D. B. and A. R. C. Markl, "Gasket Loading Conditions" Mech. Eng., 65, 1943, pp 647-66.

Roberts, E.W., "The Tribology of Sputtered Molybdenum Disulphide Films", Proc. I. Mech. E. Intl. Conf. Tribology - Friction, Lubrication and Wear Fifty Years On, London, (1-3 July, 1987, vol. 1, p. 503, Paper No. C172/87).

Robert Flitney, "Seals and Sealing Handbook", 6th edition, Elsevier, B&H, USA, 2014.

Smoley, E. M., "Sealing with Gaskets", Machine Design, 38, Oct 27, 1966, pp 172.

S. S. Goilkar, Harish Hirani, "Parametric study on balance ratio of mechanical face seal in steam environment", Tribology International, 43 (2010), pp 1180–1185.

Chapter 2

Friction, Wear and Boundary Lubrication

2.1 Friction

Friction is the tangential resistance to motion. The occurrence of friction is a part of everyday life, as it is needed to have a control on walking. In the most of machines in operation, friction is undesirable as it causes loss of energy and deteriorates performance due to heat generation. Therefore efforts (i.e., using low friction materials, lubricating surfaces, changing design to reduce sliding) are made to reduce it.

One of major misconceptions about coefficient of friction (μ) is that the value of μ is much lower than 1.0. In practice μ greater than 1.0, as shown in Table 2.1.1, has been observed. Figure 2.1.1 shows the coefficient of friction due to material adhesion/cohesion. This figure also provides a crude approximation that under thin (partial) lubrication friction coefficient reduces to ten percent of its value obtained under dry lubrication. In addition, it can be observed from this figure that coefficient of friction is defined for a pair of material, i.e., steel on indium, steel on brass, etc. Defining coefficient of friction for steel as 0.5 will be incorrect. A better material pair, to reduce adhesive friction, can be selected from Fig. 2.1.2.

Table 2.1.1 Coefficient of friction (under dry condition) between two similar metals

Metals Sliding on themselves	μ
Aluminum	1.5
Copper	1.5
Gold	2.5
Iron	1.2
Platinum	3
Silver	1.5

Adhesion (Fig. 2.1.2) increases friction. Therefore, low adhesion metal pairs must be selected to reduce frictional force. Similar material pair must be avoided as similar materials have higher tendency of adhesion. In lubricated condition, the importance of material pair from friction point of view reduces.

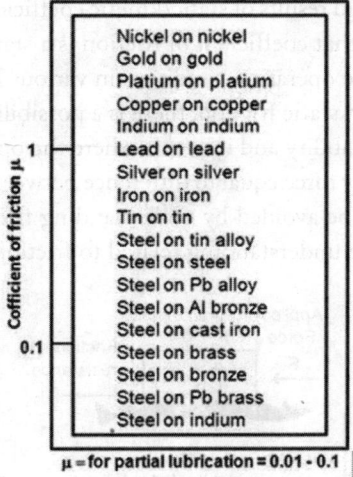

Fig. 2.1.1 Coefficient of friction for various metals

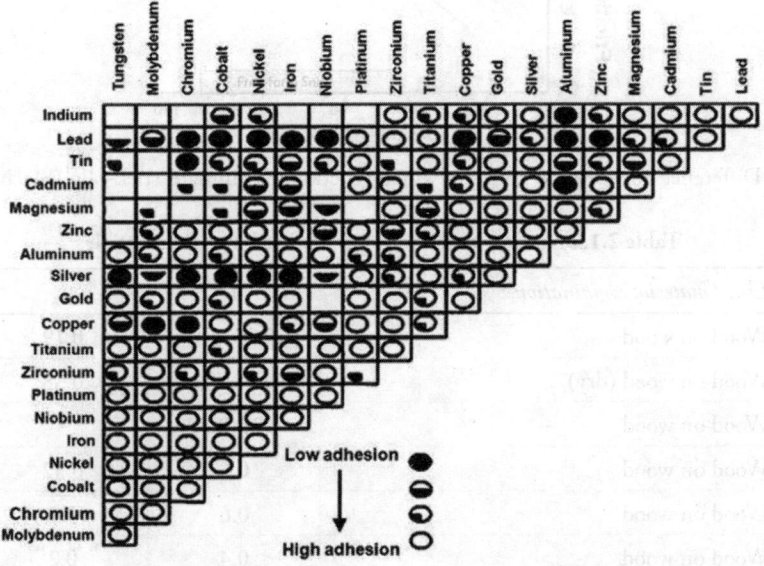

Fig. 2.1.2 Adhesive friction among various materials

2.1.1 Classification of friction

Friction can be classified as dry and lubricated frictions. Each class of friction is classified as static and kinetic frictions. To define static and kinetic frictions, let us consider a block (Fig. 2.1.3) on the surface getting pushed by a tangential force F. On application of 20 N load, the block does not move. On increasing tangential force from 20 N to 40 N (second point on the graph), still the block does not move. There is static force equilibrium between application force and friction force. On application of 50 N load, the block just starts sliding. At this point, the applied load remains equal to 50 N, but frictional resistance decreases from 50N to 40 N. In other words, static friction (50 N) is higher than kinetic friction (40 N).

Table 2.1.2 shows few published results of static/kinetic coefficient of friction. Based on the values listed in this table, it can be said that coefficient of friction is a statistical parameter. It is difficult to obtain the same value under same operating condition in various laboratory experiments. If kinetic friction is substantially lower than static friction, there is a possibility of stick slip phenomenon. The stick–slip is a type of friction instability and it occurs where the operating sliding speed of an object is very close to the speed caused by force equal to difference between static and kinetic friction forces. The stick–slip phenomenon can be avoided by understanding friction. There are a few established friction laws, which provide some understanding related to friction.

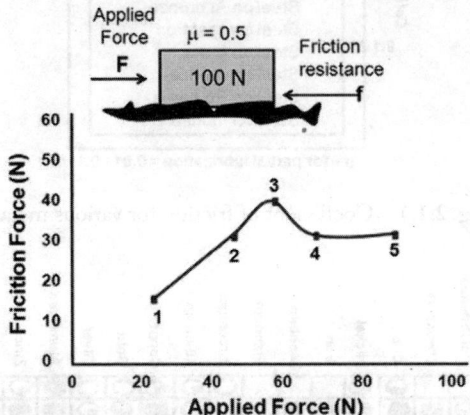

Fig. 2.1.3   Difference between the static and kinetic friction may initiate 'stick–slip' [Stachowiak, 2006]

Table 2.1.2   μ for wood–on–wood reported in various articles

Listed material combination	μ_s	μ_k
Wood on wood	0.25-0.5	0.19
Wood on wood (dry)	0.25-0.5	0.38
Wood on wood	0.30-0.70	-
Wood on wood	0.6	0.32
Wood on wood	0.6	0.5
Wood on wood	0.4	0.2

2.1.2 Laws of friction

Importance of friction motivated a number of researchers to perform experimental and theoretical studies on friction. Leonardo da Vinci (1452–1519) was one of the earliest experimenters, who made quantitative studies on friction. He measured friction by setting up simple experiments. He kept a body on a plane and slowly inclined the plane. He measured the angle of an inclined plane, where the body on the plane, started sliding. With his methods he was only able to measure static friction. Based on experimental observations, he established 'laws of friction' which are:

- The friction induced by the same weights will be of equal resistance although the contact may be of different breadths and lengths.

- Friction requires double the amount of effort if the weight be doubled. In other words
$F \alpha W$.

Leonardo defined a friction coefficient as the ratio of the friction divided by the mass of the slider. Experimentally, he found a universal friction coefficient (independent of the material) equal to 0.25.

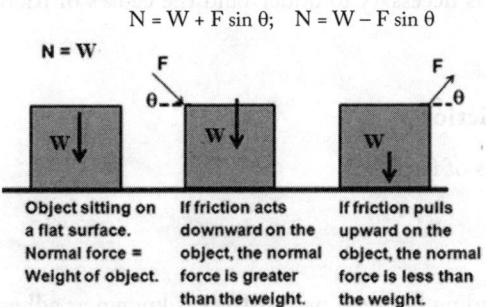

Fig. 2.1.4 Amontons' work

After Leonardo, the person who established friction laws was Guillaume Amontons (31 August 1663 – 11 October 1705), a French scientist. According to Amontons, the friction force is independent of the nominal area ($F \neq A$) of contact between two solid surfaces. The friction force is directly proportional to the normal component of the load.

$$F \alpha N$$

where N is the normal component of the load. It is interesting to note that Amontons used normal component of load compared to the load, which was considered by Leonardo da Vinci. To clarify this, he considered three cases (Fig. 2.1.4) and showed that friction force will vary as per the angle of application of load. As per Amonton, $\mu = 0.3$ for most of materials. Generally it is advisable not to define coefficient of friction for a single material. The coefficient of friction needs to be defined for pair of materials in the defined conditions (i.e., humidity, temperature, lubricant, and environment).

Based on experimental work of various researchers, following are the known laws of dry friction:

Law 1: Friction is independent of area of contact.
The amount of limiting (static) friction is independent of the area of contact between the two surfaces and the shape of the surfaces, provided that the normal reaction and surface conditions are unaltered.

Law 2: Friction is proportional to normal load.
The ratio of the limiting friction to the normal load between two surfaces is generally constant. This ratio is usually denoted by μ and known as coefficient of friction. Thus if the normal load is R, the limiting friction is μR for the given surface conditions.

Law 3: Kinetic friction is independent of magnitude of sliding velocity.
When motion takes place, the direction of friction is opposite to the direction of relative motion and is independent of magnitude of sliding velocity.

Law 4: Coefficient of static friction is slightly greater than the coefficient of kinetic friction.

Law 5: To initiate relative motion, applied force must be equal or greater than the static friction force.
Limiting (static) friction is the force which is exerted at the interface of two bodies and one body

just start sliding on another. In other words to break static equilibrium between two bodies and start relative motion, the resultant force acting on one of the bodies needs to be greater than the limiting force of friction.

It should be stressed that the above laws have been established based on experimental results and are accepted for simplicity to deal with the mathematical treatment of friction. To understand the fundamental of friction, it is necessary to understand the causes of friction, which are explained in next section.

2.1.3 Causes of dry friction

There are two main sources of friction:

- Adhesion
- Deformation

Force due to molecular binding between two surfaces is known as adhesive friction. Force required to plough asperities of harder surface through softer and deform surfaces, is treated as friction due to deformation. Apart from friction due to plastic deformation, there is possibility of friction loss due to elastic deformation. Elastic hysteresis occurs when a varying force repeatedly deforms an elastic material. The deformation produced does not completely disappear when the force is removed, and this results in energy loss on repeated deformations. Figure 2.1.5 illustrates the phenomenon of elastic hysteresis. Elastic hysteresis is more observable when the loading and unloading is done quickly than when it is done slowly. Materials such as rubber exhibit a high degree of elastic hysteresis.

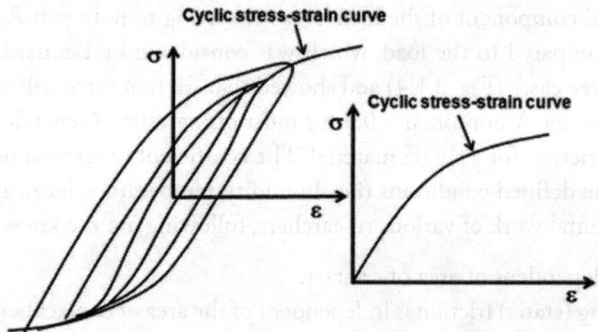

Fig. 2.1.5 Hysteresis loss

In lubricated tribo–pair case, friction due to adhesion will be negligible, while for smoother surfaces under light load conditions deformation component of friction will be negligible.

Figure 2.1.6 demonstrates the adhesion (cold weld) between two surfaces. Force F_a is required to tear the cold junction. Figure 2.1.7 demonstrates the deformation process. It shows a conical asperity approaching to a softer surface. To move upper surface relative to lower surface some force is required. Based on this discussion it can be said:

- Two friction sources: Deformation and Adhesion.
- Resulting friction force (F) is sum of two contributing (F_a and F_d) terms.
- In lubricated tribo–pair case -- negligible adhesion ($F_a \approx 0$).
- Smoother surfaces under light load conditions – negligible deformation ($F_d \approx 0$).

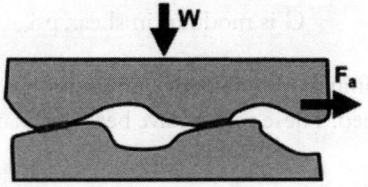
Fig. 2.1.6 Adhesion

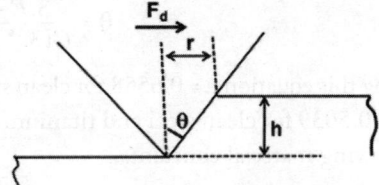
Fig. 2.1.7 Abrasion (deformation)

2.2 Theories of Dry Friction

A friction is a statistical parameter that depends on a number of variables. There is a need to understand the science of friction. To understand the effect of material pair, role of lubrication, and environmental factors, let us start with dry friction. The dry friction is also known as the solid body friction and it means that there is no coherent liquid or gas lubricant film between the two solid body surfaces. The Coulomb theory, Tomlinson theory, Bowden and Tabor theory, theory on friction due to deformation, and the theory on junction growth are explained in the following paragraphs.

1. C. A. Coulomb's theory 1781 (1736–1806):

Coulomb clearly distinguished between static and kinetic frictions. He proposed friction model as shown in Fig. 2.2.1. As per Coulomb, friction occurs due to interlocking of rough surfaces. Important points of Coulomb's work are:

- Contact at discrete points and $\mu_{static} \geq \mu_{kinetic}$.
- Friction force does not depend on contact area. $F \neq$ function (A).
- Friction force does not depend on the relative tangential velocity. $F \neq$ function (v).

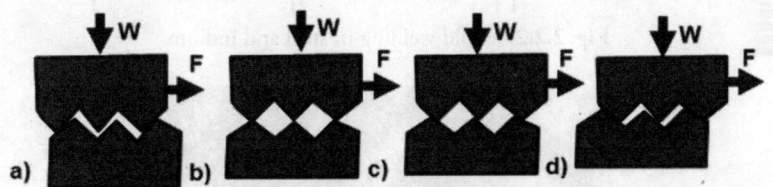

Fig. 2.2.1 Coulomb friction model

As per Coulomb, friction force is independent of sliding speed. But this law applies only to dry surfaces for a reasonable low range of sliding speeds, which depends on heat dissipation capabilities of tribo–pairs.

2. Tomlinson's theory of molecular attraction, 1929:

Tomlinson provided relation between friction coefficient and elastic properties of material involved, based on experimental study. According to Tomlinson, molecular attraction between metals causes the formation of cold weld junctions. Generally load on bearing surfaces is carried by just a few points; therefore those points are subjected to heavy unit pressure, and so probably weld together. Tomlinson suggested following friction formula:

$$\mu = 10700^* [\theta_I + \theta_{II}]^{2/3} \qquad E \text{ is young modulus, psi}$$

$$\theta = \frac{3.E + 4.G}{G(3.*E + G)} \qquad \text{G is modulus in shear, psi}$$

Using this equation μ = 0.6558 for clean steel and aluminium; μ = 0.742 for aluminium and titanium; and μ = 0.5039 for clean steel and titanium can be obtained. These values have been calculated based on following material constants.

Clean Steel	E = 30 mpsi,	G = 12 mpsi
Aluminium	E = 10 mpsi,	G = 3.6 mpsi
Titanium	E = 15.5 mpsi	G = 6.5 mpsi

Here mpsi = mega pounds per square inch

3. Bowden and Tabor Theory, 1950:

This theory is based on the fact that all surfaces are made of atoms. All atoms attract one another by attractive force. For examples, if we press steel piece over indium piece (as shown in Fig. 2.2.2) they will bind across the region of contact. This process is sometimes called 'cold welding,' since the surfaces stick together strongly without the application of heat. It requires some force to separate the two surfaces. If we now apply a sideways force to one of surfaces the junctions formed at the regions of real contact will have to be sheared if sliding is to take place. The friction force will be equal to that force. Figure 2.2.3 shows carbon graphite material adhered to stainless steel shaft.

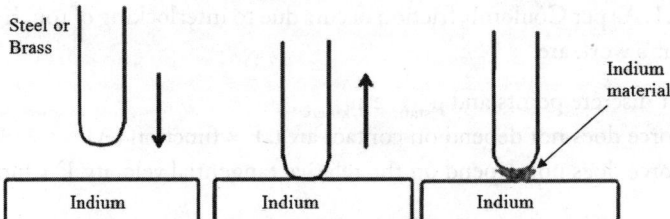

Fig. 2.2.2 Cold welding in steel and indium

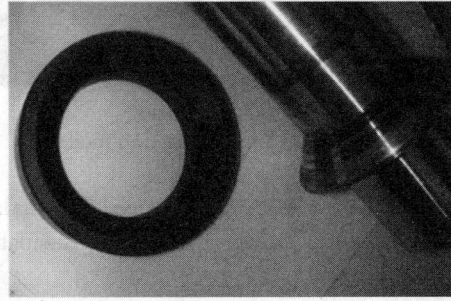

Fig. 2.2.3 Carbon graphite and stainless steel

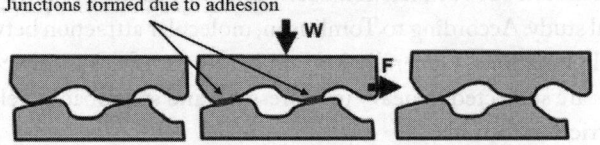

Fig. 2.2.4 Adhesion theory

Bowden and Tabor developed theory of adhesive friction. As per this theory on application of W, initial contact at some of higher asperity tips occurs. Due to high stress those asperities suffer plastic deformation and form strong adhesive bonds among asperities. Such cold formed junctions are responsible for the adhesive friction.

The real area of contact, A, can be estimated by the applied load W and hardness of the soft material, H_d. If s is shear stress of softer material, then force F_a required to break these bonds can be estimated by the formula F_a = As. The coefficient of friction due to adhesive friction is given by the ratio of friction force to applied load W. Figure 2.2.4 shows the formulation and breakage of cold junctions. The coefficient of friction due to adhesion can be explained as follows:

- Two surfaces are pressed together under load W.
- Material deforms until area of contact (A) is sufficient to support load W, A = W/H_d.
- To move the surface sideways, it must overcome shear strength of junctions with force F_a.
- $\mu_a = F_a/W = s/H_d$

In other words, shear strength (s) and hardness (H_d) of soft material decide the value of µ. This means whatever properties of the other harder pairing material, µ would not change. It is interesting to note that for most of untreated materials H_d = 3 σ_y and s = σ_y/1.7321. This means expected value of μ_a = 0.2. But for most of the material pair (shown in Fig. 2.2.5) µ is greater than 0.2. There is a huge difference between the measured values of friction coefficient and the estimated one by the theory of adhesion.

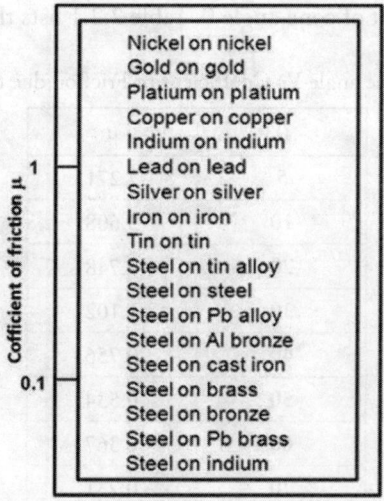

Fig. 2.2.5 Friction coefficients for various material pairs

The theory is unable to estimate different µ for steel on indium and steel on lead alloy.

4. *Friction due to deformation:*

This theory is based on the fact that contact between tribo–pairs only occurs at discrete points, where the asperities on one surface touch the other. The slope of asperities governs the friction force. Sharp edges result in more friction compared to rounded edges. Expression for the coefficient of friction can be derived based on the ploughing effect. Ploughing occurs when two bodies in contact have different hardness. The asperities on the harder surface may penetrate into the softer surface (as shown in Fig. 2.2.6) and produce grooves on it, if there is relative motion.

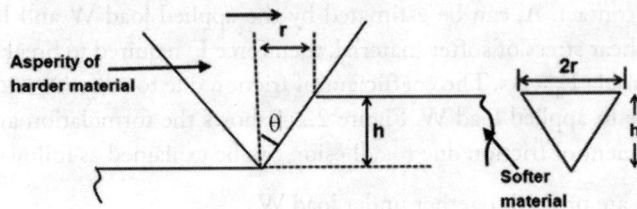

Fig. 2.2.6 Deformation theory

Contact between tribo–pairs only occurs at discrete points. Assume n similar conical asperities of hard metal in contact with flat soft metal. The vertically project area of contact to support the load is given by $A = n(0.5\,\pi r^2)$. To plough through soft material, area of contact can be expressed as $A_1 = 0.5(2r*h)$. Load can be given by $W = AH_d$; friction can be expressed as $F = A_1 H_d$ and

$$\mu_d = (F/W).$$

Substituting the equations of F and W, we get

$$\mu_d = (2/\pi)\cot\theta$$

This relation shows important of cone angle θ. Table 2.2.1 lists the μ_d for various θ values.

Table 2.2.1 Cone angle Vs Coefficient of Friction due to deformation

θ	μ
5	7.271
10	3.608
20	1.748
30	1.102
40	0.756
50	0.534
60	0.367
70	0.231
80	0.112
85	0.055

For most of the machines surfaces, angle $\theta \geq 80°$, which means $\mu_d = 0.1$. This means the derivation presented related to ploughing of soft surface with conical asperities does not provide desirable results. We can try with spherical shape of asperities. If we consider asperities on solid surfaces are spherical, vertical projected area of contact is

$$A = n(0.5 \times \pi r^2)$$

or
$$A = n[0.5 \times \pi(0.5d)^2]$$

or
$$A = n\frac{\pi d^2}{8}$$

$$W = n\frac{\pi d^2}{8} H_d$$

$$F = n\frac{2hd}{3} H_d$$

$$\mu_d = \frac{2 \times hd \times 8}{3\pi d^2} = \frac{16}{3\pi}\frac{h}{d} = \frac{16}{3\pi}\frac{h}{\sqrt{8hR}} = 0.6\sqrt{\frac{h}{R}}$$

Generally $h \ll R$; therefore $\mu_d = 0.1$. This means total μ ($= \mu_a + \mu_d$), should not exceed 0.3. From the theories related to adhesion and ploughing effects, following points can be summarized.

Adhesion,
$$\mu_a = \frac{s}{H_d}$$

Deformation by conical asperities,
$$\mu_d = \frac{2}{\pi}\cot\theta = 0.64\frac{h}{r}$$

Deformation n by spherical asperities,
$$\mu_d = 0.6\sqrt{\frac{h}{R}}$$

Coefficient of friction due to adhesion shows that friction depends on the lowest shear strength of the contact tribo–pair. Reducing shear strength and increasing the hardness reduces the coefficient of friction. The theory related to friction due to deformation by conical asperities shows the dependence of coefficient of friction on the angle of conical asperity whereas friction due to deformation by spherical asperities indicates lesser sensitivity of coefficient of friction compared to that of conical asperity. None of these theories provides reliable estimation of coefficient of friction which we observe during laboratory tests.

5. Junction growth theory:

Bowden and Tabor were motivated to think that the contact area (shown in Fig. 2.2.7) might enlarge under the additional shear force and they proposed junction growth theory. To explain their hypothesis they considered two dimensional stress system (Eq. 2.2.1) where two rough surfaces subjected to normal load W and friction force at the interface. If W force is in Y-direction and the force in X-direction is zero, then principle stresses can be expressed by Eqs. 2.2.2 and 2.2.3.

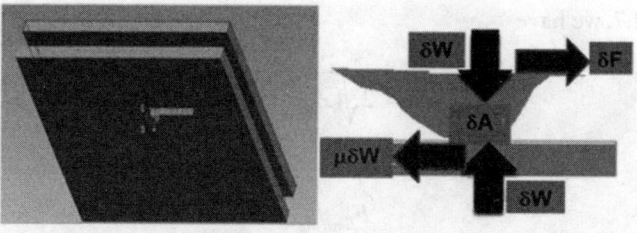

Fig. 2.2.7 Two contacting surfaces

$$\sigma_{1,2} = \frac{\sigma_x + \sigma_y}{2} \pm \sqrt{\left(\frac{\sigma_x - \sigma_y}{2}\right)^2 + \tau_{xy}^2}$$

$$\sigma_x = 0, \sigma_y = \delta W/\delta A, \tau_{xy} = \delta F/\delta A \qquad (2.2.1)$$

$$\delta A \sigma_1 = \frac{\delta W}{2} + \sqrt{\left(\frac{\delta W}{2}\right)^2 + \delta F^2} \qquad (2.2.2)$$

$$\delta A \sigma_1 = \frac{\delta W}{2} - \sqrt{\left(\frac{\delta W}{2}\right)^2 + \delta F^2} \qquad (2.2.3)$$

where,

σ_1 = first principal stress
σ_2 = second principal stress
δA = elemental area

Subtracting Eq. 2.2.3 from Eq. 2.2.2

$$\delta A(\sigma_1 - \sigma_2) = 2\sqrt{\left(\frac{\delta W}{2}\right)^2 + \delta F^2} \qquad (2.2.4)$$

If the shear strength of the softer material is $\tau_y = \dfrac{\sigma_1 - \sigma_2}{2}$, then

$$\delta A . \tau_y = \sqrt{\left(\frac{\delta W}{2}\right)^2 + \delta F^2} \qquad (2.2.5)$$

In Eq. 2.2.5 τ_y and W remain constant, which means the area of contact will increase with increasing friction force, till the force reaches its limiting value. We can state that on application of additional incremental tangential force, there will be further plastic flow at constant shear stress, resulting in an incremental contact area of A. Bowden and Tabor called this increase in the contact area as the junction growth. Assume, τ_i is shear stress of fractured interface.

$$F_{limiting} = \tau_i A_{max} \qquad (2.2.6)$$

Substituting Eq. 2.2.6 in Eq. 2.2.5, we have

$$(A_{max} \tau_y)^2 = \left(\frac{W}{2}\right)^2 + (A_{maxi} \tau_i)^2 \qquad (2.2.7)$$

Rearranging Eq. 2.2.7, we have

$$W = 2\sqrt{(A_{max})^2 \left(\tau_y^2 - \tau_i^2\right)} \qquad (2.2.8)$$

We know that

$$\mu = \frac{F_{limiting}}{W} \qquad (2.2.9)$$

Substituting Eq. 2.2.6 and Eq. 2.2.8 in Eq. 2.2.9, we have

$$\mu = \frac{\tau_i}{2\sqrt{\tau_y^2 - \tau_i^2}} \qquad (2.2.10)$$

Rearranging, we have

$$\mu = \frac{0.5}{\sqrt{\left(\frac{\tau_y}{\tau_i}\right)^2 - 1}} \qquad (2.2.11)$$

Using Eq. 2.2.11, the coefficient of friction can be calculated from the ratio τ_i/τ_y, as given in Table 2.2.2.

Table 2.2.2 Ratio of shear stresses Vs coefficient of friction

$\tau_i/\tau_y \times$	μ
1	0 005
10	0.050
20	0.102
30	0 157
40	0 218
50	0 289
60	0.375
70	0.490
80	0.667
90	1 032
99	3 509

The entries of Table 2.2.2 clearly motivates to apply thin film of low shear strength materials to the surfaces to reduce friction coefficient. Therefore, in order to reduce maintenance cost and increase life of tribo elements, interface shear strength of contacting surfaces need to be as low as possible. Interface shear strength can be reduced by:

1. Contaminations (reducing adhesion)
2. Lubrication

1. *Contamination:*

Oxide layer (encountered with metals in air), a few molecules thick on the surface (as shown in Fig. 2.2.8), can reduce the friction.

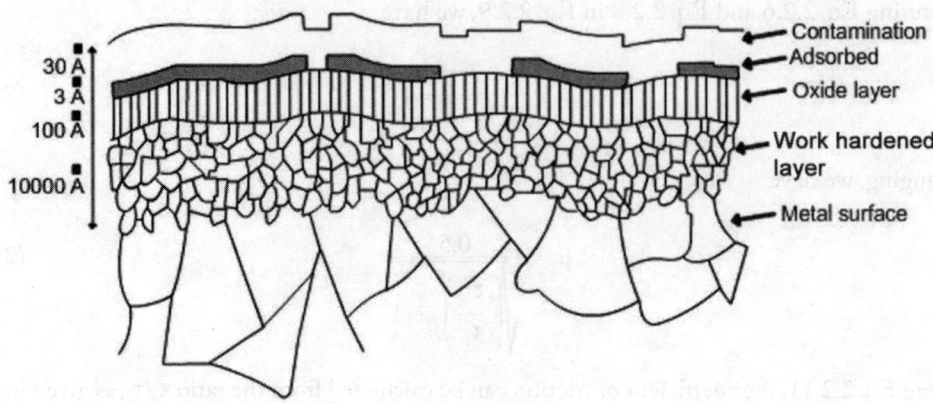

Fig. 2.2.8 Surface contamination

The contamination prevents the surfaces from sticking together strongly and restricts junction growth to occur. The formation and breakage of contamination layer may be a dynamic process; therefore, there is a possibility of variation in μ. There can be three cases:

(a) Weak (ductile) metal and weak oxide: In this case surface film breaks easily and rapid junction growth occurs, which means high value of μ; e.g. indium, gold.
(b) Weak metal and strong oxide: Transition from low to high μ as load increases (as shown in Fig. 2.2.9); e.g. copper, iron.
(c) Strong metal and strong oxide: Low μ at all loads; examples: strong steel, chromium.

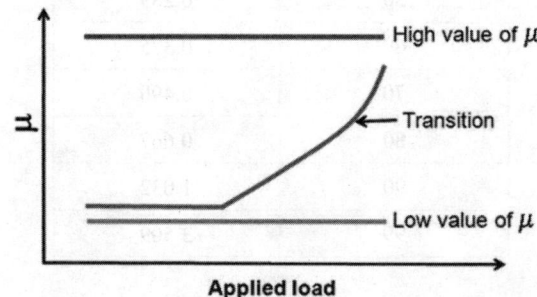

Fig. 2.2.9 Variation in Coefficient of friction due to surface contamination

2. *Lubrication:*

Lubrication can be used to reduce junction growth. To reduce junction growth, minimum value of the ratio τ_i/τ_y must be selected. Presence of liquid lubricant reduces the chances of junction growth as its shear strength is very low.

Both junction growth and ploughing affect friction, and either junction growth or ploughing may dominate friction behaviour which may vary with time. For example sliding in dry contact starts with running-in period and ploughing dominates. With time rupture/breakage of asperities takes place and surface is polished; and the effect of ploughing decreases. In this process, removal of contaminating layer may occur and possibility of increase in adhesion coefficient increases.

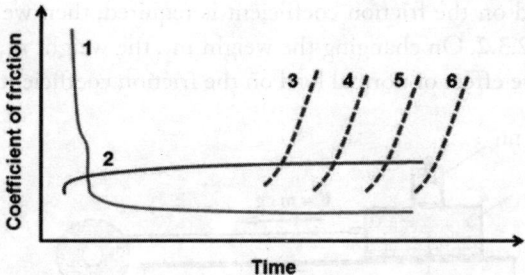

Fig. 2.2.10   Sliding friction vs. time

Variation in coefficient of friction with time is shown in Fig. 2.210. Essentially, a dry contact starts with a running-in period (as shown by 'line 1' in Fig. 2.2.10). Initially, the friction force is largely a result of ploughing of the surface by asperities. Adhesion does not play much significant role due to surface contamination. Asperity deformation takes place and affects the static coefficient of friction by polishing the surface. This is the main reason of reduction in friction coefficient, as shown in Fig. 2.2.10 by 'line 1'. Due to this reason the coefficient of friction in the initial stage is independent of the material combination. But if polishing wear process removes the contaminating layers, elements of bare surface will appear, resulting in increase in the coefficient of friction due to increased adhesion as shown by 'line 2'. In addition the coefficient of friction increases due to rapid increase in the number of wear particles entrapped between the sliding surfaces as a consequence of higher wear rates as shown by 'line 3', 'line 4', 'line 5' and 'line 6' in Fig. 2.2.10. The wear particles trapped between the surfaces cause ploughing.

2.3   Friction Measurement

Friction measurement device must be capable of supplying relative motion between two specimens, of applying a measurable normal load and of measuring the tangential resistance to motion. There are a large number of methods available and the final choice will depend largely on the exact conditions of rubbing contact under investigations.

For example, probably the simplest arrangement is the tilting plane where a specimen is placed on a flat surface which is gradually tilted until sliding starts as shown in Fig. 2.3.1. Place a block on a tilted plane and increase the angle of tilt until the block begins to slide. The tangent of the tilting angle just found is the so called 'friction angle'. This angle is related to the coefficient of friction μ, i.e., $\mu = \tan\theta$. This method is obviously unsuitable in those cases where a study of the variation of friction with continued rubbing is required but its simplicity makes it attractive in many cases.

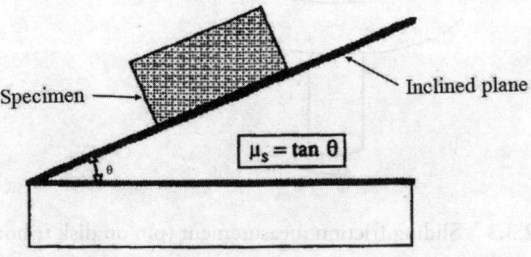

Fig. 2.3.1   Inclined plane tribometer

If effect of normal load on the friction coefficient is required, then we can use sled tribometer concept as shown in Fig. 2.3.2. On changing the weight m_1, the weight m_3 needs to be changed. A plot can be made to see the effect of normal load on the friction coefficient.

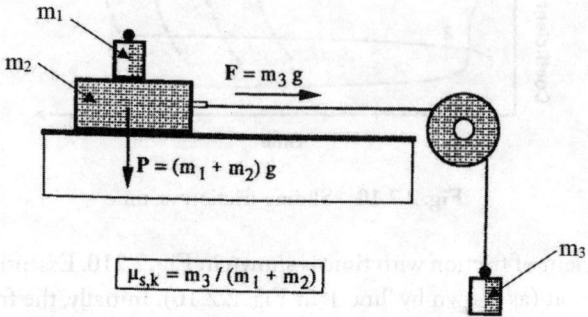

Fig. 2.3.2 Sled tribometer

If measurement of friction is required on continuous basis then a disc or a cylinder, is driven continuously, while a second specimen, nominally stationary, is loaded against it. Commonly used combinations are crossed cylinders, pin-on-cylinder or -disc, and disc-on-disc. The loading of the stationary specimen can be by simple deadweight or, if the experimental conditions demand it, by some more complicated method such as hydrostatic or magnetic loading. The measurement of the friction force is usually accomplished by mounting the nominally stationary specimen so that a very small tangential movement proportional to the frictional force occurs. This small movement is measured and recorded.

Figure 2.3.3 shows a stationary 'pin' under an applied load in contact with a rotating disc. The pin can have any shape to simulate a specific contact, but spherical tips are often used to simplify the contact geometry. Coefficient of friction is determined by the ratio of the frictional force to the loading force on the pin. Table 2.3.1 shows coefficient of friction measured using pin on disc tribometer for various materials. This table indicates statistical nature of friction coefficient, which means coefficient of friction, must be mentioned as nominal value with standard deviation.

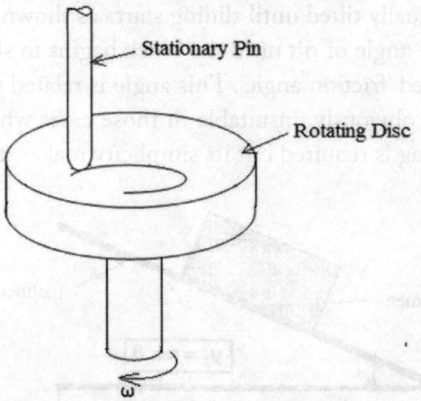

Fig. 2.3.3 Sliding friction measurement (pin on disk tribometer)

Table 2.3.1 Reproducibility of pin-on-disk friction data [Blau, 2008]

Sliding combination pin/disk	Number of tests	μ_{avg}	Standard deviation in μ
Data from round 1			
52100 Steel/52100 steel	109	0.60	0.11
Alumina/52100 steel	75	0.76	0.14
52100 Steel/alumina	64	0.60	0.12
Alumina/alumina	76	0.41	0.08
Data from round 2			
Silicon nitride/silicon nitride	83	0.70	0.21
Silicon nitride/52100 steel	83	0.80	0.22
Silicon nitride/alumina	83	0.75	0.20
52100 Steel/ silicon nitride	84	0.75	0.20
52100 Steel / 52100 steel	83	0.59	0.21

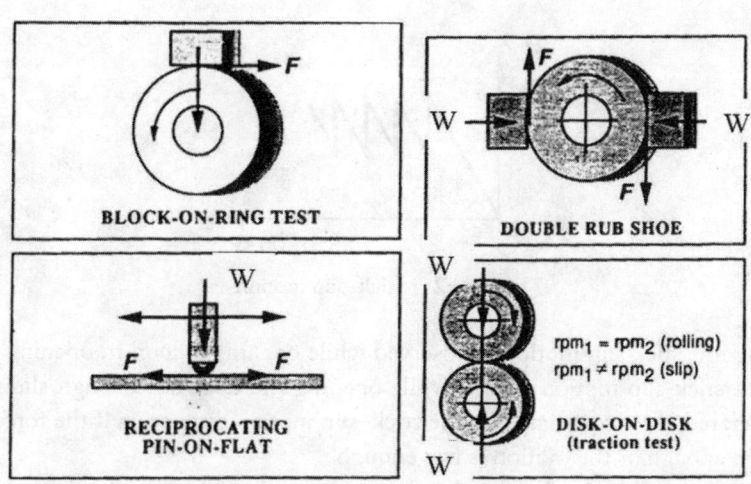

Fig. 2.3.4 Measurement using curved contact surfaces [Blau, 2008]

Figure 2.3.4 shows friction measurement in a universal tribometer. A tribometer is an instrument that measures tribological quantities, such as coefficient of friction, friction force, and wear volume, between two surfaces in contact. As shown in the figure, double rub shoe method is generally used to analyze friction in brake shoes, reciprocating pin on flat surface method is used to analyze friction behaviour between cylinder and piston rings whereas disk on disk method is used to analyze friction behaviours between two mating gears. A tribotester is the general name given to a machine or device used to perform tests of wear, friction and lubrication. Often tribotesters are extremely specific in their function and are fabricated by manufacturers who desire to test and analyze the long term performance of their products.

2.4 Stick–Slip Motion and Friction Instabilities

In a simple case of friction, if we apply a force parallel to the surface, there is no motion until the applied force is greater than the static friction force, as shown in Fig. 2.4.1. When motion begins, the force falls to correspond to the friction at the velocity of motion.

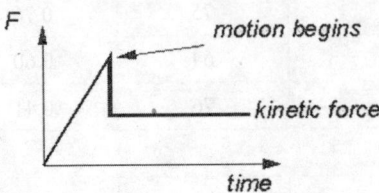

Fig. 2.4.1 Force required to overcome static friction

A common case is the stick–slip motion that is shown in Fig. 2.4.2. Possible reasons for stick–slip phenomenon can be interlocking of asperities during stick phenomenon and extent of interlocking reduces during sliding. If sliding speed is insufficient, interlocking will dominate that increases friction coefficient. To avoid this phenomenon of stick–slip motion either increase operating speed or reduce the difference between μ_s and μ_k.

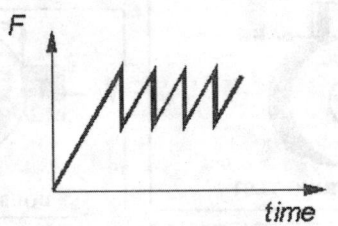

Fig. 2.4.2 Stick–slip motion

In day to day life, stick slip motion is observed while opening a door. In opening a door slowly, noise due to the stick-slip motion occurs; while opening the door fast enough sliding occurs and no noise is generated. It is common that the stick–slip motion disappears if the force in tangential direction is large enough or the motion is fast enough.

Slick–slip phenomenon is closely related to friction instability, which occurs due to large difference in the value of static and kinetic coefficient of friction. Ideally a lubricated condition having coefficient of friction equal to 0.00025 shall be preferred, but for that static coefficient of friction must of same level (i.e., 0.00025). If we assume that static coefficient of friction under lubricated conditions is equal to 0.01 and kinetic coefficient of friction is equal to 0.00025, then this lubricated contact may not be preferred, particularly for low speed applications.

Friction induced vibrations (Instability):

Decrease in coefficient of friction with velocity is shown in Fig. 2.4.3. If operating speed is in range of 200 rpm, then there is possibility of stick–slip phenomenon which may lead to friction instability.

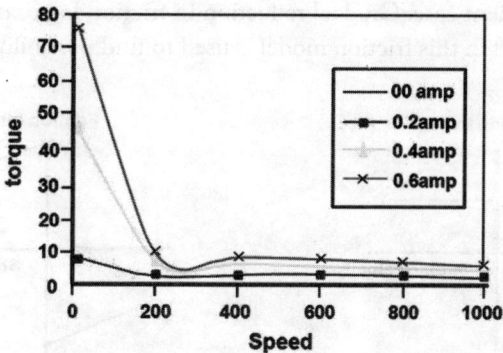

Fig. 2.4.3 Friction performance of MR (Magneto Rheological) brake

To understand the frictional instability, let us consider a block M sliding at a speed of V relative to the block A, as shown in Fig. 2.4.5. If there is difference in static and kinetic friction force, unbalance force (static friction force – kinetic friction force) causes a sudden acceleration. The velocity of M increases until the driving force falls to dynamic friction force. Eventually M comes to rest as shown in Fig. 2.4.5. Overall stick–slip behaviour of systems depends on stiffness, inertia, damping and magnitude of unbalanced force. To understand this effect mathematical model of Fig. 2.4.4 is required.

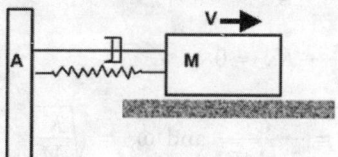

Fig. 2.4.4 Stick–slip

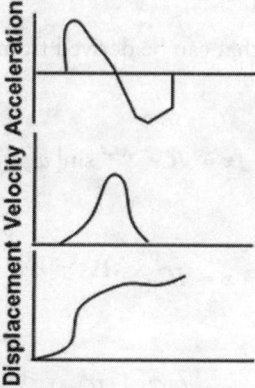

Fig. 2.4.5 Variation of vibration parameters

There are two ways to model friction:

(a) Stiction case: Instantaneous reduction in friction force as shown in Fig. 2.4.6. In this case F_s and F_c are in same vertical line. It is a hypothetical case.

(b) Negative and gradient case: Gradual reduction in friction force, as shown in Fig. 2.4.7, is a practical case. Often this friction model is used to find possibility of friction instability.

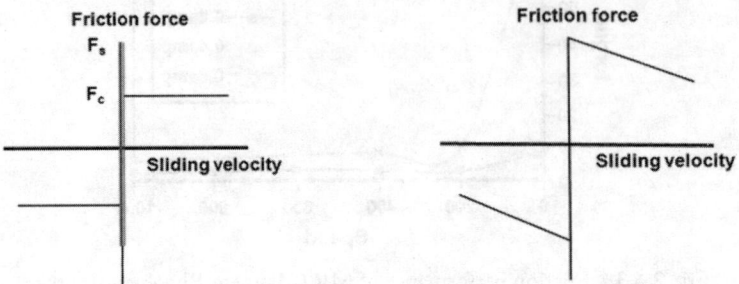

Fig. 2.4.7 Stiction case   **Fig. 2.4.8** Negative and gradient case

To understand the negative and gradient case of friction, let us consider damped vibration.

Damped vibration:

To understand friction instability let us consider system shown in Fig. 2.4.4. A mathematical model for the free vibration of this system is given in Eq. (2.4.1).

$$M\frac{d^2x}{dt^2} + C\frac{dx}{dt} + Kx = 0 \tag{2.4.1}$$

Introducing damping factor, $\varsigma = \dfrac{C}{2\sqrt{MK}}$ and $\omega_n = \sqrt{\dfrac{K}{M}}$

Substituting ζ and ω_n in Eq. 2.4.1, we have

$$\frac{d^2x}{dt^2} + 2\varsigma\omega_n\frac{dx}{dt} + \omega_n^2 x = 0 \tag{2.4.2}$$

There are three possible situations that can be derived from Eq. (2.4.2)

Case 1: Underdamped, $|\varsigma| < 1$

$$x = ICe^{-\varsigma w_n t}\sin[w_n t\sqrt{1-\varsigma^2} + \varphi] \tag{2.4.3}$$

Case 2: Overdamped, $|\varsigma| > 1$

$$x = IC_1 e^{(-\varsigma+\sqrt{\varsigma^2-1})w_n t} + IC_2 e^{(-\varsigma-\sqrt{\varsigma^2-1})w_n t} \tag{2.4.4}$$

Case 3: Critically damped, $|\varsigma| = 1$

$$x = (IC_1 + IC_2 t)e^{-\varsigma w_n t} \tag{2.4.5}$$

All three cases, namely underdamped, overdamped and critical damped, reduce vibration amplitude with time as shown in Fig. 2.2.9. But there is a possibility of negative damping ($\zeta < 0$) as shown in Fig. 2.4.10.

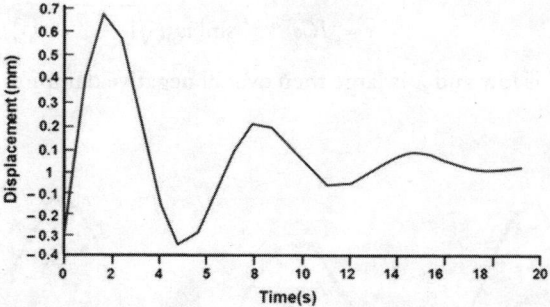

Fig. 2.4.9 Positive damping

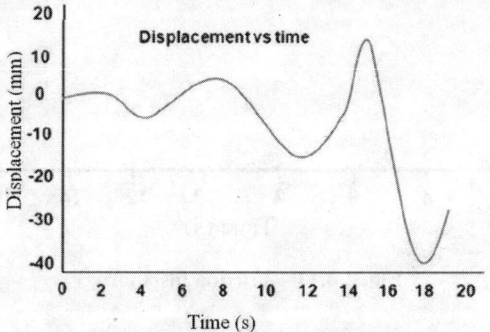

Fig. 2.4.10 Negative damping

Negative damping causes instability. If this happens due to friction, then we term it as 'Friction Instability'. To understand, we model forced damped vibrations:

$$M\frac{d^2x}{dt^2} + C\frac{dx}{dt} + Kx = F(t) \tag{2.4.6}$$

In the present case external force, F(t), is the frictional force.

$$M\frac{d^2x}{dt^2} + C\frac{dx}{dt} + Kx = -F(t) \tag{2.4.7}$$

Let us assume friction force is represented (negative and gradient case) as

$$F(t) = F_s - \lambda\frac{dx}{dt} \tag{2.4.8}$$

Substituting Eq. 2.4.7 in Eq. 2.4.8, we have

$$M\frac{d^2x}{dt^2} + C\frac{dx}{dt} + Kx = -\left(F_s - \lambda\frac{dx}{dt}\right) \tag{2.4.9}$$

Rearranging Eq. 2.4.9, we have

$$M\frac{d^2x}{dt^2} + (C - \lambda)\frac{dx}{dt} + Kx = -F_s \tag{2.4.10}$$

$$x = ICe^{-\varsigma w_n t}\sin[w_n t\sqrt{1-\varsigma^2} + \varphi] \qquad (2.4.11)$$

If system damping, C is low and λ is large then overall negative damping results, and motion may become instable.

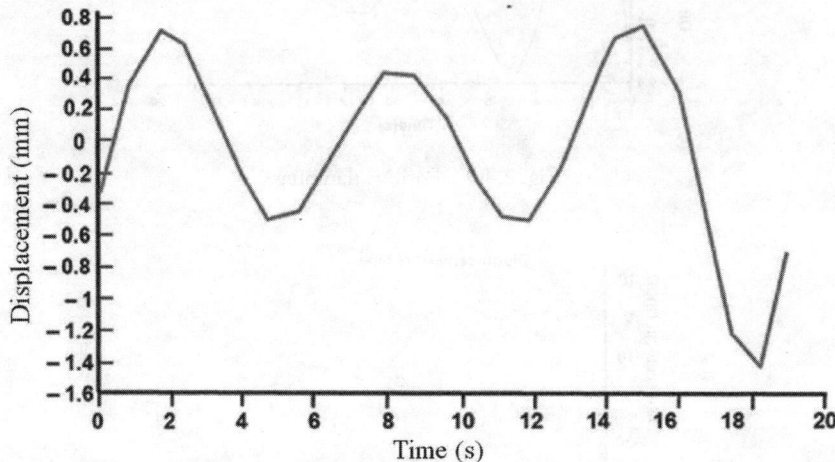

Fig. 2.4.11 Friction instability

To avoid friction instability, the following methods can be used:

- Increase the system damping (C).
- Lubricate or otherwise form a surface (additive) film to reduce the value of λ.

2.5 Wear

Material removal can be classified as either desirable or undesirable. Desirable cases of material include machining, polishing, shearing and writing with a pencil, whereas undesirable cases include almost all machine applications such as bearings, gears, cams and seals. Undesirable removal of material from one or both of the operating solid surfaces in relative motion (sliding, rolling or impact) is known as wear. Wear may also be defined as surface damage due to material displacement with no net change in volume or weight. It may occur as a natural consequence and mostly through surface interactions at asperities. Wear is strongly dominated by operating conditions and material properties. It is a system response.

Sometimes it is erroneously assumed that high friction means high wear rates. But this is not true. Interfaces with solid lubricants and polymers show relatively low friction but high wear, while ceramics show moderate friction with extremely low wear. In some isolated cases, friction and wear may be correlated. But, in general, friction and wear are two distinct system responses.

Wear can also be classified as zero wear and measurable wear.

Zero wear: Removal of material during sliding causing polishing of material surfaces may be known as 'zero wear'. It may increase the performance. In other words, zero wear is a polishing process in which the asperities of the contacting surfaces are gradually worn off until a very fine, smooth surface develops. Generally, 'polishing-in' wear is desirable for better life of tribo–pair.

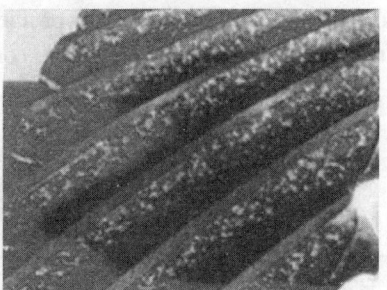

Fig. 2.5.1 (a) Zero wear of helical gear   **Fig. 2.5.1** (b) Measurable wear of helical gear

Figure 2.5.1(a) shows a polished surface of helical gear which occurs due to removal asperities from the surface. This polished surface reduces stress concentration effects and improves gear performance.

Measurable wear: Removal of material from surface that increases vibration, noise or roughness of surface, is treated as undesirable and termed as 'measureable wear'. Often wear is measured as the loss of volume/mass, but undesirable local deformation (which cannot be measured by weight loss) also falls in 'measurable wear' category. As a result of measurable wear, geometry changes and performance reduces; therefore measurable wear concept can be used to estimate the life of tribo–pair. The extent of measurable wear depends on the lubrication regime, the nature of the load, the surface hardness and roughness, and on the contaminants in the lubricating oil. A typical example of measurable wear in helical gear is shown in Fig. 2.5.1(b).

One form of measurable wear is pitting, which is a surface fatigue failure. The repeated loading of tooth surface induces contact stresses and if the contact stress exceeds the surface fatigue strength of the material wear occurs. Material in the fatigue region gets removed and a pit is formed. The pit itself will cause stress concentration and soon the pitting spreads to adjacent region till the whole surface is covered with pits. Subsequently, higher impact load resulting from pitting may cause fracture of already weakened tooth. There are two types of pitting, initial and progressive.

Figure 2.5.2 illustrates pit formation. When surface is subjected to cyclic load, sub-surface cracks may be generated. Few cracks generated within the material are shown in Fig. 2.5.2(b). Figure 2.5.2(c) shows merger of generated cracks and Fig.2.5.2 (d) shows wear particle getting detached from the surface. Such formation of pits (removal of material) comes under measurable wear.

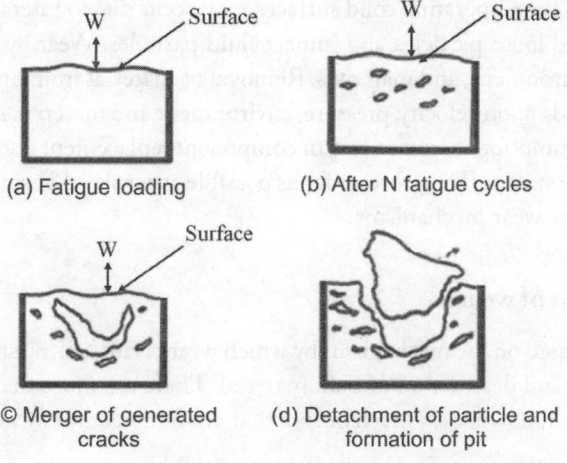

Fig. 2.5.2 Formation of pit

Many times a change in surface profile alters the optimum value of clearance and reduces load capacity of machine components. Let us consider Fig. 2.5.3 of worn out rollers. Sliding to rolling ratio for these worn out rollers increases with wear rate and usage of rolling element bearing loses its purpose.

Fig. 2.5.3 Worn out rollers

Figure 2.5.4 shows variation in bearing clearance due to abrasion of the bearing surface. With increase in bearing clearance load (1 ≤ Factor <10) capacity of the bearing decreases as shown in Fig. 2.5.5.

 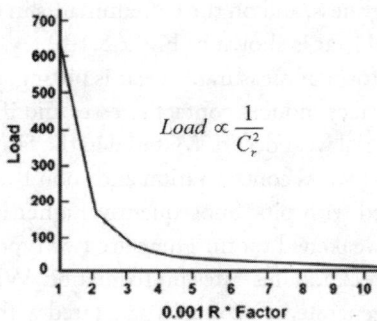

Fig. 2.5.4 Abrasion marks on bearing bore   **Fig. 2.5.5** Effect of clearance on load

Removal of material from operating solid surfaces may occur due to interaction with asperities on other solid surface, solid loose particles and impact fluid particles. Wear by solid particles depends upon load, velocity, environment, and materials. Removal of material from operating solid surface by fluid (liquid/gas) depends upon velocity, pressure, environment and material. As wear increases power losses increases, oil consumption increases, rate of component replacement also increases. Ultimately it reduces efficiency of the system. Therefore, as far as possible, wear should be minimized. To minimize wear, one must study the wear mechanisms.

2.5.1 Classification of wear

Wear can be classified based on the mechanisms by which wear occurs, i.e., plastic displacement, cutting, fatigue of surface films and destruction of bulk material. There are more than 34 wear mechanisms, but we shall discuss the following six mechanisms that are more common wear mechanisms:

1. Abrasive wear: polishing, scouring, scratching, grinding, gouging
2. Adhesive wear: galling, scuffing, and scoring

3. Corrosive wear (Chemical nature)
4. Erosive wear
5. Fatigue: de-lamination
6. Fretting wear

2.5.1.1 Abrasive wear

Abrasive wear occurs due to sliding of hard particles/asperities against the softer surfaces. Hard particle sources are: dirt, sand, metal wear particles, and loose particles.

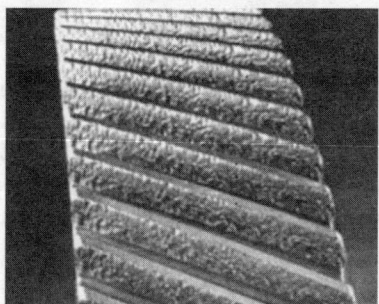

Fig. 2.5.1.1 Abrasive wear

Following are few well-known reasons of abrasive wear mechanisms:

- Scratching (Microploughing): This type of abrasive wear is characterized by short scratch-like lines in the direction of sliding. This type of damage can be stopped by removing the contaminants that caused it. Figure 2.5.1.1 shows abrasive wear of a hardened gear.
- Micro-cutting: Sharp particle or hard asperity cuts the softer surface. Cut material is removed as wear debris.
- Micro-fracture: Generally occurs in brittle, e.g. ceramic, material. Fracture of the worn surface occurs due to merging of a number of smaller cracks.
- Micro fatigue: When a ductile material is abraded by a blunt particle/asperity the worn surface is repeatedly loaded and unloaded, and failure occurs due to fatigue.
- Removal of material grains: Happens in materials (i.e., ceramics) having relatively weak grain boundaries.

Abrasive wear can be classified as two body abrasions and three body abrasions.

Two body abrasion:

This wear mechanism happens between two interacting asperities in physical contact, and one of them is harder than the other. Normal load causes penetration of harder asperities into the softer surface, thus producing plastic deformations. To slide, the material is displaced/ removed from the softer surface by combined action of micro-ploughing and micro-cutting. To quantify wear rate, Rabinowicz's equations for 2-B abrasive wear can be used.

To derive that equation, assume conical asperities indenting soft surface during traverse motion (as shown in Fig. 2.5.1.2) and all the material displaced by the cone is lost as wear debris. Total normal load W is supported by n asperities. Integer number n may have any value (i.e., 50, 150, and 500).

Here the basic assumptions are:

- All asperities can be represented by equal dimensions cones.
- All the material displaced by the conical asperity in a single pass is removed as wear particles.
- Load carried by nth asperity is $W_n = H_d \left(\frac{1}{2} \pi a^2 \right)$ where H_d is the hardness. Asperity is considered conical as shown in Fig. 2.5.1.2 so the area is multiplied by a factor of $\frac{1}{2}$.

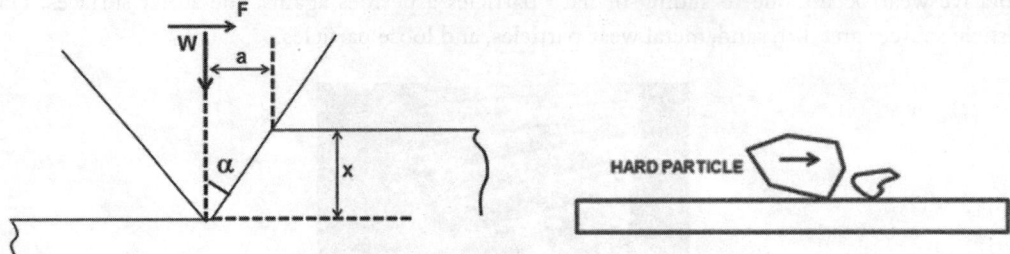

Fig. 2.5.1.2 2-body abrasion **Fig. 2.5.1.3** 3-body abrasion

Volume swept by penetrated asperity as per Fig. 2.5.1.2

$$\Delta V = axL \qquad (2.5.1)$$

$$\Rightarrow \quad \Delta V = \frac{a.a.L}{\tan \alpha} \qquad (2.5.2)$$

$$\Rightarrow \quad \Delta V = \frac{W_n.L}{0.5 H_d \pi \tan \alpha} \qquad (2.5.3)$$

Total wear is sum of the wear caused by individual asperity (same geometry of each asperity).

$$V = \frac{L \sum_{i=1}^{n} W_n}{0.5 H_d \pi \tan \alpha} \qquad (2.5.4)$$

$$\Rightarrow \quad V = \frac{LW}{0.5 H_d \pi \tan \alpha} \qquad (2.5.5)$$

$$Q = \frac{V}{L} = \frac{2}{\pi \tan \alpha} \frac{W}{H_d} \qquad (2.5.6)$$

$$Q = K \frac{W}{H_d} \qquad (2.5.7)$$

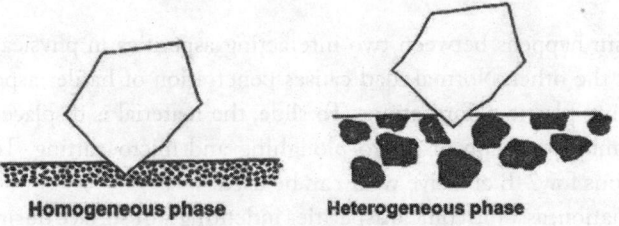

Fig. 2.5.1.4 Effect of microstructure on 3-body abrasive wear

Three body abrasion:

Three body abrasion is material removed from softer surface by hard and loose particles (Fig. 2.5.1.4), which are free to roll as well as slide over the surface, since they are not held rigidly. Same quantitative laws provided by Rabinowicz ($Q = K \dfrac{W}{H_d}$) can be used for 3-body abrasion. But due to rolling action of loose abrasive particles wear constant (K) is lower compared to 2-body abrasion. Generally K_2B = 0.005–0.05; and K_3B = 0.0005–0.005. From these values of wear constants one can conclude that wear rate is lower in three body abrasion than two body abrasion. The reduction in 3-body abrasion occurs due to energy consumed in rolling motion of free hard particles. Often 3-body abrasion occurs in sequence of other wear mechanisms. For example the hard particles may be generated locally by oxidation or during adhesive wear, and generated particles may further damage contacting surfaces by abrasion.

Shapes and size of loose particles play important role in deciding wear constant (K). For example M.R. Fluids contains loose iron particles, as shown in Fig. 2.5.1.5.

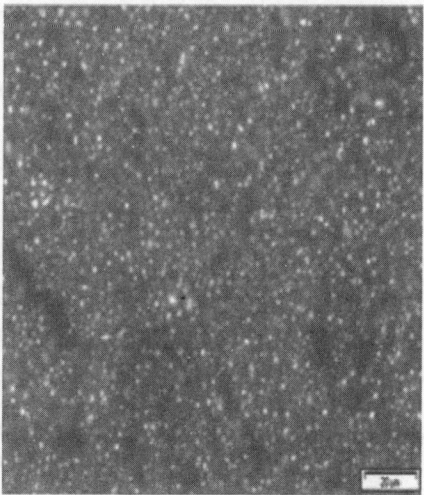

Fig. 2.5.1.5 M R Particles

M R Fluids are known as smart fluids, which vary in viscosity due to magnetic attraction (in the presence of magnetic field) among particles. If MR particles (Fig. 2.5.1.5) are spherical and relatively smaller in size compared to available clearance then abrasion by MR particles is negligible. But larger particle size and irregular shape of particles, may wear off contacting surfaces. Therefore, a good design must use spherical/regular shape of MR particles having size much smaller than provided clearance.

2.5.1.2 *Adhesive wear*

Adhesive wear is very common in metals. It depends on the mutual affinity between the materials. Let us take the example of steel and indium (Fig. 2.5.6a). When steel pin under load is pushed (Fig. 2.5.6b) in indium block, and subsequently retracted (Fig. 2.5.6c) a thin layer of indium gets transferred on the steel pin. Similar behaviour is observed by pushing brass metal in indium metal. This behaviour demonstrates the loss of indium material, which occurs due to high value of adhesive force between steel and indium. If steel pin is subjected to normal load as well as tangential load (Fig. 2.5.6a), then severe wear of indium material occurs. By introducing a thin layer of lubricant at the interface of indium and metal, the severe wear can be reduced to mild wear. Shear strength of lubricant layer is much smaller than shear strength of interface made indium and steel; therefore weak interface between steel and indium occurs which can be sheared easily and wear rate reduces to mild value.

All theories which predict wear rates start from the concept of true area of contact. It is usually assumed that the true area of contact between two real metal surfaces is determined by the elastic and plastic deformations of asperities. Severity of adhesive wear is based on the area of contact which is given by $A = W/H_d$. Here W is load applied to press one surface over other surface and H_d is hardness of soft material. This expression provides appropriate results if whole load is supported due to plastic deformation of the surface. However, for elasto–plastic deformation, the expression needs to be slightly modified as

$$A = (W/H_d)^{n1}$$

where ($2/3 < n1 < 1$). Here assumption is that higher asperities could be deformed plastically while the lower contacting asperities are subject to within elastic limits. In addition that adhesive wear will depend on the shear strength of friction junctions. If the junction is weaker than the material on either side of it, shearing occurs at the interface itself Fig. 2.5.7(a). There will be little surface damage and little wear. This situation occurs if sliding occurs within the surface oxide layer. If the junction is stronger than one of the metals, shearing will not occur at the interface but at a little distance within the softer metal [Figs. 2.5.7 (b) and (c)]. This may lead to an enormous increase in wear rate.

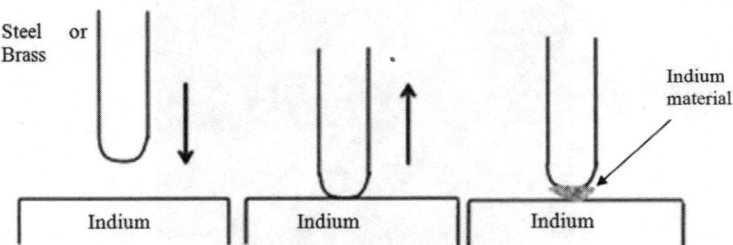

Fig. 2.5.6 Adhesive wear

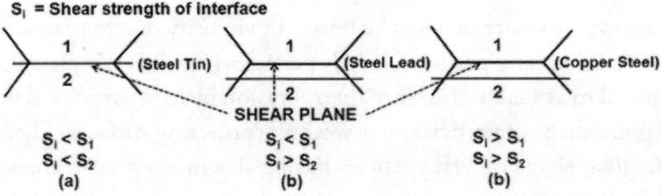

Fig. 2.5.7 Location of shear plane affect wear rate

Scoring wear, a severe form of adhesive wear, occurs due to tearing out of particles that weld together as a result of overheating (due to high contact pressure and/or high sliding velocity), permitting metal to metal contact shown in Fig. 2.5.8. After welding, sliding forces tear the metal from the surface producing a minute cavity in one surface and a projection on the other. The wear initiates microscopically; however, it progresses rapidly. Scoring is sometimes referred to as galling, seizing or scuffing.

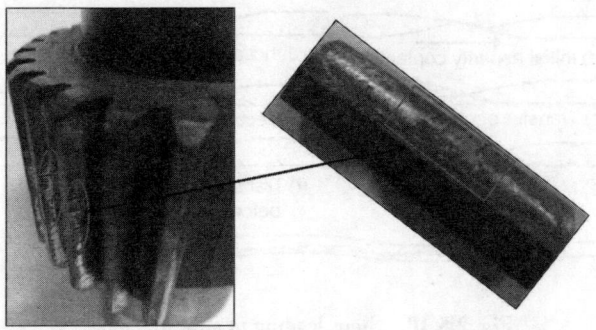

Fig. 2.5.8 Scoring

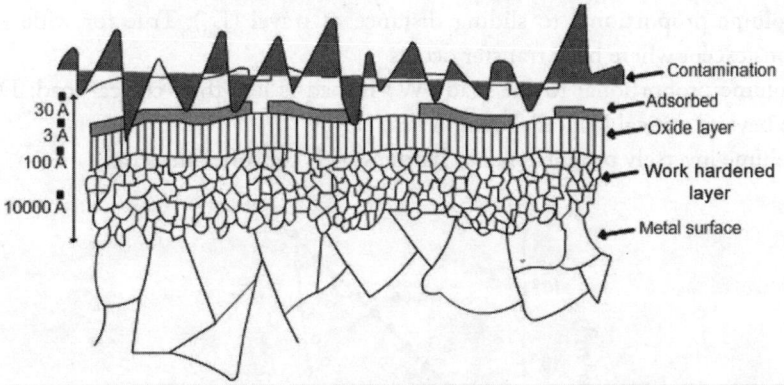

Fig. 2.5.9 Contaminant layers on metal surface

Steps leading to adhesive wear:

It is well known that macroscopically smooth surfaces are rough on micro scale. When two such surfaces are brought together, contact is made at relatively few isolated asperities. As normal load is applied, the local pressure at the asperities becomes extremely high. In the absence of surface films, surfaces would adhere but a small amount of contaminant prevents adhesion under purely normal loading. However, relative tangential motion at the interface disperses the contaminant (Fig. 2.5.9) films at the points of contact, and welding of the junctions can take place. The continued sliding causes the junctions to be sheared and new junctions to be formed. If shear occurs away from the interface, then metal is transferred from one surface to the other. With further rubbing, some of the transferred material is detached to form loose wear particles. In short the process of adhesive wear (as shown in Fig. 2.5.10) can be summarized as:

- Deformation of contacting asperities
- Removal (abrasion) of protective oxide surface film
- Formation of adhesive junctions; work hardening of metal around junction, which then becomes stronger than cohesion of soft metal
- Failure of junction by pulling out large lumps and transfer of materials
- Modification of transferred fragments
- Removal of transferred fragments and creation of loose particles

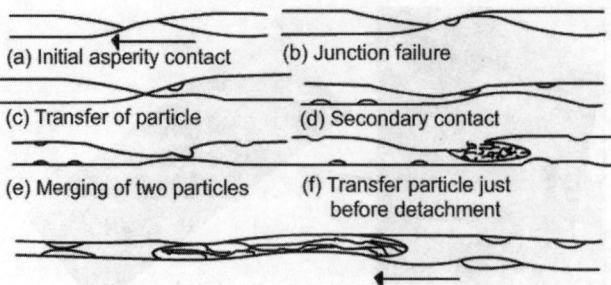

Fig. 2.5.10 Steps leading to adhesive wear

Laws of Adhesive Wear

- Wear volume proportional to sliding distance of travel (L_d); True for wide range of conditions except where back transfer occurs
- Wear volume proportional to the load (W) if load is less than critical load; Dramatic increase beyond critical load as shown in Fig. 2.5.11.
- Wear volume inversely proportional to hardness (H_d) of softer material

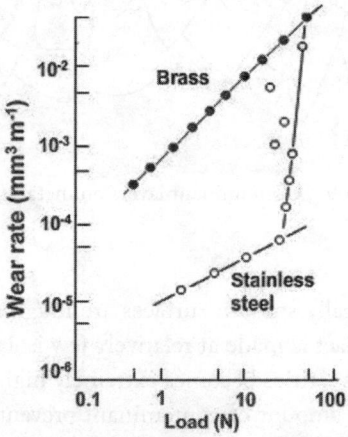

Fig. 2.5.11 Wear transition

Using these laws, wear volume is given by

$$V = \frac{K_1 W L_d}{3 H_d} \tag{2.5.8}$$

This equation is known as Archard's wear equation. The value of K_1 depends on elastic–plastic contacts, shearing of those contacts, effect of environment, mode of lubrication, etc. This expression of wear volume is a simple expression, as it does not require to estimate constant, n1 $\left(A = \left(\frac{W}{H_d} \right)^{n1} \right)$ individual shear strength of elastic and plastic junctions, effect of lubricant thickness, roughness, etc. Archard assumed that the contact between tribo–pair involves formation and breakage of junctions.

In other words, contact occurs only at asperities. The Archard model is demonstrated in Fig. 2.5.12, where cross section of asperities after plastic deformation is assumed to be circular. First sketch demonstrates the approach of junction forming asperities. Area of contact increases with sliding distance and subsequently decreases. But this process is continuous and happens among number of asperities. On average it is assumed that n asperities will be in contact at any frame of time. Wear expressed in Eq. 2.5.8 can be derived using the following sequence of equations.

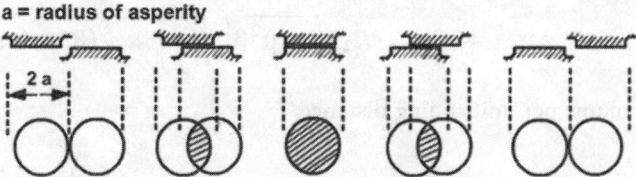

Fig. 2.5.12   Archard's wear model

$$\delta W = K_1 H_d (\pi a^2) \tag{2.5.9}$$

$$W = K_1 H_d \sum \pi a^2 \text{ for n asperities} \tag{2.5.10}$$

$$\delta V = K_2 \frac{2\pi a^3}{3} \tag{2.5.11}$$

On rearranging, we have

$$\delta v = \frac{\delta V}{2a} = \frac{K_2 \pi a^2}{3} \tag{2.5.12}$$

Considering all asperities,

$$v = K_2 \sum \frac{\pi a^2}{3} \tag{2.5.13}$$

On substituting expression of area in terms of W,

$$v = K_1 \frac{W}{3H_d} \tag{2.5.14}$$

Or,

$$V = K_1 \frac{WL}{3H_d}$$

K_1 is a dimensionless constant and expresses the probability of removing a wear particle. Factor K_1 (often referred as index of severity) represents the fraction of the friction junctions producing wear.

- $K_1 = 1$. Every junction involved in the friction process produces a wear fragment.
- $K_1 = 0.1$. One tenth of the friction junctions produce wear fragments. For clean gold surfaces K_1 is between 0.1–1. For clean–copper surfaces K_1 is between 0.1–0.01. Clean gold surfaces wear about ten times more rapidly than clean copper surfaces.
- $K_1 = 10^{-7}$ means that of the junctions responsible for friction only one in ten million produces a wear fragment.

Relation between coefficient of friction and wear constant:

Table 2.5.1 shows some data, available in the literature, related to coefficient of friction and wear rate. To establish relation between μ and K_1, Rowe proposed modified adhesion theory. In Eq. 2.5.15 k_m is constant and β is fractional surface film defect. This means β fraction of contact area is under dry lubrication, while (1- β) contact area is under lubricated condition. Here lubricated condition means shear strength of interface is smaller than shear strength of bulk material.

$$v = K_m \sqrt{1+\mu^2} \beta \left(\frac{W}{H_d} \right) \quad \text{(Rowe's Equation)} \quad (2.5.15)$$

where, v = wear volume per unit sliding distance.

$$v = K_1 \frac{W}{3H_d} \quad \text{(Archard's Equation)} \quad (2.5.16)$$

Table 2.5.1 Data related to friction coefficient and wear constant

Rubbing materials	μ	K_1
Gold on gold	2.5	0.1 to 1
Copper on copper	1.2	0.01 to 0.1
Mild steel on mild steel	0.6	0.01
Brass on herd steel	0.3	0.001
Teflon on hard steel	0.15	$2*10^{-5}$
Stainless steel on hard steel	0.5	$2*10^{-5}$
Tungsten carbide on tungsten carbide	0.35	10^{-6}
Polythene on hard steel	0.6	10^{-7}

It is interesting to compare Rowe's equation, Eq. 2.2.15, with Archard's equation, Eq. 2.5.16. There are three constants in Eq. 2.5.15 while only one constant in Eq. 2.5.16.

Equation 2.5.17 provides a modified form of wear constant K_1. In this equation 'h' represents the thickness of asperity while 'l' represents the length of asperity and P_w is the probability of wear particle formation. For spherical asperity, l = 2*h which means K_1 is equal to probability of wear particle formation. But if h is greater than radius of sphere then K_1 will be greater than P_w. Similar if h is lesser than sphere radius than K_1 will be lesser than P_w. This relation has its merits but difficulties lies in determining h, l and P_w.

$$K_1 = 2\left(\frac{h}{l}\right) P_w \quad (2.5.17)$$

In literature [Meng, 1995] there are many wear equations, but the most popular equation is Archard's equation, Eq. 2.5.2.

Some guidelines based on adhesive wear:

For longer service life or reliability of devices/machines, designers always aim for mild regime wear. It means wear particle coming out from the surfaces need to be much smaller in size. For getting this conditions dissimilar metals are usually chosen to run together as they do not weld together easily. If the metals are already at their maximum hardness, as in rolling bearing steel, no further work hardening is possible, so identical metals can be used for both elements. If severe wear behaviour cannot be avoided, such as in ore processing or earth moving equipments, routine maintenance is essential. For example, outer ring of rolling element bearings, if subjected to severe wear, then it can be rotated by few degrees to avoid wear of same localized surface. Many plastics undergo a transition from mild to severe wear as a function of sliding speed (that increase temp.) or combination of sliding and contact pressure. For better life of those plastics, load and speed conditions must be closely controlled.

Example: To find the best material for a dry journal bearing few tests were conducted on pin-on-disk machines (Fig. 2.5.13). Disk material remained AISI 1040 steel. While pin materials were: A (225), B(30), C(50), D (70), and E (100). Numbers in bracket for materials A, B,...E are surface hardness. The aim was to find the best material combination from the following experimental results. The wear on the pin can be measured with a microscope by measuring the size of wear scar.

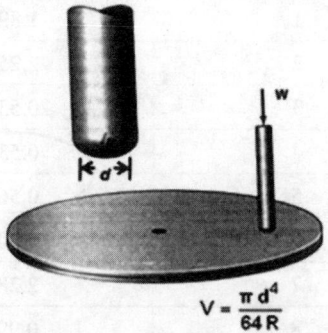

$$V = \frac{\pi d^4}{64R}$$

Fig. 2.5.13 Pin-on-disk arrangement

As per Table 2.5.2 wear scar(d) is maximum for Test 7(20.83 mm) and minimum for Test 2(8.81 mm). To find the best material following equation can be used.

Table 2.5.2 Experimental data

Test	Material	Sliding speed of disk rpm	Contact force N	Test duration min	Wear scar mm
1	A	30	100	350	9.70
2	A	30	200	100	8.81
3	A	60	200	60	16.01
4	B	30	100	480	12.63
5	B	60	100	480	15.27
6	C	60	100	430	15.89
7	D	60	200	480	20.83
8	E	60	100	240	13.02

Wear volume, $$V = K_1 \frac{WL_d}{3H_d} = \frac{\pi d^4}{64R}$$

where sliding distance, L_d = test duration * sliding speed. As R is same for all the tests, equation can be rearranged as

$$K_1 \frac{WL_d}{3H_d} = \frac{\pi d^4}{64R}$$

or,

$$K_{12345} = \frac{d^4 H_d}{time.speed.load}$$

Wear constant (K_{12345}) for various tests has been listed in Table 2.5.3. The result of test 4 is favourable as wear constant is smallest. Similar wear constant of test 5 is also very low; therefore material B may be treated as the best material. The values corresponding to material A in the table represent the transition behaviour of metal (A) from mild wear to severe wear on increase of load and speed.

Table 2.5.3 Values of wear constant

Test	K_{12345}
1	1.8971
2	2.259
3	20.5312
4	0.5301
5	0.5664
6	1.1068
7	2.2879
8	0.9978

It is important to define the 'mild wear' and 'severe wear' terms, which are based on average size of wear debris. In mild adhesive wear, small wear fragments (0.01–1 μm) mostly of metal oxides are generated. This kind of wear occurs at low contact pressure (below transition limit) and sliding velocity. Formation of black powdered oxide is typical example of mild wear. At higher velocities more oxidation replenishes losses due to break-away of oxide fragment as wear debris; therefore at higher velocities mild wear is possible.

Fig. 2.5.14   Mild wear

In some cases at higher loads, a hard surface layer (most likely martensite) is formed on carbon-steel surfaces because of high flash temperatures, followed by rapid quenching as heat is conducted into underlying bulk, and mild wear in such situation is possible. In short if oxide or contamination layers remain throughout operating time, wear will be in mild regime.

In case of severe adhesive wear due to large load, asperities break the oxide films and expose fresh metal which welds and as a result wear rate increases several hundred fold. Typical debris size range 20–200 µm of metallic particles is shown in Fig. 2.5.15.

Fig. 2.5.15 Debris in severe wear

Extreme form of severe wear is seizure, which means 'to bind' or 'fasten together'. It is a result of mutual plastic deformation of materials, as shown binding of inner ring and rollers of roller bearing in Fig. 2.5.15.

After seizure, components do not separate on their own. Manual force is required to separate the parts. In other words, after seizure tribo–pair loses its utility and cannot be used without proper reconditioning. The figure 2.5.16 clearly demonstrates the grooves made at inner ring and loss of material from roller surfaces.

The causes of seizure are:

1. Poor heat dissipation; it is related to material properties such as thermal conductivity.
2. Poor lubrication system or improper lubrication
3. Smaller clearances; it is related to improper design.
4. Installation errors; it is related to maintenance.
5. The ability of the metals to seize or to join in solid state; poor material selection

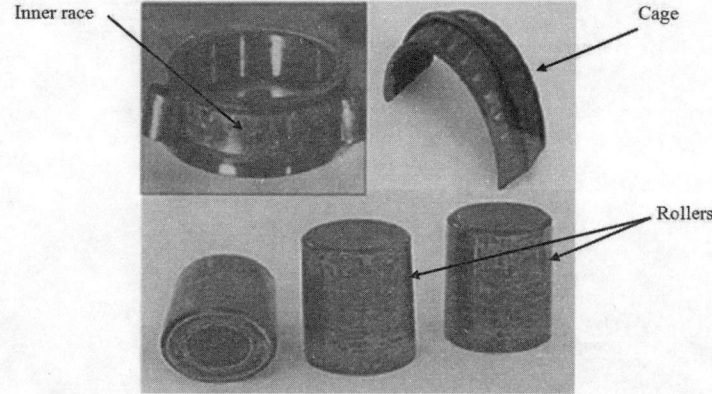

Fig.2.5.16 Seizure of rolling elements

All kind of adhesive wears can be shown in wear map as shown in Fig. 2.5.17 To plot this figure two controlling parameters, bearing pressure and sliding velocity, are used. The thick lines in this figure separate the various forms of adhesive wear. The wear constant (K_a) is expressed in numbers (i.e., 10^{-3}, 10^{-7}, 10^{-9}). Lower value of this constant means lower wear rate.

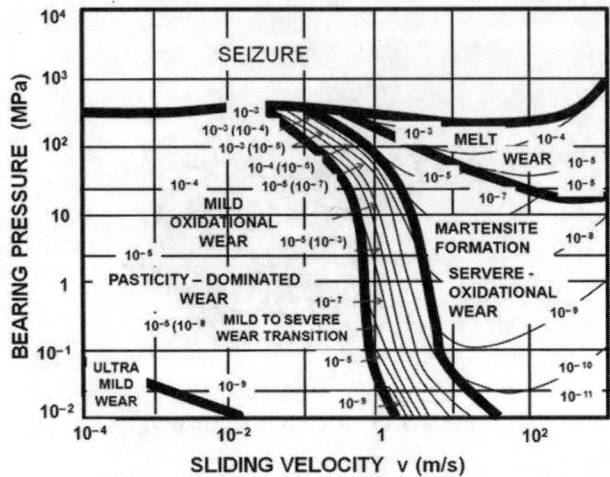

Fig. 2.5.17 Wear–mechanism map [Lim, 1998]

2.5.1.3 *Corrosive wear*

The fundamental cause of corrosive is a chemical reaction between the material and a corroding medium which can be either a chemical reagent, reactive lubricant or even air. To start wear from material surface, mechanical action is essential. In short corrosive wear can be explained as

Chemical reaction + Mechanical action = Corrosive wear

Understanding the mechanisms of corrosive wear is important to reduce this kind of wear. Let us consider a jaw coupling used for connecting shaft and motor, as shown in Fig. 2.5.18. This coupling is corroded, due to moist environment and its outer dimensions have increased. If we rub (mechanical action) this coupling with fingers, brown colour debris will get detached from the coupling surface. In other words, after chemical reactions, mechanical action is essential to initiate corrosive wear.

Fig. 2.5.18 Jaw coupling

Stages of corrosive wear:

- Sliding surfaces chemically interact with environment (humid/industrial vapour/acid)
- A reaction product (like oxide, chlorides, copper sulphide)
- Wearing away of reaction product film

The most corrosion films passivate (Fig. 2.5.19) or cease to grow beyond a certain thickness. This is favourable as corrosion process stops on its own. But most corrosion films are brittle and porous, and mechanical sliding wears away the film. The formation and subsequent loss of sacrificial (Fig. 2.5.20) film remains a continuous process. Such short life-time corrosion films are the most common form of corrosive wear.

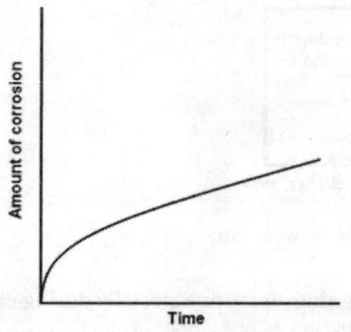

Fig. 2.5.19 Passivation of corrosion   **Fig. 2.5.20** Continuous corrosion

Debris generated by corrosive wear may increase or decrease wear rate. It is well known that the most of metal surfaces get oxide layers resulting from reactions with the environment. Thickness of these oxide layers is typically 10 microns, and they may have a protective role unless the thickness tends to grow during the cyclic contact process. If the oxide layer grows, it becomes liable to break in brittle fracture, producing wear particles. Hard, broken-off oxide particles may then profoundly affect subsequent wear life as abrasive agents. If soft, ductile debris results, however, it may form a protective lubricant layer on the surface.

2.5.1.4 *Erosive wear*

Erosive wear is caused by the impact of particles (solid/liquid) against a solid surface. For example, dust particles impacting on gas turbine blades and slurry impact on pump impeller. Erosive wear rate (V_e) is a function of:

- Particles velocity (K.E.)
- Impact angle, and
- Size of abrasive

$$V_e = K_e . A(\alpha) . (\text{particle velocity})^n . (\text{particle size})^3. \quad (2.5.15)$$

Here K_e is an empirical constant and n is a velocity exponent. This means that the relationship between wear rate and impact velocity, Eq. 2.5.15, is described by a power law. Generally n = 2–2.5 for metals; and n = 2.5–3 for ceramics.

Angle of impact (α) decides the magnitude of moment transfer. Angle between eroded surface and trajectory of particle immediately before impact may range from 0° to 90°.

- Low impact angle: cutting wear prevails, hardness resists wear.
- At large angle, fatigue wear prevails. Soft (ductile) material may be suitable.

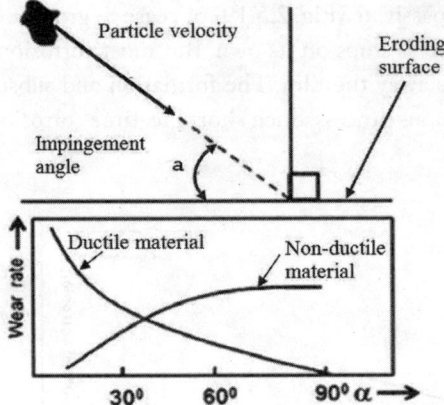

Fig. 2.5.21 Impingement angle vs. wear rate

The knowledge of angle of impingement is useful in designing air strips especially during emergency landings e.g., belly landing. Landing of an aircraft on its belly (underside without its landing gear fully extended) is termed as belly landing. During a belly landing, there is normally extensive damage to the airplane. Special tribological materials on the underside of airplane are used to reduce the friction and wear of airplane. In addition to that designing air strips to minimize the erosion shall reduce the damage to aircraft. The impingement angle is one of the most important factors and is widely recognized in literature. For ductile materials the maximum wear rate is found when the impingement angle is approximately 30°, whilst for non-ductile materials the maximum wear rate occurs when the impingement angle is normal to the surface.

Example

Pneumatic Transportation:

In Fig. 2.5.22 an elbow which carries plastic pellets is tried with steel pellets. Due to metal contact elbow got damaged (worn out).

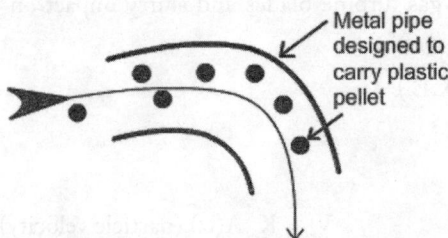

Fig. 2.5.22 Pneumatic transportation

From tribological point of view elbow must be reinforced with rubber inside the elbow to sustain the impact and absorb the additional kinetic energy.

Example

Engine particle (sand) separator:

Erosion by sand particles inside engine is a major problem. The design of helicopter engine must be modified to reduce the number of particles.

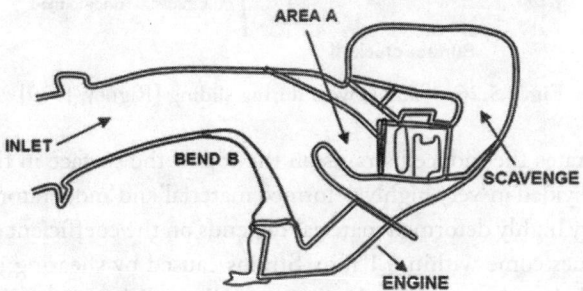

Fig. 2.5.23   Helicopter engine

As per tribology, the travelling distance of particles should be minimized. Based on this guideline modified design shown in Fig. 2.5.24 was proposed.

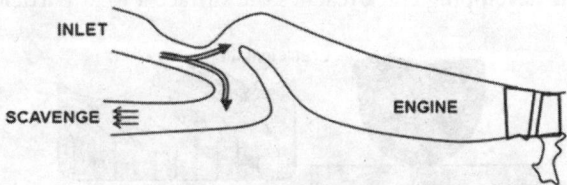

Fig. 2.5.24   Modified engine

2.5.1.5 *Fatigue wear*

Fatigue is attributed to multiple reversals (apply and release) of the contact stress, occurring due to cyclic loading, such as rolling bearings, gears, friction drives, cam and follower. Abrasive and adhesive wear involve a large contribution from fatigue. Figure 2.5.25 shows 'surface fatigue' failure of outer ring of roller bearing. At the start of bearing operation, the rolling bearings rely on smooth undamaged contacting surfaces for reliable functioning. A certain number of rolling contact cycles must elapse before surface defects are formed. Once the rolling surfaces of a bearing are pitted, its further use is prevented due to excessive vibration caused by pits passing through the rolling contact.

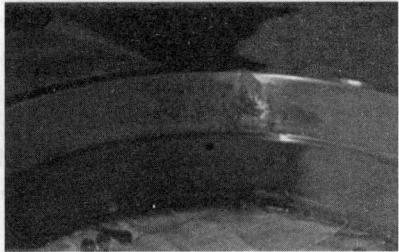

Fig. 2.5.25   Fatigue wear

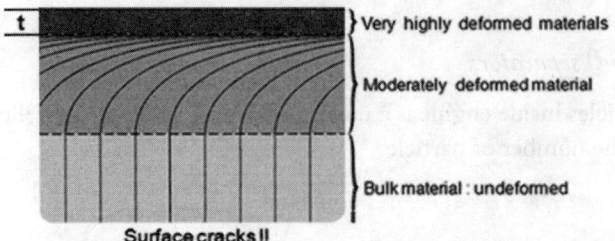

Fig. 2.5.26 Fatigue wear during sliding [Rigney, 1979]

Figure 2.5.26 illustrates the induced strains on the top of the surface in the direction of sliding. Deformed material is divided in 'very highly deformed material' and 'moderator deformation material'. The thickness 't' of 'very highly deformed material' depends on the coefficient of friction and for high value of sliding its values come within 0.1 mm. Strains caused by shearing in sliding direction are present to some depth below the surface and is the space where dislocation cells are formed. Materials vary greatly in their tendency to form dislocation cells. For example, aluminium, copper and iron have a high tendency to form dislocation cells. These dislocation cells are probable regions for void formation and crack nucleation. A primary crack originates at the surface at some weak point and propagates downward along weak planes such as slip planes or dislocation cell boundaries as shown in Fig. 2.5.27 When the developing crack reaches the surface a wear particle is released.

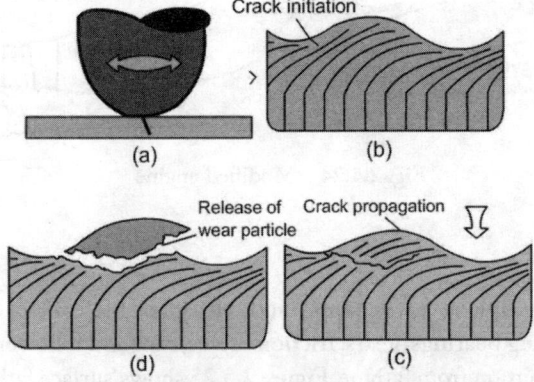

Fig. 2.5.27 Fatigue wear [Kimura, 1983]

Pure materials (without any inclusions in it) may exhibit high service life. It was found that pure copper (99.96% purity) slid against steel gives a wear rate ten times lower than any other material despite exhibiting the highest coefficient of friction of all the materials tested. On the other hand, a steel rich in carbide particles shows a low coefficient of friction and gives one of the highest wear rates. The wear rate was found to increase with inclusion of density in the material, while friction was determined by adhesion factors so that complex impure materials exhibited the lowest friction coefficient.

During rolling the local contact stresses are very high, concentrated over a small area and repetitive. Such loading occurs in rolling bearings, gears, friction drives, cams and followers, etc. Steps leading to generation of wear particles:

- Application of normal load that induces stresses at contact points

- Growth of plastic deformation per cycle
- Subsurface crack nucleation
- Expansion of crack due to reversal of stress
- Extension of crack to the surface due to traction force
- Generation of wear particles

Often large size stress raiser, such as holes, under fatigue form deep cracks and failure of machine component due to such stress raiser is termed as cracking (Fig. 2.5.28). Cracking is ultimate failure to split the component. In other words, cracking results in complete failure of the component. Causes for cracking are excessive load with vibration, loose fit and excessive impact.

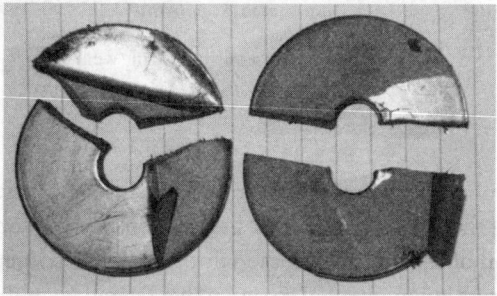

Fig. 2.5.28 Cracking

2.5.1.7 *Fretting wear*

Fretting wear was coined in 1927 by Tomlinson. It refers to small amplitude (1–300 µm) high frequency oscillatory movement mainly originated by vibration. This generally occurs in mechanical assemblies (press fit parts, rivet / bolt joints, strands of wire ropes, rolling element bearings), in which relative sliding on micron level is allowed. It is very difficult to eliminate such movements. Therefore it can be said that fretting wear and fretting fatigue are present in almost all machinery and are the cause of total failure of some otherwise robust components.

Fretting occurs wherever short amplitude (few microns) oscillating sliding between contacting surfaces (Fig. 2.5.29) is continued for a large number of cycles. The centre (Fig. 2.5.29) of the contact may remain stationary while the edges oscillate. In this process, generated wear debris remain in contacting area and contribute to accelerate (Fig. 2.5.30) the wear process by abrasion. In addition increase in local temperature further accelerates the wear process. It may be said that oscillatory movements even of 0.1 micron amplitude can cause failure of the component when the sliding is maintained for one million cycles or more.

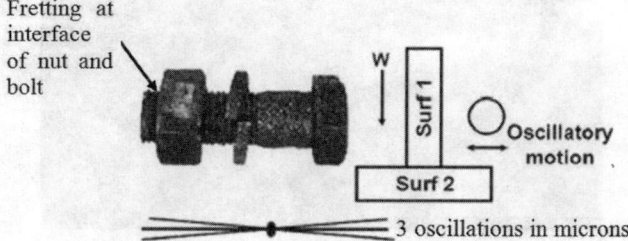

Fig. 2.5.29 Fretting wear

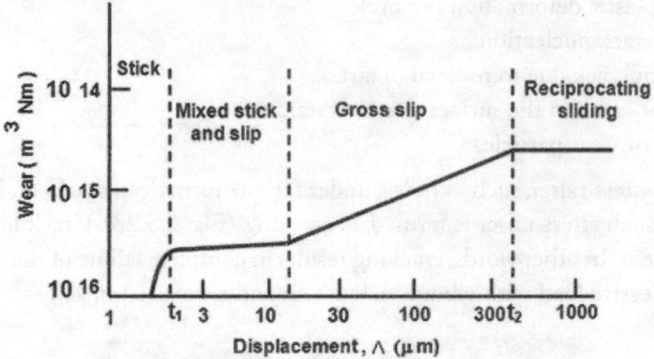

Fig. 2.5.30 Process of fretting wear

2.5.1.8 Wear analysis

Generally wear does not involve a single mechanism; therefore it is advisable to take an integrated wear analysis approach assuming the wear behaviour as a system property. In other words, wear analysis is not limited to the evaluation of the effects of materials on wear behaviour, but recommends changes in contact geometry, roughness, tolerance, and so on, so that overall favourable results can be achieved. In Ludema's words 'A scan of many wear models shows considerable incongruity. Equations have either too many undefined variables or too few variables to adequately describe the system'. Most of available equations are derived for mild wear rate of components. Therefore it can be said that to estimate wear, theoretical equations with experimental coefficients are required.

Example

Cam wear analysis:

Cam having pits on surface, as shown in the Fig. 2.5.31, was rejected because it was making noise and it was not performing intended function. It is necessary to find the cause of failure of such pitting so that in future cam service life can be improved.

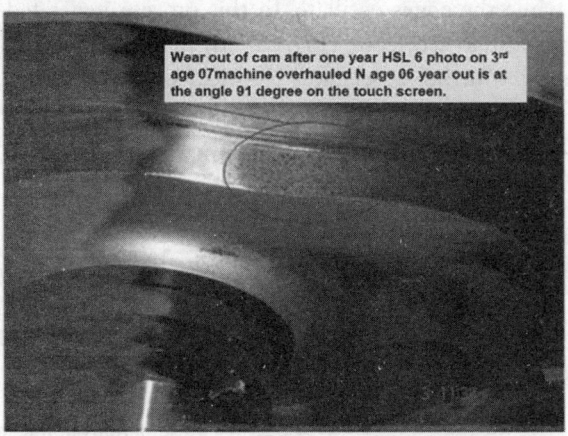

Fig. 2.5.31 Cam wear

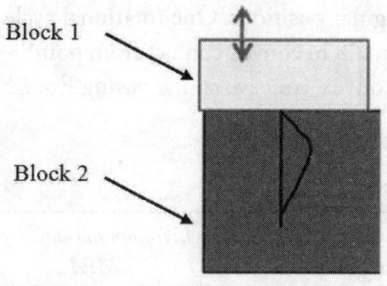 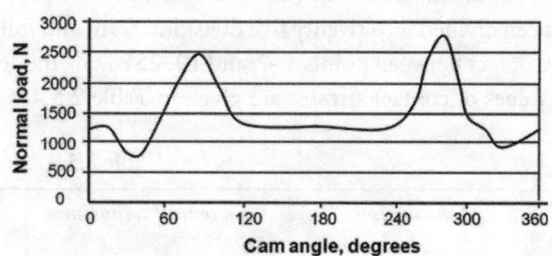

Fig. 2.5.32 Sketch to illustrate pitting   **Fig. 2.5.33** Normal load vs. cam angle

In the previous sections pitting as a fatigue wear was studied. Reversible stresses are main cause of such failure. To illustrate it, a sketch is shown (Fig. 2.5.32), in which the block 1 supported on block 2 is subjected to reversible stresses. Due to this arrangement, block 2 will experience compressive and shear stresses. The variation in magnitude of shear stress is shown by a free curve, which shows maximum shear stress below the surface. Now question comes from where reversible stresses are induced. We estimated normal compressive force applied on the cam surface and is plotted, as in Fig. 2.5.33, with the cam angle. In this figure maximum force occurs at 90 and 270 degrees, and remains constant in magnitude between 120 to 240 degrees. This means applied load on cam surface is dynamic and shall induce dynamic stresses.

Can such dynamic load be reduced?   It is very obvious to explore whether this dynamic load can be reduced or not. Variation in pressure angle with cam rotation is given in Fig. 2.5.34. Pressure angle Φ is angle between direction of motion (velocity of follower) and axis of force transmission.

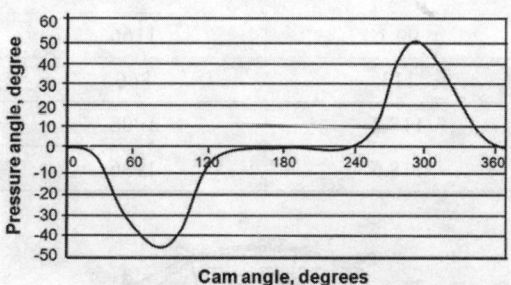

Fig. 2.5.34 Pressure angle vs. cam angle

- ø = 0 - Transmitted force is completely utilize to move the follower
- ø = 90⁰ - No motion to the follower; gross sliding

From this discussion, it can be said that redesigning cam with the maximum pressure angle lesser than 10 degrees may reduce sliding and delay pitting failure of the cam surface. But this takes long time and it is advisable to reduce contact stresses. The contact stresses depend on the cam geometry, such as

$$\sigma_c \propto \sqrt{\frac{1}{R_{\text{follower}}} \pm \frac{1}{R_{\text{cam}}}} \qquad (2.5.16)$$

In above equations, '+' sign is for convex cam geometry and '-' sign is for concave geometry. Obvious question arises, how convex and concave differentiation helps in reducing contact stresses?

Figure 2.5.35 shows cam–follower interface at various angular positions. One rotational cycle has been divided into twenty five divisions. Cam and follower remain in convex contact from point 8–18. Contact between points 1–7, and 19–25 can be modelled as convex-concave contact using Eq. 2.5.16. Values of contact stresses are given in Table 2.5.4.

Table 2.5.4 Contact stress

Cam angle	Cam contact radius, mm	Normal load on cam surface, N	Maximum normal stress, MPA
0	118.6	1995	−283
45	106.7	862	−242
60	86.2	1653	−342
75	64.4	2174	−404
90	45.7	2489	−453
105	36.5	1794	−399
120	74.4	1344	−237
240	74	1258	−229
255	74.5	1428	−244
270	38.1	2553	−473
285	55.2	2716	−461
300	79.4	1510	−329
315	99.1	1166	−283
330	113	866	−241
345	118.2	1008	−260
360	118.6	1196	−283

Compressive stress does not initiates fatigue failure.

Shear stress associated with compressive stresses causes crack formation.

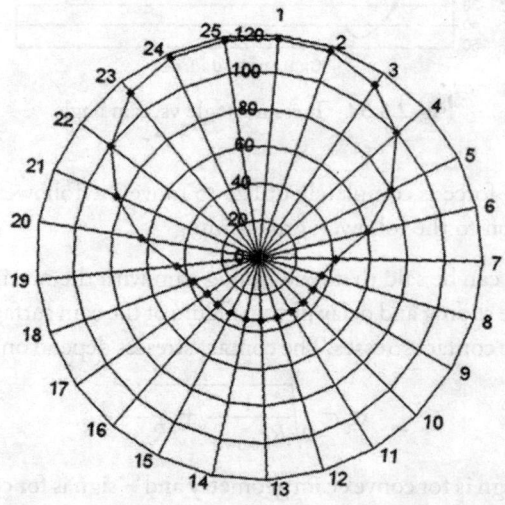

Fig. 2.5.35 Cam–follower interface at various angular positions

Transition from convex to concave contact introduces sliding. Present cam–follower mechanism is subjected to variable stresses and sliding conditions, which repeat at frequency of cam rotation. Increasing rotational speed will reduce operating life of cam-surface.

Further clearance between groove and follower (required to avoid jamming) reduces the support area. Follower contacts only one side of groove. The figure 2.5.36 shows the contact between the follower and cam. The sudden change in the velocity of roller is shown by the hatched ellipse. Sudden change in the velocity causes gross sliding at interface. Therefore cam–follower interface needs proper lubrication.

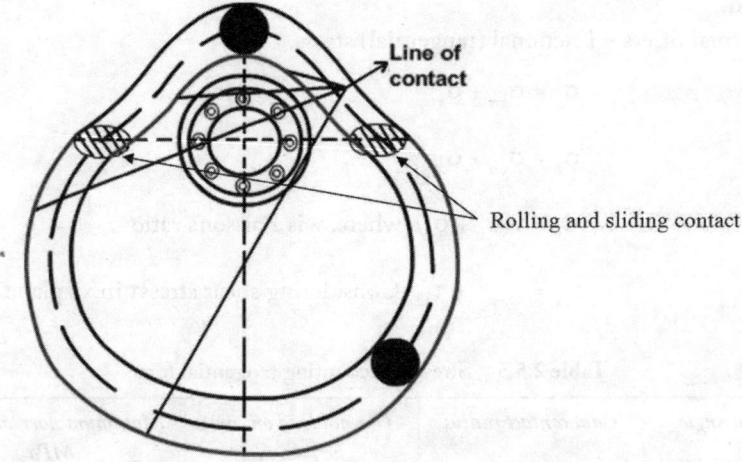

Fig. 2.5.36 Cam–follower mechanism

How sliding reduces life?

Pitting, a fatigue wear, initiates on or near the surface of component. Tangential force not only increases (Fig. 2.5.37) τ_{max} but also shifts position of τ_{max} to the surface. The pitting occurs if $\tau_{max} > S_{ys}$ where S_{ys} is the yield shear strength of the material. Total pitting life (N_f) is the sum of non-cracking life (N_O) and crack propagation life (N_{pl}). In pressure of lubrication, τ_{max} is reduced in magnitude and occurs below surface. This means lubrication delays to crack to reach at surface.

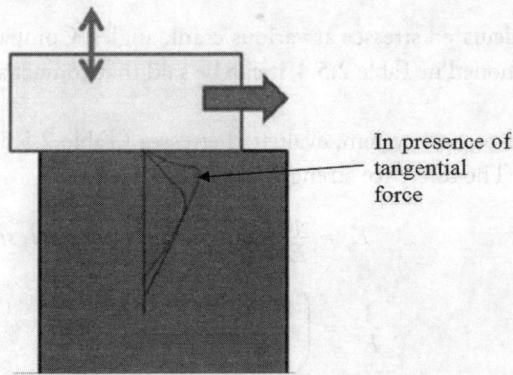

Fig. 2.5.37 Effect of tangential force

3-D stress analysis of cam–follower mechanism [Norton, 2009]:

Motion in cam–follower mechanism comprises both rolling and sliding. Frictional force causes tangential loading. Finding location of incipient crack is quite unpredictable. Therefore, it is more difficult to accurately predict the condition of stress at an expected point of failure. Therefore, the contact zone stress as a reference value to compare to material strength is used. Strength needs to be compared with the largest negative principal contact stress. In a pure rolling case, its magnitude will be equal to the maximum contact pressure. But it will be greater than that value if sliding is present. When rolling and sliding are both present, stresses due to the normal and tangential loading need to be accounted.

Stress = Normal Stress + Frictional (tangential) stress

$$\sigma_x = \sigma_{xn} + \sigma_{xt} \qquad (2.5.17)$$

$$\sigma_z = \sigma_{zn} + \sigma_{zt} \qquad (2.5.18)$$

$$\sigma_y = \nu(\sigma_x + \sigma_z) \text{ where, } \nu \text{ is Poisson's ratio} \qquad (2.5.19)$$

$$\tau_{zx} = \tau_{xz_n} + \tau_{xz_t} \text{ Considering shear stress} \tau \text{ in xz plane} \qquad (2.5.20)$$

Table 2.5.5 Stresses accounting tangential force

Cam angle	Cam contact radius	Normal load on cam surface, N	Maximum normal stress, MPa
0	118.6	1195	−344
75	64.4	2174	−491
90	45.7	2489	−551
105	36.5	1794	−485
270	38.1	2553	−574
285	55.2	2716	−560
300	79.4	1510	−400

Table 2.5.5 lists the calculated stresses at various crank angles. Comparing stress values listed in this table with those mentioned in Table 2.5.4, it can be said that contact stresses increase due to the frictional force.

To estimate the failure of cam system, evaluated stresses (Table 2.5.5) must be compared with materials reference stress. The reference strength is expressed as:

$$K_r = \frac{\pi}{E'}(\text{max normal. principal. stress})^2 \qquad (2.5.21)$$

where

$$\frac{1}{E'} = \left(\frac{1-\upsilon_1^2}{E_1} + \frac{1-\upsilon_2^2}{E_2}\right) \qquad (2.5.22)$$

and

$$\log_{10} K_r = \frac{\zeta - \log_{10} N}{\lambda} \qquad (2.5.23)$$

where
- E = Effective Young's modulus
- E_1 = Young's modulus of cam
- E_2 = Young's modulus of follower
- v_1 = Poisson's ratio of cam
- v_2 = Poisson's ratio of follower
- K_r = Reference strength
- ζ and λ are material constants as provided in Table 2.5.6.

Table 2.5.6 Material strength data for rolling with 9% sliding

Material	Rolling & 9% Sliding			
	K_r psi	Sc@1E8 psi	λ	ζ
Gray iron, Cl.20, HB 140-160	740	47000	4.09	19.72
Gray iron, Cl.30, h – t (austempered), HB 255 – 300 Phosphate coated	2510	94000	6.01	28.44
Gray iron, Cl.35, HB 225 – 255	1900	84000	8 39	35 51
Gray iron, Cl.45, HB 220 - 240	1070	65000	3.77	19.41
Nodular iron, Gr. 80 - 60 – 03 h-t HB 207 - 241	1960	93000	5.56	26.31
Nodular iron, Gr. 100-70-03 h-t HB 240 - 260	3570	122000	13.04	54.33

Table 2.5.7 Material data for pure rolling case

Material	Pure Rolling			
	K_r psi	Sc@1E8 psi	λ	ζ
Gray iron, Cl.20, HB 140 - 160	790	49000	3.83	19.09
Gray iron, CS.30. HB 200 - 220	1120	63000	4.24	20.92
Gray iron, Cl.30, h -1 (austempered), HB 255 - 300, Phosphate coated	2920	102000	5 52	27.11
Gray iron, Cl.35, HB 225 - 255	2000	86000	11 62	46.35
Gray iron, Cl.45. HB 220 -240	—	—	—	—
Nodular iron,Gr.S0 - 60 - 03 h-t, HB 207-241	2100	96000	10.09	41.53

Since finding location of incipient crack is quite unpredictable, therefore contact zone stress as a reference value to compare to material strength is used. Strength needs to be compared with largest negative principal contact stress. In a pure rolling case, its magnitude will be equal to maximum contact pressure. But it will be greater than that value if sliding is present. From maximum principal stress we can calculate value of K_r then we can calculate life in number of cycles. The calculated number of cycles will be relative. Relative reduction in life due to increase in speed is given in Table 2.5.8.

Stress vs. Cam life:

Using Eq. 2.5.21, Eq. 2.5.22, and Eq. 2.5.23 a computer code was generated to calculate the cam life at different rpm as shown in Table 2.5.8.

Table 2.5.8 Cam life for various materials

Cam Life at 60 rpm	Cam Life at 65 rpm	Cam Material	Reduction in Life
16.5 days	11.8 days	Gray iron, Cl.20, HB 140-160	28.5 %
3692 days	2351 days	Gray iron, Cl.30, h – t (austempered), Phosphate coated	36.3 %
553 days	304 days	Gray iron, Cl.35, HB 225 - 255,	45%
93 days	68 days	Gray iron, Cl.45	26.9 %
851 days	557 days	Nodular cast iron Gr. 80 – 60 – 03 h-t	34.6 %
1372628 days	566321 days	Nodular cast iron Gr. 100-70 -03 h-t HB 240 -260	58.7 %

Conclusions:

- Theoretical study shows 25–30% reduction in cam life on increasing speed from 60 rpm to 65 rpm.
- Nodular cast iron provides much higher life compared to CI 45 material. Therefore nodular cast iron will be a better choice if cam is operated at higher rpm.

To get more accurate prediction of the cam life a detailed EHL contact analysis is required.

2.5.2 Factors affecting wear

There are a number of factors which affect the wear rates of rubbing surfaces. It is difficult to separate individual affect. For example, high operating temperature at the surfaces may be due to high loads and/or large speeds. The temperature influences surface film formation and can cause changes in the surface structure and hardness. This effect may reduce or increase rate depending the load and speed conditions. Main direct factors that affect wear are:

1. Surface films:

Wear is essentially a surface phenomenon. The presence and properties of films covering rubbing surfaces have a critical effect. The wear between metals may be reduced by coating the surfaces with a thin layer of higher (than the substrate metal) wear resistance coating. The coating of hard material on a tough material provides overall economics. For example, rhodium and chromium are both hard but costly, so electroplated coatings of these metals are used to protect cylinder liners, crankshafts and similar applications experiencing extreme operating conditions.

2. Oxide films:

Most metals, during rubbing operation, are covered by a few molecular thick oxide films. There is a possibility of breakage/removal of oxide layer by rubbing, but the clean metallic surface immediately gets oxidized and formation of monomolecular layer almost instantaneously occurs. This must be taken into account in all considerations of metals rubbing under atmospheric conditions.

3. Boundary Lubrication:

Generally boundary lubrication reduces wear rate, which is explained in section 2.8 of this chapter.

4. Temperature:

The temperature of the rubbing surfaces influences the wear in three major ways. It can alter

- The properties of the rubbing materials
- The form of the surface contaminants film
- The lubricant properties

5. Load:

An increase in load generally increases wear rate. Sometimes an increase in load causes transition from mild wear to severe wear.

6. Compatibility:

Rabinowicz suggested the use of metal pairs with low metallurgical compatibility to reduce the wear rate. Here metallurgically compatible metals are those which show a high degree of mutual solubility. The compatibility ratings can be obtained from binary phase diagrams. Certainly it has been generally accepted that in selecting metals for bearings, similar metals, that is, metals with 100 per cent compatibility, should be avoided.

2.6 Theories of Wear

Most of the laws and theories of wear are based on experimental findings. Very few researches based on the quantitative law of wear are available, which can be used for general purpose tribology. Most of the available equations are very specific to a particular situation. Some of the known theories of wear are:

1. Holm's wear equation

Holm assumed that wear was an atomic transfer process that occurs at the area of contact formed by the plastic deformation of the asperities. Holm's equation for wear rate 'w' is

$$w = \frac{ZW}{H_d} \quad (2.6.1)$$

where,
- W = normal load
- H_d = hardness of the softer material
- Z = number of atoms removed per atomic encounter. The value of Z can be found from the following relation

$$Z = \frac{n_1 \cdot d}{2 \cdot a} \quad (2.6.2)$$

where,
- n_1 = atomic layers of inter planer spacing d
- a = radius of α spots (N circular regions of contact)

This wear rate equation is difficult to implement. In addition Rabinowicz showed that metallic transfer during sliding does not take place uniformly which was an assumption in Holm's hypothesis.

2. Burwell and Strang's wear equation

Using different combinations of materials Burwell and Strang performed extensive experiments and proposed the concept of transfer of wear particles during asperities interaction and gave the equation for wear material volume Q.

$$Q = \frac{K_1 W L_d}{H_d} \quad (2.6.3)$$

where,
- K_1 = probability of removing a wear particle
- L_d = sliding distance

Wear volume per unit sliding distance ν can be given by

$$\nu = \frac{K_1 W}{H_d} \quad (2.6.4)$$

3. Archard's wear equation

Archard assumed that the contact between tribo–pair involve formation and breakage of junctions. In other words contact occurs only at asperities. The Archard model is demonstrated in Fig. 2.6.1, where cross section of asperities after plastic deformation is assumed to be circular.

First sketch in the Fig. 2.6.1 demonstrates the approach of junction forming asperities. The area of contact increases with sliding distance and subsequently decreases. But this process is continuous and happens among number of asperities. On an average, it is assumed that n asperities will be in contact at any frame of time. Archard's wear equation is given by

$$Q = \frac{K_1 W L_d}{3 H_d} \quad (2.6.5)$$

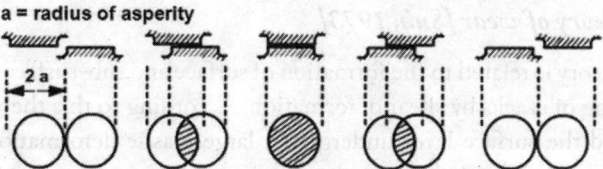

Fig. 2.6.1 Archard's wear model

The value of K_1 depends on elastic plastic contacts, shearing of those contacts, effect of environment, mode of lubrication, etc. This expression of wear volume is a simple expression, as it does not require to estimate constant n1 $\left(A = \left(\dfrac{W}{H_d}\right)^{n1} \right)$, individual shear strength of elastic and plastic junctions, effect of lubricant thickness, roughness, etc.

4. Rowe's wear equation

To establish the relation between μ and K_1, Rowe's wear equation is given in Eq. 2.6.6. In this equation k_m is constant and β is fractional surface film defect. This means β fraction of contact area is under dry lubrication, while $(1-\beta)$ contact area is under lubricated condition. Here lubricated condition means shear strength of interface is lesser than shear strength of bulk material.

$$v = K_m \sqrt{1+\mu^2}\, \beta \left(\dfrac{W}{H_d}\right) \qquad (2.6.6)$$

where v = wear volume per unit sliding distance.

5. Rabinowicz's equation for abrasive wear

The abrasive wear rate of the two–body contact has been obtained by Rabinowicz. To derive the quantitative expression for abrasive wear the asperities on the hard surface are assumed to be conical. Assuming that one asperity carries a load ΔW and will penetrate the softer surface to an extent given by

$$\Delta W = H_d . \pi . r^2 \qquad (2.6.7)$$

The projected area of the penetrating cone in the vertical plane is $r.x$. When the cone slides through a distance dl, it will sweep out a volume dv given by

$$dv = r.x.dl = r^2 \tan\theta\, dl = \dfrac{\Delta W \tan\theta . dl}{\pi H_d} \qquad (2.6.8)$$

$$\dfrac{dv}{dl} = \dfrac{\Delta W \tan\theta}{\pi H_d} \qquad (2.6.9)$$

If the contributions of all the asperities are added,

$$\dfrac{Q}{l} = \dfrac{W\, \overline{\tan\theta}}{\pi H_d} \qquad (2.6.10)$$

where $\overline{\tan\theta}$ is a weighted average of the $\tan\theta$ values of all individual cones. If we replace $\dfrac{\overline{\tan\theta}}{\pi}$ by $\dfrac{K}{3}$, Archard's wear equation results.

6. Delamination theory of wear [Suh, 1973]

The Delamination theory is related to the formation of surface and sub-surface cracks, void formation and subsequent joining of cracks by shear deformation. According to this theory wear particle is like a thin flake sheet and the surface layer undergoes a large plastic deformation. To derive the wear equation, a circular wear track produced on a pin-on-disc type of wear test and following assumptions were made with respect to this theory.

- Metal wears layer by layer and each layer consists of N sheets.
- The number of wear sheets per layer is proportional to the average number of asperities in contact.
- The rate of void and crack nucleation and the shear deformation for loose particles can be expressed in terms of sliding distance s_o.

The total wear may be expressed as

$$Q = N_1(L_d/s_{01})A_1 h_1 + N_2(L_d/s_{02})A_2 h_2 \qquad (2.6.11)$$

The subscripts 1 and 2 refer to the soft and hard metals sliding against each other

A = average area of delaminated wear sheet

h = thickness of sheet

L_d = sliding distance

The thickness h is approximated as

$$h = \frac{Gb}{4\pi(1-\upsilon)\sigma_f} \qquad (2.6.12)$$

where,

G = modulus of rigidity,

σ_f = friction stress,

b = Burgers vector, and

υ = Poisson's ratio

Average surface area A is taken to be $A = CA_r$

where, A_r = real area of contact, and
C = constant.

2.7 Approaches to Friction Control and Wear Prevention

Wear and friction cannot be fully eliminated when surfaces are in relative motion but can only be reduced to an insignificant level. Easiest way to prevent friction and wear is the lubrication. If lubrication, due to high load is insufficient to reduce friction/wear to acceptable level, a suitable change can be made in the system in some possible way so as to minimize the same.

The wear rate in a particular system depends on many factors, which can be divided into two groups:

- The structure of the mechanical system, which is constituted by the materials, making up the surfaces in relative motion, the nature of any material present at the interface and the environment.
- The operating variables such as imposed on the system during operation such as load, speed, temperature, etc.

It can be said that friction and wear can be reduced by controlling the parameters such as temperature, material of contacting surface, lubrication, reducing average load, etc.

2.8 Boundary Lubrication

'Boundary Lubrication' term was coined by the English Biologist 'Sir Hardy' in 1922. He quoted that 'very thin adsorbed layers, about 10 Å thick, were sufficient to cause two glass surfaces to slide over each other'.

In normal light load operations sliding contact under air or water, form the protective oxide layer. At relatively larger load such as occurred in steel gears, piston–rings and metal–working tools, the oxide layer is torn away, exposing the pure metal of both surfaces. These may get welded together before oxygen can reform the protective layer. To avoid such welding of metals boundary lubricants, which form a thin layer separating sliding surfaces, are required.

An obvious question arises 'How thin layer is able to separate surfaces?' The answer is, boundary lubricants can separate two surfaces. One group of boundary lubricants (i.e., fatty acids) has long chain molecules with an active end group. Representative molecules of these types are shown in Fig. 2.8.1. They consist of a hydrocarbon backbone of carbon atoms and an active end group. In fatty acids, active group is COOH, known as the carboxyl group, which when meets a solid surface attaches itself to the solid and gradually builds up a surface layer.

Some of the characteristics required for thin film lubrication are:

1. Lubricant must attach itself between contacting surfaces. In Fig. 2.8.2 boundary lubricants having long chain molecules with an active end group attached to the solid surface is shown. Longer hydrocarbon chain, will give more effective separation between solid surfaces. This brings low coefficient of friction. More number of layers reduces friction coefficient as shown in Fig. 2.8.3.
2. Boundary lubricant should be dissolvable in mineral/lubricating oils.
3. Temperature stability: It is important because increase in operating temperature may cause reduction in molecular attraction that may lead to detachment of boundary additives from surface.

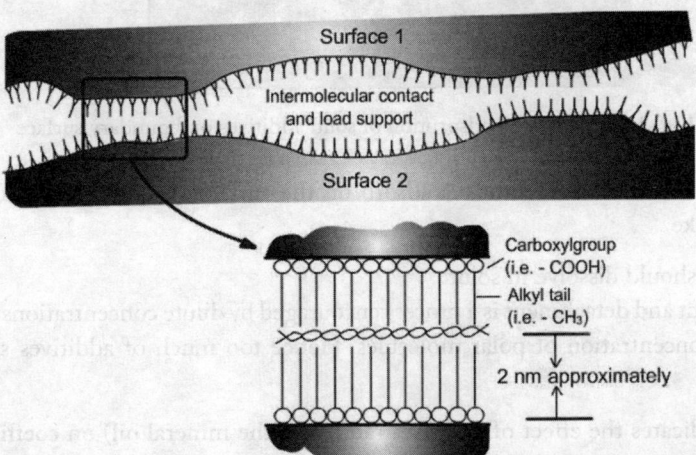

Fig. 2.8.1 Boundary lubricants 'oiliness additives'.[Stachowiak, 2006]

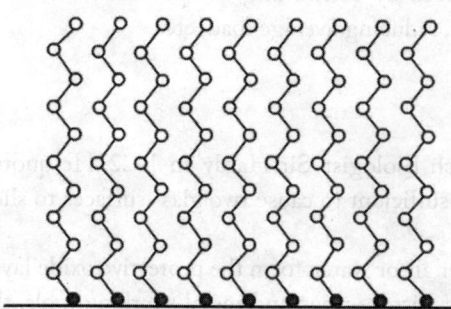

Fig. 2.8.2 Long chain boundary additives

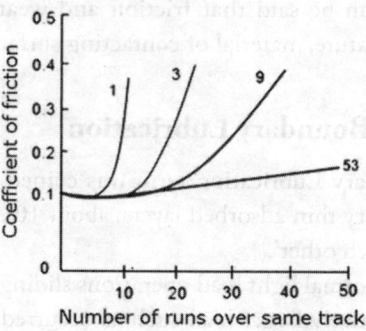

Fig. 2.8.3 Number of layers vs. friction coefficient

Mechanisms of boundary lubrication

Boundary lubricants form boundary films by physical adsorption, chemical adsorption and chemical reaction.
- Physisorption
- Chemisorption

Physisorption:

First boundary lubrication mechanism is physisorption or 'physical adsorption' (physical bonding by van der Waals' force). In this mechanism, surface active molecules of oiliness additives are attracted to surface by electrostatic (dipole) forces as shown in Fig. 2.8.4.

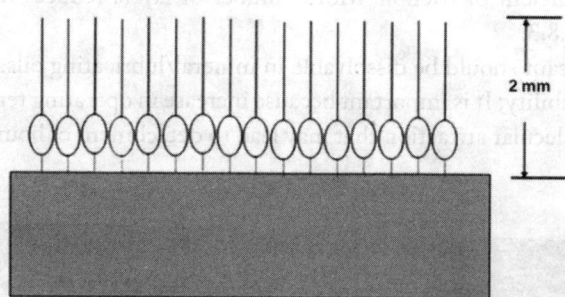

Fig. 2.8.4 Physical adsorption of solid additives on boundary surface

Energy is lowered when the additives adsorb on the surface by physical attraction. It requires some properties like

- Additives should dissolve in solute.
- Attachment and detachment is a process encouraged by dilute concentrations and hindered by high concentration of polar molecules. Hence too much of additives should not be present.

Table 2.8.1 indicates the effect of additives (added to the mineral oil) on coefficient of friction. The increase in the percentage of additives reduces the coefficient of friction, but up to a certain limit after that coefficient does not change.

Table 2.8.1 Percentage of boundary additive vs. friction coefficient

Lubricant	Friction Coefficient
Pure mineral oil	0.360
2% oleic acid in mineral oil	0.249
10% oleic acid in mineral oil	0.198
50% oleic acid in mineral oil	0.198
Pure oleic acid	0.195

As the temperature increases the viscosity reduces so that friction reduces. But when the temperature reaches critical value the friction increases, this is shown in the Fig. 2.8.5. Figure 2.8.6 indicates disruption of boundary lubricants at critical temperature which results in increase in friction coefficient as shown in Fig. 2.8.5.

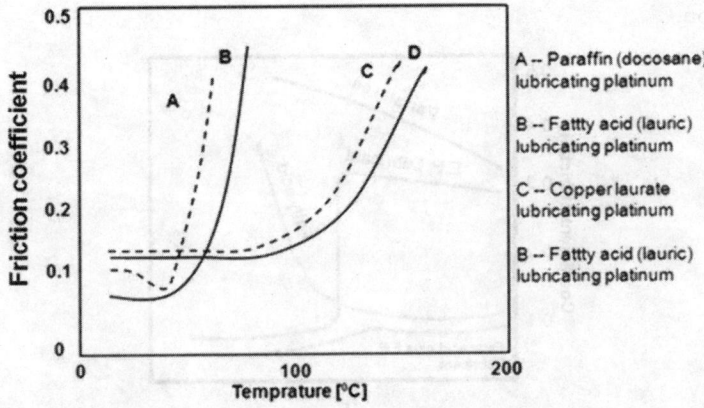

Fig. 2.8.5 Temperature vs. µ [Bowden and Tabor, 1950]

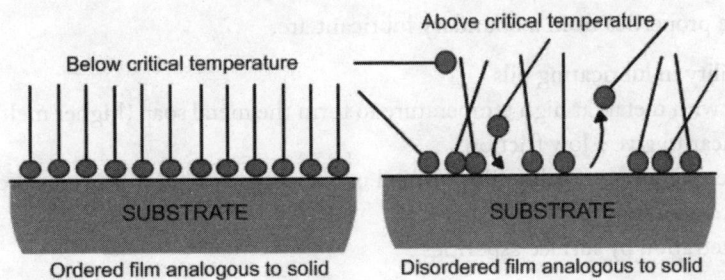

Fig. 2.8.6 Effect of temperature on adsorption [Stachowiak, 2006]

At a somewhat higher temperature physically absorbed molecules get desorbed. In other words molecules still present on the surface but lose their attachment (Fig. 2.8.6). Consequently wherever the surfaces come together the lubricant molecules are pushed away and intimate metal–metal contact is able to occur.

Chemisorption:

Chemisorption is a form of corrosion. To form a chemically bound layer three things are needed:
- Chemically active group.
- Reactive surface of material.
- Surface free from physisorbed material so that chemical reaction occurs. This creates a gap (Fig. 2.8.7) where boundary additives become ineffective.

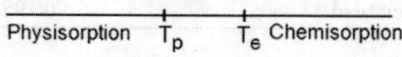

Fig. 2.8.7 Temperature gap

During each contact, the chemical layer is rubbed off the surface and has to be reformed before next contact comes round. Surface is therefore slowly worn away so the additive must be chosen with care. Figure 2.8.8 indicates effective boundary lubrication requires a combination of physisorption and chemisorption.

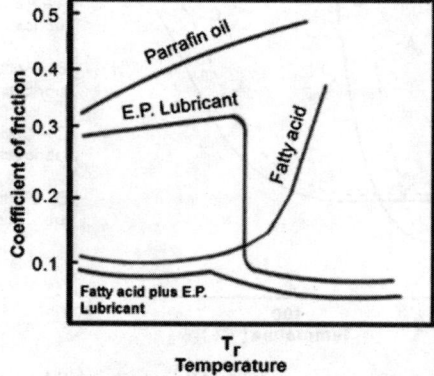

Fig. 2.8.8 Combination of physisorption and chemisorption of effective lubrication

Some desirable properties from a boundary lubricant are:
- Dissolvability in lubricating oils
- Reactivity with metals at high temperature to form the metal soap (higher melting points)
- Easy to shear to give a low friction
- High softening point of solid–film formed on substrate so that it provides protection up to a high temperature
- Resist penetration by surface asperities

Boundary (BL) vs. Extreme Pressure (EP) lubrication

Often there is doubt about BL and EP lubrication. BL is restricted to those systems where there is thermodynamic reversibility. A small change in temperature or concentration, up or down, brings a related change in film coverage. If lubricant reacts chemically (chlorine, sulphur, and phosphorus) with metal, then it must be known as EP lubricant. At low load/temperature EP lubricants are ineffective and BL remain active. Major difficulty with EP lubricants is their carcinogenic nature and environmental pollutant; therefore EP additives should be avoided as far as possible.

Comparative study:

To understand lubrication in all respect, a comparative study among dry, boundary lubricated, and fluid film lubricated has been provided.

Wear rate vs. time (Fig. 2.8.9a):

As the time increases the wear rate decreases and remains constant up to certain time, then again increases for dry lubrication. For boundary lubrication the wear rate decreases up to certain time then decreases or increases depending on the improvement in surface smoothness. If surface smoothness occurs, boundary lubrication turns out to fluid film lubrication. This means the wear rate decreases, otherwise the wear rate increases. For fluid film lubrication the wear rate drastically decreases, then remains constant up to certain limit and then increases.

Wear rate vs. load (Fig. 2.8.9b):

For dry lubrication, as the load increases, the wear rate increases. For boundary lubrication, the wear rate increases and the rate of increase in the wear rate is lower than that of the dry lubrication. For fluid film lubrication also the wear rate increases, but the rate of increase in the wear rate is initially lower. Same process does occur with increase in the temperature.

Wear rate vs. temperature (Fig. 2.8.9c):

Wear rate increases for all three lubrication mechanisms with increase in temperature. One of the common elements in all machines is spur gear. Spur gear generally operates under boundary lubrication regime. To understand the effect of time and load on the wear rate of spur gears, an experimental (Fig. 2.8.10) study was performed.

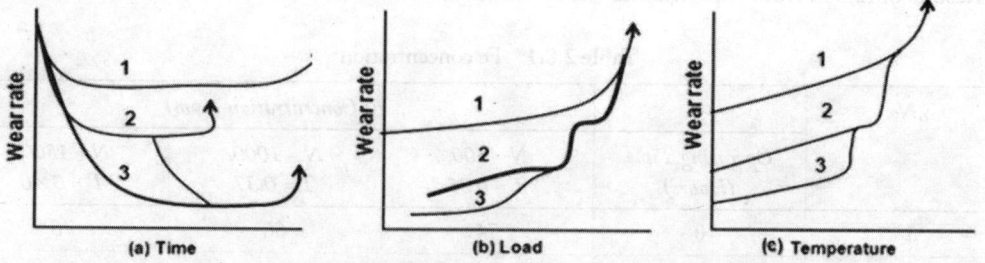

Fig. 2.8.9 Comparative study among (1) dry, (2) boundary and (3) hydrodynamic lubrication mechanisms

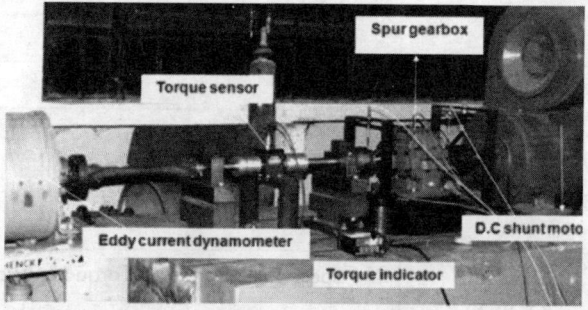

Tp = 22, Tg = 44, Module = 2.5 mm

Fig. 2.8.10 Online monitoring of spur gears

Geometric data related to single stage gearbox (Fig. 2.8.10) are given below:

- Pinion speed (N_p = 3000 rpm)
- Gear speed (N_g = 1500 rpm)
- Pitch circle diameter of pinion (D_p = 55 mm)
- Pitch circle diameter of gear (D_g = 110 mm)
- No. of teeth on pinion (T_p = 22)
- No. of teeth on gear (T_g = 44)
- Module = 2.5 mm

To measure the wear rate, online sensors, as shown in Fig. 2.8.11, were used.

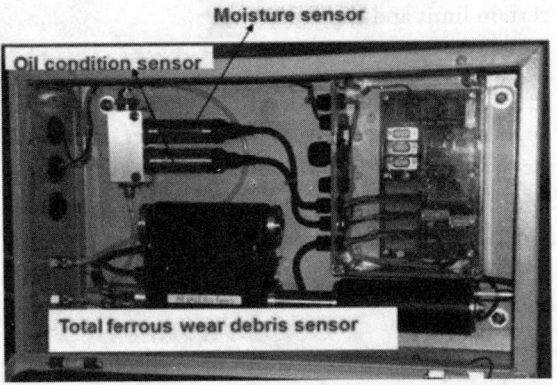

Fig. 2.8.11 Online sensor instrument

Results obtained from experimental study are tabulated in Tables 2.8.1–2.8.4:

Table 2.8.1 Fe concentration

S.No.		Fe Concentration (ppm)		
	Operating Time (Hours)	N = 500, T = 0.35	N = 1000, T = 0.37	N = 1500, T = 0.40
1	0	74	66	76
2	1	-	-	93
3	2	84	68	95
4	3	-	-	111
5	4	77	71	121
6	6	73	73	-
7	8	69	74	-
8	10	65		
Lubricating oil: 80W–90	R.H. 45° to 55°	N = speed in RPM	T = Torque in N/m	R.H. = Relative Humidity

Results listed in Table 2.8.1 show running in behaviour of spur gear at speed 500 rpm. But on changing speed from 500 to 1500 rpm, the Fe concentration increases. This concludes that running in behaviour occurs at every speed. Results of Table 2.8.2 illustrate that on changing lubricating oil (using fresh oil), the Fe concentration (compared to results shown in Table 2.8.1) decreases.

Experimental Results after changing oil:

Table 2.8.2 Changing operating condition changes the dynamics of B.L.L

S. No.	Operating time (hours)	Fe Concentration (pprn) R.H. 38% to 52%		
		$N = 500, T = 0.35$	$N = 2000, T = 0.42$	$N = 2500, T = 0.45$
1	1	43	89	219
2	2	47	136	283
3	3	30	141	344
4	4	32	143	615
5	5	34	130	540
6	6	35	124	658
7	7	36	120	

Table 2.8.3 and Table 2.8.4 show wear rate versus time. Both of these tables indicate reduction in wear rate with time.

Table 2.8.3 Particle size vs. time at N = 500 rpm

S. No.	Particle Size (μm)	Time (hours)–500 rpm			
		$t = 2$	$t = 4$	$t = 6$	$t = 8$
1	5 – 15	18895523	17284428	14598593	7657957
2	15 – 25	403509	388530	364556	130996
3	25 – 50	58218	43026	50805	18934
4	50 – 100	1853	1544	3552	462
5	> 100	154	0	0	0

Table 2.8.4 Particle size vs. time at N = 2000 rpm

S.No,	Particle Size (μm)	Time (hours) –2000			
		$t = 3$	$t = 4$	$t = 6$	$t = 7$
1	5-15 ^im	19075570	19638570	14932115	12883162
2	15 - 25 |Jm	409160	453867	337724	266367
3	25-50 pm	59576	80657	74741	43051
4	50 - 100 fjm	1235	5089	5714	2009
5	> 100 pm	0	0	0	0

Table 2.8.5 indicates reduction in the wear rate on increasing load from zero load to very low load. This means some load is required to reduce the vibration of gears and reduce the wear rate.

Table 2.8.5 Fe concentration vs. load

S. No.	Operating time (hours)	N = 200, T = 5	N = 200, T = 8.5	N = 600 T = 20	N = 800, T = 28	N = 200, T = 5	N = 800, T = 28	N = 1000, T = 35	N = 1000, T = 0
1	1	152	100	67	45	-	-	-	-
2	1.5	-	-	-	-	-	-	-	33
3	2	154	94	68	42	-	-	-	-
4	2.5	138	93	65	41	33	30	33	-
5	3	135	90	63	40	-	-	-	30
6	4	130	85	62	40	-	-	-	-
7	4.5	-	-	-	-	-	-	-	30
8	5	127	85	61	39	30	32	30	-
9	6	102	67	56	35	-	-	-	29
10	7	104	69	54	37	-	-	-	-
11	7.5	107	69	56	36	32	32	33	32
12	8	-	-	53	34	-	-	-	-
13	9	-	-	53	33	-	-	-	-
14	10	-	-	52	32	-	-	-	-
15	11	-	-	47		-	-	-	-

Frequently Asked Questions

Q.1. Instead of providing lubrication between two rubbing surfaces, is it advisable to improve the surface finish of the two surfaces to reduce friction?

Ans. Friction occurs due to ploughing (due to surface roughness) and adhesion (due to surface smoothness). This means improving surface finish would definitely help in reducing the friction caused by ploughing and micro–cutting, but it will increase friction caused by molecular attraction between the tribo–pair. Lubricant reduces molecular attraction between tribo–surfaces; therefore, lubrication (in form of solid / liquid /gaseous) is essential to reduce friction. In addition liquid lubricant has some other advantages also like cooling, transferring of wear debris etc. Also achieving high surface finish would be very costly as compared to providing lubrication.

Q.2. Since coefficient of friction is dependent on various parameters such as temperature, surface roughness and hardness, how can we quantify the coefficient for a particular material in general?

Ans. Coefficient of friction is a system property. It may be defined for a pair of materials for particular environment condition. In other words, coefficient of friction cannot be quantified for a material and may vary significantly.

Q.3. Is there any other alternative to reduce the adhesion component of friction without lubrication?

Ans. No. Lubrication is a generic term. Reducing chemical activities between tribo–surface may be treated as lubrication. In other words, choosing appropriate surface coating or surface treatment, which reduces chemical affinity between tribo–material and provides wear-resistant surface, is part of lubrication mechanisms.

Q.4. What is friction instability and how it is related to stick–slip process?

Ans. Friction instability generally occurs due to large difference in the value of static and kinetic coefficients of friction. Similarly, stick–slip phenomenon occurs due to huge variation in static and kinetic coefficients of friction. Ideally lubricated condition having kinetic coefficient of friction equal to 0.00025 shall be preferred, but there is a possibility of high value of static coefficient of frictions. If we assume that static coefficient of friction under lubricated conditions is equal to 0.01 and kinetic coefficient of friction is equal to 0.00025, then this lubricated contact may not be preferred. This huge difference between static and kinetic friction coefficients provide negative damping and initiate a 'stick–slip' process. Instantaneous speed of sliding object does not remain close to the average sliding speed of that object and friction coefficient decreases as velocity increases.

Q.5. What is the difference between static and kinetic coefficients of friction and what is the practical significance of these two terms?

Ans. The coefficient of friction is defined as a ratio of tangential force to normal force. Tangential force required to start sliding between two contacting surfaces is termed as static friction force. The ratio of static friction force to normal force is expressed as static coefficient of friction. The coefficient of static friction is typically (but not necessary) larger than the coefficient of kinetic friction.

Kinetic friction: Kinetic friction force acts at interface of two relatively sliding surfaces. The kinetic frictional resistance is almost constant over a wide range of sliding speeds. The kinetic coefficient is typically less than the coefficient of static friction. Often starting force is calculated based on static friction while energy loss is calculated based on kinetic friction force.

Q.6. Why are the junctions formed at the region of contact between two surfaces?

Ans. All surfaces are made of atoms. All atoms attract one another by attractive force. For example, if we press steel piece over indium piece they will bind across the region of contact. This process is sometimes called 'cold welding or junction' since the surfaces stick together without the application of heat. It requires some force to separate the two surfaces. If we now apply a sideways force to one of the surfaces, the junctions formed at the regions of real contact will have to be sheared if sliding is to take place. The force to do this is the frictional force. Due to high stress those asperities suffer plastic deformation, which permits strong adhesive bonds among asperities. Such cold formed junctions are responsible for the adhesive friction.

Q.7. What is ploughing effect?

Ans. In a tribo–pair (two bodies in contact having relative sliding) harder surface may penetrate into the softer surface and produce grooves on it. This is called ploughing effect.

Q.8. Out of spherical asperity and conical asperity, which is more significant for increase in the coefficient of friction and why?

Ans. Conical asperities are much more severe than spherical asperity for increase in coefficient of friction. Conical asperity has comparatively more ploughing effect. The tip of the conical asperities may break, further adding to the debris which would further add to the friction.

Q.9. How does lubrication help in reducing the coefficient of friction?

Ans. Lubricants reduce the friction between sliding surfaces by reducing shear strength of interface. The interface shear strength in the presence of lubricant is generally very low compared to shear strength of any of material in tribo–contact. In addition lubricant fills the surface cavities and making the surfaces smoother, which helps one surface to slide over another easily and thus reducing friction.

Q.10. What are the various methods to reduce friction instability?

Ans. The best alternative to reduce friction instability is to choose the appropriate lubrication mechanism that maintains negligible difference in values of static and kinetic coefficients of friction.

Q.11. While designing a mechanical system should the selection of material for a tribological pair be based on the coefficient of friction or on the basis of functional requirement?

Ans. Functional importance of material is the primary requirement. It plays the most crucial role in the material selection. Friction can be handled by choosing pairing materials, surface treatment, coating, lubricant, etc. Right choice is 'overall economic solution', which may require few practical trails, before finalizing complete design.

Q.12. In a tribological pair wearing of softer material takes place at a much faster rate, so is it advisable to use materials of same hardness?

Ans. Choosing same hardness of tribo–pair shall reduce abrasive wear, but adhesive, corrosion, fatigue and fretting wear may continue. Therefore material and corresponding lubricant selection must be able to minimize all kinds of wear phenomena.

Q.13. What is wear constant and how it is related to coefficient of friction?

Ans. As per Archard's wear equation, wear volume rate (m^3/s) is proportional to normal load (N), relative speed (m/s) and inversely proportional to hardness (N/m^2). Wear constant (non-dimensional) is constant of proportionality relating wear volume rate with load, relative speed and hardness. There is no established relation between friction-coefficient and wear rate, but reduction in friction-coefficient means lesser resistant to relatively moving surface and as a result lesser wear rate.

Q.14. What are the main criteria for classification of wear and what are various types of wear?

Ans. Failure mechanisms causing wear are used to classify wear. For example wear caused by formation and rupture of adhesive junctions is known as 'adhesive wear'. Wear caused by abrasion of hard surface on soft surface is termed as 'abrasive wear'. Formation of brittle layer due to chemical action and removal of that layer by mechanical action is called 'corrosive wear'. Similarly formation of subsurface cracks and rapid propagation of those cracks to surface due to friction is named as' fatigue wear'. There are more than 2000 wear equations to describe the wear behaviour. There are more than 35 mechanisms detailed as wear sources. This means there are more than 35 types of wear.

Q.15. In a tribological system where two tribo–surfaces are moving at relatively high speed, should the design of such system involve the use of similar metals or dissimilar metals to prevent adhesive wear and seizure?

Ans. Whether low or high speed, molecular attraction, between similar materials will be higher so resultant friction shall be higher. Therefore, from tribological point of view similar materials shall never be recommended. However if overall cost (i.e., inventory cost and manufacturing processing cost) of dissimilar metals is much higher compared to savings in friction loss, then designers may select appropriate lubricant for similar materials tribo–pair.

Q.16. What is the difference between zero wear and measurable wear and how does zero wear increases performance?

Ans. Removal of material which causes polishing of material surfaces may be known as zero wear. In zero wear process, the asperities of the contacting surfaces are gradually worn off. On other hand removal of material from surface that increases vibration; noise or surface roughness may be treated as measureable wear. Zero wear is desirable for better life of tribo–pair. Due to zero wear significant reduction in friction, noise and vibration can be achieved which increases the overall performance of the system.

Q.17. In a tribo–pair interaction, why does contact occur only at asperities?

Ans. Every engineered surface is rough on nanometre to micro meter level. In absence of full film lubrication (thick lubrication), when surfaces come in contact, asperities(peaks on the surfaces) interact. During the interaction tip of the asperities may break-off (reduce roughness) or scratch opposite surface (increase roughness).However asperity contact can be avoided by developing appropriate lubricant layer between tribo–surfaces.

Q.18. What is seizure and what causes it in a tribological system?

Ans. Seizure means 'to bind' or 'fasten together'. Seizure is an extreme form of adhesive wear, which occurs as a result of mutual plastic deformation of materials. In ordinary cases after seizure components do not get separated on their own. Manual force is required to separate the parts. In other words, after seizure tribo–pair lose its utility and cannot be used without proper reconditioning. Causes for seizure are:
- Poor heat dissipation.
- Poor lubrication or improper lubrication.
- Smaller clearances.
- Installation errors.
- The tendency of the metals to form strong bond in solid state.

Q.19. What are MR fluids? What are their usage in tribological system?

Ans. MR Fluids stand for magneto rheological fluid. Apparent viscosity of MR fluid increases with increasing its exposure to magnetic field. Increase in viscosity is only restricted by magnetic saturation limit. Due to regulation of apparent viscosity, MR fluid is known as a smart fluid. There are a number of usages of MR fluids. It can be used to make wear-free brakes.

Q.20. What is the optimum value of impingement angle to minimize erosive wear?

Ans. To minimize the erosive wear, it is important to estimate material (surface on which particles are going to strike) hardness. Low value of impact angle results cutting wear, which means material hardness help to reduce wear rate. On other hand large impact angle causes fatigue wear, therefore soft (ductile) material may be suitable.

Q.21. Are there any general changes/improvements that can be incorporated in the design of mechanical system to eliminate different types of wear?

Ans. Different types of wear can be eliminated by following these guidelines:
- Selecting right lubricant and ensuring optimum quantity of lubricant.
- Confirming surface finish of tribo surfaces to reduce friction, vibration, noise.
- Right material selection with appropriate surface hardness.
- Increasing heat dissipation from the system to avoid excessive operating temperatures.
- Replacing the lubricant to remove the debris after periodic intervals.

Q.22. Does suspension system in an automobile help in reducing or eliminating fretting wear?
Ans. Suspension system in an automobile does help in reducing fretting wear by damping the vibration and shocks generated by the interaction of tires and road surface.

Q.23. Is lubrication useful in reducing fretting wear?
Ans. Fluid lubricant reduces friction and takes away wear debris from source. The removal of debris helps to control wear within mild regime. Therefore lubricants are useful in reducing fretting wear.

Q.24. How can compliant mechanism be useful in reducing wear due to friction?
Ans. Compliant mechanisms are joint-less mechanisms. Due to the absence of any joints, sliding becomes zero and hence there is no wear other than fatigue wear.

Q.25. Can the use of tribology knowledge be made in metal removing operations in machine tools?
Ans. Understanding any material interaction where one material slides or rubs over another requires knowledge of tribological interactions. For example, in metal-forming operations, friction increases tool wear, which in turn increases the required power to work a piece. In addition, loss of tolerance accuracy occurs, which demands the frequent replacement of tool. This increases the manufacturing cost. On implementing tribology manufacturing cost may decrease.

Q.26. What specific design changes are implemented on piston rings and oil rings in order to reduce wear?
Ans. Ring wear can be reduced by designing appropriate profile so that full film lubrication is made. But excessive lubricant may enter in combustion chamber and cause incomplete combustion and as result pollution. Therefore minimizing wear is achieved by appropriate selection of (chemically inert) material and appropriate design with suitable hardness of piston rings.

Q.27. Is there any method in tribology to separate the wear debris from the lubricants?
Ans. Separation mechanism of wear debris from the lubricants varies from application to application. For example in ferrography wear particles are separated by magnetic (or similar) arrangement. To recirculate the same lubricant in machine filtering system is used to separate wear particles. For example in IC engine filtration is widely used to remove particles from oil in the oil pump. The size of debris particle that would be filtered depends upon the mesh size of the filter membrane.

Q.28. If a tribological pair is brought up to the hardness and surface finish of the order of slip gauge, would it help in reducing friction and wear?
Ans. Excessive smooth surface minimizes the abrasion but provides many locations for adhesive wear. Sufficiently high hardness minimizes the plastic deformation and as result minimizes adhesive wear. This means abrasive and adhesive wear and friction can be minimized by very good surface hardness and roughness, but other wear mechanisms (pitting, fretting, corrosive, etc.) may persist.

Q.29. Is it possible to predict the fatigue life of a mechanical component like a passenger car chassis or an I-beam in a commercial vehicle chassis due to fatigue wear?
Ans. The fatigue life of any mechanical component can be predicted by simulating the real life conditions inside a test chamber with the help of artificially generated operating conditions. After extensive testing, curve fit equations may be generated, which may be used later to estimate the fatigue life.

Q.30. What is fluid film lubrication? What is the difference between hydrostatic and hydrodynamic lubrication?
Ans. Fluid film lubrication is a generic term used for full film lubrication, in which solid surfaces are completely separated by fluid and wear is negligible. Fluid may be gas or liquid. Liquid

lubrication is more common compared to gaseous lubrication. Hydrostatic and hydrodynamic lubrications are two mechanisms of liquid lubrications. In hydrostatic lubrication external pressure is supplied to separate two surfaces, while in hydrodynamic lubrication relative velocity between two surfaces is used to generate liquid pressure between two surfaces. Hydrostatic lubrication is generally costlier compared to hydrodynamic lubrication.

Q.31. How transition between various regimes of lubrication occurs on the basis of increasing load?

Ans. On increasing load on the tribo–surfaces fluid film between solids decreases. If fluid film is sufficiently thick and solid surfaces are rigid (no elastic deformation), fluid film lubrication hydrodynamic (hydrostatic), aerodynamic (aerostatic) occurs between surfaces. On increase normal load, fluid pressure increases, viscosity thickening occurs and to some extend solid surfaces deform. This form of lubrication is known as 'Elastohydrodynamic lubrication'. In this lubrication regime elastic deformation of the contacting surfaces increase the load bearing area whereby the viscous resistance of the lubricant becomes capable of bearing the load. On further increase in load direct contact between asperities occur and friction increases. Due to relative motion among asperities wear of surface occur. This regime is called boundary lubrication. To minimize wear under boundary lubrication boundary/extreme pressure additives are used.

Q.32. What is the effect on wear rate with respect to increase in temperature and speed?

Ans. Wear rate generally increases with increase in temperature. On increasing speed, wear rate may increase or decrease depending the lubrication regime. If tribo–pair is under mixed lubrication, then increasing speed shall reduce the wear rate. But if increasing speed in mixed/boundary lubrication causes high temperature rise due to insufficient thermal dissipation, then wear rate may increase. During 'running in' period increasing speed causes increase in wear rate.

Multiple Choice Questions

Q.1. Which one of the following statements is not true about friction?
  (a) Friction is tangential resistance to motion.
  (b) Friction is dependent upon the surface of the content.
  (c) Friction is greater on rough surface.
  (d) Friction does not decrease with lubrication.

Q.2. Coefficient of friction is independent of
  (a) Temperature   (b) Surface Roughness
  (c) Hardness   (d) Area of contact.

Q.3. Phenomenon of stick-slip occurs because of
  (a) There is a large difference between static and kinetic coefficients of friction.
  (b) Additional force is applied to move the object.
  (c) Coefficient of friction changes.
  (d) Lubrication is applied on the surfaces.

Q.4. Adhesion component of dry friction is negligible in which of the following?
  (a) In high temperature surfaces   (b) Lubricated tribo pair
  (c) Rough surfaces   (d) Surface with sharp asperities

Q.5. Cold weld between two surfaces happens because of
   (a) Excessive lubrication
   (b) Adhesion between two surfaces
   (c) No load on surfaces
   (d) Low temp on area of contact

Q.6. As per the theory of friction due to deformation which of the following statements is not true?
   (a) Slope of asperities govern the friction force.
   (b) Sharp asperities causes more friction compared to round or spherical asperities.
   (c) Asperities on one surface interact with the asperities on the other surface.
   (d) An asperity of softer surface causes ploughing on the harder surface.

Q.7. The formation of junction growth can be reduced by
   (a) Lubrication of the surfaces.
   (b) Increasing the surface finish of the rubbing surfaces.
   (c) Surface treatment of the surfaces to increase hardness.
   (d) All of above

Q.8. Easy deformation of asperities causes
   (a) Increase in friction.
   (b) Decrease in friction.
   (c) No change.
   (d) Cannot be predicted.

Q.9. Ploughing effect causes
   (a) Piercing and penetration of the soft surface by the asperities of the hard surface.
   (b) Increase in friction.
   (c) Both (a) & (b)
   (d) None of these.

Q.10. Coefficient of friction due to rolling is generally
   (a) Greater than coefficient of sliding friction
   (b) Less than coefficient of sliding friction.
   (c) Equal to sliding friction.
   (d) Cannot say.

Q.11. Which of the following are the major contributors to rolling friction?
   (a) Micro-slip effect within the contact area.
   (b) Elastic hysteresis of the contacting materials.
   (c) Plastic deformation of the materials & adhesion effects in the contact.
   (d) All of the above.

Q.12. Which one of the following is true with respect to ball bearing?
   (a) Sliding occurs between cage and balls.
   (b) Lubricants such as grease are used to reduce friction within ball bearing.
   (c) Both (a) and (b).
   (d) There is no sliding and only rolling motion involved between cage and balls.

Q.13. If an automobile tire is not filled up to the optimum pressure level it means:
   (a) There would be less hysteresis loss.
   (b) Rolling friction would be lower.
   (c) More controllability because of higher hysteresis loss.
   (d) None of these.

Q.14. To avoid the phenomenon of stick slip due to friction instability which of the following is the right approach?
  (a) Increase the operation speed.
  (b) Decrease the operation speed.
  (c) Operation speed does not have any effect on the stick slip process.
  (d) Increase the difference between static and kinetic coefficient of friction.

Q.15. In a mechanical system, negative damping due to friction instability causes
  (a) Increase in amplitude over a period of time.
  (b) Decrease in amplitude over a period of time.
  (c) Amplitude remains unchanged with time.
  (d) None of these.

Q.16. Zero wear increases performance because:
  (a) It causes polishing of surface and hence increases surface finish.
  (b) Size of surface asperities increases.
  (c) It removes lubrication from the surface.
  (d) It increases load bearing capacity of the surface.

Q.17. Which of the following is not true about measurable wear?
  (a) Measurable wear is undesirable.
  (b) It causes vibration and noise.
  (c) Measurable wear makes the surface rough.
  (d) It increases performance.

Q.18. Which of the following is not true about pitting on the gear surface?
  (a) It is a surface fatigue failure.
  (b) It occurs due to repeated loading of the tooth surface.
  (c) It occurs because contact stress exceeds than the surface fatigue strength of the material.
  (d) It increases performance of the system.

Q.19. With increase in bearing clearance the load capacity of the bearing
  (a) Increases   (b) Decreases
  (c) Does not change (d) Can't say.

Q.20. Which among the following is not an adhesive wear mechanism?
  (a) Galling   (b) Scoring
  (c) Scuffing   (d) Polishing

Q.21. Compared to the shear strength of the tribo surfaces, the shear strength of the lubricant should be:
  (a) Greater   (b) Lesser
  (c) Equal   (d) Cannot say

Q.22. As per Archard's wear equation wear volume in adhesive wear is independent of
  (a) Sliding distance of travel.   (b) Load
  (c) Hardness of the soft material. (d) Temperature.

Q.23. Seizure refers to
  (a) Binding and fastening together of the material due to plastic deformation.
  (b) Cracking on the surface.
  (c) Significant wear on the surface.
  (d) None of these.

Q.24. Causes of seizure are
(a) Poor heat dissipation.
(b) Poor lubrication.
(c) Smaller clearances.
(d) All of above.

Q.25. The thickness of the oxide layer formed on the surface is dependent upon
(a) Time required of the rupture the oxide layer.
(b) Time available to re-oxidise.
(c) Rate of formation of oxide layer.
(d) All of the above.

Q.26. Scratching is a form of
(a) Abrasive wear.
(b) Adhesive wear.
(c) Corrosive wear.
(d) Fatigue.

Q.27. Wear rate is lesser in 3-body abrasion as compared to 2-body abrasion because:
(a) Energy is consumed in rolling motion of free hard particles.
(b) Only spherical asperities are involved in 3-body abrasion.
(c) Size of the asperities is smaller in 3-body abrasion.
(d) None of these.

Q.28. The property of MR fluid is
(a) Their viscosity can vary due to magnetic attraction among particle.
(b) Viscosity is constant and does not change.
(c) MR fluids have very high viscosity.
(d) None of these.

Q.29. Which of the following represents correct sequence of corrosive wear?
(i) Sliding surfaces chemically interact with environment.
(ii) A reaction product (oxide, chloride).
(iii) Wearing away of reaction product film.

(a) (iii), (ii), (i)
(b) (ii), (iii), (i)
(c) (i), (iii), (ii)
(d) (i), (ii), (iii)

Q.30. Erosive wear is a function of
(a) Particle velocity.
(b) Impact angle.
(c) Size of abrasive.
(d) All of above.

Answers

Q.1. (d)	Q.2. (d)	Q.3. (a)	Q.4. (b)	Q.5. (b)	Q.6. (d)
Q.7. (d)	Q.8. (b)	Q.9. (c)	Q.10. (b)	Q.11. (d)	Q.12. (c)
Q.13. (c)	Q.14. (a)	Q.15. (a)	Q.16. (a)	Q.17. (d)	Q.18. (d)
Q.19. (b)	Q.20. (d)	Q.21. (b)	Q.22. (d)	Q.23. (a)	Q.24. (d)
Q.25. (d)	Q.26. (a)	Q.27. (a)	Q.28. (a)	Q.29. (d)	Q.30. (d)

References

Blau PJ. Friction Science and Technology: From Concepts to Applications, Second Edition. CRC Press; 2008.

F.P. Bowden and D. Tabor, The Friction and Lubrication of Solids, Part 1, Clarendon Press, Oxford, 1950.

D.A. Rigney and J.P. Hirth, Plastic Deformation and Sliding Friction of Metals, Wear, Vol. 53, 1979, pp. 345- 370.

Meng HC, Ludema KC. Wear models and predictive equations: Their form and content. Wear 1995;181-183:443–57.

Norton RL. Cam Design and Manufacturing Handbook. Industrial Press Inc.; 2009.

Stachowiak GW, Batchelor AW. Engineering Tribology. First edition. UK: Butterworth Heinemann; 2006.

Suh, Nam P, The Delamination theory of Wear, Wear, 25, 1, 1973, pp 111-124.

Y. Kimura, Mechanisms of Wear – the Present State of Our Understanding, Transactions JSLE, Vol. 28, 1983, pp. 709-714.

Chapter 3

Lubrication of Bearings

A bearing is defined as a load support allowing relative motion between two surfaces. Figure 3.1 illustrates a few possible configurations of bearings. The first two concepts (elastomer and flexible strips) are useful for small amplitude oscillations. Bearing assembly with continuous relative motion, such as rotational motion, requires solid lubricants, boundary lubricants, rolling elements and associate lubricants, fluid film (hydrostatic, aerostatic, hydrodynamic, aerodynamic) lubrication or magnetic/electric field to provide desirable performance.

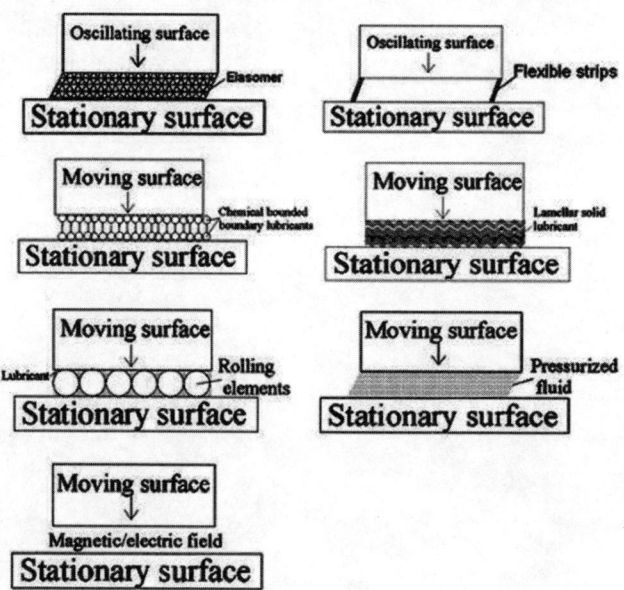

Fig. 3.1 Various concepts to separate two solid surfaces

Electric and magnetic fields are very good solutions, but require gigantic dimensions to compete with any mechanism of fluid film lubrication. In the fluid film lubrication regime, the sliding surfaces are completely separated by a film of liquid or gaseous lubricant. It is important to note that design of sliding bearings started in the nineteenth century. Over the period of time, sliding speed and imposed load have increased and permitted film thickness (separation between two solid surfaces) has reduced. In other words, designs of sliding bearings are getting more and more aggressive with lesser margin for errors. The desire for lower power loss, lower oil consumption and improved reliability requires better understanding of fluid mechanics.

3.1 Mechanics of Fluid Flow

A fluid is a substance, i.e., gases, liquids and easy flowing solids, which flows and changes its relative position without a separation from the bulk mass. If a thin film of fluid is provided between two relatively moving surfaces, significant reduction in friction and wear is expected. Primary reason for this reduction is load sharing by lubricant layer. In normal conditions, the real area of contact between tribo–pair (relatively moving surfaces) is hardly ten percent of the apparent (projected) area. With such low contact area, stress–state always exceeds the elastic limit point of all known solids. It results in plastic deformation of asperities in ductile materials, while generates cracks in brittle materials and as a result the relative motion between surfaces causes wear. By introducing a thin fluid film the redistribution of load happens, the severity of normal stresses reduces and in addition low shear force is required to induce relative motion. The fluid film lubrication can be classified as:

Hydrodynamic (Aerodynamic):

- In a hydrodynamic/ aerodynamic lubrication mechanism, a fluid is drawn into the region between the tribo–surfaces by the virtue of relative velocity. This mechanism will be called as hydrodynamic if the introduced fluid is liquid; and aerodynamic mechanism if the drawn fluid is gaseous. It is important to note that aerodynamic requires very sophisticated manufacturing of tribo–pair, and can sustain 1–2 % load compared to hydrodynamic lubrication mechanisms.
- Converging wedge shaped geometry, as shown in Fig. 3.1.1(a), is essential for this type of lubrication.
- Viscosity of lubricant plays an important role to support the load and reduce the friction.

Squeeze Film:

- Fluctuation in load and/or relative speed, as shown in Fig. 3.1.1(b), generates squeeze film action.
- Viscosity of lubricant plays an important role.

Hydrostatic (Aerostatic):

- External pressure of fluid, as shown in Fig. 3.1.1 (c), needs to be supplied to generate hydrostatic/ aerostatic lubrication.

Fig. 3.1.1(a) Converging wedge shape geometry

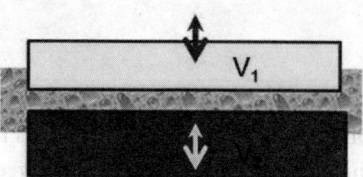
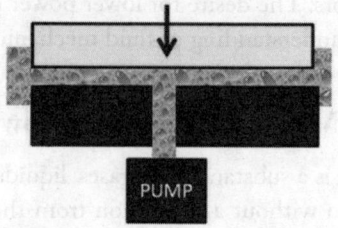

Fig. 3.1.1 (b) Squeeze lubrication **Fig. 3.1.1** (c) Hydrostatic lubrication

Every mechanism (hydrodynamic/ squeeze/ hydrostatic) of fluid film lubrication is suitable for a particular set of operating and environmental conditions. For example, hydrostatic lubrication that separates two surfaces by an external pressure source is suitable for extremely high load carrying capacity at low (or zero) speed. This mechanism finds its application in large telescopes, radar tracking units, machine tools and gyroscopes. In reverse, aerostatic (here aero is not restricted to air, but nitrogen and helium are also used) works well in light load applications even in odd temperature conditions (-200–2000 °C). As pressure is supplied by external sources, these mechanisms require closed feedback lubricant system. For example, if the applied load is reduced, the film thickness (separation between tribo–pair) will increase. Similarly if load is added to the moving surface, the film thickness will decrease. To maintain the film thickness within specified range, the active control system is essential. The requirement of external pressure supply and control system makes the bearing very costly. To compensate the cost, often a hybrid concept of hydrodynamic + hydrostatic or aerodynamic + aerostatic is used to achieve the best of both the mechanisms of fluid film lubrications.

3.1.1 Theory of hydrodynamic lubrication

The theory of hydrodynamic lubrication relates to pressure generation in fluid film due to the motion of fluids. The fluid may be liquid or gas. In the case of gas as separating fluid, lubrication mechanism is termed as aerodynamic lubrication. To quantify the pressure generation, mathematical modelling imitating the realistic lubrication mechanisms are required.

The basic purpose of lubrication is to separate two sliding surfaces with a film of lubricant, which can be sheared easily without causing any damage to the surfaces. The lubricant film must be of sufficient thickness to ensure no mechanical contact (so no wear) occurring between the opposing surfaces. This type of lubrication, which occurs in the most of journal and thrust bearings, is known as full fluid lubrication. The mechanism of full film lubrication may be hydrostatic (aerostatic), squeeze film or hydrodynamic (aerodynamic). In the present chapter only hydrodynamic lubrication mechanism has been detailed.

The hydrodynamic lubrication mechanism is often referred as 'the ideal form of lubrication', as fluid is pressurized due to relative velocity and lubricant viscosity; and solid surfaces are prevented from coming into contact (as shown in Fig. 3.1.1.1) that eliminates the chances of wear. In this lubrication mechanism interacting surfaces are separated by a relatively thick fluid film (Fig. 3.1.1.2) and the influence of asperities is neglected (i.e., the surfaces can be assumed as smooth surfaces). In this mechanism, nearly converging geometry of the fluid gets pressurized and separates the tribo–surfaces. The minimum separation (h_{min}) between the surfaces, which is a critical design criterion, is a function of relative velocity (U), lubricant viscosity (η) and applied load (W). The expression $h_{min} \alpha$ $(\eta U/W)^{1/2}$ generally guides preliminary design of fluid film bearings. It is worth noting that lubricant viscosity plays an important role in the hydrodynamic lubrication. Therefore, any liquid (i.e., water, oil, alcohol, liquid refrigerant, molten metal, etc.,) having desirable viscosity can be used as lubricant.

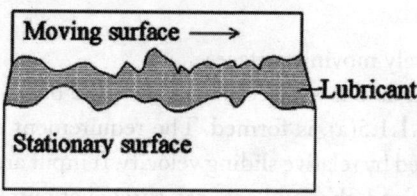

Fig. 3.1.1.1 Hydrodynamic lubrication   Fig. 3.1.1.2 Shearing of lubricant

Before starting mathematical expressions related to hydrodynamic lubrication, let us understand the experiments performed by Tower, who was first to realize the importance of hydrodynamic lubrication.

Tower's experiments:

Tower performed experiments on a partial arc bearing by imposing load on bearing cap and rotating journal assembly as shown in the Fig. 3.1.1.3a. Lower part of the journal was immersed in lubricating oil and friction resistance on bearing was obtained by measuring friction moment acting on bearing cap. He found reduction in friction in the presence of liquid lubricants. In subsequent experiments, a hole of 0.5 inches (1.27 cm) was drilled in the centre of the bearing cap (Fig. 3.1.1.3b) and the shaft was rotated, an outflow of oil took place through that hole. To prevent the outflow of oil a wooden plug was driven in the hole but on restarting (rotating shaft) the plug was slowly forced out. This proved to be a turning point in the history of lubrication. Following are the observations made by Tower:

- Very small (~0.001) coefficient of friction in the presence of lubricant
- Increase in frictional resistance on increasing sliding speed; this was different from the Coulomb's frictional resistance, which did not depend on the sliding velocity.
- Decrease in friction resistance on increase in operating temperature

The Tower's experimental study motivated a number of researchers to understand the mechanism of hydrodynamic lubrication applied to journal bearing.

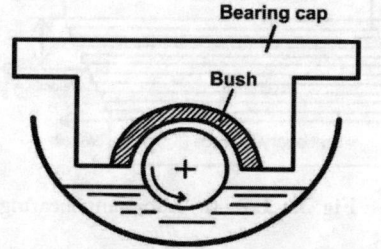 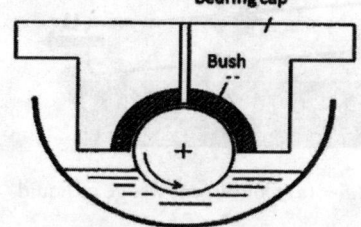

Fig. 3.1.1.3 (a) Sketch of Tower's test setup   Fig. 3.1.1.3 (b) Sketch of Tower's test setup with hole

A typical geometry of journal bearing in a polar coordinate system (r, θ, z) with $\theta = 0°$ being aligned with the line of centres, is shown in Fig. 3.1.1.4. In this figure the subscript 'b' is used for bearing and 'j' for journal. In journal bearing, a fluid wedge, due to eccentric position of journal, is generated between shaft outer diameter and bearing bore. Being incompressible, the fluid is pressurized to support the load and keep the two surfaces apart. Based on available literature, it can be said that characteristic features of hydrodynamic lubrication are:

- Very low coefficient of friction ($\mu \sim 0.001$)
- In the ideal case, there will be no wear of relatively moving surfaces
- The geometry of the surfaces must be such that as one surface moves over the other, a convergent wedge of liquid, as shown in Fig. 3.1.1.5(a), is formed. The requirement of convergent wedge comes to restrict the flow caused by relative sliding velocity. If input area is larger than output area and sliding velocity is fixed, the fluid gets pressurized at output and separates the surfaces.
- In hydrodynamic lubrication, the whole of the friction arises from shearing (Fig. 3.1.1.5(b)) of the lubricant film. In other words, reducing the viscosity of the oil reduces the friction. However, the minimum film thickness between sliding surfaces places a limit to the lowest possible viscosity. As lubricant viscosity plays an important role, it is necessary to understand the viscosity of commercially available oils and the effect of temperature, pressure and relative velocity on the lubricant viscosity.

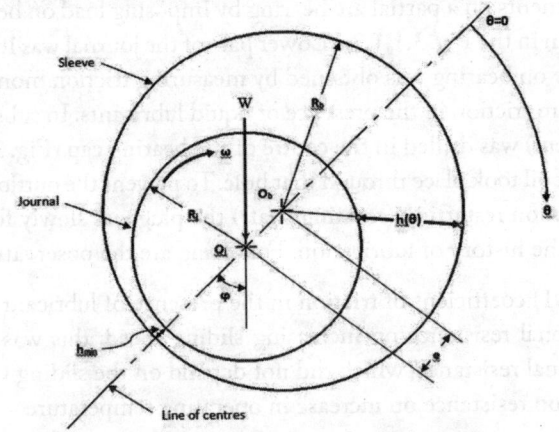

Fig. 3.1.1.4 Journal bearing

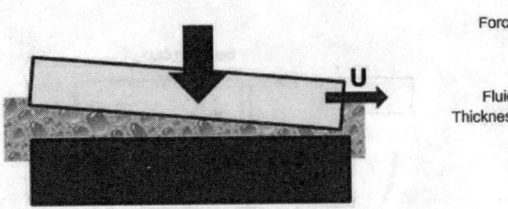

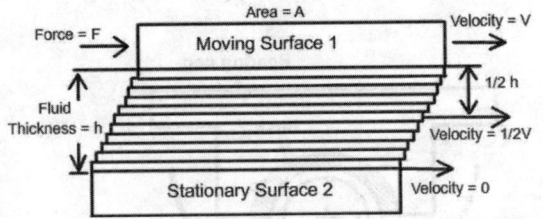

Fig. 3.1.1.5 (a) Convergent wedge of liquid   **Fig. 3.1.1.5** (b) Lubricant shearing

3.1.2 Lubricant Viscosity

Viscosity is a fluid property, which resists the flow of fluid due to internal friction between fluid molecules. Figure 3.1.2.1 illustrates the flow of three liquids under the same gravitational force. Higher flow rate of lesser viscous lubricant (left hand side lubricant) compared to lower flow rate of more viscous lubricant (right hand side lubricant) is shown in this figure.

Lubrication of Bearings 125

Fig. 3.1.2.1 Viscosity comparison

Lubricant Viscosity is classified as 'dynamic viscosity' and 'kinematic viscosity'. The dynamic viscosity, also known as absolute viscosity, is the ratio of the shear stress to the shear rate of flowing fluid. The unit of dynamic viscosity in SI units is Pascal.sec (Pa.s) and in CGS units it is centipoise. The kinematic viscosity can be expressed by the dynamic viscosity divided by fluid-density. The SI unit of kinematic viscosity is square meter per second, but its CGS unit centistokes is more widely accepted. Table 3.1.1.1 lists the viscosity range for various machine elements. Lesser viscosity, which can bear the applied load, is preferred to reduce the friction.

Table 3.1.1.1 Typical operating viscosity ranges

Lubricants	Viscosity range, cSt
Light duty instruments (i.e., clock)	5-20
Roller bearings	10-300
Plain bearings	20-500
Gears	50-1000

Table 3.1.2.1 ISO viscosity grades

ISO Viscosity Grade Numbers International Organization for Standardization (ISO)	Viscosity Grade Ranges Centistokes at 40 °C	
	Minimum	Maximum
2	1.98	2.42
3	2.88	3.52
5	4.14	5.06
7	6.12	7.48
10	9.0	11.0
15	13.5	16.5
22	19.8	24.2
32	28.8	35.2
46	41.4	50.6
68	61.2	74.8
100	90	110

Note: 1 Centistoke = 10^{-6} m²/s

To standardize lubricant oils, International Organization for Standardization (ISO) has setup grading system. In this system (Table 3.1.2.1), the kinematic viscosity of oil is expressed in centistokes. It is worth noting that there is a tolerance band for viscosity. For example, viscosity of ISO 10 varies in the range of 9–11. To analyze tribo–pair using ISO 10, one must consider the worst case (either viscosity = 9 or viscosity = 11). If tolerance band is not defined, then 10% variation may be considered.

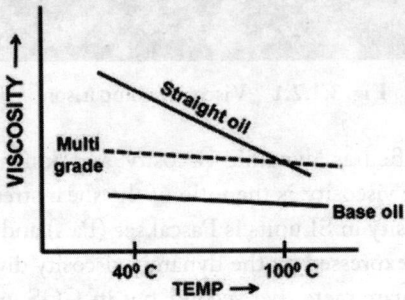

Fig. 3.1.2.4 Variation in oil viscosity with temperature

3.1.2.1 Viscosity–temperature relation

It is important to note that lubricant viscosity is a strong function of temperature as shown in Fig. 3.1.2.4. The oil viscosity must be mentioned with temperature. For all liquids, the viscosity decreases with increase $\eta = \eta_{ref}\, e^{\frac{b}{T+\theta}}$ in the temperature, but the rate of decrease varies considerably as shown in Fig. 3.1.2.4. To model this there are following three viscosity–temperature relations (Stachowiak and Batchelor, 2006):

$$\eta = \eta_{ref}\, e^{-\beta.(T-T_{ref})} \qquad (3.1.2.1)$$

Vogel's equation: $\qquad \eta = \eta_{ref}\, e^{\frac{b}{T+\theta}} \qquad (3.1.2.2)$

Walther's equation: $\log\log\left[\dfrac{\eta}{\rho} + 0.6\right] = C1 - C2\,(\log T) \qquad (3.1.2.3)$

Where ρ is density of oil, η is the dynamic viscosity, T is the absolute temperature and C1, C2 are constants.

Vogel's equation is the most accurate among all three relations and it very useful in engineering calculations. Walther's equation forms the basis of the ASTM viscosity–temperature chart (Stachowiak and Batchelor, 2006). For better evaluation of the relationship between viscosity and temperature, a comparator 'Viscosity Index'(ASTM standard, 2010) is used. It is a measure of the variation in kinematic viscosity due to change in the temperature between 40 °C and 100 °C. Higher viscosity index indicates lesser sensitivity of lubricant viscosity toward temperature.

Note: *The viscosity index is not defined and shall not be reported for oils with kinematic viscosity of lesser than 2.0 mm²/s at 100 °C (ASTM standard, 2010)*

Figure 3.1.2.5 shows lubrication regimes for three oils with low, medium and high viscosity indices. Merits of oil with high viscosity index (low sensitivity toward temperature) are clearly illustrated in the Fig. 3.1.2.5. The High VI oil is preferable in the operating range and in the left side of the operating range and Low VI oil is preferable beyond the right hand side of operating range. One way to increase viscosity index is by adding polymers in mineral oils (i.e., low VI oils). At cold temperatures the polymers are coiled up and allow the oil to flow as their low numbers indicate. As the oil warms up, the polymers begin to unwind into long chains that prevent the oil from thinning as much as it normally would. Such oils are known as multi-grade oils and are represented by two numbers as listed in last three rows of Table 3.1.2.3. It is worthy to note that the two numbers indicate two separate grades: first number indicates the base oil used in multi-grade oil and second grade indicates the viscosity achieved at higher operating temperature by additives mixed in the base oil. A multi-grade oil having lesser numerical value as the first number performs better in cold conditions. Similarly multi-grade oil having higher the second number provides better protection at higher temperatures.

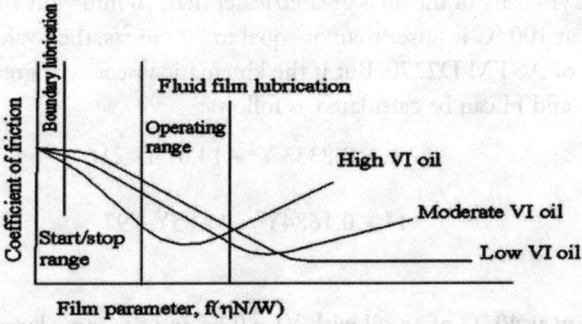

Fig. 3.1.2.5 Stribeck diagram shows viscosity index of few commonly used lubricating oils.

Table 3.1.2.3 Variation of viscosity with temperature for commonly used engine oil

SAE grade	ISO Grade	Viscosity in cSt 40 °C 100 °C 130 °C			VI
10W	32	32.6	5.57	3.20	107
20W	68	62.3	8.81	5.01	118
SAE 30	100	100	11.9	6.25	110
SAE 40	150	140	14.7	8.0	102
5W-20	46	138	6.92	4.17	140
10W-30	68	66.4	10.2	5.7	135
10W-40	100	77.1	14.4	8.4	193

To estimate the VI, viscosity of oil at temperature 37.78 °C (100 °F) is compared with viscosity of two oils, (viscosity H of Pennsylvania crude oil having VI = 100 and viscosity L of gulf coast oil having VI = 0), at 37.78 °C. Only condition is to select two oils which have same viscosity at 98.89 °C (210 °F), as shown in Fig. 3.1.2.6.

In the present time, viscosity index is calculated based on kinematic viscosities determined at 40 and 100 °C, as calculated VI obtained from these temperature are virtually the same as those obtained from temperature at 37.78 and 98.89 °C (ASTM standard, 2010).

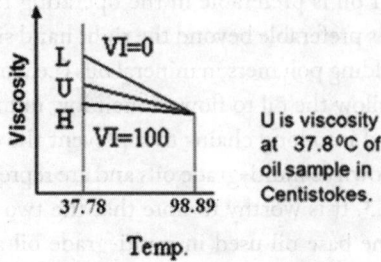

Fig. 3.1.2.6 Viscosity index

To calculate VI of a given oil, ASTM standard D2270 is used. As per this standard, first step is to find whether kinematic viscosity of the oil is greater/lesser than 70 mm²/s at 100 °C. If the kinematic viscosity of the sample at 100 °C is lesser than or equal to 70 mm²/s, then values of L and H can be extracted from Table 1 of ASTM D2270. But if the kinematic viscosity is greater than 70 mm²/s at 100 °C, the values of L and H can be calculated as follows:

$$L = 0.8353\, Y^2 + 14.67\, Y - 216 \qquad (3.1.2.4)$$

$$H = 0.1684 Y^2 + 11.85 Y - 97 \qquad (3.1.2.5)$$

where,

L = kinematic viscosity at 40 °C of an oil with VI = 0 having the same kinematic viscosity at 100 °C as the oil whose viscosity index is to be calculated, mm²/s.

Y = kinematic viscosity, at 100 °C, of the oil whose viscosity index is to be calculated, mm²/s.

H = kinematic viscosity at 40 °C of an oil with VI = 100 having the same kinematic viscosity at 100 °C as the oil whose viscosity index is to be calculated, mm²/s.

If U (kinematic viscosity in mm²/s at 40 °C of the oil whose viscosity index is to be calculated) is greater than H, the viscosity index VI of the oil can be calculated as follows:

$$VI = 100\left[\frac{L-U}{L-H}\right] \qquad (3.1.2.6)$$

In case U < H, then Eq. (3.1.2.6) cannot be used, instead VI of that oil shall be calculated using following equation:

$$VI = \left[\left(\left(anti\log N_v\right) - 1\right) / 0.00715\right] + 100 \qquad (3.1.2.7)$$

where $N_v = \left(\log H - \log U\right) / \log Y$

Problem 1: Kinematic viscosities of oil at 40 °C and 100 °C are 136.75 mm²/s and 13.05 mm²/s, respectively. Find the VI of the oil.

Ans. As the kinematic viscosity of the given sample at 100 °C is lesser than 70 mm²/s, Table 1 of ASTM D2270 shall be used. As per that Table, L = 233.45 and H = 122.2. As in the present case U_{40} > H, VI of the oil shall be estimated by substituting the viscosities values in Eq. (3.1.2.6),

$$VI = \left[(233.45 - 136.75)/(233.45 - 122.2)\right] \times 100 = 86.92$$

Rounding to the nearest whole number VI = 87.
It is worth noting that Eq. 3.1.2.7, as illustrated below, provides the same results of VI= 87

$$N_v = \left[\frac{\log(122.2) - \log(136.75)}{\log(13.05)}\right] = -0.0438$$

Substituting the value of N_v in Eq. 3.1.2.7, we have

$$VI = \left[\frac{anti\log(-0.0438) - 1}{0.00715}\right] + 100$$

$$VI = \left[\frac{0.9189 - 1}{0.00715}\right] + 100$$

$$VI = 86.58$$

$$VI = 87$$

Problem 2: Measured kinematic viscosities of the commercial (SM-120) oil at 40 °C and 100 °C are 125.07 mm²/s and 12.48 mm²/s, respectively. Estimate the VI of the oil.

Ans. As U_{100} is lesser than 70 mm²/s, the viscosity index of SM-120 shall be estimated using Table 1 of ASTM D2270. In this case H = 114 and L = 215.1. The viscosity index is to be calculated using the Eq. 3.1.2.6,

$$VI = \left[(215.1 - 125.07)/(215.1 - 114)\right] \times 100 = 89.05$$

Rounding to the nearest whole number VI = 89.
Similarly this VI = 89 can be obtained using Eq. 3.1.2.7

$$N_v = \left[\frac{\log(114) - \log(125.07)}{\log(12.48)}\right] = -0.0367$$

Substituting the value of N_v in Eq. 3.1.2.7, we have

$$VI = \left[\frac{anti\log(-0.0367) - 1}{0.00715}\right] + 100$$

$$VI = \left[\frac{0.9189 - 1}{0.00715}\right] + 100$$

$$VI = 88.66$$

$$VI = 89$$

3.1.2.2 Viscosity–shear rate relation

In hydrodynamic lubrication mechanism, the fluid film thickness is largely affected by the viscosity of lubricant. Increase in viscosity can increase the film thickness, but also increases the friction loss. In ideal case of Newtonian liquid, friction is given by

$$\text{Friction} = \text{shear stress} \times \text{area}$$

$$\text{Friction} = (\textit{Viscosity} \times V/h) \times \text{area}$$

In this case viscosity remains constant, irrespective of lubricant velocity. This behaviour is represented by straight line as shown in Fig. 3.1.2.2.1 with n = 1. For non-Newtonian fluids (n>1 or n<1), viscosity of lubricant depends on the relative velocity and this effect (viscosity–shear rate relation) must be accounted. One way to model such behaviour is to express shear stress as:

$$\tau_i = \tau_p + \eta \left(\frac{du}{dh}\right)^n$$

Newtonian behaviour is a special case where $\tau_p = 0$ and n = 1.

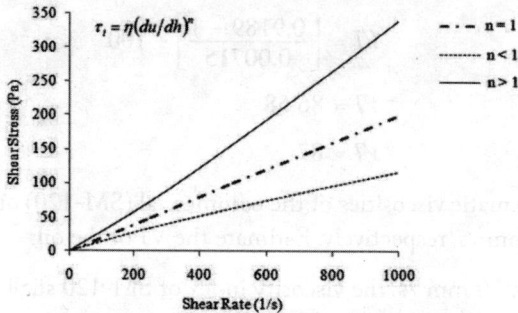

Fig. 3.1.2.2.1 Viscosity–shear rate relationship

Multi-grade oils are the preferable choice from temperature point of view, but multi-grade oils exhibit a non-Newtonian behaviour due to dependence of viscosity on the rate of shear. Higher relative velocity of tribo–pairs causes shear thinning effect, as shown in Fig. 3.1.2.2.2. Such a behaviour is represented by expression

$$\eta = \eta_1 \frac{K + \eta_2 \gamma}{K + \eta_1 \gamma} \tag{3.1.2.8}$$

where, γ is shear rate, K is shear stability parameter and η is dynamic or absolute viscosity.

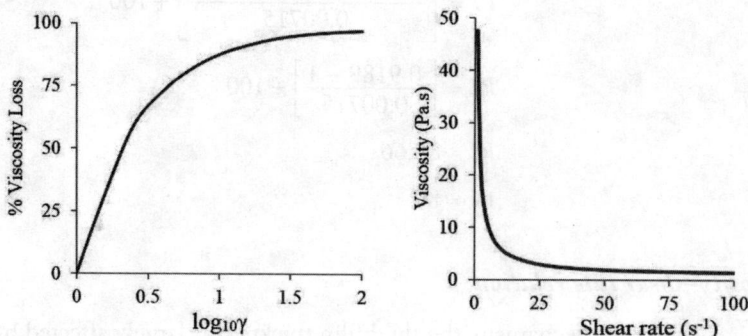

Fig. 3.1.2.2.2 Shear thinning effect of multi-grade oils

To understand the role of shear stability parameter K, let us consider two mono-grade oils (η_1 = 0.0111 Pa.s; η_2 = 0.0063 Pa.s), and two multi-grade oils, viscosities of which can be expressed using Eq. 3.1.2.8. The shear stability parameter for oil A is 1500 Pa, and K for oil B is 20000 Pa. These four oils (as listed in Table 3.1.2.4) were tested to evaluate the performance of a journal bearing operating at 5000 rpm, under radial load of 21110 N (Hirani *et al.* 2001). The results of minimum film thickness and maximum pressure are listed in Table 3.1.2.4.

Table 3.1.2.4 Shear stability of multi-grade oils

Different oils	Minimum film thickness (μm)	Maximum pressure (MPa)
Oil with μ_1	3.60	73.0
Multigrade oil A	2.82	79.9
Multigrade oil B	2.98	74.3
Oil with μ_2	2.78	81.5

Table 3.1.2.4 shows multi-grade oil with a high value of K performs better than the oil with low value of K.

3.1.2.3 Viscosity–pressure relation

Inherent assumption in hydrodynamic lubrication mechanism is that tribo–surfaces are perfectly rigid and retain their geometric shape during operation. But sometimes the generated hydrodynamic fluid pressure is high and it can deform the tribo–surfaces elastically. In the presence of such high fluid pressure, lubricant viscosity also increases. This mechanism is termed as Elasto–Hydrodynamic Lubrication (EHL) and it provides the lowest friction. As the name suggests this lubrication mechanism utilizes: (1) elastic deformation and (2) hydrodynamics. Therefore, to estimate the film thickness, EHL requires simultaneous/sequential solution of fluid and solid mechanics equations. The simplest way to analyze EHL is: assume film thickness, estimate pressure using hydrodynamic equations, evaluate elastic deformation of surfaces, modify film thickness and iterate. The iteration continues until the modified film thickness distribution matches with the new film thickness distribution. One of the notable points of EHL compared to hydrodynamic lubrication is the negligible effect of load on minimum film thickness $h_{min} \propto W^{-0.075}$ and significant effect of relative velocity on film thickness $h_{min} \propto U^{0.68}$. EHL mostly occurs in rolling element bearings, gears, and cam–follower contact.

A well-known expression, relating the viscosity with pressure given by Roelands (Wang et al., 2005), is expressed as

$$\eta = \eta_o \left((\ln \eta_o + 9.67) \left\{ \left(1 + 5.1 \times 10^{-9} p \right)^Z - 1 \right\} \right) \quad (3.1.2.9)$$

where η_0 is the viscosity at atmospheric zero pressure, p is the pressure and Z is the pressure–viscosity index which is a lubricant constant. It is commonly taken as 0.6 for conventional mineral oils and falls in the range of 0.4–0.8 for most synthetic oils, as given in Table 3.1.2.5.

Another relation, given by Barus (Hamrock, 1994) is expressed as

$$\eta = \eta_0 e^{\alpha P} \quad (3.1.2.10)$$

Here α is the pressure-viscosity coefficient. For most of the lubricating oils, α remain in the range of 10^{-8} to $2*10^{-8}$ /Pa.

Table 3.1.2.5 Representative values of viscosity–pressure index Z [Roelands, 1966]

Material	Z
Mineral oil, typical value	0.60
Synthetic Oils	
PAO synthetic hydrocarbon	0.45
Diester	0.47
Polyol ester	0.48
Polymethylsiloxane	0.49
Castor oil	0.43
Rapeseed	0.42

Table 3.1.2.6 lists the effect of pressure on viscosity using Eq. 3.1.2.10 assuming α as 10^{-8} /Pa. When pressure increases from atmospheric to higher pressure, the viscosity of the lubricant also increases very significantly. As may be observed from Table 3.1.2.5, if pressure reaches to 100 MPa, the viscosity multiplier increases to about 2.72 times. Barus' relation (Eq. 3.1.2.10) can be used for fluid pressure lesser than 200 MPa.

Table 3.1.2.6 Viscosity variation with pressure

Pressure (Pa)	Viscosity variation or multiplier
1.e5	1.0010
1.e6	1.0101
1.e7	1.1052
1.e8	2.7183

3.1.3 Mechanism of pressure development in lubricant film

The hydrodynamic action is often desirable in two types of bearing: journal bearings and thrust bearings. These bearings are known for their economic way of pressure generation and low coefficient of friction (~0.001–0.01). To understand the process leading to generation of a pressure in a converging oil film let us take an example of parallel plate as shown in Fig. 3.1.3.1 and use fluid mechanics concepts.

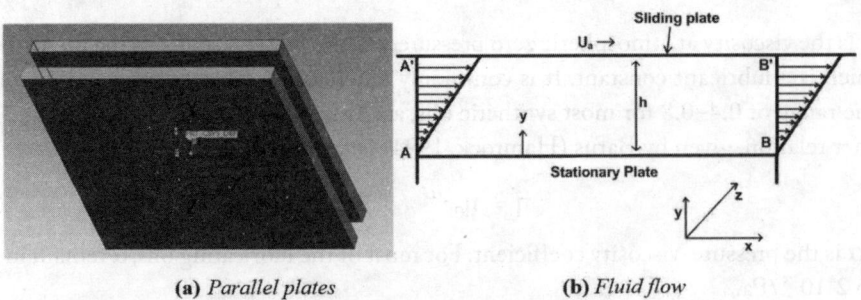

(a) *Parallel plates*   (b) *Fluid flow*

Fig. 3.1.3.1 Lubrication between parallel plates

Considering two parallel plates, AB and A'B', separated by fluid film of thickness h in Y-direction and assuming plate A'B' is sliding at velocity u_a relative to plate AB. No pressure will develop in between these parallel plates.

Now assuming plate AB is inclined at angle α and film thickness h is function of coordinate x. Due to inclination, as shown in Fig. 3.1.3.2, exit area (BB') will be smaller compared to entrance area (AA'). To conserve the mass flow rate, a positive pressure gradient will be generated at exit and negative pressure gradient will be generated at entrance as shown in Fig. 3.1.3.3.

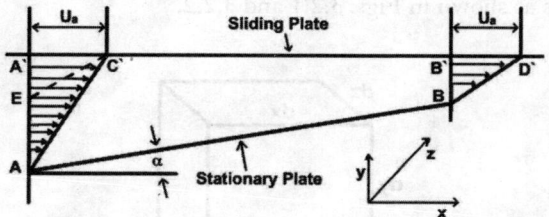

Fig. 3.1.3.2 Lubrication between inclined plates

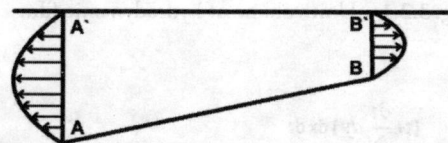

Fig. 3.1.3.3 Positive pressure gradient at exit and negative pressure gradient at entrance

The negative pressure gradient at entrance, decreases the flow rate. Similarly the positive pressure gradient at exit increases the flow rate. In other words, to support load positive pressure is developed between inclined relatively moving surfaces. To quantify the contribution of angle of inclination, the converging wedge between relatively moving plates, as shown in Fig. 3.1.3.4, may be assumed in a number of ways. Fluid film pressure within these geometries can be explained using Reynolds' equation.

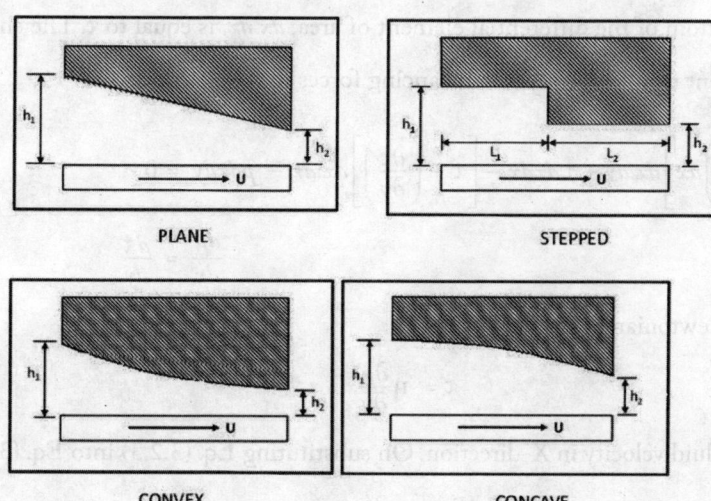

Fig. 3.1.3.4 Formation of converging wedge

3.2 Reynolds' Equation and its Limitations

In 1886, Reynolds derived an equation to estimate the pressure distribution for 'Fluid Film Lubrication'. Quantification of fluid film lubrication can be made by solving Reynolds' Equation, which provides fluid film pressure as a function of coordinates and time. Reynolds' equation helps to predict hydrodynamic, squeeze, and hydrostatic film mechanisms.

To model the pressure as a function of angle of inclination and film thickness, let us consider a differential fluid element dx, dz and of depth dy, which has certain forces acting on it, neglecting gravity and inertia forces as shown in Figs. 3.2.1 and 3.2.2.

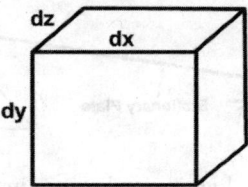

Fig. 3.2.1 Unit volume of hydrodynamic film

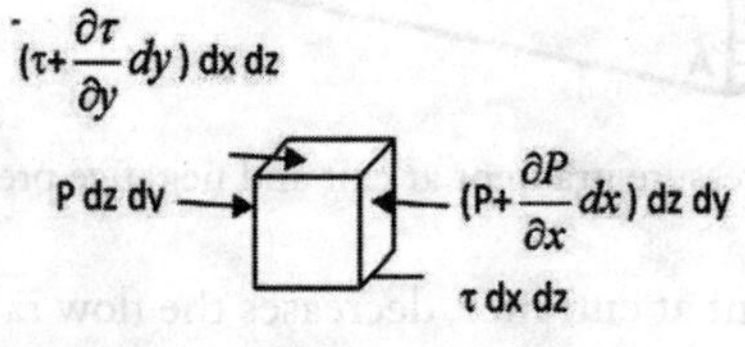

$$\left(\tau + \frac{\partial \tau}{\partial y} dy\right) dx\, dz$$
$$P\, dz\, dy \longrightarrow \longleftarrow \left(P + \frac{\partial P}{\partial x} dx\right) dz\, dy$$
$$\tau\, dx\, dz$$

Fig. 3.2.2 Fluid element subjected to pressure and viscous forces

Pressure on the left face of area $dz.dy$ is p. Pressure on the right face is $p + \left(\dfrac{\partial p}{\partial x}\right) dx$. The shear stress on the bottom of the differential element of area, $dx.dz$, is equal to τ. The shear stress on the top of the element is $\tau + \left(\dfrac{\partial \tau}{\partial y}\right) dy$. On balancing forces in X-direction:

$$\left[p + \left(\frac{\partial p}{\partial x}\right) dx\right] dz.dy + \tau\, dzdx - \left[\tau + \left(\frac{\partial \tau}{\partial y}\right)\right] dz.dx - p\, dzdy = 0 \quad (3.2.1)$$

Or

$$\frac{dp}{dx} = \frac{d\tau}{dy} \quad (3.2.2)$$

Assuming, Newtonian behaviour of fluid

$$\tau = \eta \frac{\partial u}{\partial y} \quad (3.2.3)$$

where, u is the fluid velocity in X-direction. On substituting Eq. (3.2.3) into Eq. (3.2.2)

$$\Rightarrow \quad \frac{\partial P}{\partial x} = \eta \frac{\partial^2 u}{\partial y^2} \quad (3.2.4)$$

Similarly, on force balance in Z-direction,

$$\frac{\partial P}{\partial z} = \eta \frac{\partial^2 w}{\partial y^2} \qquad (3.2.5)$$

Equation (3.2.5) is based on the following assumptions:

- Negligible inertia terms
- Negligible pressure gradient in the direction of film thickness (Y- direction)
- Newtonian behaviour of fluid

To find flow velocity (u) in the X-direction, integrate $\frac{\partial P}{\partial x} = \eta \frac{\partial^2 u}{\partial y^2}$ two times. On integrating first time

$$\eta \frac{\partial u}{\partial y} = \frac{\partial P}{\partial x} y + C_1 \qquad (3.2.6)$$

On integrating second time

$$\eta u = \frac{\partial P}{\partial x} \frac{y^2}{2} + C_1 y + C_2 \qquad (3.2.7)$$

Assuming no slip at liquid solid boundary, this means
y = 0, u = U; and y = h, u = 0 as shown in Fig. 3.1.3.4.
Substituting these boundary conditions in Eq. (3.2.7), constants C_1 and C_2 can be obtained.

$$\eta U = \frac{\partial P}{\partial x} \frac{(0)^2}{2} + C_1 (0) + C_2 \qquad (3.2.8)$$

$$\Rightarrow \qquad \eta U = C_2 \qquad (3.2.9)$$

$$0 = \frac{\partial P}{\partial x} \frac{h^2}{2} + C_1 h + C_2$$

$$\Rightarrow \qquad -\frac{\eta U}{h} - \frac{\partial P}{\partial x} \frac{h}{2} = C_2 \qquad (3.2.10)$$

On substituting C_1 and C_2 in Eq. (3.2.7), we get

$$u = \left(\frac{y^2 - yh}{2\eta}\right) \frac{\partial P}{\partial x} + \left(1 - \frac{y}{h}\right) U \qquad (3.2.11)$$

In Eq. (3.2.11) there are two velocity terms, 'shear flow' and 'flow due to pressure gradient (Poiseuille flow)'. The Poiseuille flow term retards the fluid flow at entrance as shown in Fig. 3.2.3. This pressure term boasts flow at exit as indicated in Fig. 3.2.3.

Similarly flow velocity in Z-dir,

$$w = \left(\frac{y^2 - yh}{2\eta}\right) \frac{\partial P}{\partial z} + \left(1 - \frac{y}{h}\right) W_z \qquad (3.2.12)$$

where, W_z is the relative velocity of sliding surfaces in Z-direction. This is a relation of velocity in terms of pressure. To find out the pressure in terms of velocity, one additional equation i.e., the conservation of mass or continuity equation (Eq. 3.2.13), is required.

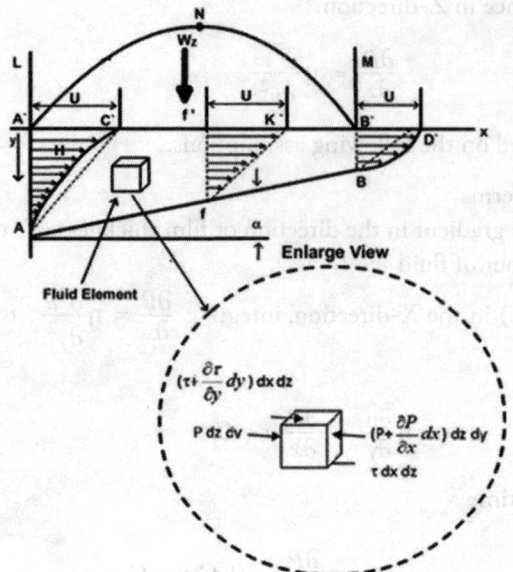

Fig. 3.2.3 Pressure profile between inclined plates

Continuity equation for incompressible fluid can be expressed as:

$$\frac{\partial u}{\partial x} + \frac{\partial v}{\partial y} + \frac{\partial w}{\partial z} = 0 \qquad (3.2.13)$$

Assumptions:

1. Negligible inertia terms
2. Negligible pressure gradient in the direction of film thickness
3. Newtonian fluid
4. Constant value of viscosity
5. No slip at liquid solid boundary
6. Neglecting angle of inclination for coordinate system
7. Incompressible flow

Integrating Eq. 3.2.13 in Y-direction from y = 0 to y = h,

$$\int_0^h \frac{\partial u}{\partial x} dy + \int_0^h dv + \int_0^h \frac{\partial w}{\partial z} dy = 0 \qquad (3.2.14)$$

$$u = \left(\frac{y^2 - yh}{2\eta}\right)\frac{\partial P}{\partial x} + (U_1 - U_2)\frac{y}{h} + U_2 \qquad (3.2.15)$$

$$w = \left(\frac{y^2 - yh}{2\eta}\right)\frac{\partial P}{\partial z} + (W_{1z} - W_{2z})\frac{y}{h} + W_2 \qquad (3.2.16)$$

Using Leibnitz rule,

$$\int_a^b \frac{\partial u(y,x)}{\partial x} dy = \frac{d}{dx}\int_a^b u\, dy - u(b,x)\frac{db}{dx} + u(a,x)\frac{da}{dx} \qquad (3.2.17)$$

Using
$$\int_0^h \frac{\partial u(x,y)}{\partial x}dy = \frac{\partial}{\partial x}\int_0^h u(x,y)dy - u(x,h)\frac{\partial h}{\partial x} \quad (3.2.18)$$

$$\int_0^h \frac{\partial u}{\partial x}dy = \frac{\partial}{\partial x}\int_0^h u\,dy - (U_1)\frac{\partial h}{\partial x} \quad (3.2.19)$$

$$\int_0^h u\,dy = \frac{1}{2\eta}\left[\frac{y^3}{3} - \frac{y^2 h}{2}\right]_0^h \frac{\partial P}{\partial x} + \frac{(U_1 - U_2)}{h}\left[\frac{y^2}{2}\right]_0^h + U_2 h \quad (3.2.20)$$

$$= \frac{1}{2\eta}\left[-\frac{h^3}{6}\right]\frac{\partial P}{\partial x} + \frac{(U_1 - U_2)}{h}\frac{h^2}{2} + U_2 h \quad (3.2.21)$$

$$= -\frac{h^3}{12\eta}\frac{\partial P}{\partial x} + \frac{h}{2}(U_1 - U_2) + U_2 h \quad (3.2.22)$$

$$= -\frac{h^3}{12\eta}\frac{\partial P}{\partial x} + \frac{h}{2}(U_1 + U_2) \quad (3.2.23)$$

Substituting Eq. 3.2.23 in Eq. 3.2.19, we get

$$\int_0^h \frac{\partial u}{\partial x}dy = \frac{\partial}{\partial x}\left[-\frac{h^3}{12\eta}\frac{\partial P}{\partial x} + \frac{h}{2}(U_1 + U_2)\right] - (U_1)\frac{\partial h}{\partial x} \quad (3.2.24)$$

$$\int_0^h \frac{\partial u}{\partial x}dy = \frac{\partial}{\partial x}\left[-\frac{h^3}{12\eta}\frac{\partial P}{\partial x} + \frac{h}{2}(U_1 + U_2 - 2U_1)\right] \quad (3.2.25)$$

$$\int_0^h \frac{\partial u}{\partial x}dy = \frac{\partial}{\partial x}\left[-\frac{h^3}{12\eta}\frac{\partial P}{\partial x} + \frac{h}{2}(U_2 - U_1)\right] \quad (3.2.26)$$

$$\int_0^h \frac{\partial w}{\partial z}dy = -\frac{\partial}{\partial z}\left(\frac{h^3}{12\eta}\right)\frac{\partial P}{\partial z} - \frac{1}{2}\frac{\partial(W_{1z} - W_{2z})h}{\partial z} \quad (3.2.27)$$

Substituting Eq. 3.2.26 and Eq. 3.2.27 in equation 3.2.14 and rearranging, we get

$$-\frac{\partial}{\partial x}\left\{\frac{h^3}{12\eta}\frac{\partial P}{\partial x}\right\} - \frac{1}{2}\frac{\partial}{\partial x}\left\{(U_1 - U_2)h\right\} + (V_h - V_0) - \frac{\partial}{\partial z}\left\{-\frac{h^3}{12\eta}\frac{\partial P}{\partial z}\right\} - \frac{1}{2}\frac{\partial}{\partial z}\left\{(W_{1z} - W_{2z})h\right\} = 0 \quad (3.2.28)$$

$$\frac{\partial}{\partial x}\left\{\frac{h^3}{12\eta}\frac{\partial P}{\partial x}\right\} + \frac{\partial}{\partial z}\left\{\frac{h^3}{12\eta}\frac{\partial P}{\partial z}\right\} = \frac{1}{2}\frac{\partial}{\partial x}\left\{(U_2 - U_1)h\right\} + (V_h - V_0) + \frac{1}{2}\frac{\partial}{\partial z}\left\{(W_{2z} - W_{1z})h\right\} \quad (3.2.29)$$

Generalized Reynolds' equation for liquid lubricant under isothermal conditions for geometry, shown in Fig. 3.2.3, is given by

$$\frac{\partial}{\partial x}\left(\frac{h^3}{\eta}\frac{\partial p}{\partial x}\right) + \frac{\partial}{\partial z}\left(\frac{h^3}{\eta}\frac{\partial p}{\partial z}\right) = 6\left[\frac{\partial}{\partial x}(U_1 + U_2)h + \frac{\partial}{\partial z}(W_{1z} + W_{2z})h + 2\left\{\left(V_h - U_1\frac{\partial h}{\partial x} - W_{1z}\frac{\partial h}{\partial z}\right) - V_0\right\}\right] \quad (3.2.30)$$

For hydrostatic (explained in chapter 5) case the equation would reduce to:

$$\frac{\partial}{\partial x}\left(\frac{h^3}{\eta}\frac{\partial p}{\partial x}\right) + \frac{\partial}{\partial z}\left(\frac{h^3}{\eta}\frac{\partial p}{\partial z}\right) = 0 \quad (3.2.31)$$

For hydrodynamic case this equation will be reduced to:

$$\frac{\partial}{\partial x}\left(\frac{h^3}{\eta}\frac{\partial p}{\partial x}\right) + \frac{\partial}{\partial z}\left(\frac{h^3}{\eta}\frac{\partial p}{\partial z}\right) = 6\left[\frac{\partial}{\partial x}(U_2 - U_1)h + \frac{\partial}{\partial z}(W_{2z} - W_{1z})h\right] \quad (3.2.32)$$

For squeeze film case (explained in Chapter 5) this equation will be reduce to:

$$\frac{\partial}{\partial x}\left(\frac{h^3}{\eta}\frac{\partial p}{\partial x}\right) + \frac{\partial}{\partial z}\left(\frac{h^3}{\eta}\frac{\partial p}{\partial z}\right) = 12(V_h - V_0) \quad (3.2.33)$$

Some of the limitations of Reynolds' equation are that:

- inertia forces are neglected;
- the constant value of lubricant viscosity is accounted;
- compressibility of the lubricant is ignored;
- the lubricants are assumed to be Newtonian; and
- the variation of pressure across the film is assumed to be negligible, that is, $\frac{\partial P}{\partial y} = 0$

3.3 Idealized Bearings

The Reynolds' equation has been expressed in Eq. 3.2.32. To design a hydrodynamic bearing, one requires solution of this partial differential equation. Since it is very difficult to get a closed form solution of Reynolds' equation, assumptions are made to simplify (idealize) the bearing by neglecting lubricant flow in one (X-direction or Z-direction) of the directions. The closed form solution obtained for idealized bearing provides an understanding of geometric parameters and their effect on bearing performance. To start with 'plane' geometry has been detailed in sections 3.3.1 and 3.3.2. Sections 3.3.3 and 3.3.4 cover cylindrical geometry of the bearing.

3.3.1 Infinitely long plane fixed sliders

To support load by hydrodynamic bearing, there is need of converging wedge, as shown in Fig. 3.1.3.4. To form fluid wedge between two infinitely long planes, there are two possibilities as shown in Fig. 3.3.1. The fixed taper shown in Fig. 3.3.1(a) can be machined into either the stationary (top surface of Fig. 3.3.1a) or moving face of the bearing.

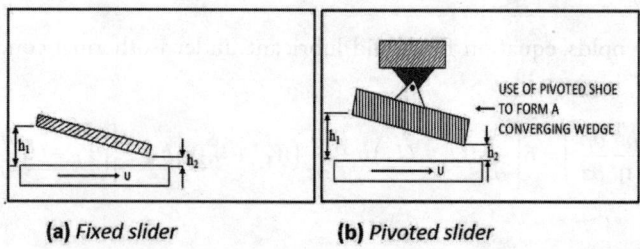

(a) *Fixed slider* (b) *Pivoted slider*

Fig. 3.3.1 Slider bearing

Now let us consider a long fixed slider bearing as shown in Fig. 3.3.2 where the length B of the slider is much larger than the dimension L (generally B>5L). In such cases, it is considered that there is no flow of lubricant in the Z-direction; no pressure drop in the Z-direction $\left(\dfrac{dp}{dz} = 0\right)$ and no variation in film thickness along Z-direction $\dfrac{dh}{dz} = 0$. Implementing these assumptions in Eq. 3.2.32,

$$\frac{\partial}{\partial x}\left(\frac{h^3}{\eta}\frac{\partial p}{\partial x}\right) = 6\left[\frac{\partial}{\partial x}(U)h\right] \tag{3.3.1}$$

Due to this simplification, Eq. 3.3.1 can be represented in definite differential form:

$$\frac{dp}{dx} = 6U\eta\left[\frac{h + const}{h^3}\right] \tag{3.3.2}$$

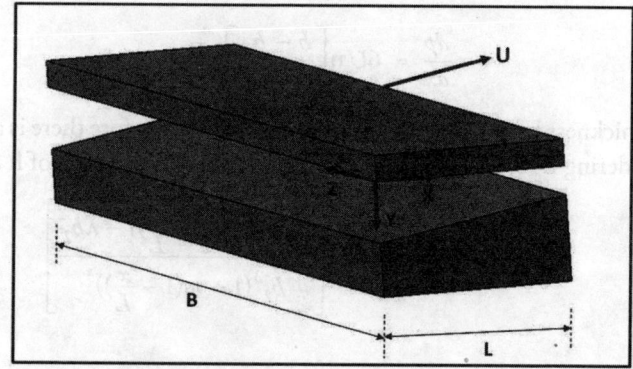

Fig. 3.3.2 A long fixed slider

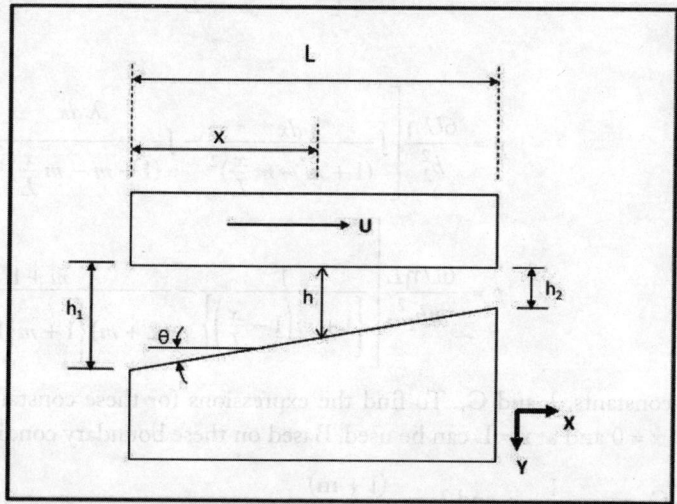

Fig. 3.3.3 Oil wedge in a long hydrodynamic slider bearing

Figure 3.3.3 shows the formation of oil wedge in a hydrodynamic slider bearing. Film thickness, h, at any value of x can be represented by

$$h = h_2 + (L - x)\tan\theta$$

$$\Rightarrow h = h_2 + (L - x)(h_1 - h_2)/L$$

$$\Rightarrow h = h_2[1 + m(1 - x/L)] \quad (3.3.3)$$

Where m is aspect ratio and expressed as $m=(h_1-h_2)/h_2$. Generally the value of the aspect ratio m ranges from 0.5 to 2.0. The value of θ is extremely small of the order of $1/2°$.

The maximum value of fluid pressure can be obtained by equating pressure gradient equal to zero,

$$\left.\frac{dp}{dx}\right|_m = 6U\eta\left[\frac{h_m + const}{h_m^3}\right] = 0$$

Here h_m is film thickness at location of maximum pressure. On substitution of value of constant in Eq. 3.3.2

$$\frac{dp}{dx} = 6U\eta\left[\frac{h - h_m}{h^3}\right] \quad (3.3.4)$$

In Eq. 3.3.3 film thickness has been expressed in terms of h_2; therefore there is a need to express h_m in terms of h_2. Considering $h_m = \lambda h_2$ where $\lambda > 1$. On substituting expression of h and h_m in Eq. 3.3.4

$$\frac{dp}{dx} = 6U\eta\left[\frac{h_2(1 + m(1 - \frac{x}{L})) - \lambda h_2}{h_2^3(1 + m(1 - \frac{x}{L}))^3}\right]$$

On rearranging,

$$\frac{dp}{dx} = \frac{6U\eta}{h_2^2}\left[\frac{1}{(1 + m - m\frac{x}{L})^2} - \frac{\lambda}{(1 + m - m\frac{x}{L})^3}\right] \quad (3.3.5)$$

On integrating

$$p = \frac{6U\eta}{h_2^2}\left[\int\frac{dx}{(1 + m - m\frac{x}{L})^2} - \int\frac{\lambda\, dx}{(1 + m - m\frac{x}{L})^3} + C_1\right]$$

$$p = \frac{6U\eta L}{mh_2^2}\left[\frac{1}{\{1 + m(1 - \frac{x}{L})\}} - \frac{m+1}{(2+m)\{1 + m(1 - \frac{x}{L})\}^2} + C_1\right]$$

There are two constants, λ and C_1. To find the expressions for these constants, two boundary conditions $p = 0$ at $x = 0$ and at $x = L$ can be used. Based on these boundary conditions:

$$C_1 = -\frac{1}{2 + m}; \text{ and } \lambda = \frac{(1 + m)}{2(2 + m)}$$

On substituting

$$p = \frac{6U\eta L}{mb_2^2}\left[\frac{1}{\left\{1+m\left(1-\frac{x}{L}\right)\right\}} - \frac{m+1}{(2+m)\left\{1+m\left(1-\frac{x}{L}\right)\right\}^2} - \frac{1}{2+m}\right] \quad (3.3.6)$$

Load capacity of slider fixed pad can be obtained by integrating Eq. 3.3.6 with respect to dx and dz. As Eq. 3.3.6 does not contain any Z-term, therefore integration can be simplified as:

$$W = \frac{6U\eta LB}{mb_2^2}\int_0^L \left[\frac{1}{\left\{1+m\left(1-\frac{x}{L}\right)\right\}} - \frac{m+1}{(2+m)\left\{1+m\left(1-\frac{x}{L}\right)\right\}^2} - \frac{1}{2+m}\right]dx$$

$$W = \frac{6U\eta LB}{mb_2^2}\left[\frac{L}{m}\log(1+m) - \frac{L}{m+2} - \frac{L}{2+m}\right]$$

Or,
$$W = \frac{6U\eta L^2 B}{m^2 b_2^2}\left[\log(1+m) + \frac{2m}{m+2}\right] \quad (3.3.7)$$

Equation 3.3.7 provides an expression of load capacity in terms of a number of geometric and operational parameters. The load capacity increases with increasing relative velocity (U), viscosity (η) and bearing dimensions (B and L). Decreasing minimum film thickness (h_2) increases load capacity drastically. Hypothetically load capacity will be infinity if film thickness h_2 approaches to zero. However, there is a limit on reducing the minimum film thickness which comes due to roughness of plane surfaces. The value of h_2 must be greater than five times of composite roughness ($\sqrt{(\text{rms of surface 1})^2 + (\text{rms of surface 2})^2}$) of bearing surfaces. Variation in load capacity with increase in value of 'm' is shown in Fig. 3.3.4. It is worth noting that with increase in lubricant viscosity (η) load capacity of the bearing (Eq. 3.3.7) increases, but limitation on viscosity comes from the friction loss. The friction force (Eq. 3.3.9) can be estimated by integrating the shear stress on the moving surface, as expressed in Eq. 3.3.8.

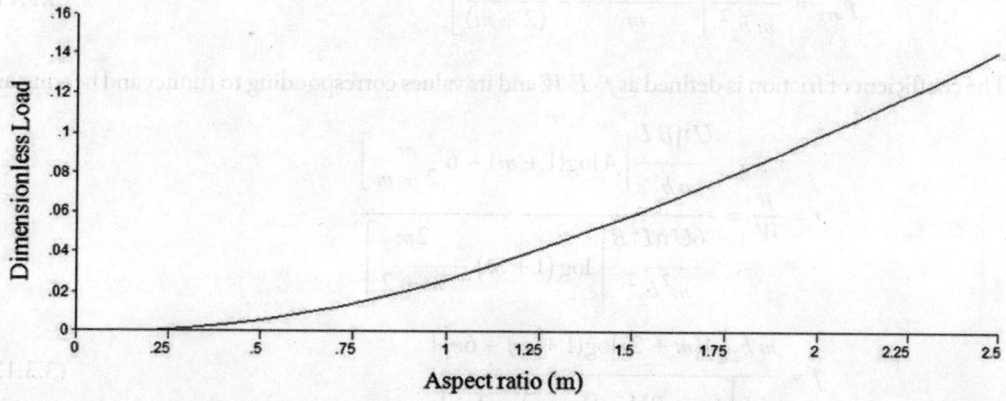

Fig. 3.3.4 Values of dimensionless load plotted against aspect ratio (m)

$$\tau = \frac{h}{2}\frac{dp}{dx} + \frac{\eta U}{h} \qquad (3.3.8)$$

$$F = \int_0^L \int_0^B \tau \, dz \, dx \qquad (3.3.9)$$

On substituting expressions from Eqs. 3.3.8, 3.3.5 and 3.3.3;

$$F = \int_0^L \int_0^B \left(\frac{h_2[1 + m(1 - x/L)]}{2} \frac{6U\eta}{h_2^2} \left[\frac{1}{(1 + m - m\frac{x}{L})^2} - \frac{2(2 + m)}{(1 + m - m\frac{x}{L})^3} \right] \right.$$

$$\left. + \frac{\eta U}{h_2[1 + m(1 - x/L)]} \right) dz \, dx$$

Or $\qquad F = \dfrac{U\eta B}{h_2} \displaystyle\int_0^L \left[\dfrac{4}{(1 + m - m\frac{x}{L})} - \dfrac{3\frac{(1+m)}{(2+m)}}{(1 + m - m\frac{x}{L})^2} \right] dx$

Or $\qquad F = \dfrac{U\eta B L}{m\, h_2} \left[4\log(1 + m) - 6\dfrac{m}{2 + m} \right] \qquad (3.3.10)$

Apart from friction force and load capacity, often 'average fluid pressure' and coefficient of friction are also needed. Average pressure can be estimated using

$$p_{avg} = \frac{1}{L}\int_0^L p \, dx$$

Or $\qquad p_{avg} = \dfrac{1}{L}\displaystyle\int_0^L \dfrac{6U\eta L}{mh_2^2}\left[\dfrac{1}{\left\{1 + m\left(1 - \frac{x}{L}\right)\right\}} - \dfrac{m+1}{(2+m)\left\{1 + m\left(1 - \frac{x}{L}\right)\right\}^2} - \dfrac{1}{2+m} \right] dx$

$$p_{avg} = \frac{6U\eta L}{m\, h_2^2}\left[\frac{\log(1+m)}{m} - \frac{2}{(2+m)} \right] \qquad (3.3.11)$$

The coefficient of friction is defined as $f = F/W$ and its values corresponding to runner and bearing are

$$f = \frac{F}{W} = \frac{\dfrac{U\eta B L}{m\, h_2}\left[4\log(1+m) - 6\dfrac{m}{2+m} \right]}{\dfrac{6U\eta L^2 B}{m^2 h_2^2}\left[\log(1+m) - \dfrac{2m}{m+2} \right]}$$

Or $\qquad f = \dfrac{m\, h_2 \left[4(m+2)\log(1+m) - 6m \right]}{6L\left[(m+2)\log(1+m) - 2m \right]} \qquad (3.3.12)$

To observe the variation of 'f' against 'm', let us consider $h_2 = 0.001*L$ and plot the variation as shown in Fig. 3.3.5. This figure indicates that the minimum value of coefficient of friction occurs at m = 1.5.

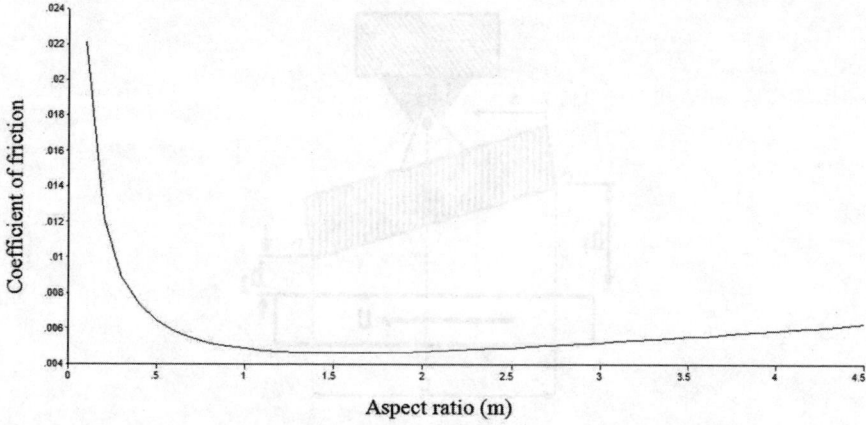

Fig. 3.3.5 Values of coefficient of friction plotted against aspect ratio (m)

3.3.2 Infinitely long plane pivoted sliders

In the previous section, it was observed that the minimum value of film thickness affects load and coefficient of friction. For certain value of 'm', the minimum value of friction coefficient occurs, which means fixed inclination of slider can provide desirable results only in the narrow range of operating conditions. It is desirable to provide flexibility related to the minimum film thickness and 'm', which can be achieved by pivoting the slider instead of fixing it. With pivoted slider, the value of 'm' and the minimum film thickness h_2 can freely be adjusted by itself and pivoted slider can be used for wide range of design parameters.

The pivot should be located at a point through which the resultant load passes, such as

$$W \bar{x} = B \int_0^L px\,dx \qquad (3.3.13)$$

where, $\bar{x}$ = distance of edge from centre of pressure.

Substituting expressions of W and p from Eq. 3.3.7 and Eq. 3.3.6, respectively

$$\frac{6U\eta L^2 B}{m^2 h_2^2}\left[\log(1+m) - \frac{2m}{m+2}\right]\bar{x}$$

$$= B\int_0^L \frac{6U\eta L}{mh_2^2}\left[\frac{1}{\left\{1+m\left(1-\frac{x}{L}\right)\right\}} - \frac{m+1}{(2+m)\left\{1+m\left(1-\frac{x}{L}\right)\right\}^2} - \frac{1}{2+m}\right]x\,dx$$

Or, $\dfrac{L}{m}\left[(m+2)\log(1+m) - 2m\right]\bar{x} = \displaystyle\int_0^L\left[\dfrac{(m+2)}{\left\{1+m\left(1-\frac{x}{L}\right)\right\}} - \dfrac{m+1}{\left\{1+m\left(1-\frac{x}{L}\right)\right\}^2} - 1\right]x\,dx$

Or, $\quad\bar{x} = \dfrac{2(3+m)(1+m)\ln(1+m) - m(6+5m)}{2m\big[(2+m)\ln(1+m) - 2m\big]} L \quad$ (3.3.14)

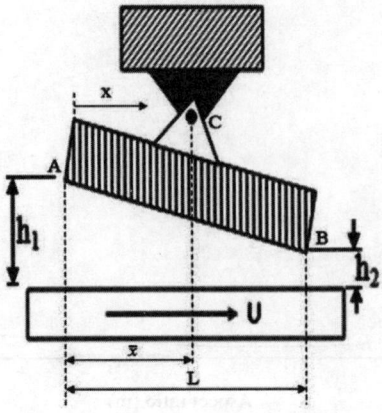

Fig. 3.3.6 Analysis of pivoted slider bearing

It can be said that location of centre of pressure depends on m and L. For example, for m=2, $\bar{x}$ = 0.6047 L. With increase in the value of m, the location of centre of pressure drifts towards trailing edge (i.e., distance from the left corner of the pad increases). It is worth noting that pivot must be of finite dimension and manufacturing accuracy is not always practical; therefore $\bar{x}$ must be defined in finite range (i.e., 0.55–0.62).

3.3.3 Infinitely long journal bearings

In an infinitely long journal bearing, length of the bearing is too large compare to its diameter, as shown in Fig. 3.3.7. In such a situation pressure gradient in Z-direction will be negligible and Eq. 3.2.32 will be modified as:

$$\dfrac{d}{dx}\left(\dfrac{h^3}{\eta}\dfrac{\partial p}{\partial x}\right) = 6\dfrac{d}{dx}(U_2 - U_1)h \quad (3.3.16)$$

Fig. 3.3.7 Infinitely long journal bearing

Since the cross-section of the journal bearing is circular, it would be beneficial to use polar coordinates, which means, $x = R\theta$. On substituting this polar coordinate in Eq. 3.3.16, and assuming relative speed as constant equal to U,

$$\frac{d}{d\theta}\left(\frac{h^3}{\eta}\frac{\partial p}{\partial \theta}\right) = 6UR\frac{dh}{d\theta} \qquad (3.3.17)$$

Figure 3.3.8 shows a circular geometry of a full journal bearing with an exaggerated clearance between the journal and the bearing. The oil film thickness at any angular location θ can be expressed as

$$h = c + e\cdot\cos\theta \qquad (3.3.18)$$

where, $c = R_B - R_J$ (indicate R_B and R_J in figure) and e = eccentricity (distance between O_B and O_J). The film thickness h (as shown in Fig. 3.3.8 as distance between point Q and A) expressed in Eq. 3.3.18, can be explained by using the law of sines

$$\frac{\sin\alpha}{e} = \frac{\sin\theta}{R_B}$$

$$\Rightarrow \cos\alpha = \sqrt{\left(1 - \frac{e^2}{R_B^2}\sin^2\theta\right)}$$

and expressing length $O_J Q$ as $h + R_J = e\cos\theta + R_B\sqrt{\left(1 - \frac{e^2}{R_B^2}\sin^2\theta\right)}$ and assuming $R_B \gg e$.

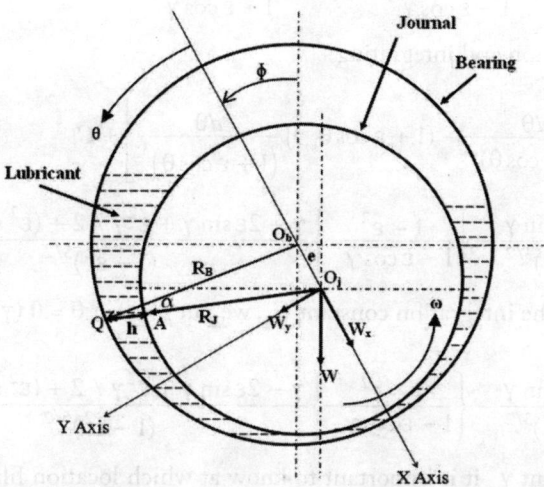

Fig. 3.3.8 Coordinate system and force components in a journal bearing

One of the very important parameters in journal bearing design is the eccentricity ratio, 'ε', which is expressed as ε = e/c. Extreme values of eccentricity (ε = 0 and ε = 1) shall be avoided. If journal and bearing are concentric (ε = 0) then load capacity of hydrodynamic journal bearing will be zero, which means there is no use of the bearing. Similarly ε = 1 forces shaft to contact bearing inner surface and rubbing occurs which should be completely avoided. Using eccentricity ratio, film thickness is expressed as

$$h = c(1 + \varepsilon\cos\theta) \qquad (3.3.19)$$

with constraints 0 < ε < 1

Integrating Eq. 3.3.17 with the condition that at $h = h_m$ at $\frac{dp}{d\theta} = 0$

$$\frac{dp}{d\theta} = 6\eta UR\left(\frac{h - h_m}{h^3}\right) \qquad (3.3.20)$$

On integrating Eq. 3.3.20 with respect to θ

$$p = \frac{6\eta UR}{c^2}\left[\int\frac{d\theta}{(1 + \varepsilon\cos\theta)^2} - \frac{h_m}{c}\int\frac{d\theta}{(1 + \varepsilon\cos\theta)^3}\right] + C_1 \qquad (3.3.21)$$

where, C_1 is the constant of integration. There are two constants in Eq.3.3.21 and to determine those constants, two boundary conditions related to fluid pressure are required. Those conditions are known as 'pressure boundary condition' and shall be discussed in the present section, but before that it is important to integrate and evaluate the integrals of Eq. 3.3.21. Let us consider

$$\cos\gamma = \frac{\varepsilon + \cos\theta}{1 + \varepsilon\cos\theta} \qquad (3.3.22)$$

This Sommerfeld substitution is important as it provides same important limits as $\theta = 0 \Rightarrow \gamma = 0$; $\theta = \pi \Rightarrow \gamma = \pi$; and $\theta = 2\pi \Rightarrow \gamma = 2\pi$.

By rearranging $\cos\theta = \dfrac{\cos\gamma - \varepsilon}{1 - \varepsilon\cos\gamma}$; $\sin\theta = \dfrac{\sqrt{1 - \varepsilon^2}\sin\gamma}{1 + \varepsilon\cos\gamma}$ and $d\theta = \dfrac{\sqrt{1 - \varepsilon^2}.d\gamma}{1 - \varepsilon\cos\gamma}$

Using above substitution and integrating

$$p = \frac{6\eta UR}{c^2}\left[\int\frac{d\theta}{(1 + \varepsilon\cos\theta)^2} - (1 + \varepsilon\cos\theta_m)\int\frac{d\theta}{(1 + \varepsilon\cos\theta)^3}\right] + C_1$$

$$p = \frac{6\eta UR}{c^2}\left[\frac{\gamma - \varepsilon\sin\gamma}{(1 - \varepsilon^2)^{3/2}} - \left\{\frac{1 - \varepsilon^2}{1 - \varepsilon\cos\gamma_m}\right\}\frac{\gamma - 2\varepsilon\sin\gamma + \varepsilon^2\gamma/2 + (\varepsilon^2\sin 2\gamma)/4}{(1 - \varepsilon^2)^{5/2}}\right] + C_1$$

In order to evaluate the integration constant C_1, we put $p = 0$ at $\theta = 0$ ($\gamma = 0$). We get $C_1 = 0$ and the equation becomes

$$p = \frac{6\eta UR}{c^2}\left[\frac{\gamma - \varepsilon\sin\gamma}{(1 - \varepsilon^2)^{3/2}} - \left\{\frac{1 - \varepsilon^2}{1 - \varepsilon\cos\gamma_m}\right\}\frac{\gamma - 2\varepsilon\sin\gamma + \varepsilon^2\gamma/2 + (\varepsilon^2\sin 2\gamma)/4}{(1 - \varepsilon^2)^{5/2}}\right] \qquad (3.3.23)$$

To get second constant γ_m it is important to know at which location film will rupture and fluid pressure will be zero. There are three pressure boundary conditions: Full Sommerfeld, Half Sommerfeld and Reynolds' conditions that can provide solutions.

Full Sommerfeld boundary condition

In the full-Sommerfeld boundary condition, in addition to $p = 0$ at $\theta = 0$ ($\gamma = 0$), $p(0) = p(2\pi)$ is also used. This pressure boundary conditions provides

$$p(2\pi) = \frac{6\eta UR}{c^2}\left[\frac{2\pi}{(1 - \varepsilon^2)^{3/2}} - \left\{\frac{1 - \varepsilon^2}{1 - \varepsilon\cos\gamma_m}\right\}\frac{2\pi + \pi\varepsilon^2}{(1 - \varepsilon^2)^{5/2}}\right]$$

$$\Rightarrow \qquad 0 = \frac{6\eta\pi UR}{(1 - \varepsilon^2)^{3/2}c^2}\left[2 - \frac{2 + \varepsilon^2}{(1 - \varepsilon\cos\gamma_m)}\right]$$

$$\Rightarrow \qquad 1 + \frac{\varepsilon^2}{2} = 1\,\varepsilon \cos \gamma_m$$

$$\Rightarrow \qquad \cos \gamma_m = -\frac{\varepsilon}{2} \qquad (3.3.24)$$

If $\cos\gamma$ and $\sin\gamma$ are replaced by the corresponding relations in θ, then it is found that

$$p = \frac{6\eta UR}{c^2}\left[\frac{\varepsilon \sin\theta(2 + \varepsilon\cos\theta)}{(2 + \varepsilon^2)(1 + \varepsilon\cos\theta)^2}\right] \text{ with corresponding } \cos\theta_m = \frac{-3\varepsilon}{2 + \varepsilon^2} \qquad (3.3.25)$$

Equation 3.3.25 can be written in non-dimensional form as

$$\bar{p} = \frac{pc^2}{6\eta UR} = \frac{\varepsilon \sin\theta(2 + \varepsilon\cos\theta)}{(2 + \varepsilon^2)(1 + \varepsilon\cos\theta)^2} \qquad (3.3.26)$$

Figure 3.3.9 provides variation of $\bar{p}$ vs θ. As shown in Fig. 3.3.9, θ measures the angular position from the position of maximum film thickness and the minimum film thickness occurs at $\theta = \pi$. The positive pressure occurs in the range of $\theta = 0$ to π. The negative pressure is developed when θ varies from π to 2π. The pressure distribution is skewed symmetrically.

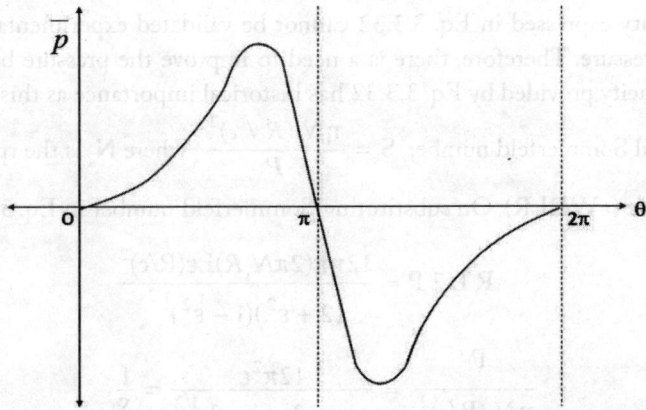

Fig. 3.3.9 Pressure distribution for full Sommerfeld solution in a journal bearing

Using $\cos\theta_m = \frac{-3\varepsilon}{2 + \varepsilon^2}$ the maximum pressure value (where $\frac{dp}{d\theta} = 0$) can be estimated. The non-dimensional maximum pressure is then given as

$$\bar{p}_m = \frac{3\varepsilon(4 - 5\varepsilon^2 + \varepsilon^4)^{\frac{1}{2}}(4 - \varepsilon^2)}{2(2 + \varepsilon^2)(1 - \varepsilon^2)^2} \qquad (3.3.27)$$

Having known the pressure, load components can be found. The components of the resultant load along and perpendicular to the line of centres (as shown in Fig. 3.3.8) are determined.

$$W_x = W \cos\varphi = -L\int_0^{2\pi} p \cos\theta . R . d\theta \qquad (3.3.28)$$

$$W_x = W\sin\varphi = L\int_0^{2\pi} p\sin\theta \cdot R\,d\theta \qquad (3.3.29)$$

Here L = the axial length of the journal and p does not vary with L.
φ = attitude angle, which is angle between the load vector and the line of centres.

Integrating Eqs. 3.3.28 and 3.3.29 require Sommerfeld substitution. It is important to note that due to positive and negative pressure (as shown in Fig. 3.3.3), resultant load capacity in X-direction will be zero.

$$W_x = W\cos\varphi = 0 \qquad (3.3.30)$$

The other load component W_y may be evaluated as

$$W_y = W\sin\varphi = \frac{12\eta\pi UL\varepsilon(R/c)^2}{(2+\varepsilon^2)(1-\varepsilon^2)^{1/2}} \qquad (3.3.31)$$

Thus the total load is

$$W = W_y = \frac{12\eta\pi UL\varepsilon(R/c)^2}{(2+\varepsilon^2)(1-\varepsilon^2)^{1/2}} \qquad (3.3.32)$$

The load capacity expressed in Eq. 3.3.32 cannot be validated experimentally as liquid cannot sustain negative pressure. Therefore, there is a need to improve the pressure boundary conditions. However, load capacity provided by Eq. 3.3.32 has historical importance as this equation gave birth to non-dimensional Sommerfeld number, $S = \dfrac{\eta N_s (R/c)^2}{P}$, where N_s is the rps of the journal and P is average pressure (=W/2LR). On substituting Sommerfeld number in Eq. 3.3.32

$$RL2P = \frac{12\eta\pi(2\pi N_s R)L\varepsilon(R/c)^2}{(2+\varepsilon^2)(1-\varepsilon^2)^{1/2}}$$

Rearranging,

$$\frac{P}{\eta N_s(R/c)^2} = \frac{12\pi^2\varepsilon}{(2+\varepsilon^2)(1-\varepsilon^2)^{1/2}} = \frac{1}{S}$$

Or

$$S = \frac{(2+\varepsilon^2)(1-\varepsilon^2)^{1/2}}{12\pi^2\varepsilon} \qquad (3.3.33)$$

The expression of Sommerfeld, provided in Eq. 3.3.33 is important historically. It MUST NOT BE used to evaluate the eccentricity ratio.

Half-Sommerfeld boundary condition

As seen in Fig. 3.3.9, the solution of idealized full journal bearing using full-Sommerfeld boundary conditions leads to skewed symmetrical pressure distribution. In practice liquid film cannot sustain negative pressure and it ruptures. To account this phenomenon, half-Sommerfeld condition (also known as Gumbel condition) is used to simulate the pressure profile in journal bearing. As per this pressure boundary condition (as shown in Fig. 3.3.10),

(i) P = 0 at θ = 0 and P = 0 at π ≤ θ ≤ 2π. (3.3.34)

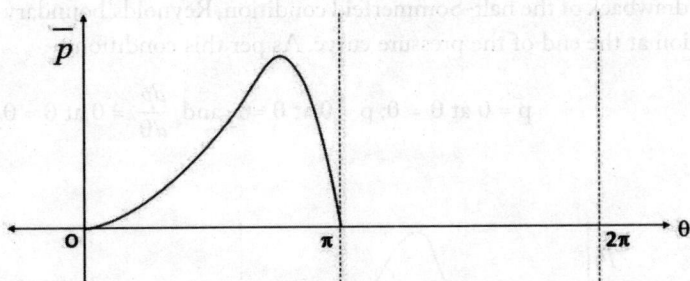

Fig. 3.3.10 Pressure distribution using half-Sommerfeld boundary condition

The pressure distribution as obtained with full-Sommerfeld boundary condition, can be applied here for the region θ = 0 to π. Thus

$$p = \begin{cases} \dfrac{6\eta UR}{c^2} \left[\dfrac{\varepsilon \sin\theta(2 + \varepsilon\cos\theta)}{(2 + \varepsilon^2)(1 + \varepsilon\cos\theta)^2} \right] & \text{for } 0 \leq \theta \leq \pi \\ 0 & \text{for } \pi < \theta \leq 2\pi \end{cases} \quad (3.3.35)$$

The pressure distribution with half-Sommerfeld solution is shown in Fig. 3.3.10. The load components W_x and W_y can be evaluated in the same way.

$$W_x = W\cos\varphi = -LR \int_0^\pi p\cos\theta\, d\theta \quad (3.3.36)$$

$$W_y = W\sin\varphi = LR \int_0^\pi p\sin\theta\, d\theta \quad (3.3.37)$$

On integrating Eq. 3.3.36 and Eq. 3.3.37, we have

$$W_x = 12\eta UL \left(\dfrac{R}{c}\right)^2 \cdot \dfrac{\varepsilon^2}{(2+\varepsilon^2)(1-\varepsilon^2)}$$

$$W_y = 6\eta UL \left(\dfrac{R}{c}\right)^2 \cdot \dfrac{\pi\varepsilon}{(2+\varepsilon^2)(1-\varepsilon^2)^{1/2}}$$

The resultant load is given by

$$W = \sqrt{W_x^2 + W_y^2} = \dfrac{6\eta UL (R/c)^2 \cdot \varepsilon \left[\pi^2 - \varepsilon^2(\pi^2 - 4)\right]^{1/2}}{(2+\varepsilon^2)(1-\varepsilon^2)} \quad (3.3.38)$$

The attitude angle, φ, is expressed as

$$\varphi = \tan^{-1}\left(\dfrac{W_y}{W_x}\right) = \tan^{-1}\left(\dfrac{\pi\sqrt{1-\varepsilon^2}}{2\varepsilon}\right) \quad (3.3.39)$$

It may be noted here that Eq. 3.3.38 provides reasonable good estimation of bearing load capacity. However, assumption of film pressure $p = 0$ in the range of $\pi < \theta \leq 2\pi$ may not be correct and one needs to account Reynolds' pressure boundary condition.

Reynolds' boundary condition

To overcome the drawback of the half-Sommerfeld condition, Reynolds' boundary condition account mass flow condition at the end of the pressure curve. As per this condition:

$$p = 0 \text{ at } \theta = 0;\ p = 0 \text{ at } \theta = \theta_2 \text{ and } \frac{dp}{d\theta} = 0 \text{ at } \theta = \theta_2 \qquad (3.3.40)$$

where $\theta_2 > \pi$

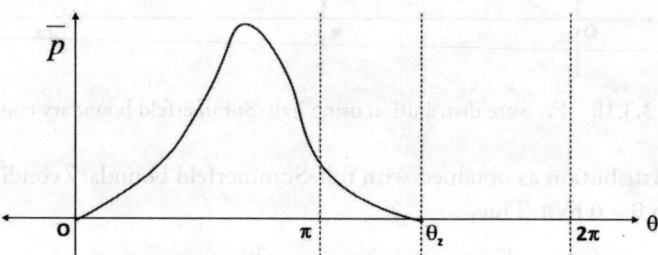

Fig. 3.3.11 Pressure distribution in a journal bearing using Reynolds' boundary conditions

This boundary condition takes care of the film rupture at $\theta = \theta_2$ as shown in Fig. 3.3.11. Using $p = 0$ at $\theta = 0$, we can get similar equation as Eq. 3.3.23

$$p = \frac{6\eta UR}{c^2}\left[\frac{\gamma - \varepsilon \sin\gamma}{(1-\varepsilon^2)^{3/2}} - \left\{\frac{1-\varepsilon^2}{1-\varepsilon\cos\gamma_2}\right\}\frac{\gamma - 2\varepsilon\sin\gamma + \varepsilon^2\gamma/2 + (\varepsilon^2\sin 2\gamma)/4}{(1-\varepsilon^2)^{5/2}}\right] \qquad (3.3.41)$$

where, $\cos\gamma = \dfrac{\varepsilon + \cos\theta}{1 + \varepsilon\cos\theta}$ and γ_2 corresponds to θ_2.

Using $p = 0$ at $\gamma = \gamma_2$ in Eq. 3.3.41, we have

$$\varepsilon[\sin\gamma_2\cos\gamma_2 - \gamma_2] + 2[-\gamma_2\cos\gamma_2 + \sin\gamma_2] = 0 \qquad (3.3.42)$$

For a particular ε, γ_2 can be determined from Eq. 3.3.42 and hence θ_2 can be found.

The load components can be given as

$$W_x = -3\eta UL\left(R/c\right)^2 \frac{\varepsilon(1-\cos\gamma_2)^2}{(1-\varepsilon^2)(1-\varepsilon\cos\gamma_2)}$$

$$W_y = 6\eta UL\left(R/c\right)^2 \frac{(\sin\gamma_2 - \gamma_2\cos\gamma_2)^2}{(1-\varepsilon^2)^{1/2}(1-\varepsilon\cos\gamma_2)}$$

The total load is given by

$$W = \frac{3\eta UL\left(R/c\right)^2}{(1-\varepsilon^2)^{1/2}(1-\varepsilon\cos\gamma_2)}\left[\frac{\varepsilon^2(1-\cos\gamma_2)^4}{1-\varepsilon^2} + 4(\sin\gamma_2 - \gamma_2\cos\gamma_2)^2\right]^{1/2} \qquad (3.3.43)$$

The attitude angle ϕ is given as

$$\varphi = \tan^{-1}\left[\frac{2(1-\varepsilon^2)^{1/2}\cdot(\sin\gamma_2 - \gamma_2\cos\gamma_2)}{\varepsilon(1-\cos\gamma_2)^2}\right] \qquad (3.3.44)$$

3.3.4 Infinitely short journal bearings

If the bearing is infinitely short or narrow i.e., L << 2*R_J (journal diameter) as shown in Fig.3.3.12, the pressure gradient in the X-direction can be neglected $\frac{\delta p}{\delta z} >> \frac{\delta p}{\delta x}$. The governing equation in this case can be reduced to

$$\frac{\partial}{\partial z}\left(h^3 \frac{\partial p}{\partial z}\right) = 6\eta U \frac{\partial h}{\partial x} \qquad (3.3.45)$$

In polar coordinates with $x = R\theta$ and $dx = R.d\theta$,

$$\frac{\partial}{\partial z}\left(h^3 \frac{\partial p}{\partial z}\right) = 6\frac{\eta U}{R}\frac{\partial h}{\partial \theta} \qquad (3.3.46)$$

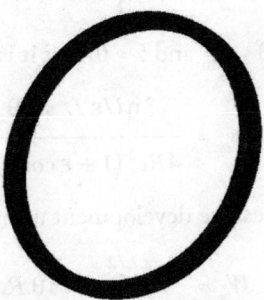

Fig. 3.3.12 Infinitely short journal bearing

Assuming no misalignment, the film thickness is a function of θ

$$h = c(1 + \varepsilon \cos\theta)$$

Integrating Eq. 3.3.46 twice with respect to z, we have

$$p = \frac{6\mu\omega}{h^3} \cdot \frac{dh}{d\theta} \cdot \frac{z^2}{2} + C_1 z + C_2$$

where, C_1 and C_2 are the constants of integration and can be evaluated by substituting the boundary conditions p = 0 at z = ± L/2 as shown in Fig.3.3.13 where Z-coordinate is chosen from the bearing centre.

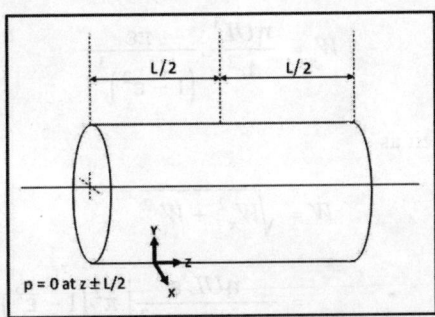

Fig. 3.3.13 Boundary conditions for evaluating pressure distribution

Using the boundary conditions, the pressure distribution is given by

$$p = \frac{3\eta U}{Rc^2}\left(\frac{L^2}{4} - z^2\right) \cdot \frac{\varepsilon \sin\theta}{(1+\varepsilon\cos\theta)^3} \quad (3.3.47)$$

The pressure distribution is parabolic in Z-direction. Equation 3.3.47 is applicable for $\theta = 0$ to $\theta = \pi$ range, as the pressure becomes negative for $\theta > \pi$. The maximum pressure can be obtained by differentiating Eq. 3.3.47 with respect to θ and then equating to zero, such as

$$\left.\frac{\partial p}{\partial \theta}\right|_{\theta = \theta_m} = 3\frac{\eta U}{R}\left(\frac{L^2}{4} - z^2\right)\frac{\partial}{\partial \theta}\left(\frac{\varepsilon \sin\theta}{(1+\varepsilon\cos\theta)^3}\right) = 0$$

$$\theta_m = \cos^{-1}\left[\frac{1-\sqrt{(1+24\varepsilon^2)}}{4\varepsilon}\right] \quad (3.3.48)$$

The maximum pressure occurs at $\theta = \theta_m$ and $z = 0$, and it is given by

$$p_m = \frac{3\eta U \varepsilon L^2 \sin\theta_m}{4Rc^2(1+\varepsilon\cos\theta_m)^3} \quad (3.3.49)$$

The load components from the pressure development using half-Sommerfeld assumption are

$$W_x = -2\int_0^\pi \int_0^{L/2} p\cos\theta \cdot R \cdot d\theta \cdot dz$$

$$= -\frac{\eta U L^3 \varepsilon}{2c^2}\int_0^\pi \frac{\sin\theta\cos\theta}{(1+\varepsilon\cos\theta)^3}d\theta \quad (3.3.50)$$

$$W_y = 2\int_0^\pi \int_0^{L/2} p\sin\theta \cdot R \cdot d\theta \cdot dz$$

$$= -\frac{\eta U L^3 \varepsilon}{2c^2}\int_0^\pi \frac{\sin^2\theta}{(1+\varepsilon\cos\theta)^3}d\theta \quad (3.3.51)$$

Using Sommerfeld table of integration [Booker, 1965],

$$W_x = -\frac{\eta U L^3}{c^2} \cdot \frac{\varepsilon^2}{(1-\varepsilon^2)^2}$$

$$W_y = \frac{\eta U L^3}{4c^2} \cdot \frac{\pi\varepsilon}{(1-\varepsilon^2)^{3/2}}$$

The load capacity W is given as

$$W = \sqrt{W_x^2 + W_y^2}$$

$$= \frac{\eta U L^3 \varepsilon}{4c^2(1-\varepsilon^2)^2}\left[\pi^2(1-\varepsilon^2)+16\varepsilon^2\right]^{1/2} \quad (3.3.52)$$

The attitude angle ϕ is defined as

$$\varphi = \tan^{-1}\left(-\frac{W_y}{W_x}\right) = \tan^{-1}\left(\frac{\pi}{4} \cdot \frac{(1-\varepsilon^2)^{1/2}}{\varepsilon}\right) \qquad (3.3.53)$$

The load bearing capacity of infinitely long bearing (Eq. 3.3.38) can be compared to infinitely short bearing (Eq. 3.3.52) considering half-Sommerfeld pressure condition. The ratio of load capacities will be:

$$\text{Ratio} = \frac{W_{\text{(Short bearing)}}}{W_{\text{(Long bearing)}}} = \left(\frac{L}{R}\right)^2 \frac{(2+\varepsilon^2)}{24(1-\varepsilon^2)} \sqrt{\frac{[16\varepsilon^2 + \pi^2(1-\varepsilon^2)]}{[\pi^2 - \varepsilon^2(\pi^2 - 4)]}} \qquad (3.3.54)$$

As we can see in Eq. 3.3.54, the load ratio are functions of (L/R) and ε. Figure 3.3.14 provides a comparison of load ratio (Eq. 3.3.54) versus eccentricity ratio for L = R. It is interesting to note that at eccentricity ratio 0.9, load capacity predicted by short bearing (Eq. 3.3.52) becomes more than that predicted by long bearing approximation (Eq. 3.3.38).

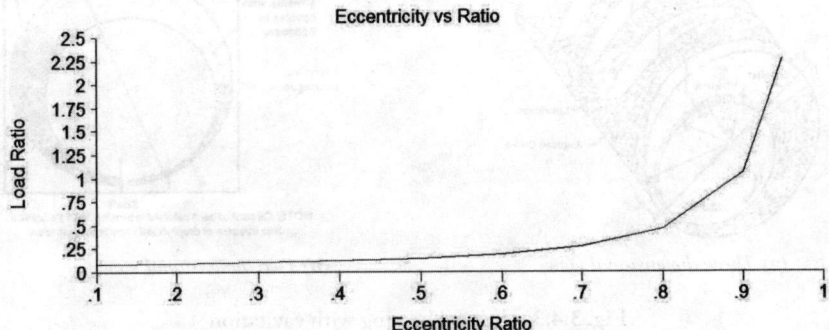

Fig. 3.3.14 Load ratio vs. Eccentricity ratio

NOTE: The infinitely short bearing theory overestimates the load carrying capacity for all ε and should be used for L/D lesser than 0.5. The infinitely long bearing theory provides a much greater estimate for finite bearings particularly for L/D ratio greater than 2.

3.4 Journal Bearings

Low cost, simple manufacturing, zero wear and very low coefficient of friction are main features of hydrodynamic journal bearings. Two journal bearings, made of solid rubber and lubricated by sea water are shown in Fig. 3.4.1. These bearings are known as long bearing as length to diameter ratio is greater than two. These bearings are full (360°) cylindrical bearings. In practice nearly half of cylindrical bearing supports the loads. To observe this, a journal bearing made of acrylic material as shown in Fig. 3.4.2 was fabricated. The cavitation in upper half of the bearing was observed. The observed cavitation is sketched in Fig. 3.4.3.

Fig. 3.4.1 Long journal bearings

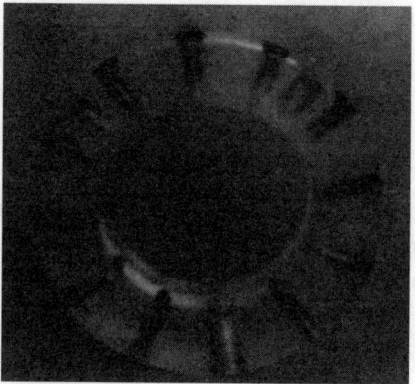

Fig. 3.4.2 Acrylic journal bearing with temperature measuring copper rivets

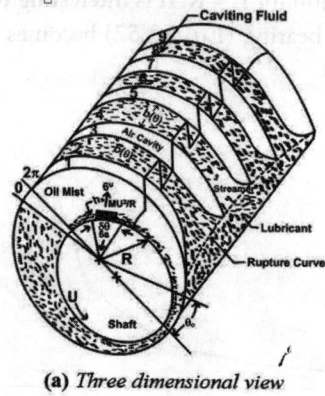

(a) *Three dimensional view*

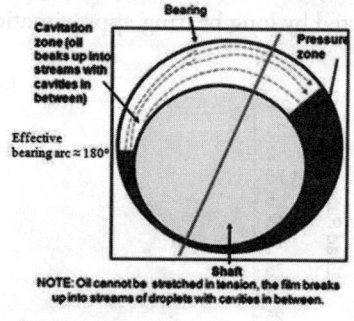

(b) *Two dimensional view*

Fig. 3.4.3 Journal bearing with cavitation

The eccentric position of journal in a hydrodynamic bearing forms a converging-diverging clearance space. Due to unavailability of sufficient liquid lubricant, liquid-streamers separated by gas/vapour space are formed in the divergent clearance space, as indicated in Fig. 3.4.3. This 'cavitation' phenomenon cannot be appropriately predicted using the Reynolds' equation alone, or associating it with Gumbel/Reynolds' boundary condition. The Gumbel (half Sommerfeld) condition does not satisfy the continuity of flow at film rupture. It considers only half bearing, the cavitated zone is neglected and recirculating flow is neglected (as shown in below Fig. 3.4.4).

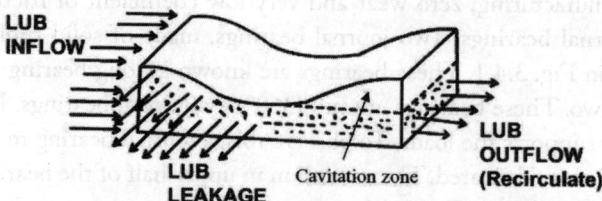

Fig. 3.4.4 Recirculation of lubricant indicating side leakage and cavitation

Even though prediction of load capacity is fairly reasonable (as cavitated pressure does not contribute to load carrying capacity), but oil flow and power loss cannot accurately be predicted using this boundary condition.

The Reynolds' boundary condition (also known as Swift–Striber boundary condition) represents rupture interface in a reasonable way but does not satisfy the principle of mass conservation when the film is re-established. To implement the mass conservation principle at the rupture as well as reformation, Elrod (Elrod, 1981) utilized Jakobsson–Floberg–Olsson (JFO) cavitation boundary condition. To implement mass conservation a switch function, g, as depicted by following equation is required.

$$g = \begin{cases} 1 & \text{in full film region} \\ 0 & \text{in cavitated region} \end{cases}$$

The cavitation pressure is slightly lesser than ambient pressure, and to account this density term is used in the Reynolds' equation, such as:

$$\frac{\partial(\rho h)}{\partial t} + \frac{\partial}{\partial x}\left(\frac{\rho h u}{2} - \frac{\rho h^3}{12\mu}\frac{\partial P}{\partial x}\right) + \frac{\partial}{\partial z}\left(-\frac{\rho h^3}{12\mu}\frac{\partial P}{\partial z}\right) = 0 \qquad (3.4.1)$$

The term $\alpha_c = \rho/\rho_c$ is used to account the fluid compressibility due to presence of air bubbles/gas in the lubricant. Here ρ_c is the density of lubricant at cavitation pressure.

$$\alpha_c = \begin{cases} \rho/\rho_c & \text{in full film region, } \alpha_c \geq 1 \\ h/h_c & \text{in cavitated region, } \alpha_c < 1 \end{cases}$$

To identify the cavitation ($\alpha_c < 1$) and full film ($\alpha_c \geq 1$) regions in hydrodynamic bearing, α_c needs to be expressed in terms of fluid pressure as

$$P = P_c + g\beta_c (\log \alpha_c) \qquad (3.4.2)$$

where, β_c is bulk modulus (in pascal).

As value of α remains close to 1.0, the above expression can be approximated as

$$P = P_c + g\beta_c (\alpha_c - 1) \qquad (3.4.3)$$

Now incorporating α_c and g in Reynolds' equation (Eq. 3.4.1),

$$\frac{\partial}{\partial t}(\alpha_c h) + \frac{\partial}{\partial x}\left(\frac{\alpha_c h u}{2} - \frac{g\beta_c h^3}{12\mu}\frac{\partial \alpha_c}{\partial x}\right) + \frac{\partial}{\partial z}\left(-\frac{g\beta_c h^3}{12\mu}\frac{\partial \alpha_c}{\partial z}\right) = 0 \qquad (3.4.4)$$

The value of g = 1 makes Eq. 3.4.4 an elliptic partial differential equation (PDE), while g = 0 makes Eq. 3.4.4 a hyperbolic PDE. A robust convergent solution needs a central-finite-difference-scheme for elliptical PDE and upwind-finite-difference scheme for hyperbolic PDE. This can be achieved by adopting a type difference scheme, such as:

$$\frac{\partial}{\partial x}\left(\frac{h u \alpha_c}{2}\right)_i = \frac{u}{4\Delta x}\left[g_{i+\frac{1}{2}}(\alpha_c h)_{i+1} + \left(2 - g_{i+\frac{1}{2}} - g_{i-\frac{1}{2}}\right)(\alpha_c h)_i - \left(2 - g_{i-\frac{1}{2}}\right)(\alpha_c h)_{i-1}\right]$$

(3.4.5)

$$\frac{\partial}{\partial x}\left(-\frac{\beta_c h^3}{12\mu}g\frac{\partial \alpha_c}{\partial x}\right)_i$$

$$= -\frac{1}{(\Delta x)^2}\left[h_{i+1/2}^3 g_{i+1}\left(\alpha_{c(i+1)}-1\right)-\left(h_{i+1/2}^3+h_{i-1/2}^3\right)g_i\left(\alpha_{ci}-1\right)+h_{i-1/2}^3 g_{i-1}\left(\alpha_{c(i-1)}-1\right)\right]$$

(3.4.6)

$$\frac{\partial}{\partial z}\left(-\frac{\beta_c h^3}{12\mu}g\frac{\partial \alpha_c}{\partial z}\right)_j = -\frac{\beta_c h_i^3}{12\mu(\Delta z)^2}\left[g_{j+1}\left(\alpha_{c(j+1)}-1\right)-2g_j\left(\alpha_{c(j+1)}-1\right)+g_{j-1}\left(\alpha_{c(j-1)}-1\right)\right]$$

(3.4.7)

This formulation considers the phenomenon of cavitation in journal bearing, but its solution takes appreciable computational efforts. To solve these equations numerically it is necessary to find the location of journal centre in the clearance circle, which is described in the next section.

3.4.1 Locating journal position

In a hydrodynamic journal bearing, journal floats and gets located based on the operating parameters (i.e., load, speed, viscosity, etc.). Figure 3.4.5 shows three positions of journal in bearing. First position is corresponding to negligible load and relatively high speed. In this situation journal centre coincides ($\varepsilon = 0$) with bearing centre. Third position corresponds to high load and negligible speed. To find middle position (to find ε and ϕ) for given speed and load, numerical solution of Reynolds' equation and integration of fluid pressure over effective area is required.

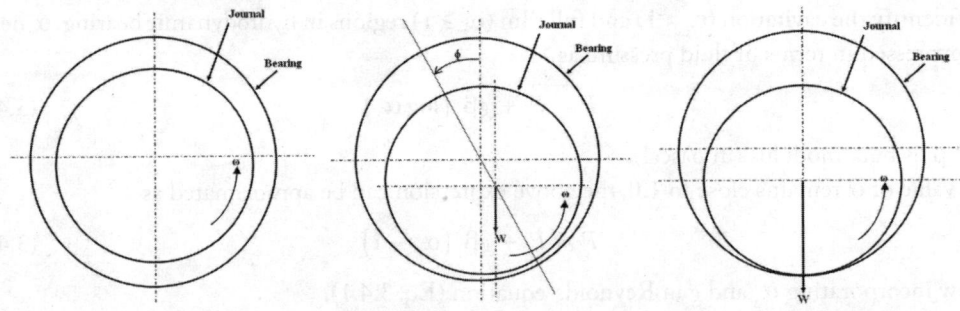

(a) $\varepsilon = 0, \varphi = \pi/2, W = 0$ (b) $0 < \varepsilon < 1, 0 < \varphi < \pi/2, 0 < W < W_{max}$ (c) $\varepsilon = 1, \varphi = 0, W = $ Max Value

Fig. 3.4.5 Locations of journal in bearing

The lubricant flow rate, required to reduce bearing starvation, depends on the feeding geometry and eccentricity ratio of the rotating shaft. This topic is explained in the next section.

3.4.2 Lubricant supply in bearing

Effective full film (hydrodynamic) lubrication can be achieved by continuously feeding lubricant (as pressure greater than ambient pressure) through oil groove/hole arrangement as shown in Fig. 3.4.6. In other words, bearings are provided with feed hole and/or groove to get lubricant.

Lubrication of Bearings 157

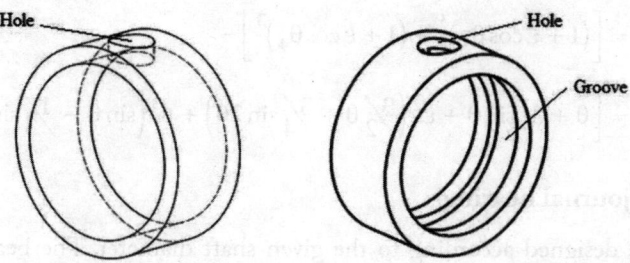

Fig. 3.4.6 Hole and groove arrangement in bearing

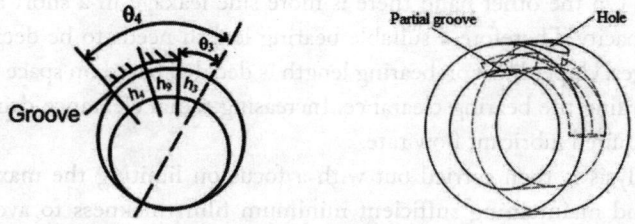

Fig. 3.4.7 Partial oil groove

Oil can be supplied to journal bearing by oil hole (Fig. 3.4.6), partial groove (Fig. 3.4.7) or full groove (Fig. 3.4.6) arrangement depends on the required quantity of oil and load direction. If applied load changes its direction then partial to full groove arrangement is essential to avoid starvation. Martin and Lee (Martin and Lee, 1983) published curve fit equations to estimate the required flow rate of lubricant.

3.4.2.1 *Flow rate under supply pressure, P_S, in bearing with oil hole (Martin, 1998)*

$$Q = \frac{0.675 h_g^3 P_{supply}}{\eta} \left[\frac{d_h}{L + 0.4}\right]^{1.75} \tag{3.4.8}$$

where d_h is the diameter of the hole ($d_h < L/2$) and h_g is the film thickness at the groove mid position.

3.4.2.2 *Flow rate under supply pressure P_S with full oil groove (Martin, 1998)*

$$Q = \frac{C_d^3 P_{supply} \pi D (1 + 1.5\varepsilon^2)}{24\eta(L - a_g)} \tag{3.4.9}$$

where, a_g is the axial length of groove

3.4.2.3 *Flow rate under supply pressure P_S with partial groove (Martin, 1998)*

$$Q = \frac{C_d^3 \cdot P_{supply}}{8\eta} \left[(f_1/6)\left(\frac{5}{4} - \frac{1}{4}\left(\frac{a}{L}\right)\right)\left(\frac{L}{a_g} - 1\right)^{-1/3} + \frac{f_2 D}{6L\left(1 - \frac{a}{L}\right)} \right] \tag{3.4.10}$$

where,
$$f_1 = \left[(1 + \varepsilon \cos \theta_3)^3 + (1 + \varepsilon \cos \theta_4)^3\right] \quad (3.4.11)$$

$$f_2 = \left[\theta + 3\varepsilon \sin \theta + \varepsilon^2 \left(\tfrac{3}{2}\theta + \tfrac{3}{4}\sin 2\theta\right) + \varepsilon^3 \left(\sin \theta - \tfrac{1}{3}\sin^3 \theta\right)\right]_{\theta_3}^{\theta_4} \quad (3.4.12)$$

3.4.3 Design of journal bearings

Journal bearings are designed according to the given shaft diameter. The bearing length is then decided based on the required load carrying capacity. A longer bearing (L/D ratio >1) is able to provide a higher load carrying capacity, but this also increases the chances of edge loading due to shaft misalignment. On the other hand there is more side leakage in a short bearing that sharply reduces the load capacity. Therefore a suitable bearing length needs to be decided based on these conflicting advantages. Upper limit on bearing length is decided based on space limitation. The next is the decision regarding the bearing clearance. Increasing radial clearance decreases load capacity and increases the required lubricant flow rate.

The bearing analysis is then carried out with a focus on limiting the maximum pressure and temperature rise, and maintaining sufficient minimum film thickness to avoid wear of bearing surface. The temperature rise occurs due to shearing of the viscous fluid, which can be explained using Petroff's equation.

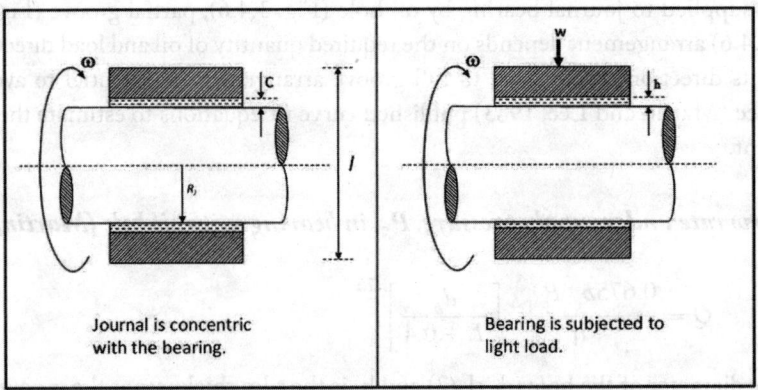

Fig. 3.4.8 Journal rotating inside a bearing

3.4.3.1 Petroff's equation

Petroff was the first to estimate friction in a journal bearing assuming the journal to be concentric with the bearing. Even though this method CANNOT BE used for predicting friction due to inherent wrong assumptions, but it retains its importance because it defines dimensionless parameter 'Sommerfeld number' that is widely used in ascertaining the operative lubrication regime. To derive Petroff's equation, let us consider a journal bearing, as shown in Fig. 3.4.8.

The velocity at the surface of the journal is given by,

$$U = 2\pi R_j N_s \quad (3.4.13)$$

According to Newton's law of viscosity, tangential force required to shear lubricant layer between the journal and the bearing in the circumferential direction,

$$F = F = \left(\eta \frac{U}{h}\right) A \qquad (3.4.14)$$

Here A = area of journal surface = $2\pi R_J L$.
For concentric shaft position, h = C = radial clearance between journal and bearing.
Substituting the above values in Eq. 3.4.13, we have

$$F = \eta \frac{(2\pi R_J L)(2\pi R_J N_s)}{C} = \frac{4\pi^2 R_J^2 L \eta N_s}{C} \qquad (3.4.15)$$

The unit bearing pressure p acting on the bearing, shown in Fig. 3.4.8, is given by

$$p = \frac{W}{projected.area.of.bearing} = \frac{W}{2R_J L} \qquad (3.4.16)$$

or, $\qquad W = 2p R_J L \qquad (3.4.17)$

The frictional force is given by (μW), where μ is the coefficient of friction between journal and bearing. From Eqs. 3.4.15 and 3.4.17, we get

$$\frac{4\pi^2 R_J^2 L \eta N}{C} = \mu(2pR_J L) \qquad (3.4.18)$$

Rearranging

$$\mu = \left(2\pi^2\right)\left(\frac{R_J}{C}\right)\left(\frac{\eta N_s}{p}\right) \qquad (3.4.19)$$

Equation 3.4.19 is Petroff's equation which provides an estimation of friction coefficient in journal bearing. Petroff's equation provides a first approximation of the friction loss in a journal bearing; it indicates two important dimensionless parameters namely, $\left(\frac{R_J}{C}\right)$ and $\left(\frac{\eta N}{p}\right)$ that govern the coefficient of friction.

3.4.3.2 *Raimondi and Boyd method*

To design a journal bearing, it is necessary to estimate the fluid film pressure, which requires solution of Reynolds' equation. To solve the Reynolds' equation, infinitely long bearing approximation and infinitely short bearing approximation have been explained in previous subsections. As both the approximations are applicable for narrow range of L/D ratio, it is necessary to solve the Reynolds' equation using numerical method, such as finite difference method. The pressure solution using finite difference technique requires appreciable computational efforts. To deal with this problem, Raimondi and Boyd (Raimondi and Boyd, 1958) developed design charts, which are explained in the present section.

Raimondi and Boyd (Raimondi and Boyd, 1958) solved Reynolds' equation (Eq. 3.2.32) using iterative numerical approach. An average value of the lubricant temperature was used in the numerical computation scheme considering the inlet and outlet lubricant temperature for the determination of the lubricant dynamic viscosity. After the determination of hydrodynamic pressure, the load carrying capacity, minimum film thickness, lubricant flow and friction can be estimated using the charts. In the Raimondi and Boyd method (Raimondi and Boyd, 1958), the performance of the bearing is determined by the use of five dimensionless parameters, which are related to the Sommerfeld number and are given in Table 3.4.3.1.

Table 3.4.3.1 Dimensionless parameters in Raimondi and Boyd method

S. No.	Dimensionless Parameter	Description	Determination of performance parameters
1.	$\dfrac{h_0}{c}$	Minimum film thickness variable	Minimum film thickness, h_0
2.	$\left(\dfrac{r_J}{c}\right)f$	Coefficient of friction variable	Coefficient of friction, f
3.	$\dfrac{Q}{r_J cNL}$	Flow variable	Lubricant flow rate, Q
4.	$\dfrac{Q_L}{Q}$	Flow ratio variable	Side leakage, Q_L
5.	$\left(\dfrac{P}{p_{max}}\right)$	Maximum film pressure ratio	Maximum film pressure, p_{max}

In the Raimondi and Boyd method, the lubricant viscosity at operating temperature is to be used. The easiest method to find operating temperature is to calculate an average value of lubricant temperature by considering the inlet temperature and 50% of the expected temperature rise during the bearing operation, as expressed in Eq. 3.4.20.

$$T_{average} = T_{inlet} + \frac{\Delta T_{rise}}{2} \qquad (3.4.20)$$

Initially, the dynamic viscosity (Fig. 3.4.9) is determined for temperature calculated using Eq. 3.4.20. The temperature rise of the lubricant is determined using the Raimondi and Boyd chart 6 (Fig. 3.4.10). The average temperature is calculated using Eq. 3.4.20 and compared with initial average temperature. If average temperature is different, then the dynamic viscosity at this average temperature is determined. This dynamic viscosity is then used to repeat calculations for Sommerfeld number and temperature rise variable. The procedure is repeated until the convergence is obtained in the dynamic viscosity and temperature rise values.

After freezing the value of operating dynamic viscosity and corresponding Sommerfeld number, the non-dimensional 'minimum film thickness $\left(\dfrac{h_o}{C}\right)$' value is read from chart 1 (Fig. 3.4.11). Using this value the minimum film thickness is determined. This is compared with the composite surface roughness, $\sigma = \sqrt{R_{q,bearing}^2 + R_{q,journal}^2}$ of the journal bearing. In order to ensure that the operative regime of the journal bearing is hydrodynamic and there is no asperity contact, the minimum film thickness must be 5 times of the composite thickness. If this is not achieved then the necessary changes in bearing design (i.e., lubricant viscosity, length, clearance) must be made.

The position of minimum film thickness is determined from chart 2 (Fig. 3.4.12). The coefficient of friction is determined from the value of coefficient of friction variable $\left(\dfrac{r_J}{c}\right) f$ obtained from chart 3 (Fig. 3.4.13). The friction torque and power loss are estimated using the following equations:

$$T = f \cdot W \cdot r_J$$

and
$$H = T \cdot (2 \cdot \pi \cdot N)$$

The total volumetric flow rate is determined by finding the value of flow variable $\dfrac{Q}{r_J c N L}$ from chart 4 (Fig. 3.4.14). The side flow is calculated by finding the value of flow ratio (ratio of side flow to total flow) $\dfrac{Q_s}{Q}$, from chart 5 (fig. 3.4.15). The maximum pressure is evaluated by finding the value of maximum pressure ratio, $\dfrac{P}{P_{max}}$ from chart 7 (Fig. 3.4.16). The position of maximum film pressure is determined from chart 8 (Fig. 3.4.17). The location of terminating position θ_{p0} is estimated from chart 9 (Fig. 3.4.18).

Example 3: Determine the performance parameters (minimum film thickness, coefficient of friction, power loss, side flow, location and magnitude of the maximum fluid pressure) of a full journal bearing having following specifications:

Journal radius (R_J) = 25 mm, bearing length (L) = 50 mm, rotational speed of journal = 1440 rpm, load (W) = 1000 N, lubricant oil SAE40, and radial clearance (c) = 0.00005 m. The lubricant supply temperature is 70°. An initial assumption of expected temperature rise is 8 °C. The density and specific heat of the lubricant is 887 kg/m³ and 1800 J/kg/°C respectively.

Solution: As a first step, average temperature of the lubricant is determined using the initial values.

$$T_{average} = T_{inlet} + \dfrac{\Delta T_{rise}}{2}$$

$$T_{average} = 70 + (8/2) = 74\ °C$$

The dynamic viscosity (η) of lubricant, SAE40 at 74 °C, determined from the viscosity–temperature chart (Fig. 3.4.9), is 20 mPa.s.

The Sommerfeld number corresponding to this dynamic viscosity is given by:

$$S = \dfrac{\eta N}{p}\left(\dfrac{R_J}{C}\right)^2 = \left(\dfrac{0.02 \cdot (1440/60)}{1000/(2 \cdot 0.025 \cdot 0.050)}\right)\left(\dfrac{0.025}{0.00005}\right)^2 = 0.3$$

Corresponding to $\dfrac{L}{D} = 1$ and Sommerfeld number 0.3, we obtain a value of temperature rise variable of 25 using chart 6 (Fig. 3.4.10).

162 *Fundamentals of Engineering Tribology with Applications*

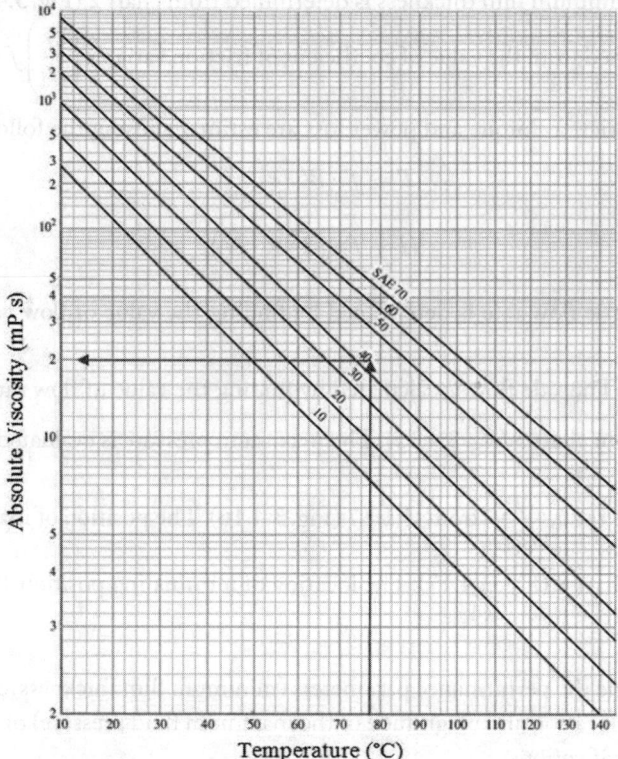

Fig. 3.4.9 Viscosity–temperature chart

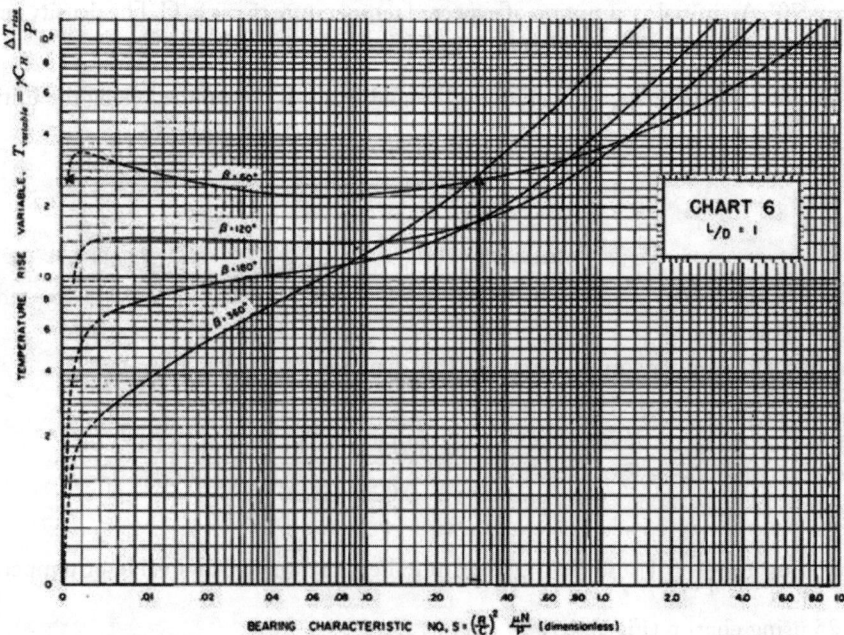

Fig. 3.4.10 Temperature rise variable [Raimondi and Boyd, 1958] Chart 6

Determine the temperature rise using the temperature rise variable given by the following expression:

$$T_{variable} = \gamma C_H \frac{\Delta T_{rise}}{P} \qquad (3.4.21)$$

or

$$\Delta T = \frac{T_{variable} \cdot P}{\gamma \cdot C_H} = 6.61 \,°C$$

The average temperature is now estimated

$$T_{average} = T_{inlet} + \frac{\Delta T_{rise}}{2} = 70 + \frac{6.61}{2} = 73.3°.$$

The dynamic viscosity is now determined at this temperature from the viscosity–temperature chart (Fig. 3.4.9) to be 0.022 Pa.s. The Sommerfeld number corresponding to this dynamic viscosity of 0.022 Pa.s is given by:

$$S = \frac{\eta N}{p}\left(\frac{R_J}{C}\right)^2 = \left(\frac{0.022 \cdot (1440/60)}{1000/(2 \cdot 0.025 \cdot 0.050)}\right)\left(\frac{0.025}{0.00005}\right)^2 = 0.33$$

Corresponding to $\frac{L}{D} = 1$ and Sommerfeld number 0.33, we obtain a value of temperature rise variable of 30 using chart 6 (Fig. 3.4.10).

The temperature rise is then determined using the temperature rise variable given by the following expression:

$$T_{variable} = \gamma C_H \frac{\Delta T_{rise}}{P}; \quad \text{or} \quad \Delta T = \frac{T_{variable} \cdot P}{\gamma \cdot C_H} = 7.92 \,°C$$

The average temperature is now estimated

$$T_{average} = T_{inlet} + \frac{\Delta T_{rise}}{2} = 70 + \frac{7.92}{2} = 73.96°.$$

This average temperature is close to 74 °C; therefore the iterations are stopped here and the dynamic viscosity is now determined at this temperature from the viscosity–temperature chart (Fig. 3.4.9) to be 0.02 Pa.s. The further calculations are based on this dynamic viscosity.

Corresponding to $\frac{L}{D} = 1$ and Sommerfeld number 0.3, we obtain a value of $\frac{h_0}{c} = 0.64$ from chart 1 (Fig. 3.4.11). The minimum film thickness (0.64*radial clearance) is determined to be 32 micro-meters. The angular location of the minimum film thickness is determined from chart 2 (Fig. 3.4.12) corresponding to S = 0.3 and L/D=1 which comes out to be 65°.

The coefficient of friction variable, determined from chart 3 (Fig. 3.4.13), is $\left(\frac{r_J}{c}\right)f = 6.25$, from which the coefficient of friction is determined to be 0.0125. The friction torque on the journal is $T = f \cdot W \cdot r_J = 0.0125 \cdot 1000 \cdot 0.025 = 0.3125$ Nm. The power loss is estimated to be $H = T \cdot (2 \cdot \pi \cdot N)$.

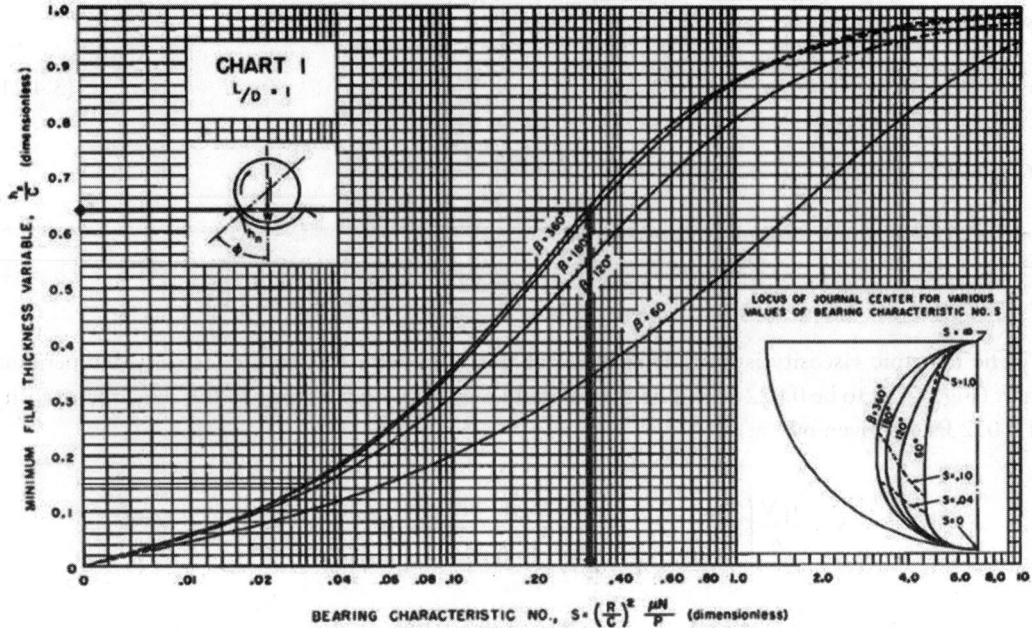

Fig. 3.4.11 Minimum film thickness parameter [Raimondi and Boyd, 1958] Chart 1

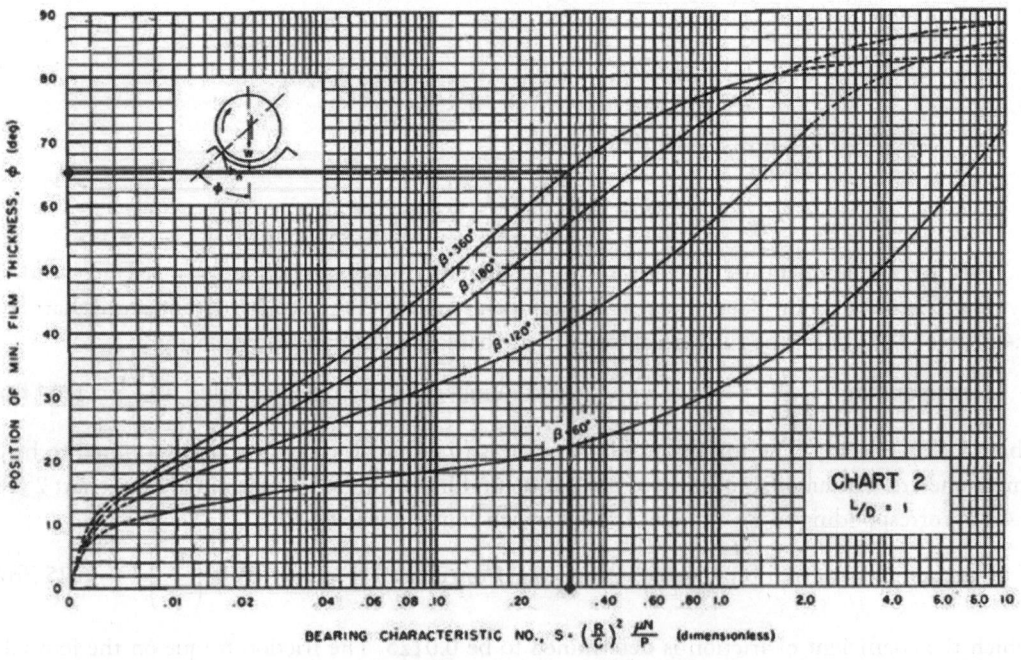

Fig. 3.4.12 Position of minimum film thickness [Raimondi and Boyd, 1958] Chart 2

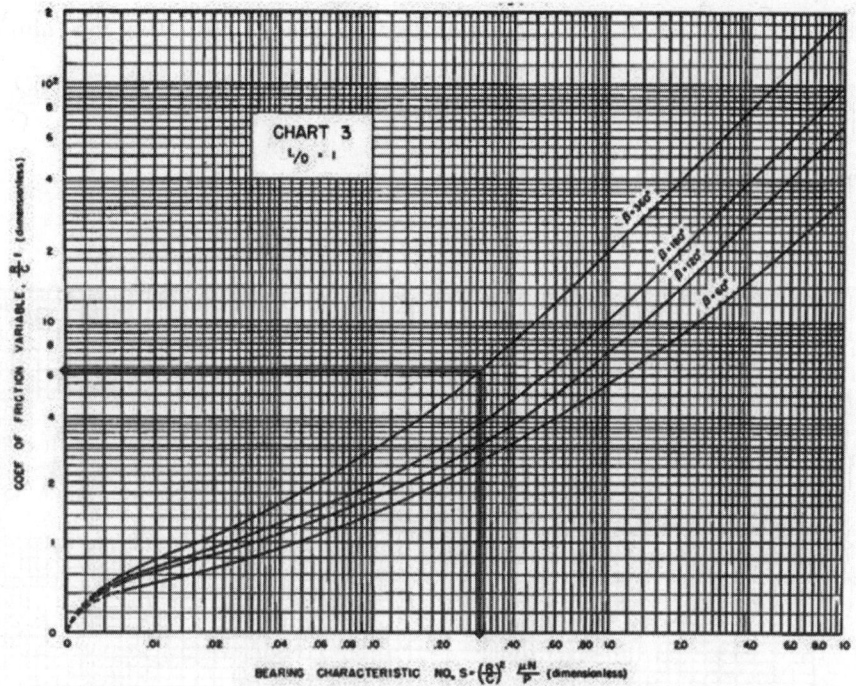

Fig. 3.4.13 Coefficient of friction variable [Raimondi and Boyd, 1958] Chart 3

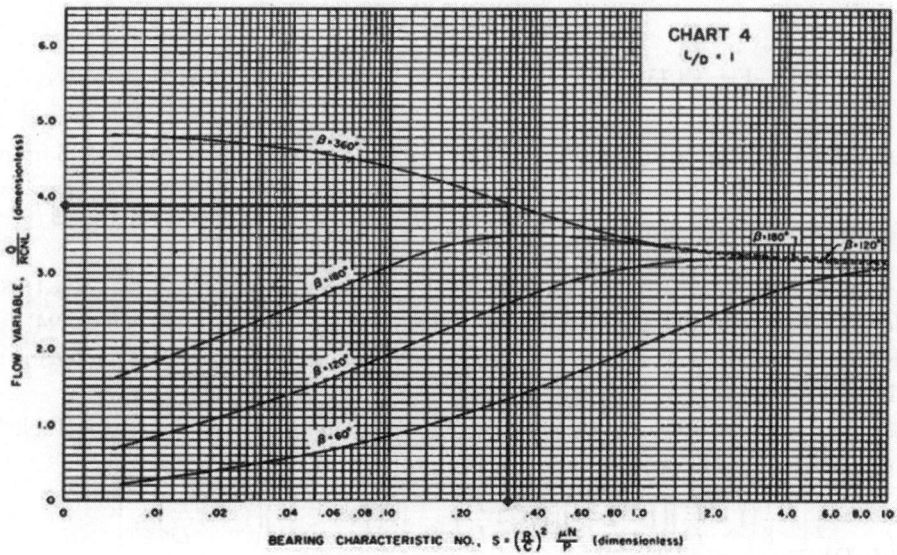

Fig. 3.4.14 Flow variable [Raimondi and Boyd, 1958] Chart 4

The lubricant flow can be determined by finding the value of flow variable from chart 4 (Fig.3.4.14). The value of this variable $\dfrac{Q}{r_J cNL}$ is 3.9. From this the total volumetric flow rate is determined as

$$Q = 3.9 \cdot 25 \cdot 0.05 \cdot (1440/60) \cdot 50 = 5850 \text{ mm}^3/s.$$

The flow ratio (ratio of side flow to total flow), determined from chart 5 (Fig. 3.4.15) is $\dfrac{Q_s}{Q} = 0.475$.

From this, the side flow can be determined as

$$Q_s = 0.475 \cdot 5850 = 2778.75 \text{ mm}^3/s.$$

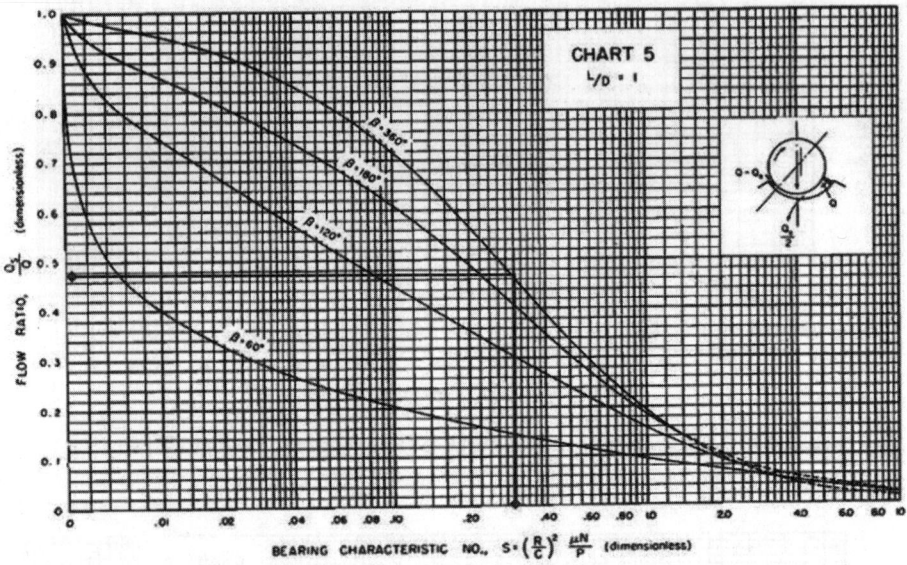

Fig. 3.4.15 Flow ratio [Raimondi and Boyd, 1958] Chart 5

The maximum pressure ratio, determined from chart 7 (Fig. 3.4.16), is $\dfrac{P}{P_{max}} = 0.48$ from which the maximum pressure is determined to be $P_{max} = \dfrac{0.4}{0.48} = 0.833 \text{ }MPa$.

The position of the maximum film pressure determined from chart 8 (Fig. 3.4.17) is 15.8°.
The location of terminating position determined from chart 9 (Fig. 3.4.18) is $\theta_{p0} = 94°$.

The Raimondi and Boyd charts given in this example are for a full journal bearing having an L/D ratio of 1. For L/D ratios of ∞, 1, ½, and ¼, Table 3.4.3.2 may be used.

For L/D ratios other than ∞, 1, ½, and ¼, the interpolation may be required using Table 3.4.3.2, following expression may be used:

$$y = \frac{1}{(L/D)^2}\left[-\frac{1}{8}\left(1-\frac{L}{D}\right)\left(1-2\frac{L}{D}\right)\left(1-4\frac{L}{D}\right)y_\infty + \frac{1}{3}\left(1-2\frac{L}{D}\right)\left(1-4\frac{L}{D}\right)y_1\right]$$

$$+ \frac{1}{(L/D)^2}\left[-\frac{1}{4}\left(1-\frac{L}{D}\right)\left(1-4\frac{L}{D}\right)y_{\frac{1}{2}} + \frac{1}{24}\left(1-\frac{L}{D}\right)\left(1-2\frac{L}{D}\right)y_{\frac{1}{4}}\right]$$

where $y_\infty, y_1, y_{\frac{1}{2}}$ and $y_{\frac{1}{4}}$ are the variables corresponding to L/D ratios of ∞, 1, ½, and ¼.

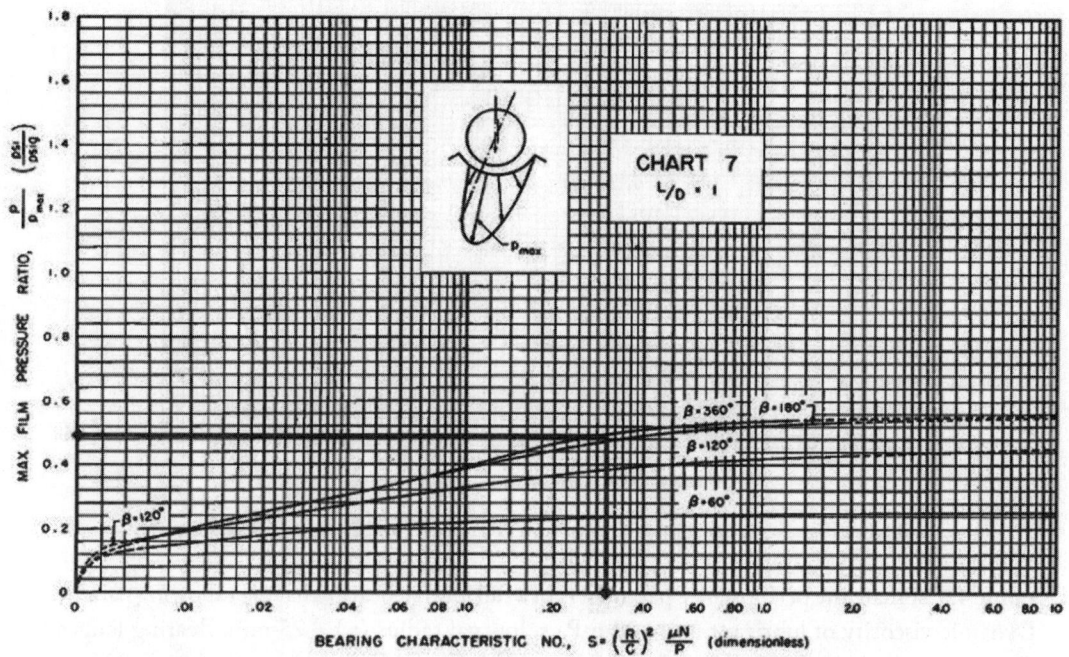

Fig. 3.4.16 Maximum film pressure ratio [Raimondi and Boyd, 1958] Chart 7

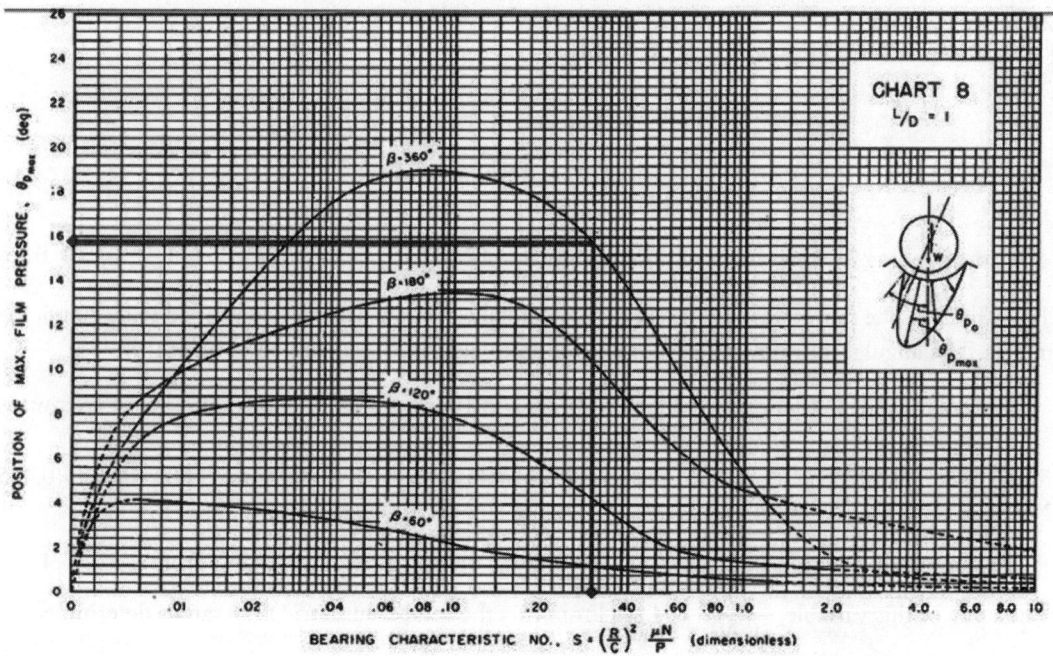

Fig. 3.4.17 Position of maximum film pressure [Raimondi and Boyd, 1958] Chart 8

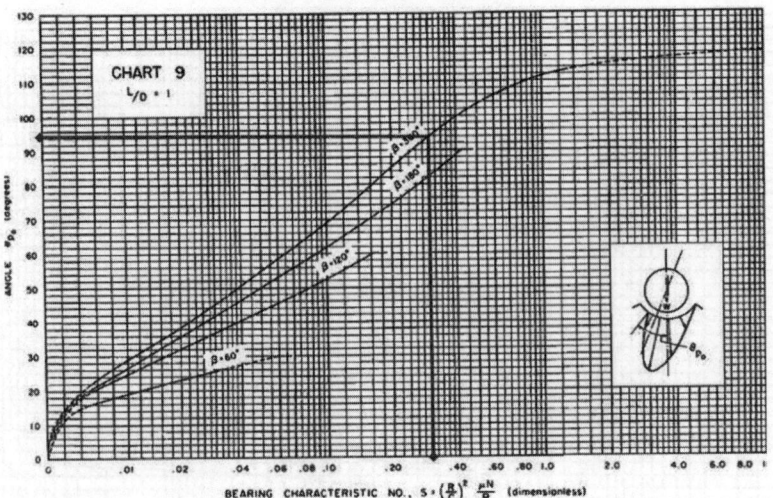

Fig. 3.4.18 Terminating position of film [Raimondi and Boyd, 1958] Chart 9

Example 4: Estimate the performance parameters of a full journal bearing having following parameters:
Dynamic viscosity of lubricant (η) = 20 mPa.s, Journal radius (r_J) = 25 mm, Bearing length (L) = 25mm, Journal speed =3062.4 rpm, Load (W) = 1000 N, and Radial clearance (c) = 0.00005 m.

Solution: Since the L/D ratio is 0.5 therefore table 3.4.3.2 will be used for determining the bearing performance parameters.

First step is to find the Sommerfeld number,

$$S = \frac{\eta N}{p}\left(\frac{R_J}{C}\right)^2 = \left(\frac{0.02 \cdot (3062.4 / 60)}{1000/(2 \cdot 0.025 \cdot 0.025)}\right)\left(\frac{0.025}{0.00005}\right)^2 = 0.319$$

Corresponding to $\frac{L}{D}$ = 0.5 and Sommerfeld number 0.319, we obtain a value of $\frac{h_0}{c}$ = 0.4 from Table 3.4.3.2. The minimum film thickness (0.4*radial clearance) is thus determined to be 20 micrometers. The angular location of the minimum film thickness is determined to be 48.14°.

The coefficient of friction variable is $\left(\frac{r_J}{c}\right)f$ = 8.10 from which the coefficient of friction is found to be 0.0162. The friction torque on the journal is T = $f \cdot W \cdot r_J$ = 0.0162 · 1000 · 0.025 = 0.405 Nm. The power loss is estimated as $H = T \cdot (2 \cdot \pi \cdot N)$ = 129.9 W.

The lubricant flow can be determined by finding the value of flow variable from Table 3.4.3.2. The value of this variable $\frac{Q}{r_J cNL}$ is 4.85. From this the total volumetric flow rate is determined as

$$Q = 4.85 \cdot 25 \cdot 0.05 \cdot (3062.4 / 60) \cdot 25 = 7735.8 \, mm^3/s.$$

The flow ratio (ratio of side flow to total flow) is $\frac{Q_s}{Q}$ = 0.730. From this, the side flow can be determined as Q_s = 0.730 · 7735.8 = 5641 mm^3/s.

The maximum pressure ratio is $\dfrac{P}{P_{max}} = 0.365$ from which the maximum pressure is determined to be $P_{max} = \dfrac{0.8}{0.365} = 2.191\ MPa$.

The position of the maximum film pressure is determined to be $17.1°$.
The location of terminating position is determined to be $\theta_{p0} = 64.3°$.

Table 3.4.3.2 Performance characteristics of full journal bearing [Raimondi and Boyd, 1958]

L/D	ε	$h_{n/c}$	θ_A	α/β	s	φ	$\dfrac{R_J}{c}f$	$\dfrac{Q}{R_J cNL}$	$\dfrac{Q_s}{Q}$	$\dfrac{J\gamma c\Delta t}{p}$	$\dfrac{P}{P_{max}}$	$\theta_{P_{max}}$	θ_{P0}
∞	0	1.0	0	------	∞	(70.92)	∞	"	0	∞	------	0	(149.38)
	0.1	0.9	"	.308	.240	69.10	4.80	3.03	"	19.9	.826	0.0	137
	.2	.8	"	.314	.123	67.26	2.57	2.83	"	11.4	.814	5.6	128
	.4	.6	"	.328	.0626	61.94	1.52	2.26	"	8.47	.764	14.4	107
	.6	.4	"	.349	.0389	54.31	1.20	1.56	"	9.73	.667	20.8	86
	.8	.2	"	.383	.0210	42.22	.961	.760	"	15.9	.495	21.5	58.8
	.9	.1	"	.412	.0115	31.62	.756	.411	"	23.1	.358	19	44
	.97	.03	"	------	------	------	------	------	"	------	------	----	------
	1.0	0	"	------	0	0	0	0	"	∞	0	--0	0
1	0	1.0	0	------	∞	(85)	∞	--	0	∞	------	0	(119)
	0.1	0.9	"	.279	1.33	79.5	26.4	3.37	.150	106	.540	3.5	113
	.2	.8	"	.294	.631	74.02	12.8	3.59	.280	52.1	.529	9.2	106
	.4	.6	"	.325	.264	63.10	5.79	3.99	.497	24.3	.484	16.5	91.2
	.6	.4	"	.360	.121	50.58	3.22	4.33	.680	14.2	.415	18.7	72.9
	.8	.2	"	.399	.0446	36.24	1.70	4.62	.842	8.00	.313	18.2	52.3
	.9	.1	"	.426	.0188	26.45	1.05	4.74	.919	5.16	.247	13.9	37.3
	.97	.03	"	.457	.00474	15.47	.514	4.82	.973	2.61	.152	7.1	20.5
	1.0	0	"	------	0	0	0	------	1.0	0	0	0	0
1/2	0	1.0	0	------	∞	(88.5)	∞	--	0	∞	------	0	(107)
	0.1	0.9	"	.273	4.31	81.62	85.6	3.43	.173	343	.523	5.8	99.2
	.2	.8	"	.292	2.03	74.94	40.9	3.72	.318	164	.506	11.9	92.5
	.4	.6	"	.329	.779	61.45	17.0	4.29	.552	68.6	.441	16.9	78.8
	.6	.4	"	.366	.319	48.14	8.10	4.85	.730	33.0	.365	17.1	64.3
	.8	.2	"	.408	.0923	33.31	3.26	5.41	.874	13.4	.267	15.3	44.2
	.9	.1	"	.434	.0313	23.66	1.60	5.69	.939	6.66	.206	11	33.8
	.97	.03	"	.462	.00609	13.75	.610	5.88	.980	2.56	.126	3.8	19.1
	1.0	0	"	-----	0	0	0	------	1.0	0	0	0	0
1/4	0	1.0	0	------	∞	(89.5)	∞	--	0	∞	------	0	(99)
	0.1	0.9	"	.271	16.2	82.31	322	3.45	.180	1287	.515	7.4	98.9
	.2	.8	"	.291	7.57	75.18	153	3.76	.330	611	.489	13.5	85
	.4	.6	"	.331	2.83	60.86	61.1	4.37	.567	245	.415	17.4	70
	.6	.4	"	.370	1.07	46.72	26.7	4.99	.746	107	.334	16.4	55.5
	.8	.2	"	.414	.261	31.04	8.80	5.60	.884	35.4	.240	11.5	39.7
	.9	.1	"	.439	.0736	21.85	3.50	5.91	.945	14.1	.180	8.6	27.8
	.97	.03	"	.466	.0101	12.22	.922	6.12	.984	3.73	.108	4	17.7
	1.0	0	"	------	0	0	0	------	1.0	0	0	0	0

Tables 3.4.3.3, 3.4.3.4 and 3.4.3.5 may be used for a partial arc bearings having extent of 180°, 120° and 60°, respectively.

Table 3.4.3.3 Performance characteristics of 180° arc journal bearing [Raimondi and Boyd, 1958]

L/D	ε	$h_{n/c}$	θ_A	α/β	s	φ	$\frac{R}{C}f$	$\frac{Q}{RCNL}$	$\frac{Q_s}{Q}$	$\frac{J\gamma c \Delta t}{p}$	$\frac{P}{P_{max}}$	$\theta_{P_{max}}$	θ_{P0}
∞	0	1.0	0	.50	∞	90	∞	--	0	∞	-----	0	90
	0.1	0.9	17.00	.50	.347	72.90	3.55	3.04	"	14.7	.778	1.6	90
	.2	.8	28.60	.50	.179	61.32	2.01	2.80	"	8.99	.759	3.4	90
	.4	.6	40.00	.50	.0898	49.99	1.29	2.20	"	7.34	.700	7.7	90
	.6	.4	46.90	.50	.0523	43.15	1.06	1.52	"	8.71	.607	12.3	71.8
	.8	.2	56.70	.50	.0253	33.35	.859	.767	"	14.1	.459	13.7	51.2
	.9	.1	64.20	.501	.0128	25.57	.681	.380	"	22.5	.337	13.3	36.8
	.97	.03	74.65	.50	.00384	15.43	.416	.119	"	44.0	.190	9.1	20.7
	1.0	0	90	.50	0	0	0	0	"	∞	0	0	0
1	0	1.0	0	.500	∞	90	∞	--	0	∞	------	0	90
	0.1	0.9	11.500	.500	1.40	78.50	14.1	3.34	.139	57.0	.525	4.5	90
	.2	.8	21.000	.500	.670	68.93	7.15	3.46	.252	29.7	.513	5.9	90
	.4	.6	34.167	.500	.278	58.86	3.61	3.49	.425	16.5	.466	10.8	80.8
	.6	.4	45.000	.502	.128	44.67	2.28	3.25	.572	12.4	.403	13.4	66.8
	.8	.2	58.000	.498	.0463	32.33	1.39	2.63	.721	10.4	.313	12.9	48.5
	.9	.1	66.000	.499	.0193	24.14	.921	2.14	.818	9.13	.244	11.3	35.2
	.97	.03	75.584	.499	.00483	14.57	.483	1.60	.915	6.96	.157	8.2	20
	1.0	0	90	.500	0	0	0	------	1.0	0	0	0	0
1/2	0	1.0	0	.500	∞	90	∞	-	0	∞	-----	0	90
	0.1	0.9	10.000	.500	4.38	79.97	44.0	3.41	.167	177	.518	5.4	90
	.2	.8	17.800	.500	2.06	72.14	21.6	3.64	.302	87.8	.499	9.1	90
	.4	.6	32.000	.500	.794	58.01	9.96	3.93	.506	42.7	.438	13.7	71.9
	.6	.4	45.000	.500	.321	45.01	5.41	3.93	.665	25.9	.365	14.1	59.1
	.8	.2	59.000	.498	.0921	31.29	2.54	3.56	.806	15.0	.273	12.1	43.9
	.9	.1	67.200	.500	.0314	22.80	1.38	3.17	.886	9.80	.208	10.4	31.5
	.97	.03	76.50	.499	.00625	13.63	.581	2.62	.951	5.30	.132	6.9	18.1
	1.0	0	90	.500	0	0	0	-------	1.0	0	0	0	0
1/4	0	1.0	0	.500	∞	90	∞	-	0	∞	------	0	90
	0.1	0.9	9.000	.498	16.3	81.40	163	3.44	.176	653	.513	6.3	90
	.2	.8	16.300	.500	7.60	73.70	79.4	3.71	.320	320	.489	12.4	80.9
	.4	.6	31.000	.500	2.84	58.99	35.1	4.11	.534	146	.417	15.8	70.5
	.6	.4	45.000	.500	1.08	44.96	17.6	4.25	.698	79.3	.336	14.8	53.6
	.8	.2	59.300	.502	.263	30.43	6.88	4.07	.837	36.5	.241	11.4	39
	.9	.1	68.900	.498	.0736	21.43	2.99	3.72	.905	18.4	.180	9.3	27.3
	.97	.03	77.680	.500	.0104	12.28	.877	3.29	.961	6.46	.110	6.3	17.5
	1.0	0	90	.500	0	0	0	------	1.0	0	0	0	0

Lubrication of Bearings   171

Table 3.4.3.4   Performance characteristics of 120° arc journal bearing, [Raimondi and Boyd, 1958]

L/D	ε	$h_{n/c}$	θ_A	α/β	s	φ	$\frac{R}{C}f$	$\frac{Q}{RCNL}$	Q_s/Q	$\frac{J\gamma c\Delta t}{p}$	$\frac{P}{P_{max}}$	$\theta_{P_{max}}$	θ_{P0}
∞	0	1.0	30	.500	∞	90	∞	–	0	∞	-----	0	60
	0.1	.9007	53.300	.500	.877	66.69	6.02	3.02	"	25.1	.610	0.4	60
	.2	.8	67.400	.500	.431	52.60	3.26	2.75	"	14.9	.599	0.9	60
	.4	.6	81.000	.500	.181	39.02	1.78	2.13	"	10.5	.566	2.4	60
	.6	.4	87.300	.500	.0845	32.67	1.21	1.47	"	10.3	.509	5.1	60
	.8	.2	93.200	.500	.0328	26.80	.853	.759	"	14.1	.405	8.2	44.2
	.9	.1	98.500	.500	.0147	21.51	.653	.388	"	21.2	.311	8.7	32.9
	.97	.03	106.15	.500	.00406	13.86	.399	.118	"	42.4	.199	6.6	19.6
	1.0	0	120	.500	0	0	0	0	"	∞	0	0	0
1	0	1.0	30	.500	∞	90	∞	–	0	∞	-----	0	60
	0.1	.9024	47.500	.500	2.14	72.43	14.5	3.20	.0876	59.5	.427	1.1	60
	.2	.8	62.000	.498	1.01	58.25	7.44	3.11	.157	32.6	.420	1.3	60
	.4	.6	76.000	.500	.385	43.98	3.60	2.75	.272	19.0	.396	3.2	60
	.6	.4	84.500	.499	.162	35.65	2.16	2.24	.384	15.0	.356	6.5	60
	.8	.2	92.600	.500	.0531	27.42	1.27	1.57	.535	13.9	.290	8.6	43.6
	.9	.1	98.667	.500	.0208	21.29	.855	1.11	.657	14.4	.233	8.5	32.5
	.97	.03	106.50	.500	.00498	13.49	.461	.694	.812	14.0	.162	6.3	19.3
	1.0	0	120	.500	0	0	0	-----	1.0	0	0	0	0
1/2	0	1.0	30	.500	∞	90	∞	–	0	∞	-----	0	60
	0.1	.9034	45.000	.500	5.42	74.99	36.6	3.29	.124	149	.431	1.2	60
	.2	.8003	56.650	.500	2.51	63.38	18.1	3.32	.225	77.2	.424	2.4	60
	.4	.6	72.000	.499	.914	48.07	8.20	3.15	.386	40.5	.389	4.8	60
	.6	.4	81.500	.500	.354	38.50	4.43	2.80	.530	27.0	.336	8.1	53.4
	.8	.2	92.000	.500	.0973	28.02	2.17	2.18	.684	19.0	.261	9.0	40.5
	.9	.1	99.000	.500	.0324	21.02	1.24	1.70	.787	15.1	.203	8.2	30.4
	.97	.03	107.00	.500	.00631	13.00	.550	1.19	.899	10.6	.136	6.0	18.2
	1.0	0	120	.500	0	0	0	------	1.0	0	0	0	0
1/4	0	1.0	30	.500	∞	90	∞	π	0	∞	-----	0	60
	0.1	.9044	43.000	.500	18.4	76.97	124	3.34	.143	502	.456	3.0	60
	.2	.8011	54.000	.500	8.45	65.97	60.4	3.44	.260	254	.438	4.8	60
	.4	.6	68.833	.500	3.04	51.23	26.6	3.42	.442	125	.389	8.4	60
	.6	.4	79.600	.500	1.12	40.42	13.5	3.20	.599	75.8	.321	10.4	48.3
	.8	.2	91.560	.500	.268	28.38	5.56	2.67	.753	42.7	.237	9.4	35.8
	.9	.1	99.400	.500	.0743	20.55	2.63	2.21	.846	25.9	.178	7.8	26.9
	.97	.03	108.00	.499	.0105	12.11	.832	1.69	.931	11.6	.112	5.5	17.4
	1.0	0	120	.500	0	0	0	-----	1.0	0	0	0	0

Table 3.4.3.5 Performance characteristics of 60° arc journal bearing, [Raimondi and Boyd, 1958]

L/D	ε	$h_{n/c}$	θ_A	α/β	s	φ	$\dfrac{R}{C}f$	$\dfrac{Q}{RCNL}$	$\dfrac{Q_s}{Q}$	$\dfrac{J\gamma c \Delta t}{p}$	$\dfrac{P}{P_{max}}$	$\theta_{P_{max}}$	θ_{P0}
∞	0	1.0	60	.500	∞	90	∞	–	0	∞	----	0	30
	0.1	.9191	84.00	.502	5.75	65.91	19.7	3.01	"	82.3	.337	0.16	30
	.2	.8109	101.00	.502	2.66	48.91	10.1	2.73	"	46.5	.336	0.18	30
	.4	.6002	118.00	.501	.931	31.96	4.67	2.07	"	28.4	.329	0.25	30
	.6	.4	126.80	.500	.322	23.21	2.40	1.40	"	21.5	.317	0.54	30
	.8	.2	132.60	.500	.0755	17.39	1.10	.722	"	19.2	.287	1.7	30
	.9	.1	135.06	.500	.0241	14.94	.667	.372	"	22.5	.243	3.2	25.5
	.97	.03	139.14	.500	.00495	10.89	.372	.115	"	40.7	.163	4.2	16.9
	1.0	0	150	.500	0	0	0	0	"	∞	0	0	0
1	0	1.0	60	.500	∞	90	∞	--	0	∞	-----	0	30
	0.1	.9212	82.00	.501	8.52	67.92	29.1	3.07	.0267	121	.252	0.3	30
	.2	.8133	99.00	.501	3.92	50.96	14.8	2.82	.0481	67.4	.251	0.3	30
	.4	.6010	166.00	.500	1.34	33.99	6.61	2.22	.0849	39.1	.247	0.54	30
	.6	.4	125.50	.499	.450	24.56	3.29	1.56	.127	28.2	.239	0.95	30
	.8	.2	131.60	.501	.101	18.33	1.42	.883	.200	22.5	.220	2.2	30
	.9	.1	134.67	.500	.0309	15.33	.822	.519	.287	23.2	.192	3.5	25.5
	.97	.03	139.10	.500	.00584	10.88	.422	.226	.469	30.5	.139	4.2	16.9
	1.0	0	150	.500	0	0	0	-----	1.0	0	0	0	0
1/2	0	1.0	60	.500	∞	90	∞	--	0	∞	-----	0	30
	0.1	.9223	81.00	.500	14.2	69.00	48.6	3.11	.0488	201	.239	0	30
	.2	.8152	97.50	.498	6.47	52.60	24.2	2.91	.0883	109	.239	0.03	30
	.4	.6039	113.00	.500	2.14	37.00	10.3	2.38	.160	59.4	.233	0.45	30
	.6	.4	123.00	.500	.695	26.98	4.93	1.74	.236	40.3	.225	1.0	30
	.8	.2	130.40	.500	.149	19.57	2.02	1.05	.350	29.4	.201	2.2	30
	.9	.1	134.09	.500	.0422	15.91	1.08	.664	.464	26.5	.172	3.8	25.4
	.97	.03	139.22	.499	.00704	10.85	.490	.329	.650	27.8	.122	4.2	16.6
	1.0	0	150	.500	0	0	0	-----	1.0	0	0	0	0
1/4	0	1.0	60	.500	∞	90	∞	--	0	∞	-----	0	30
	0.1	.9251	78.50	.499	35.8	71.55	121	3.16	.0666	499	.251	0	30
	.2	.8242	91.50	.500	16.0	58.51	58.7	3.04	.131	260	.249	0.1	30
	.4	.6074	109.00	.500	5.20	41.01	24.5	2.57	.236	136	.242	0.5	30
	.6	.4	119.80	.501	1.65	30.14	11.2	1.98	.346	86.1	.228	1.5	30
	.8	.2	128.30	.500	.333	21.70	4.27	1.30	.496	54.9	.195	3.2	30
	.9	.1	133.10	.500	.0844	16.87	2.01	.894	.620	41.0	.159	4.3	23.7
	.97	.03	139.20	.500	.0110	10.81	.713	.507	.786	29.1	.107	4.1	15.9
	1.0	0	150	.500	0	0	0	-----	1.0	0	0	0	0

The method of using charts given by Raimondi and Boyd is helpful in predicting the bearing performance for a given set of specified operating conditions. By varying the inputs, the change in bearing performance can be determined. However, there is possibility of graphical error in reading the values from the charts. It requires referring to many charts and the process becomes cumbersome.

Further, optimization of journal bearing using Raimondi and Boyd charts is very burdensome. It is worth noting that charts are available only for L/D ratio equal to 1.0, for all other L/D ratios Tables 3.4.3.2–3.4.3.5 and corresponding interpolation methods are required. The next section explains an improved method for the performance analysis of journal bearing.

3.4.3.3 Improved method of journal bearing

An improved and rapid method for bearing performance evaluation was proposed by Hirani et al. [Hirani et al., 1997]. They presented a design procedure in tabular form for prediction of load capacity, attitude angle, power loss, flow and maximum temperature. Their method is based on hybridization of pressure profiles (Eq. 3.4.22) predicted by short and long bearing approximation shown in Figs. 3.4.19 and 3.4.20. As indicated in Fig. 3.4.19, pressure gradient in axial (Z-direction) is zero, which is only justifiable if L/D ratio is very high. In practice such long bearings do not exist and there is a need to modify the pressure distribution estimated by long bearing approximation. Similarly there is a need to modify pressure distribution estimated by short bearing approximation.

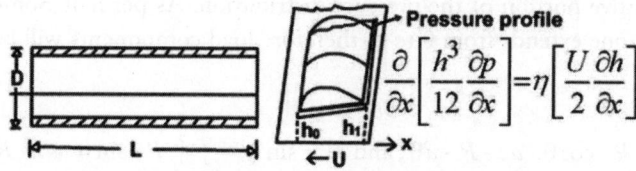

Fig. 3.4.19 Long static bearing

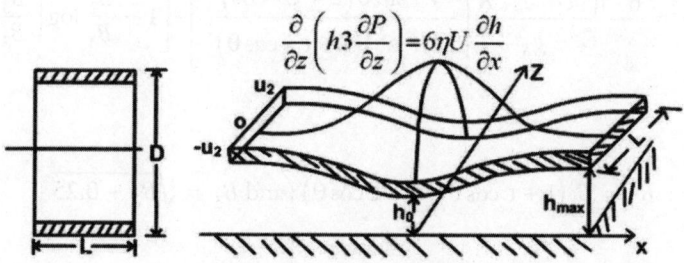

Fig. 3.4.20 Short static bearing

$$\frac{1}{p} = \frac{g_0}{p_0} + \frac{g_S}{p_\infty} \qquad (3.4.22)$$

where p_∞ is fluid pressure estimated by long bearing approximation (Eq. 3.3.35) and p_0 is fluid pressure estimated by short bearing approximation (Eq. 3.3.47). Here $g_0 \left(= f(\varepsilon, L/D)\right)$ and $g_S \left(= f(\varepsilon)\right)$ are correction factors given by $g_0 = 1 + \varepsilon \left(\frac{L}{D}\right)^{1.2} \left[e^{\varepsilon^5} - 1\right]$ and $g_S = e^{(1-\varepsilon)^3}$. Substituting these expressions in Eq. 3.4.22, the pressure profile is given by:

$$p = \frac{12 \cdot \eta \cdot U \cdot R}{C^2 \cdot g_0}\left(\frac{L}{D}\right)^2 \frac{\left\{\frac{1}{4}-\left(\frac{Z}{L}\right)^2\right\}\frac{\varepsilon \cdot \sin\theta}{(1+\varepsilon\sin\theta)^3}}{1+\frac{2 \cdot g_s\left(\frac{L}{D}\right)^2(2+\varepsilon^2)}{g_0(1+\varepsilon\cos\theta)(2+\varepsilon\cos\theta)}\left\{\frac{1}{4}-\left(\frac{Z}{L}\right)^2\right\}} \quad (3.4.23)$$

The total load supported by bearing is obtained by integrating the pressure components normal and parallel to the line of centres over the whole film, as expressed in following two equations,

$$W \cdot \cos\varphi = -\int_A p \cdot \cos\theta \cdot dA \quad (3.4.24)$$

$$W \cdot \sin\varphi = \int_A p \cdot \sin\theta \cdot dA \quad (3.4.25)$$

For complete films, this integration is to be carried out over the entire bearing. For bearings with ruptured films (Fig. 3.4.3), occurring under negative pressure, the integration is to be carried out only over the positive portion of the pressure distribution. As per half-Sommerfeld condition, the positive pressure zone extends from 0 to π, therefore load components will be calculated by the following equation:

$$W \cdot \cos\varphi = -\int_0^\pi \int_{-\frac{L}{2}}^{\frac{L}{2}} P \cdot \cos\theta \cdot dz \cdot R \cdot d\theta, \text{ and } W \cdot \sin\varphi = \int_0^\pi \int_{-\frac{L}{2}}^{\frac{L}{2}} P \cdot \sin\theta \cdot dz \cdot R \cdot d\theta \quad (3.4.26)$$

The load component along the z direction can be simply carried out analytically and is given as

$$I_z = \int_{-\frac{L}{2}}^{\frac{L}{2}} P \cdot dz = \frac{6 \cdot \eta \cdot U \cdot L \cdot R}{C^2 \cdot g_s}\left[\frac{\varepsilon \cdot \sin\theta(2+\varepsilon \cdot \cos)}{(2+\varepsilon^2)(1+\varepsilon \cdot \cos\theta)^2}\right] \cdot \left[1-\frac{B_2}{B_3}\log\left(\frac{B_3+0.5}{B_3-0.5}\right)\right]$$

$$(3.4.27)$$

where

$$B_1 = g_0\left(\frac{D}{L}\right)^2; \quad B_2 = B_1(1+\varepsilon\cos\theta)(2+\varepsilon\cos\theta); \text{ and } B_3 = \sqrt{B_2+0.25}$$

Therefore,

$$W \cdot \cos\varphi = -\int_0^\pi I_z R\cos\theta \cdot d\theta \text{ and } W \cdot \sin\varphi = \int_0^\pi I_z R\sin\theta \cdot d\theta \quad (3.4.28)$$

Integrating Eq. 3.4.28 in closed form is not possible; therefore numerical integration shall be used. One of such integration method is Weddle's rule. The Weddle's formula is given by:

$$\int_a^b y \cdot dx = \frac{b-a}{20}\left[y_1 + 5y_2 + y_3 + 6y_4 + y_5 + 5y_6 + y_7\right] \quad (3.4.29)$$

where, $y_1,\ldots y_7$ are seven equidistant ordinates of mutual distance $(b-a)/6$. The accuracy of the above expression for journal bearing depends on the eccentricity ratio. For journal bearings, intervals of 30° can be chosen if eccentricity ratio is lesser than 0.8. However at high eccentricity ratios, due to peak of oil pressure shifting towards $\theta = \pi$, the interval of 30° gives erroneous results. Therefore to minimize the error in evaluating the integrals at high eccentricity ratios above expression must be combined with expression given below:

$$\int_0^\pi f(x) \cdot dx = \int_0^{\frac{\pi}{2}} f(x) \cdot dx + \int_{\frac{\pi}{2}}^{\frac{3\pi}{4}} f(x) \cdot dx + \int_{\frac{3\pi}{4}}^{\pi} f(x) \cdot dx \quad (3.4.30)$$

The accuracy of this method is comparable [Hirani et al., 1997] with many time consuming techniques such as thermo–hydrodynamic (THD). To simplify the complete design procedure, Hirani et al. [Hirani et al., 1997] presented all performance parameters (location of maximum pressure, oil leakage, attitude angle, load capacity, friction force, temperature rise, maximum temperature and maximum pressure) in Table 3.4.3.6.

Using Table 3.4.3.6 the supply and re-circulating flow along with the location of maximum pressure can be evaluated step by step using expressions 1–7. Integrations in expressions 10–11 can be determined by using Weddle's formula (Eq. 3.4.29). Since the effective temperature is unknown at the beginning; therefore it can be evaluated by iterating steps 16–17 along with a viscosity–temperature relation given by the following expression $\eta_{eff} = \eta_{in} \cdot e^{-\beta(T_{eff} - T_m)}$ until viscosity converges, however, other viscosity temperature relations, such as Vogel's equation (Eq. 3.1.2.2):

$$\eta = \eta_{ref} \, e^{\frac{b}{T+\theta}} \quad (3.1.2.2)$$

or Walther's equation (Eq. 3.1.2.3):

$$\log \log \left[\frac{\eta}{\rho} + 0.6 \right] = C1 - C2(\log T) \quad (3.1.2.3)$$

may also be used.

In expressions 13–15 load capacity, friction force and temperature rise are evaluated without the inclusion of viscosity, to reduce computational effort, and assigned to symbols W_η, F_η, and temperature rise Δt_η, respectively. Once effective temperature and corresponding effective viscosity are determined, load capacity, friction force, the maximum temperature and maximum pressure can be evaluated using steps 18–21. By using the proposed design table, with the exception of supply flow, which is marginally over estimated, the prediction of other parameters has been found to be very good.

Table 3.4.3.6 Design table for journal bearing

	Design expressions
1.	$\Lambda = \frac{L}{D}; \quad g_s = e^{(1-\varepsilon)3}; \quad g_0 = 1 + \varepsilon \cdot \Lambda^{1.2}\left[e^{\varepsilon^5} - 1\right]$
2.	$B_1 = \frac{g_0}{2 \cdot g_s \cdot \Lambda^2 \cdot (2+\varepsilon^2)}; \quad B_{21} = B_1 \cdot (1+\varepsilon)(2+\varepsilon); \quad B_{22} = B_1 \cdot (1-\varepsilon)(2-\varepsilon)$
3.	$B_{3j} = \sqrt{B_{2j} + 0.25}; \quad B_{4j} = \frac{B_{2j}}{B_{3j}} \log\left(\frac{B_{3j} + 0.5}{B_{3j} - 0.5}\right) : j = 1,2$

Contd.

Contd.

4.	$\theta_{0\max} = \cos^{-1}\left(\dfrac{1-\sqrt{1+24\cdot\varepsilon^2}}{4\cdot\varepsilon}\right);\quad \theta_{S\max} = \cos^{-1}\left(\dfrac{-3\cdot\varepsilon}{2+\varepsilon^2}\right)$		
5.	$\Theta = \dfrac{\Lambda^2 \cdot g_s}{3\cdot g_0} \dfrac{(12+\varepsilon^2)^4 \left(12\cdot\varepsilon^2 + (2+\varepsilon^2)\left(1-\sqrt{1+24\cdot\varepsilon^2}\right)\right)}{\varepsilon^2\cdot(1-\varepsilon^2)^2(2-\varepsilon^2)\left(9-\sqrt{1+24\cdot\varepsilon^2}\right)^2}$		
6.	$\theta_{p\max} = \theta_{S\max} + (\theta_{0\max} - \theta_{S\max})\left(\dfrac{1}{1-\Theta}\right)$		
7.	$Q_{leakage} = Q_p + UCL\varepsilon\left[1 - \dfrac{1}{g_0}\Lambda^2\left\{B_{21}(1-B_{41}) + B_{22}(1-B_{42})\right\}\right]$ $Q_{rec} = \dfrac{UCL(1-\varepsilon)}{2} + \dfrac{UCL\Lambda^2}{g_0}\left[B_{22}(1-B_{42})\right]$		
8.	$H = 1 + \varepsilon\cdot\cos\theta;\quad B_{23} = B_1\cdot H(1+H);\quad B_{33} = \sqrt{B_{23}+0.25};\quad B_{43} = \dfrac{B_{23}}{B_{33}}\log\left(\dfrac{B_{33}+0.5}{B_{33}-0.5}\right)$		
9.	$I_{z\eta} = \dfrac{6ULR}{C^2 g_s}\left[\dfrac{\varepsilon\cdot\sin\theta(1+H)}{(2+\varepsilon^2)H^2}\right]((1-B_{43}))$	10.	$W_{\varepsilon\eta} = -\int_0^\pi I_{z\eta}\cdot R\cdot\cos\theta\cdot d\theta$
11.	$W_{\varphi\eta} = \int_0^\pi I_{z\eta}\cdot R\cdot\sin\theta\cdot d\theta$	12.	$\varphi = \tan^{-1}\left(\dfrac{W_{\varphi\eta}}{W_{\varepsilon\eta}}\right)$
13.	$W_\eta = \sqrt{W_{\varphi\eta}^2 + W_{\varepsilon\eta}^2}$	14.	$F_\eta = \dfrac{ULR\pi}{c\sqrt{1-\varepsilon^2}}\left(\dfrac{2+\varepsilon}{1+\varepsilon}\right) + \dfrac{C\varepsilon W_{\varphi\eta}}{2R}$
15.	$\Delta_{t\eta} = \dfrac{\varepsilon F_\eta U}{\rho_0 C_0 Q_{leakage}}$	16.	$\Delta_t = \eta\Delta t_\eta$
17.	$T_{eff} = T_{in} + \left[2 + \dfrac{Q_{rec}}{Q_{leakage}}\right]\Delta_t$	18.	$W = \eta W_\eta;\quad F = \eta F_\eta$
19.	$H_{p\max} = 1 + \varepsilon\cos\theta_{p\max};\quad B_{21\max} = 4\cdot B_1 H_{p\max}(1+H_{p\max})$		
20.	$P_{\max} = \dfrac{3\cdot\eta UR\Lambda^2}{C^2 g_0}\dfrac{\varepsilon\sin\theta_{p\max}}{H_{p\max}^3}\dfrac{B_{21\max}}{1+B_{21\max}}$	21.	$T_{\max} = T_{eff} + \dfrac{Q_{leakage}}{Q_{rec}}\Delta_t$

To understand the use of the Table 3.4.3.6, example 3 is considered. The following specifications of the journal bearing are used: Journal radius (R_J) = 25mm, bearing length (L) =50 mm, rotational speed of journal = 1440 rpm, load (W) = 1000 N, lubricant oil SAE40, and radial clearance (c) =

0.00005 m. The supply and re-circulating flow and location of maximum pressure are determined using expressions 1–7. The Weddle's formula (Eq. 3.4.29) is used to evaluate expressions 10–11. The load capacity, friction force and maximum pressure are determined using expressions 18–21.

The results obtained by the improved method of Hirani et al. [Hirani et al., 1997] are compared with the Raimondi and Boyd method [Raimondi and Boyd, 1958] and tabulated in Table 3.4.3.7 for comparison. As is observed from Table 3.4.3.7, the parameters obtained by the two methods are found to be in close agreement with each other, with the exception of supply flow, which is marginally over estimated using the method of Hirani et al. [Hirani et al., 1997].

Similarly, the results obtained by the improved method of Hirani et al. [Hirani et al., 1997], for example 4, are compared with the Raimondi and Boyd method [Raimondi and Boyd, 1958] and tabulated in Table 3.4.3.8 for comparison.

As is observed from Tables 3.4.3.7 and 3.4.3.8, the parameters obtained by the two methods are found to be in close agreement with each other. One of the main advantages of using the improved method of Hirani et al. [Hirani et al., 1997] is the accurate determination of the bearing performance without referring to charts and/or tables. This improved method may be programmed using any programming language, and this will automate the evaluation of journal bearing performance. The listing of the complete program in MATLAB is given at the end of this chapter.

Table 3.4.3.7 Comparison of results for example 3

Parameter	Raimondi & Boyd method (1958)	Hirani et al. method (1997)
h_{min} (μm)	35	30.22
Location of h_{min}	65°	65.94°
Coefficient of friction	0.0125	0.0110
Power loss (w)	47.12	41.72
Side leakage (mm³/s)	2778	2955
P_{max} (MPa)	0.833	0.7925
Temperature Rise (°C)	3.96	3.4975

Table 3.4.3.8 Comparison of results for example 4

Parameter	Raimondi and Boyd method using table (1958)	Hirani et al. method (1997)
h_{min} (μm)	20	19.77
Location of h_{min}	48.14°	48.49
Coefficient of friction	0.0162	0.0128
Power loss (w)	129	102
Side leakage (mm³/s)	5641	5652
P_{max} (MPa)	2.191	2.146

Frequently Asked Questions

Q.1. What is the difference between dynamic viscosity and kinematic viscosity?

Ans. Dynamic viscosity is the ratio of the shear stress to the shear rate of flowing fluid, whereas kinematic viscosity is equal to the dynamic viscosity divided by density. The usage of two terms is based on the measurement of the viscosity. It is easy to measure kinematic viscosity even with low cost equipment, but the measurement of dynamic viscosity requires good quality rheometer.

The unit of dynamic viscosity in SI units is pascal and in CGS units it is centipoises. The SI unit of kinematic viscosity is square meter per second, but CGS unit centistokes is more widely accepted.

Q.2. How do we measure dynamic and kinematic viscosities?

Ans. Kinematic viscosity is generally measured by counting the time it takes for the liquid to travel through the orifice of a capillary under the force of gravity. Different sizes of capillary tubes are available for fluids of varying viscosity. The time taken by the fluid to flow through the capillary tube can be converted directly to a kinematic viscosity using a simple calibration constant which is unique for each capillary tube. One of the commonly referred standards for performing kinematic viscosity measurements is ASTM D445.

To measure dynamic (absolute) viscosity, rheometer is required, in which rotational movement is provided to fluid and resisting torque is measured. Commonly referenced standards for testing absolute viscosity are ASTM D2983, D6080.

Q.3. What is the difference between shear thickening and shear thinning fluids?

Ans. A fluid that behaves according to Newton's law, whose viscosity is independent of the stress, is said to be Newtonian. Gases, water and many common liquids can be considered to be Newtonian in ordinary conditions. There are fluids which do not follow Newtonian Law and are termed as non-Newtonian fluids. The shear thickening and shear thinning fluids belong to non-Newtonian fluids. The viscosity increases with the rate of shear rate for shear thickening liquids, while for shear thinning liquids, viscosity decreases with the rate of shear rate.

Q.4. Which equation, relating viscosity with operating temperature, is the most widely used?

Ans. In literature there are several viscosity temperature relations available such as Reynolds, Walther, Slotte, Vogel, etc. One of the commonly used viscosity–temperature equation is Walther's equation which has been approved by ASTM.

Q.5. What can a tribologist do to counter the effects of variation of viscosity of oil/lubricant with respect to variation in temperature?

Ans. For better evaluation of the relationship between viscosity and temperature, a comparator 'viscosity index' is used. High viscosity index indicates less sensitivity of viscosity toward temperature. As viscosity is a very influencing parameter in designing fluid film lubricated tribo–pair, it is important to keep this parameter within certain limits. Use of multi-grade oil is a way to reduce sensitivity of viscosity-temperature. Therefore the most oils on shelf today are multi-grade oils, such as 10W30 or 20W50.

Q.6. What are idealized bearings?

Ans. Idealized bearings are those bearings which have very smooth surface finish and dimension in one of the direction is negligible (i.e., infinitely long bearing, infinitely small bearing). In idealized bearings the flow of lubricant is laminar and the lubricant follows Newtonian behaviour.

Q.7. What is the practical significance of pivoted slider bearing?

Ans. In a plane slider bearing maximum load can be obtained for a definite angle of inclination when dimensions of the bearing, viscosity of lubricant, minimum film thickness are held constant. Also for a given velocity, viscosity and minimum film thickness there is a definite value of angle of inclination which gives minimum friction. Hence a bearing with a fixed inclination will give satisfactory performance on a narrow range of operating conditions. If the fixed (zero tangential velocity) surface of the bearing is pivoted, bearing can have variable angle of inclination and hence a wide range of operating conditions. This is the main practical significance of pivoted slider bearing.

Q.8. What is the purpose of Sommerfeld substitutions while determining pressure distribution in an infinitely long journal bearing?

Ans. It is difficult to integrate and to evaluate the integrals required for determining the pressure distribution in infinitely long journal bearing. Sommerfeld substitutions are well known and provide easy solution to find the close form expressions for pressure distributions. Hence in order to find the pressure solution Sommerfeld substitutions are used.

Q.9. Why does cavitation occur in the divergent zone of the operating journal bearing?

Ans. The eccentric position of journal in a hydrodynamic bearing forms a converging–diverging clearance space. Due to unavailability of sufficient liquid lubricant, liquid–streamers separated by gas/vapour space are formed in the divergent clearance space. Hence the phenomenon of cavitation occurs.

Q.10. What is the effect of cavitation on the load bearing capacity of the journal bearing?

Ans. The cavitated pressure or the pressure developed in the cavitated zone does not contribute to the load bearing capacity of the journal bearing. However to find the fluid pressure and its extent, following universal equation (Eq. 3.4.4) that describes both the full film and cavitation regions can be used.

Q.11. Describe the significance of eccentricity ratio ε with respect to journal bearing.

Ans. If $\varepsilon = 0$ it means there is zero load on the bearing and the journal does not have any weight. In this situation journal centre coincides with bearing centre. If $\varepsilon = 1$ it means there is a very high load on the bearing. If $0 < \varepsilon < 1$, it means the journal bearing was designed to account appropriate speed and load.

Q.12. What effect does the length of the bearing and the radial clearance have on the overall design of the bearing?

Ans. Increasing bearing length increases load capacity of the bearing. Upper limit on bearing length is decided based on space limitation. Increasing radial clearance decreases load capacity and increases lubricant flow rate.

Q.13. What is the significance of Raimondi and Boyd method for designing journal bearing?

Ans. The method is useful in obtaining quick determination of the bearing performance but its accuracy is limited and needs many number of charts containing the dimensionless parameters which are time intensive.

Q.14. Why do we use Petroff's equation and what are the assumptions on which the Petroff's equation is based on?

Ans. Petroff's equation is used to determine the coefficient of friction in journal bearings. Petroff's equation finds its importance because it helps in defining a group of dimensionless parameters $\left(\dfrac{R_J}{C}\right)$ and $\left(\dfrac{\eta N}{p}\right)$ that govern the frictional properties of bearing. Petroff's equation can be used as a first approximation of the friction loss in a journal bearing. The equation would provide exact and precise results for $\varepsilon = 0$ and reasonably good results for values of ε up to 0.5. Following assumptions form the basis of Petroff's equation:

(i) The journal is concentric with the bearing.
(ii) The bearing is subjected to light load.

Q.15. What are the important parameters for designing a journal bearing?

Ans. In the initial design phase of journal bearing, following parameters play a crucial role:
- Unit bearing pressure
- Temperature rise
- Length to diameter ratio
- Radial clearance
- Minimum oil-film thickness

Q.16. Describe the reasons for selecting an L/D ratio of unity.

Ans. The load capacity of a journal bearing is dependent on the L/D ratio. For a given diameter of a journal a longer bearing will have more load carrying capacity but may result in edge loading due to shaft misalignment. But a shorter bearing causes more oil leakage from sides resulting in excessive pressure losses and reducing the load carrying capacity.

Multiple Choice Questions

Q.1. Increase in radial clearance of a journal bearing would:
(a) Increase the load bearing capacity. (b) Decrease the load bearing capacity.
(c) Decrease the supply pressure. (d) Increase the supply pressure.

Q.2. Sommerfeld number S is defined as:
(a) $S = \dfrac{\eta N}{P}\left(\dfrac{R}{C}\right)$ (b) $S = \dfrac{\eta N}{P}\left(\dfrac{C}{R}\right)^2$
(c) $S = \dfrac{\eta N}{P}\left(\dfrac{R}{C}\right)^2$ (d) $S = \dfrac{\eta N}{P}\left(\dfrac{C}{R}\right)$

Q.3. Findings of Tower's experiment are:
(a) Detection of fluid film pressure in journal bearing.
(b) Reduction in coefficient of friction in presence of lubricant.
(c) Increase in frictional resistance on increase in sliding speed.
(d) All of the above.

Q.4. Radial clearance in a journal bearing is defined as follows:
(a) $R_B - R_J$ (b) $R_B + R_J$
(c) $(R_B - R_J)^2$ (d) $(R_B + R_J)^2$

Q.5. The SI unit of kinematic viscosity is
(a) m³/sec (b) sec⁻¹
(c) m² (d) m²/sec

Q.6. Which of the following statement is true?
  (a) The kinematic viscosity is equal to the dynamic viscosity divided by density.
  (b) The dynamic viscosity is equal to the kinematic viscosity divided by density.
  (c) The kinematic viscosity is equal to the dynamic viscosity multiplied by density.
  (d) The dynamic viscosity is equal to the kinematic viscosity multiplied by density.

Q.7. The CGS units of dynamic viscosity and kinematic viscosity are respectively:
  (a) Centistokes and centipoise.
  (b) Centipoise and centistokes.
  (c) Pascal and centistokes
  (d) Centistokes and Pascal

Q.8. The similarity between Bingham fluid and Newtonian fluid is:
  (a) Both have linear shear stress and shear strain relationship.
  (b) The relationship between shear stress and shear strain is not linear.
  (c) The relationship between shear stress and shear strain varies exponentially.
  (d) The relationship between shear stress and shear strain varies logarithmically.

Q.9. Reynolds' equation for hydrodynamic lubrication, assuming journal rotation in X-direction and film thickness in Y-direction, is expressed as:

  (a) $\dfrac{\partial}{\partial x}\left(h^3 \dfrac{\partial p}{\partial x}\right) + \dfrac{\partial}{\partial z}\left(h^3 \dfrac{\partial p}{\partial z}\right) = 6U\eta^3 \dfrac{\partial h}{\partial x}$

  (b) $\dfrac{\partial}{\partial x}\left(h^3 \dfrac{\partial p}{\partial x}\right) + \dfrac{\partial}{\partial z}\left(h^3 \dfrac{\partial p}{\partial z}\right) = 6U\eta \dfrac{\partial h}{\partial x}$

  (c) $\dfrac{\partial}{\partial x}\left(h^3 \dfrac{\partial p}{\partial y}\right) + \dfrac{\partial}{\partial y}\left(h^3 \dfrac{\partial p}{\partial y}\right) = 0$

  (d) $\dfrac{\partial}{\partial x}\left(h^3 \dfrac{\partial p}{\partial x}\right) + \dfrac{\partial}{\partial y}\left(h^3 \dfrac{\partial p}{\partial y}\right) = 6U\eta \dfrac{\partial h}{\partial x}$

Q.10. Which of the following are the limitations of Reynolds' equation?
  (a) Inertia forces are not considered in Reynolds' equation.
  (b) The viscosity of the lubricant is considered constant.
  (c) Variation of pressure across the film thickness is assumed to be negligible.
  (d) All of the above.

Q.11. In a journal bearing the eccentricity is defined as
  (a) The ratio of the radii of the journal and the bearing.
  (b) The square of the ratio of the radii of the journal and the bearing.
  (c) The distance between the centres of the journal and the bearing.
  (d) The ratio of the radii of the journal to difference between radius of journal and bearing.

Q.12. In a journal bearing, the oil film thickness is represented by the relation.
  (a) $h = c + e^2 \cos\theta$
  (b) $h = e \cdot \cos\theta$
  (c) $h = c^2 + e \cdot \cos\theta$
  (d) $h = c + e \cdot \cos\theta$

Q.13. In a journal bearing where 'c' is the clearance and 'e' the eccentricity, the eccentricity ratio 'ε' is defined as
  (a) $\varepsilon = \left(\dfrac{e}{c}\right)^2$
  (b) $\varepsilon = e/c$
  (c) $\varepsilon = \left(\dfrac{e}{c}\right)^{1/2}$
  (d) $\varepsilon = \left(\dfrac{c}{e}\right)^2$

Q.14. Cavitation phenomenon in a hydrodynamic journal bearing generally happens in the
  (a) Convergent zone.
  (b) Divergent zone.
  (c) Both (a) and (b).
  (d) Cavitation does not happen in the journal bearing.

Q.15. In a journal bearing generally how much portion of the bearing supports the load?
  (a) $1/4^{th}$ of the bearing.
  (b) ½ portion of the bearing.
  (c) $3/4^{th}$ portion of the bearing.
  (d) 100% of the bearing supports the load.

Q.16. In a hydrodynamic journal bearing does the cavitated pressure contribute to the load bearing capacity?
  (a) Yes
  (b) No
  (c) To a certain extend.
  (d) Only 50% of the cavitated pressure contributes to the load bearing capacity.

Q.17. For a journal bearing operating in hydrodynamic lubrication regime, the eccentricity depends on:
  (a) Load
  (b) Operating speed
  (c) Viscosity
  (d) All of the above.

Q.18. In a hydrodynamic journal bearing if the centre of the journal and the bearing coincides it means
  (a) $\varepsilon < 1$
  (b) $\varepsilon > 1$
  (c) $\varepsilon = 0$
  (d) $0 < \varepsilon < 1$

Q.19. In a hydrodynamic journal bearing if $\varepsilon = 1$, it means
  (a) Journal is subjected to very high load and the operating speed is negligible.
  (b) Journal is subjected to very light load and the operating speed is very high.
  (c) Journal is subjected to no load and the operating speed is very high.
  (d) Journal is subjected to no load and the operating speed is moderate.

Q.20. With respect to design of journal bearing, increasing the length of the bearing would
  (a) Increase linearly the load bearing capacity of the bearing.
  (b) Decrease linearly the load bearing capacity of the bearing.
  (c) Increase quadratically the load bearing capacity of the bearing.
  (d) Decrease quadratically the load bearing capacity of the bearing.

Q.21. Increase in radial clearance of the bearing would
  (a) Decrease the load bearing capacity.
  (b) Increase the lubricant flow rate.
  (c) Both (a) and (b)
  (d) None of these.

Q.22. Raimondi and Boyd method
  (a) is a finite element method to evaluate bearing performance.
  (b) utilizes the charts to evaluate the bearing performance.
  (c) considers localized lubricant temperature.
  (d) considers effect of pressure on viscosity.

Q.23. The Petroff's equation is used to determine
  (a) Operating temperature in journal bearing.
  (b) Viscosity of lubricant in journal bearing.
  (c) Minimum film thickness in journal bearing.
  (d) Coefficient of friction in journal bearing.

Q.24. The Petroff's equation is derived using the assumption that the:
 (a) Journal is concentric with the bearing.
 (b) Bearing is subjected to light load.
 (c) Both (a) and (b).
 (d) None of these.

Q.25. The two dimensionless parameter used in Petroff's equation are
(where R_J = radius of journal in mm; C = radial clearance in mm; η = absolute or dynamic viscosity in N-s/mm²; N = journal speed in rev/sec; p = unit bearing pressure)
 (a) $\left(\dfrac{R_J}{C}\right)^2$ and $\left(\dfrac{\eta N}{p}\right)$
 (b) $\left(\dfrac{R_J}{C}\right)$ and $\left(\dfrac{\eta N}{p}\right)$
 (c) $\left(\dfrac{R_J}{C}\right)$ and $\left(\dfrac{\eta N}{p}\right)^2$
 (d) $\left(\dfrac{R_J}{C}\right)^3$ and $\left(\dfrac{\eta N}{p}\right)$

Q.26. Which of the following statements is true?
 (a) While designing journal bearing low ratio of (C/R_J) are used for small bearings and high ratio of (C/R_J) are used for large bearings.
 (b) Decreasing the eccentricity ratio increases the load carrying capacity of the journal bearing
 (c) Radial clearance should be small to provide the necessary velocity gradient.
 (d) An increase in radial clearance increases the load bearing capacity of the bearing.

Q.27. Some of the important parameters of journal bearing design is/are:
 (a) Length to diameter ratio
 (b) Temperature rise
 (c) Radial clearance
 (d) All of the above

Q.28. For a journal bearing where excessive heat is generated which is more preferred?
 (a) A journal bearing with a short length.
 (b) A journal bearing with a long length.
 (c) Anyone between (a) and (b) can be used.
 (d) None of these.

Answers

Q.1. (b)	Q.2. (c)	Q.3. (d)	Q.4. (a)	Q.5. (d)	Q.6. (a)
Q.7. (b)	Q.8. (a)	Q.9. (b)	Q.10. (d)	Q.11. (c)	Q.12. (d)
Q.13. (b)	Q.14. (b)	Q.15. (b)	Q.16. (b)	Q.17. (d)	Q.18. (c)
Q.19. (a)	Q.20. (a)	Q.21. (c)	Q.22. (b)	Q.23. (d)	Q.24. (c)
Q.25. (b)	Q.26. (a)	Q.27. (d)	Q.28. (a)		

References

ASTM standard (2010). ASTM D2270-10 Standard practice for calculating Viscosity Index from Kinematic Viscosity at 40° and 100°C.

Booker, J F., 1965, A Table of the Journal Bearing Integral, Transactions of ASME, Journal of Basic Engineering, Vol. 87, pp 533-535.

Elrod, H. G. 1981. A Cavitation Algorithm. *Journal of Lubrication Technology*. *103*(3): 350.

Gopinath K., Mayuram M. M. 2014. NPTEL Web course on Machine Design, Module 5 Sliding Contact Bearings

Hamrock, B. J. 1994. *Fundamentals of Fluid Film Lubrication*. McGraw Hill.

Hirani, H., Athre, K., and Biswas, S. 2001. A Simplified Mass Conserving Algorithm for Journal Bearing under Large Dynamic Loads. *International Journal of Rotating Machinery*. *7*(1): 41–51.

Hirani H., Rao T. V. V. L. N., Athre K., Biswas S. 1997. Rapid Performance Evaluation of Journal Bearings. *Tribology International*. *30*(11): 825–834.

Martin, F. A. 1998. Oil flow in plain steadily loaded journal bearings: realistic predictions using rapid techniques. *Proceedings of the Institution of Mechanical Engineers, Part J: Journal of Engineering Tribology*. *212*(6): 413–425.

Martin, F. A., and Lee, C. S. 1983. Feed-Pressure Flow in Plain Journal Bearings. *A S L E Transactions*. *26*(3): 381–392.

Raimondi, A. A., and Boyd, J. 1958. A Solution for the Finite Journal Bearing and its Application to Analysis and Design : III, *1*(1): 194–209.

Roelands, C. J. A. 1966. *Correlational Aspects of the Viscosity · Temperature · Presure Relationship of Lubricating Oils*. Delft University of Technology, Netherlands.

Stachowiak, G. W., and Batchelor, A. W. 2006. *Engineering Tribology* (First edit.). UK: Butterworth Heinemann.

Wang, J., Hashimoto, T., Nishikawa, H., and Kaneta, M. 2005. Pure rolling elastohydrodynamic lubrication of short stroke reciprocating motion. *Tribology International*. *38*: 1013–1021.

Program Listing in MATLAB for Problem 3

```
%For the rapid evaluation of parameters of journal bearing without
    temperature effect, following are variables:
% L=Length of bearing (m)
% lam = (L/D)- Bearing Length to Bearing Diameter (D) Ratio,
% rc = (R/C)- Bearing Radius (R) to radial clearance Ratio,
% w = (2*pi*N/60)- angular velocity (N is the rotational speed) (rad)
% U = R*w- linear velocity (m/sec)
% eta= Viscosity of lubricant (Pas)
% ecc= eccentricity ratio
% n = Number of iterations
% Fapp = Applied load (N)
% roc = row * C0
% C0 = specific heat capacity of lubricant (J/kg.°C)
% row = density of lubricant, (kg/m^3)
% hmin= Minimum film thickness
% fi=Location of attitude angle
%delT= Temperature Rise
%Qleak=Side Leakage
%F=Frictional Force
% COF=Coefficient of friction
%T=Torque
%Ploss=Power Loss

% variables to be given as Inputs:
L=0.05;
```

```
R=0.025;
C=0.00005;
row=887 ;
C0=1800;
eta=0.020;
N=1440;
n=1000;
roc=row*C0;
D=2*R;
lam=L/D;
w=2*pi*N/60;
U=R*w;
n=100;
eccc(2)=0.1;
Fapp=1000;
W(1)=0;

for i=1:n

ecc=eccc(i+1);

gs = exp((1-ecc)^3);
g0 = 1+ecc*(lam.^1.2)*(exp(ecc^5)-1);
const1 = (6*U*L*R*eta*ecc)/((C^2*gs)*(2+ecc^2));
B1 = g0/(2*gs*(lam.^2)*(2+ecc^2));
for I = 2:24
    theta = 0 + (I-1)*pi/24;
    H = 1+ecc*cos(theta);
    const2 = sin(theta)*(1+H)/(H^2);
    B23 = B1*H*(1+H);
    B33 = sqrt(B23+0.25);
    B43 = (B23/B33)*log((B33+0.5)/(B33-0.5));
    IZN=const1*const2*(1-B43);
    IC(I)=IZN*R*cos(theta);
    IS(I)=IZN*R*sin(theta);
end
WC=-pi/80*(5*IC(2)+IC(3)+6*IC(4)+IC(5)+5*IC(6)+2*IC(7)+5*IC(8)+IC(9)+6*IC(1
    0)+IC(11)+5*IC(12)+2*IC(13)+5*IC(14)+IC(15)+6*IC(16)+IC(17)+5*IC(18)+2*
    IC(19)+5*IC(20)+IC(21)+6*IC(22)+IC(23)+5*IC(24));
WS=pi/80*(5*IS(2)+IS(3)+6*IS(4)+IS(5)+5*IS(6)+2*IS(7)+5*IS(8)+IS(9)+6*IS(10)+
    IS(11)+5*IS(12)+2*IS(13)+5*IS(14)+IS(15)+6*IS(16)+IS(17)+5*IS(18)+2*IS(19-
    )+5*IS(20)+IS(21)+6*IS(22)+IS(23)+5*IS(24));
W(i+1)= sqrt(WC*WC+WS*WS);
fi=(atand(WS/WC));
F=eta*U*L*R*pi*((2+ecc)/(1+ecc))/(C*sqrt(1-ecc^2))+C*ecc*WS/2*R;

Ecc=(Fapp-W(i+1))/Fapp;

if Ecc>1e-6
    eccc(i+2)=eccc(i+1)+0.2*(Fapp-W(i+1))/Fapp;
```

```
    else if Ecc <0
         eccc(i+2)=eccc(i+1)+0.2*(-Fapp+W(i+1))/Fapp;
    else
         break
         end
end
  Eccentricity=eccc(i);

end

  if ecc<1

hmin=C*(1- Eccentricity)
fi

B22=B1*(1-ecc)*(2-ecc);
B32=sqrt(B22+0.25);
B42=B22*log((B32+0.5)/(B32-0.5))/B32;
Qrec=U*C*L*(1-ecc)*0.5+U*C*L*lam^2*B22*(1-B42)/g0;
Qp=0;
B21=B1*(1+ecc)*(2+ecc);
B31=sqrt(B21+0.25);
B41=B21*log((B31+0.5)/(B31-0.5))/B31;
Qleak=U*C*L*eccc(i)*(1-lam^2*(B21*(1-B41)+B22*(1-B42))/g0)

delT=eccc(i)*F*U/(roc*Qleak)

F=eta*U*L*R*pi*((2+ecc)/(1+ecc))/(C*sqrt(1-ecc^2))+C*ecc*WS/2*R;
COF=F/W(i)

T=COF*W(i)*R
Ploss=T*2*pi*N/60

tetsmax=acos(-3*ecc/(2+ecc^2));
tetomax= acos(1-sqrt(1+24*ecc^2)/(4*ecc));
teta=lam^2*gs/(3*g0)*(12+ecc^2)^4*(12*ecc^2+(2+ecc^2)*(1-sqrt(1+24*ecc^2)))/
     (ecc^2*(1-ecc^2)^2*(2-ecc^2)*(9-sqrt(1+24*ecc^2))^2);
tetpmax=tetsmax+(tetomax-tetsmax)*(1/(1-teta));
Hpmax=1+ecc*cos(tetpmax);
B21max=4*B1*Hpmax*(1+Hpmax);
Pmax=3*eta*U*R*lam^2*ecc*sin(tetpmax)/(C^2*g0*Hpmax^3)*(B21max/(1+B21max))

      else
         S = 'Eccentricity more than 1';
         disp(S)
    end
```

Chapter 4

Hydrodynamic Thrust Bearing

4.1 Introduction

An axial load is termed as a thrust. There are many types of machinery where an end thrust is produced either from a reaction (i.e., ship's propeller) or as a dead load from a vertical shaft. The bearings, those are capable of carrying this type of load, are named as thrust bearings. A typical thrust bearing has been shown in Fig. 4.1, in which the thrust bearing consists of a rotating disk, fixed to the shaft, and a mating disk, fixed to the stationary hollow cylinder (shaded part). At first glance it appears that both the disks are parallel to each other, but in reality one of the disks has four to six inclined pads. The inclined pads form the required wedge (as explained in chapter 3, section 3.1.1). Due to relative motion between the disks, lubricating oil is sucked into wedge shaped zone over the taper land and forced out through reducing downstream clearance due to the generated hydrodynamic fluid pressure. The thrust load from the shaft is transferred to the stationary bearing part through the hydrodynamic oil film.

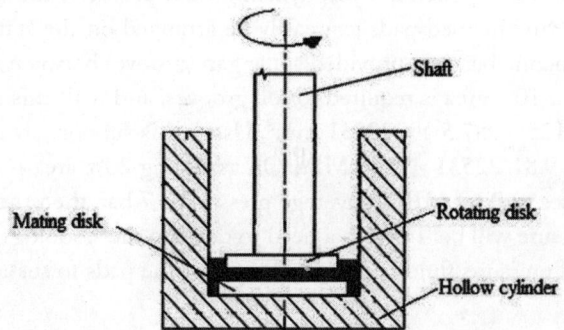

(a) Schematic of shaft, rotating disk, mating disk and hollow cylinder

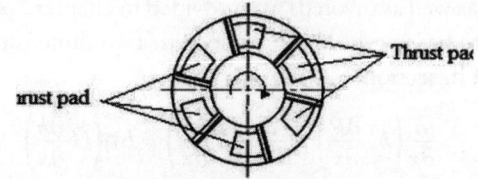

(b) Schematic of thrust pads attached to the stationary disk

Fig. 4.1 Thrust bearing supporting vertical shaft

188 *Fundamentals of Engineering Tribology with Applications*

A group of bearing pads, working on hydrodynamic lubrication mechanism (as explained in chapter 3 in sections 3.3.1), are arranged with oil lubricating grooves to form a hydrodynamic thrust bearing (Fig. 4.1) that can support axial load. Based on the fixed and variable inclinations, these bearings are classified as 'fixed pad thrust bearing' and 'tilting pad thrust bearing', respectively. Hydrodynamic simplicity and compact design of hydrodynamic fixed pad thrust bearings allow these bearings to operate reliably for long period of time. However, these fixed pad bearings perform well only if load, speed and other operating conditions are well known in advance.

In tilting pad thrust bearing, pads are arranged supported in a carrier ring with pivot so that each pad is free to tilt and able to generate appropriate hydrodynamic pressure. These bearings were invented by Michell in 1905 and independently by Kingsbury in 1910. Let us consider two examples to realize the importance of thrust bearings analysis.

Example 1: A vertical shaft of 100 mm diameter is rotating at 1440 rpm in a footstep hydrodynamic thrust bearing. Assume surfaces of shaft end and bearing are separated by a lubricant (viscosity 40 mPa.s) of 25 μm film. Estimate the power loss.

This is a hypothetical example, as parallel surfaces do not form wedge and hydrodynamic pressure cannot be generated. However, this assumption is made to obtain an initial estimate of the power loss. The power loss is calculated by multiplying 'friction force' with 'velocity', which can be represented as $\int_{0}^{r=0.05} \left(\eta \frac{\omega r}{h}\right)\left(2\pi r \, dr\right) r\omega$. The power loss comes out to be 357 watts. This power loss is on higher side. Question arises can we increase film thickness or decrease oil viscosity? Answer can only be provided after suitable analysis of the bearing.

Example 2: Assume 1400 kg load of vertical shaft, rotating at a speed of 1800 rpm, being transferred to a bearing. Outer and inner diameters of the rotating plate (Fig. 4.1a) are 250 mm and 175 mm respectively. Using these data, it can be estimated that average relative sliding speed V, is greater than 20 m/s. This sliding speed can generate hydrodynamic fluid pressure, and for that purpose wedge formation is important. Six inclined pads can easily be arranged on the stationary disk (Fig. 4.1b) to form thrust hydrodynamic bearing, provided some gap (groove) between the pads is maintained. Assuming a minimum of 10% area is required for oil grooves, and with this assumption the bearing support area, $A = 0.9\pi(125^2 - 87.5^2) = 22531$ mm^2. Using this support area, average fluid pressure is given as $P_{av} = 1400 * 9.81/22531 \approx 0.61$ MPa. On assuming 20% area is occupied by oil groove, average pressure increases to 0.69 MPa. If average pressure is 7 bar, then there is need to find what the maximum fluid pressure will be. There is a need to develop methodology to predict the required oil flow, maximum fluid pressure, fluid friction and taper of the pads to sustain the applied load.

4.2 Pressure Distribution

Analysis of single pad (fixed as well as pivoted) was provided in chapter 3, section 3.3.1. In the present chapter, analysis for complete bearing has been presented. Two dimensional Reynolds equation for static application (as derived in section 3.2) is given as:

$$\frac{\partial}{\partial x}\left(h^3 \frac{\partial P}{\partial x}\right) + \frac{\partial}{\partial z}\left(h^3 \frac{\partial P}{\partial z}\right) = 6\eta\left(U \frac{\partial h}{\partial x}\right) \tag{4.2.1}$$

Solving this partial differential equation using analytical approach is not possible. One requires numerical method to solve the Eq. 4.2.1. To reduce the numerical error, often non-dimensionalization of variables *[Khonsari and Booser, 2008]* is preferred. On using non-dimensional parameters $\bar{x} = \frac{x}{X}$, $\bar{z} = \frac{z}{Z}$, $\bar{h} = \frac{h}{C}$ Eq. 4.2.1 becomes

$$\frac{C^3}{X^2}\frac{\partial}{\partial \bar{x}}\left(\bar{h}^3 \frac{\partial P}{\partial \bar{x}}\right) + \frac{C^3}{Z^2}\frac{\partial}{\partial \bar{z}}\left(\bar{h}^3 \frac{\partial P}{\partial \bar{z}}\right) = 6\eta\left(U\frac{C}{X}\frac{\partial \bar{h}}{\partial \bar{x}}\right) \quad (4.2.2)$$

or

$$\frac{C^3}{6\eta U X^2}\frac{\partial}{\partial \bar{x}}\left(\bar{h}^3 \frac{\partial P}{\partial \bar{x}}\right) + \frac{C^3}{6\eta U X^2}\frac{X^2}{Z^2}\frac{\partial}{\partial \bar{z}}\left(\bar{h}^3 \frac{\partial P}{\partial \bar{z}}\right) = \left(\frac{C}{X}\frac{\partial \bar{h}}{\partial \bar{x}}\right) \quad (4.2.3)$$

On substituting non-dimensional pressure $\left(\bar{p} = \frac{C^3 P}{6\eta U X^2}\right)$, Eq. 4.2.3 becomes,

$$\frac{\partial}{\partial \bar{x}}\left(\bar{h}^3 \frac{\partial \bar{p}}{\partial \bar{x}}\right) + \frac{X^2}{Z^2}\frac{\partial}{\partial \bar{z}}\left(\bar{h}^3 \frac{\partial \bar{p}}{\partial \bar{z}}\right) = \left(\frac{C}{X}\frac{\partial \bar{h}}{\partial \bar{x}}\right) \quad (4.2.4)$$

Two most common approximations are 'short bearing' and 'long bearing' as discussed in section 3.3.1, which can also be used

if $\frac{X}{Z} = 10$,

$$\frac{\partial}{\partial \bar{x}}\left(\bar{h}^3 \frac{\partial \bar{p}}{\partial \bar{x}}\right) + 100\frac{\partial}{\partial \bar{z}}\left(\bar{h}^3 \frac{\partial \bar{p}}{\partial \bar{z}}\right) = \left(\frac{C}{X}\frac{\partial \bar{h}}{\partial \bar{z}}\right) \quad (4.2.5)$$

if $\frac{X}{Z} = 0.1$,

$$\frac{\partial}{\partial \bar{x}}\left(\bar{h}^3 \frac{\partial \bar{p}}{\partial \bar{x}}\right) + 0.01\frac{\partial}{\partial \bar{z}}\left(\bar{h}^3 \frac{\partial \bar{p}}{\partial \bar{z}}\right) = \left(\frac{C}{X}\frac{\partial \bar{h}}{\partial \bar{x}}\right) \quad (4.2.6)$$

Equation 4.2.5 illustrates that pressure gradient in X-direction, $\frac{\partial}{\partial \bar{x}}\left(\bar{h}^3 \frac{\partial \bar{p}}{\partial \bar{x}}\right)$, is negligible, while Eq. 4.2.6 shows that pressure gradient in Z-direction, $0.01\frac{\partial}{\partial \bar{z}}\left(\bar{h}^3 \frac{\partial \bar{p}}{\partial \bar{z}}\right)$, is negligible. In other words, effect of terms in the Reynolds' equation depends on the geometry of the bearings. For most of the bearings $0.1 < \frac{X}{Z} < 10$ and numerical methods are required to solve the Reynolds' equation and estimate the fluid pressure. To solve Eq. 4.2.4 for tribo-surfaces (Fig. 4.1), the surface can be discretized in number of nodes, as presented in Fig. 4.2.

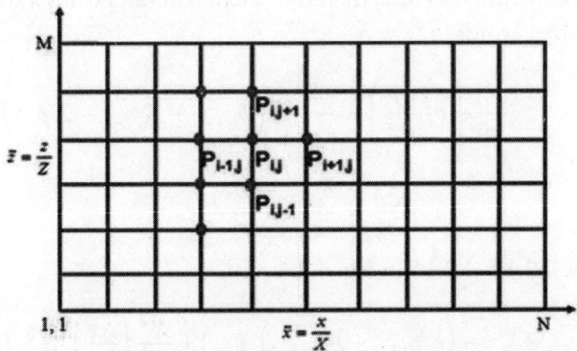

Fig. 4.2 Discretization of surface

Taylor series can be utilized to approximate the partial derivatives of Eq. 4.2.4, such as:

$$\overline{p}(\overline{x} + \Delta\overline{x}) = \overline{p}(\overline{x}) + \Delta\overline{x}\frac{\partial \overline{p}}{\partial \overline{x}} + \frac{1}{2}(\Delta\overline{x})^2 \frac{\partial^2 \overline{p}}{\partial \overline{x}^2} + \ldots \qquad (4.2.7)$$

$$\overline{p}(\overline{x} - \Delta\overline{x}) = \overline{p}(\overline{x}) - \Delta\overline{x}\frac{\partial \overline{p}}{\partial \overline{x}} + \frac{1}{2}(\Delta\overline{x})^2 \frac{\partial^2 \overline{p}}{\partial \overline{x}^2} + \ldots \qquad (4.2.8)$$

On subtracting Eq. 4.2.8 from Eq. 4.2.7,

$$\overline{p}(\overline{x} + \Delta\overline{x}) - \overline{p}(\overline{x} - \Delta\overline{x}) = 2\Delta\overline{x}\frac{\partial \overline{p}}{\partial \overline{x}} \qquad (4.2.9)$$

This means,
$$\frac{\partial \overline{p}}{\partial \overline{x}} = \frac{\overline{p}(\overline{x} + \Delta\overline{x}) - \overline{p}(\overline{x} - \Delta\overline{x})}{2\Delta\overline{x}} \qquad (4.2.10)$$

Similarly
$$\frac{\partial \overline{p}_{i,j}}{\partial \overline{x}} = \frac{\overline{p}_{i+1,j} - \overline{p}_{i-1,j}}{2\Delta\overline{x}} \qquad (4.2.11)$$

and
$$\frac{\partial}{\partial \overline{x}}\left(\frac{\partial \overline{p}_{i,j}}{\partial \overline{x}}\right) = \frac{\left(\frac{\partial \overline{p}}{\partial \overline{x}}\right)_{i+0.5,j} - \left(\frac{\partial \overline{p}}{\partial \overline{x}}\right)_{i-0.5,j}}{\Delta \overline{x}} \qquad (4.2.12)$$

Using Eq. 4.2.11 in Eq. 4.2.12

$$\frac{\partial}{\partial \overline{x}}\left(\frac{\partial \overline{p}_{i,j}}{\partial \overline{x}}\right) = \frac{\overline{p}_{i+1,j} - 2\overline{p}_{i,j} + \overline{p}_{i-1,j}}{(\Delta\overline{x})^2} \qquad (4.2.13)$$

Using this kind of Taylor series approximation, we get

$$\frac{\partial}{\partial \overline{x}}\left(\overline{h}^3 \frac{\partial \overline{p}}{\partial \overline{x}}\right) = \frac{\overline{h}^3_{i+0.5,j} \cdot \overline{p}_{i+1,j} + \overline{h}^3_{i-0.5,j} \cdot \overline{p}_{i-1,j} - \left(\overline{h}^3_{i+0.5,j} + \overline{h}^3_{i-0.5,j}\right) \cdot \overline{p}_{i,j}}{(\Delta\overline{x})^2} \qquad (4.2.14)$$

Similarly
$$\frac{\partial}{\partial \overline{z}}\left(\overline{h}^3 \frac{\partial \overline{p}}{\partial \overline{z}}\right) = \frac{\overline{h}^3_{i,j+0.5} \cdot \overline{p}_{i,j+1} + \overline{h}^3_{i,j-0.5} \cdot \overline{p}_{i,j-1} - \left(\overline{h}^3_{i,j+0.5} + \overline{h}^3_{i,j-0.5}\right) \cdot \overline{p}_{i,j}}{(\Delta\overline{z})^2}$$

As film thickness is only function of x, therefore there will not be any variation in film thickness in the Z-direction. In other words $\overline{h}_{i,j} = \overline{h}_{i,j-0.5} = \overline{h}_{i,j+0.5}$,

$$\Rightarrow \qquad \frac{\partial}{\partial \overline{z}}\left(\overline{h}^3 \frac{\partial \overline{p}}{\partial \overline{z}}\right) = \frac{\overline{h}^3_{i,j}}{(\Delta\overline{z})^2}\left(\overline{p}_{i,j+1} + \overline{p}_{i,j-1} - 2\cdot\overline{p}_{i,j}\right) \qquad (4.2.15)$$

$$\frac{\partial \overline{h}}{\partial \overline{x}} = \frac{\overline{h}_{i+1,j} - \overline{h}_{i-1,j}}{2\Delta\overline{x}} \qquad (4.2.16)$$

On substituting Eqs. 4.2.14–4.2.16 in Eq. 4.2.4

$$\left(\overline{h}^3_{i+0.5,j} \cdot \overline{p}_{i+1,j} + \overline{h}^3_{i-0.5,j} \cdot \overline{p}_{i-1,j} - \left(\overline{h}^3_{i+0.5,j} + \overline{h}^3_{i-0.5,j}\right) \cdot \overline{p}_{i,j}\right) + \frac{X^2}{Z^2}\left(\frac{\Delta\overline{x}}{\Delta\overline{z}}\right)^2 \cdot \overline{h}^3_{i,j}\left(\overline{p}_{i,j+1} + \overline{p}_{i,j-1} - 2\overline{p}_{i,j}\right)$$

$$= \frac{C}{X}\frac{\Delta\overline{x}}{2}\left(\overline{h}_{i+1,j} - \overline{h}_{i-1,j}\right)$$

On rearranging

$$\left(\bar{h}_{i+0.5,j}^3 \cdot \bar{P}_{i+1,j} + \bar{h}_{i-0.5,j}^3 \cdot \bar{P}_{i-1,j}\right) + \frac{X^2}{Z^2}\left(\frac{\Delta \bar{x}}{\Delta \bar{z}}\right)^2 \cdot \bar{h}_{i,j}^3 \left(\bar{P}_{i,j+1} + \bar{P}_{i,j-1}\right) - \frac{C}{X}\frac{\Delta \bar{x}}{2}\left(\bar{h}_{i+1,j} - \bar{h}_{i-1,j}\right)$$

$$= \left(\bar{h}_{i+0.5,j}^3 + \bar{h}_{i-0.5,j}^3 + \frac{2X^2}{Z^2}\left(\frac{\Delta \bar{x}}{\Delta \bar{z}}\right)^2 \bar{h}_{i,j}^3\right) \cdot \bar{P}_{i,j}$$

Or,

$$\bar{P}_{i,j} = \left(\frac{\bar{h}_{i+0.5,j}^3}{\left(\bar{h}_{i+0.5,j}^3 + \bar{h}_{i-0.5,j}^3 + \frac{2X^2}{Z^2}\left(\frac{\Delta \bar{x}}{\Delta \bar{z}}\right)^2 \bar{h}_{i,j}^3\right)} \cdot \bar{P}_{i+1,j} + \frac{\bar{h}_{i-0.5,j}^3}{\left(\bar{h}_{i+0.5,j}^3 + \bar{h}_{i-0.5,j}^3 + \frac{2X^2}{Z^2}\left(\frac{\Delta \bar{x}}{\Delta \bar{z}}\right)^2 \bar{h}_{i,j}^3\right)} \cdot \bar{P}_{i-1,j}\right)$$

$$+ \frac{\frac{X^2}{Z^2}\left(\frac{\Delta \bar{x}}{\Delta \bar{z}}\right)^2 \bar{h}_{i,j}^3}{\left(\bar{h}_{i+0.5,j}^3 + \bar{h}_{i-0.5,j}^3 + \frac{2X^2}{Z^2}\left(\frac{\Delta \bar{x}}{\Delta \bar{z}}\right)^2 \bar{h}_{i,j}^3\right)} \left(\bar{P}_{i,j+1} + \bar{P}_{i,j-1}\right)$$

$$- \frac{\Delta \bar{x}}{2} \cdot \frac{C}{X} \frac{\left(\bar{h}_{i+1,j} - \bar{h}_{i-1,j}\right)}{\left(\bar{h}_{i+0.5,j}^3 + \bar{h}_{i-0.5,j}^3 + \frac{2X^2}{Z^2}\left(\frac{\Delta \bar{x}}{\Delta \bar{z}}\right)^2 \bar{h}_{i,j}^3\right)}$$

$$\bar{P}_{i,j} = A_{i,j} \cdot \bar{P}_{i,j+1} + B_{i,j} \cdot \bar{P}_{i,j-1} + C_{i,j} \cdot \bar{P}_{i+1,j} + D_{i,j} \cdot \bar{P}_{i-1,j} + E_{i,j} \quad (4.2.17)$$

where,

$$A_{i,j} = \frac{\frac{X^2}{Z^2}\left(\frac{\Delta \bar{x}}{\Delta \bar{z}}\right)^2 \bar{h}_{i,j}^3}{\left(\bar{h}_{i+0.5,j}^3 + \bar{h}_{i-0.5,j}^3 + \frac{2X^2}{Z^2}\left(\frac{\Delta \bar{x}}{\Delta \bar{z}}\right)^2 \bar{h}_{i,j}^3\right)}$$

$$B_{i,j} = \frac{\frac{X^2}{Z^2}\left(\frac{\Delta \bar{x}}{\Delta \bar{z}}\right)^2 \bar{h}_{i,j}^3}{\left(\bar{h}_{i+0.5,j}^3 + \bar{h}_{i-0.5,j}^3 + \frac{2X^2}{Z^2}\left(\frac{\Delta \bar{x}}{\Delta \bar{z}}\right)^2 \bar{h}_{i,j}^3\right)}$$

$$C_{i,j} = \frac{\bar{h}_{i+0.5,j}^3}{\left(\bar{h}_{i+0.5,j}^3 + \bar{h}_{i-0.5,j}^3 + \frac{2X^2}{Z^2}\left(\frac{\Delta \bar{x}}{\Delta \bar{z}}\right)^2 \bar{h}_{i,j}^3\right)}$$

$$D_{i,j} = \frac{\bar{h}_{i-0.5,j}^3}{\left(\bar{h}_{i+0.5,j}^3 + \bar{h}_{i-0.5,j}^3 + \frac{2X^2}{Z^2}\left(\frac{\Delta \bar{x}}{\Delta \bar{z}}\right)^2 \bar{h}_{i,j}^3\right)}$$

$$E_{i,j} = -\frac{\Delta \bar{x}}{2} \cdot \frac{C}{X} \frac{\left(\bar{h}_{i+1,j} - \bar{h}_{i-1,j}\right)}{\left(\bar{h}_{i+0.5,j}^3 + \bar{h}_{i-0.5,j}^3 + \frac{2X^2}{Z^2}\left(\frac{\Delta \bar{x}}{\Delta \bar{z}}\right)^2 \bar{h}_{i,j}^3\right)}$$

Equation 4.2.17 can be solved for $i = 1, N$ and $j = 1, M$ with appropriate boundary conditions by writing sequence of equations, such as

$$\overline{p}_{1,1} = A_{1,1} \cdot \overline{p}_{1,2} + B_{1,1} \cdot \overline{p}_{1,0} + C_{1,1} \cdot \overline{p}_{2,1} + D_{1,1} \cdot \overline{p}_{0,1} + E_{1,1}$$

As most of values of nodal pressures ($p_{i,j}$) are unknown assuming all the beginning nodes other than boundary nodes as zero, therefore iterative loop is employed to estimate the fluid pressure

$$\overline{p}_{1,1}^{k+1} = A_{1,1} \cdot \overline{p}_{1,2}^{k} + B_{1,1} \cdot \overline{p}_{1,0}^{k+1} + C_{1,1} \cdot \overline{p}_{2,1}^{k} + D_{1,1} \cdot \overline{p}_{0,1}^{k+1} + E_{1,1}$$

Where k is iteration number. Process of iterations will be repeated till a convergence criterion is satisfied.

$$\frac{\left| \left(\sum_{i=1}^{n} \sum_{j=1}^{m} \overline{p}_{i,j} \right)^{k+1} - \left(\sum_{i=1}^{n} \sum_{j=1}^{m} \overline{p}_{i,j} \right)^{k} \right|}{\left| \left(\sum_{i=1}^{n} \sum_{j=1}^{m} \overline{p}_{i,j} \right)^{k+1} \right|} \leq \varepsilon \qquad (4.2.18)$$

The value of ε is very small. To understand how to estimate the pressure using the above mentioned procedure, let us assume the number of nodes in the X-direction, (N) as 25, the number of nodes in Z-direction, (M) as 25, and the number of iterations as $k = 30$. This means that 25*25*30=18750 calculations are required. The number of calculation can be reduced if transverse (Z) length of surface is much lesser than tangential(X) length. In that case pressure gradient in X-direction will be negligible and number of iterations required to converge the solution will be lesser.

if $\dfrac{X}{Z} = 10$, $\qquad \dfrac{\partial}{\partial \overline{x}} \left(\overline{h}^3 \dfrac{\partial \overline{p}}{\partial \overline{x}} \right) + 100 \dfrac{\partial}{\partial \overline{z}} \left(\overline{h}^3 \dfrac{\partial \overline{p}}{\partial \overline{z}} \right) = \left(\dfrac{C}{X} \dfrac{\partial \overline{h}}{\partial \overline{x}} \right)$

In such case number of 25 * 15 = 375 may be sufficient to provide solution. Therefore, there is a reduction in number of iterations i.e., 18750/375 = 50 times. Let us consider one example to comprehend this concept.

Example: Assume a thrust pad of 10×100 mm dimensions. Leading and trailing film thicknesses are 0.04 and 0.02 mm respectively. Sliding speed is 20 m/s. Viscosity of oil is 10 mPa.s. Find pressure distribution.

As $\dfrac{X}{Z} = 0.1$, the second term of equation $\dfrac{\partial}{\partial \overline{x}} \left(\overline{h}^3 \dfrac{\partial \overline{p}}{\partial \overline{x}} \right) + 0.01 \dfrac{\partial}{\partial \overline{z}} \left(\overline{h}^3 \dfrac{\partial \overline{p}}{\partial \overline{z}} \right) = \dfrac{C}{X} \dfrac{\partial \overline{h}}{\partial \overline{x}}$, can be neglected,

$$\dfrac{\partial}{\partial \overline{x}} \left(\overline{h}^3 \dfrac{\partial \overline{p}}{\partial \overline{x}} \right) = \dfrac{C}{X} \dfrac{\partial \overline{h}}{\partial \overline{x}} \qquad (4.2.19)$$

The average separation between stationary and rotating surfaces is $C = \dfrac{h_{min} + h_{max}}{2} = 0.03$ mm.

To find expression of film thickness in terms of x, the following expression can be used,

$$h = h_{max} - (h_{max} - h_{min}) \dfrac{x}{X}$$

$$h = 0.04 - 0.02\bar{x}$$

On non-dimensionalization,

$$\bar{h} = \frac{h}{C} = \frac{2}{3}(2 - \bar{x})$$

$$\Rightarrow \frac{\partial \bar{h}}{\partial \bar{x}} = -\frac{2}{3}$$

Substituting this term in Eq. 4.2.19

$$\frac{\partial}{\partial \bar{x}}\left(\bar{h}^3 \frac{\partial \bar{p}}{\partial \bar{x}}\right) = -\frac{C}{X}\frac{2}{3}$$

On substituting from Eq. 4.2.14

$$\frac{\bar{h}_{i+0.5}^3 \cdot \bar{p}_{i+1} + \bar{h}_{i-0.5}^3 \cdot \bar{p}_{i-1} - \left(\bar{h}_{i+0.5}^3 + \bar{h}_{i-0.5}^3\right) \cdot \bar{p}_i}{(\Delta \bar{x})^2} = -\frac{0.03}{10}\frac{2}{3} = -0.002 \quad (4.2.20)$$

Rearranging the equation,

$$\bar{h}_{i+0.5}^3 \cdot \bar{p}_{i+1} + \bar{h}_{i-0.5}^3 \cdot \bar{p}_{i-1} - \left(\bar{h}_{i+0.5}^3 + \bar{h}_{i-0.5}^3\right) \cdot \bar{p}_i = -0.002(\Delta \bar{x})^2$$

Taking step size, $\Delta \bar{x} = 0.1$

$$\bar{h}_{i+0.5}^3 \cdot \bar{p}_{i+1} + \bar{h}_{i-0.5}^3 \cdot \bar{p}_{i-1} - \left(\bar{h}_{i+0.5}^3 + \bar{h}_{i-0.5}^3\right) \cdot \bar{p}_i = -0.00002 \quad (4.2.21)$$

The values of $\bar{x}$ are listed in Table 4.1. The film thickness is maximum at the entrance and minimum at the exit due to the convergence of the shape of the bearing.

Table 4.1 Values of $\bar{x}$ at different nodes

X Dimension	Node number	Non-dimensional film thickness				
		$\bar{h}_i$	$\bar{h}_{i-0.5}$	$\bar{h}_{i+0.5}$	$\bar{h}_{i-0.5}^3$	$\bar{h}_{i+0.5}^3$
0.00	1	1.33				
0.10	2	1.27	1.30	1.23	2.20	1.88
0.20	3	1.20	1.23	1.17	1.88	1.59
0.30	4	1.13	1.17	1.10	1.59	1.33
0.40	5	1.07	1.10	1.03	1.33	1.10
0.50	6	1.00	1.03	0.97	1.10	0.90
0.60	7	0.93	0.97	0.90	0.90	0.73
0.70	8	0.87	0.90	0.83	0.73	0.58
0.80	9	0.80	0.83	0.77	0.58	0.45
0.90	10	0.73	0.77	0.70	0.45	0.34
1.00	11	0.67				

The pressure at any equation can be given as,

$$\bar{p}_i = \frac{\bar{h}_{i+0.5}^3 \cdot \bar{p}_{i+1} + \bar{h}_{i-0.5}^3 \bar{p}_{i-1} + 0.00002}{\left(\bar{h}_{i+0.5}^3 + \bar{h}_{i-0.5}^3\right)} \qquad (4.2.22)$$

Table 4.2 shows the pressure values at different iterations. At entrance and exit, the pressure at each node is equal to zero. The pressure continuously increases with the number of iterations as depicted in Fig. 4.3. The solution is terminated if error is lesser than ε.

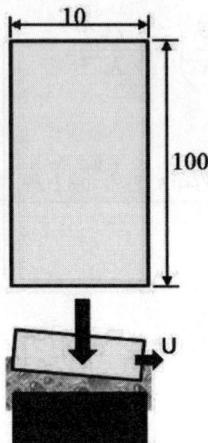

Fig. 4.3 Thrust pad of 10 × 100 mm dimensions

The pressure ($\bar{p}$) values as listed in Table 4.2 are non-dimensional pressures. To get the dimensional pressure (P), the following equation is used.

$$P = \frac{6\eta U X^2 \bar{p}}{C^3} \qquad (4.2.23)$$

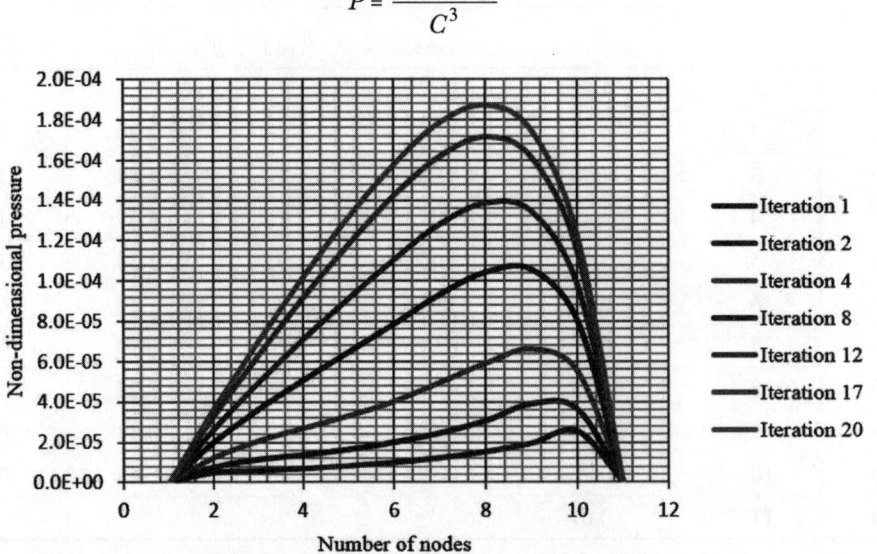

Fig. 4.4 Pressure profile at different nodes of the bearing

Table 4.2 Pressure values at different iterations

Iteration	0th	1st	2nd	3rd	4th	5th	6th	7th	8th	9th	10th	11th	12th	13th
Entrance Condition	0	0	0	0	0	0	0	0	0	0	0	0	0	0
	0	4.9E-06	7.6E-06	1.0E-05	1.2E-05	1.4E-05	1.6E-05	1.8E-05	2.0E-05	2.2E-05	2.3E-05	2.5E-05	2.6E-05	2.8E-05
	0	5.8E-06	1.2E-05	1.6E-05	2.1E-05	2.5E-05	2.9E-05	3.3E-05	3.6E-05	4.0E-05	4.3E-05	4.6E-05	4.9E-05	5.2E-05
	0	6.8E-06	1.4E-05	2.1E-05	2.7E-05	3.3E-05	3.9E-05	4.5E-05	5.0E-05	5.6E-05	6.1E-05	6.6E-05	7.1E-05	7.5E-05
	0	8.2E-06	1.7E-05	2.5E-05	3.3E-05	4.1E-05	4.9E-05	5.7E-05	6.4E-05	7.1E-05	7.8E-05	8.5E-05	9.1E-05	9.7E-05
	0	1.0E-05	2.0E-05	3.0E-05	4.0E-05	5.1E-05	6.0E-05	7.0E-05	7.9E-05	8.7E-05	9.5E-05	1.0E-04	1.1E-04	1.2E-04
	0	1.2E-05	2.5E-05	3.7E-05	5.0E-05	6.1E-05	7.3E-05	8.3E-05	9.3E-05	1.0E-04	1.1E-04	1.2E-04	1.3E-04	1.4E-04
	0	1.5E-05	3.1E-05	4.6E-05	5.9E-05	7.2E-05	8.3E-05	9.5E-05	1.0E-04	1.1E-04	1.2E-04	1.3E-04	1.4E-04	1.5E-04
	0	1.9E-05	3.9E-05	5.3E-05	6.6E-05	7.7E-05	8.8E-05	9.7E-05	1.1E-04	1.1E-04	1.2E-04	1.3E-04	1.3E-04	1.4E-04
	0	2.5E-05	3.6E-05	4.8E-05	5.5E-05	6.3E-05	6.9E-05	7.5E-05	8.0E-05	8.5E-05	9.0E-05	9.4E-05	9.8E-05	1.0E-04
Exit condition	0	0	0	0	0	0	0	0	0	0	0	0	0	0
Summation of all pres.	0	1.1E-04	2.0E-04	2.9E-04	3.6E-04	4.4E-04	5.1E-04	5.7E-04	6.3E-04	6.9E-04	7.5E-04	8.0E-04	8.5E-04	8.9E-04
Error			4.6E-01	3.0E-01	2.2E-01	1.7E-01	1.4E-01	1.1E-01	9.7E-02	8.4E-02	7.3E-02	6.5E-02	5.8E-02	5.2E-02

Example: Assume a thrust pad of 100 × 10 mm dimensions. Leading and trailing film thicknesses are 0.04 and 0.02 mm, respectively. Sliding speed is 20 m/s. Viscosity of oil is 10 mPa.s. Find the pressure distribution.

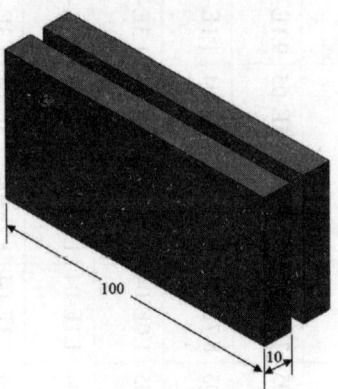

Fig. 4.5 Thrust pad of 10 × 100 mm dimensions

In the present example $\frac{X}{Z} = 10$ which means the first term of equation $\frac{\partial}{\partial \bar{x}}\left(\bar{h}^3 \frac{\partial \bar{P}}{\partial \bar{x}}\right) + 100 \frac{\partial}{\partial \bar{z}}\left(\bar{h}^3 \frac{\partial \bar{P}}{\partial \bar{z}}\right)$
$= \frac{C}{X} \frac{\partial \bar{h}}{\partial \bar{x}}$ can be neglected.

Or, $\quad 100 \frac{\partial}{\partial \bar{z}}\left(\bar{h}^3 \frac{\partial \bar{P}}{\partial \bar{z}}\right) = \frac{C}{X} \frac{\partial \bar{h}}{\partial \bar{x}}$

$$C = \frac{h_{max} + h_{min}}{2} = 0.03 \text{ mm}$$

Film thickness, $\quad h = h_{max} - \frac{h_{max} - h_{min}}{X} \bar{x} = 0.04 - 0.02\bar{x}$

$$\bar{h} = \frac{h}{c} = \frac{2}{3}(2 - \bar{x})$$

Or, $\quad \frac{\partial \bar{h}}{\partial \bar{x}} = -\frac{2}{3}$

$$100 \frac{\partial}{\partial \bar{z}}\left(\bar{h}^3 \frac{\partial \bar{p}}{\partial \bar{z}}\right) = -0.0002$$

$$\frac{\partial}{\partial \bar{z}}\left(\frac{\partial \bar{p}}{\partial \bar{z}}\right) = -\frac{0.0002}{100 \bar{h}^3}$$

$$\frac{\bar{P}_{i,j+1} + \bar{P}_{i,j-1} - 2\bar{P}_{i,j}}{(\Delta \bar{z})^2} = -\frac{-0.000002}{\bar{h}_i^3}$$

On rearranging

$$\bar{P}_{i,j} = \frac{1}{2}\left[\bar{P}_{i,j+1} + \bar{P}_{i,j-1} + \frac{-0.000002}{\bar{h}_i^3}(\Delta \bar{z})^2\right]$$

Taking step size, $\Delta \bar{x} = \Delta \bar{z} = 0.1$, which means i = 1, 2, ..., 11 and j = 1, 2, ..., 11

Non-dimensional pressure can be evaluated using the procedure described in the previous example. The pressure profile at the different nodes is shown in Fig. 4.6

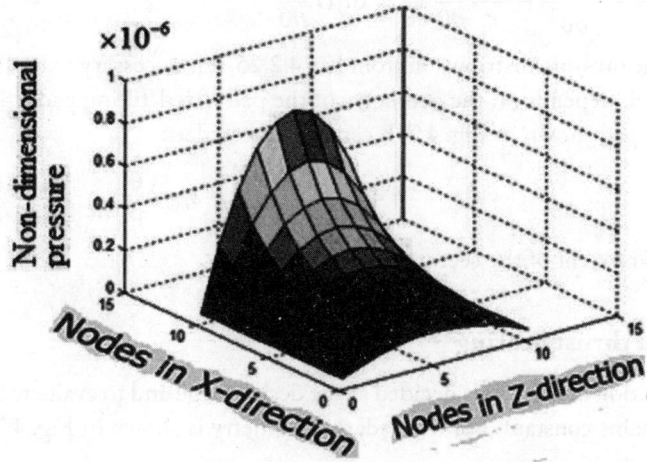

Fig. 4.6 Non-dimensional pressure profile at different nodes in X- and Z- directions

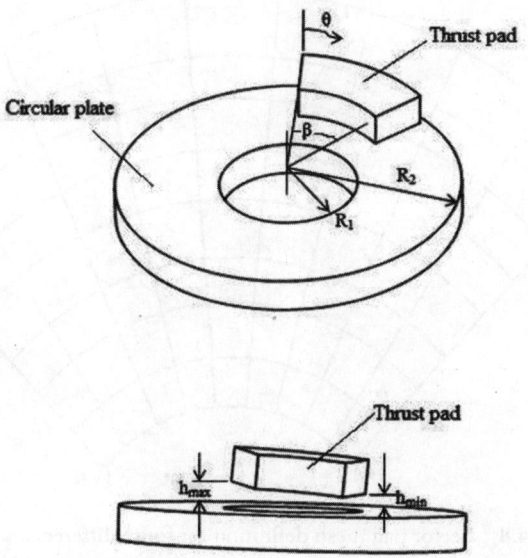

Fig. 4.7 Geometry of sector pad bearing

In the previous two examples, solutions of Reynolds' equation, expressed in rectangular coordinates, were described. To estimate the load capacity of thrust bearing shown in Fig. 4.1, there is a need to solve Reynolds' equation expressed in polar coordinate system (Eq. 4.2.24).

$$\frac{\partial}{\partial r}\left(rh^3 \frac{\partial p}{\partial r}\right) + \frac{1}{r}\frac{\partial}{\partial \theta}\left(h^3 \frac{\partial p}{\partial \theta}\right) = 6\eta U \frac{\partial h}{\partial \theta} \qquad (4.2.24)$$

Each pad of the bearing is in shape of a sector as depicted in Fig. 4.7. The pad has angular span of β and it is bounded by the inner and outer radii of R_1 and R_2 respectively. The film thickness h is function of independent variable θ. On simplifying Eq. 4.2.24 by differentiating,

$$h^3 \frac{\partial p}{\partial r} + rh^3 \frac{\partial^2 p}{\partial r^2} + \frac{h^3}{r} \frac{\partial^2 p}{\partial \theta^2} + \frac{3h^2}{r} \frac{\partial h}{\partial \theta} \frac{\partial p}{\partial \theta} = 6\eta U \frac{\partial h}{\partial \theta} \quad (4.2.25)$$

To determine the pressure distribution from Eq. 4.2.25, it is necessary to model the film thickness (h) expression, which depends on the geometry of the pad (fixed/tilting pad).

For example for pad shown in Fig. 4.7, h can be expressed as:

$$h = h_{max} - (h_{max} - h_{min}) \frac{\theta}{\beta} \quad (4.2.26)$$

Here β is angular extent of the sector shaped pad.

4.2.1 Fixed pad thrust bearing

In fixed pad, inclination of the pad is decided at the design stage and to evaluate bearing performance the inclination remains constant. For the pad, the geometry is shown in Fig. 4.7,

$$\frac{dh}{d\theta} = 0 - \frac{(h_{max} - h_{min})}{\beta} \quad (4.2.27)$$

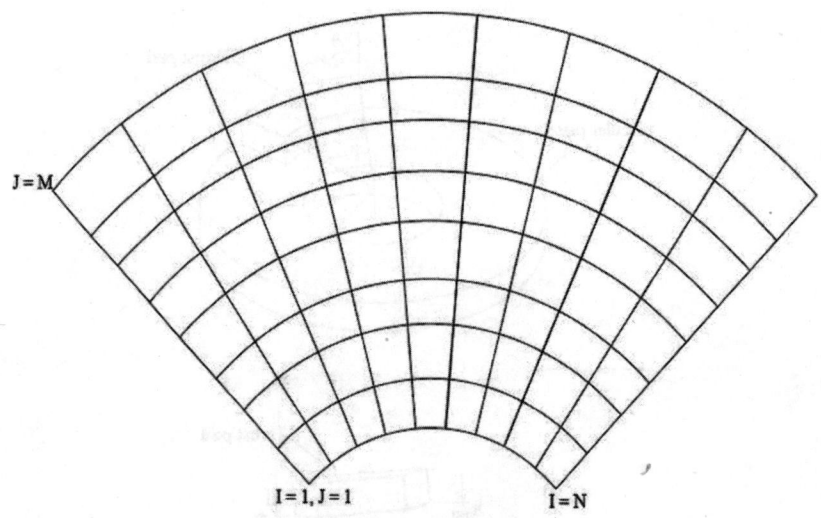

Fig. 4.8 Sector pad mesh definition for finite difference scheme

Even though the values of h_{max} and h_{min} may change depending on the operational conditions (i.e., load, speed, viscosity), but difference in these values ($h_{max} - h_{min}$) remain constant. On substituting this expression in Eq. (4.2.25)

$$h^3 \frac{\partial p}{\partial r} + rh^3 \frac{\partial^2 p}{\partial r^2} + \frac{h^3}{r} \frac{\partial^2 p}{\partial \theta^2} + \frac{3h^2}{r}\left(-\frac{(h_{max} - h_{min})}{\beta}\right)\left(\frac{\partial p}{\partial \theta}\right) = 6\eta U \left(-\frac{(h_{max} - h_{min})}{\beta}\right)$$

On rearranging

$$h^3 \frac{\partial p}{\partial r} + rh^3 \frac{\partial^2 p}{\partial r^2} + \frac{h^3}{r}\frac{\partial^2 p}{\partial \theta^2} - \left(\frac{3(h_{max} - h_{min})}{\beta}\right)\left(\frac{h^2}{r}\frac{\partial p}{\partial \theta}\right) = -\left(\frac{6\eta U(h_{max} - h_{min})}{\beta}\right) \quad (4.2.28)$$

To solve Eq. 4.2.28, finite difference scheme as sketched in Fig. 4.8 can be used. As per Fig. 4.8, I represents θ and J represent r. Limit of θ varies from 0 to β, while limit of r varies from R_1 to R_2.

$$h^3 \frac{\partial p}{\partial r} = h_i^3 \frac{p_{i,j+1} - p_{i,j-1}}{2(\Delta r)}$$

$$rh^3 \frac{\partial^2 p}{\partial r^2} = r_j h_i^3 \frac{p_{i,j+1} + p_{i,j-1} - 2p_{i,j}}{(\Delta r)^2}$$

$$\frac{h^3}{r}\frac{\partial^2 p}{\partial \theta^2} = \frac{h_i^3}{r_j}\frac{p_{i+1,j} + p_{i-1,j} - 2p_{i,j}}{(\Delta \theta)^2}$$

$$\left(\frac{3(h_{max} - h_{min})}{\beta}\right)\left(\frac{h^2}{r}\frac{\partial p}{\partial \theta}\right) = \left(\frac{3(h_{max} - h_{min})}{\beta}\right)\frac{h_i^2}{r_i}\frac{p_{i+1,j} - p_{i-1,j}}{\Delta \theta}$$

On substituting these finite difference terms in Eq. 4.2.28

$$h_i^3 \frac{p_{i,j+1} - p_{i,j-1}}{2(\Delta r)} + r_j h_i^3 \frac{p_{i,j+1} + p_{i,j-1} - 2p_{i,j}}{(\Delta r)^2} + \frac{h_i^3}{r_j}\frac{p_{i+1,j} + p_{i-1,j} - 2p_{i,j}}{(\Delta \theta)^2}$$

$$-\left(\frac{3(h_{max} - h_{min})}{\beta}\right)\frac{h_i^2}{r_i}\frac{p_{i+1,j} - p_{i-1,j}}{\Delta \theta} = -\left(\frac{6\eta U(h_{max} - h_{min})}{\beta}\right)$$

(4.2.29)

On rearranging the Eq. 4.2.29,

$$r_j h_i^3 \frac{2p_{i,j}}{(\Delta r)^2} + \frac{h_i^3}{r_j}\frac{2p_{i,j}}{(\Delta \theta)^2} = \left(\frac{6\eta U(h_{max} - h_{min})}{\beta}\right) + h_i^3 \frac{p_{i,j+1} - p_{i,j-1}}{2(\Delta r)} + r_j h_i^3 \frac{p_{i,j+1} + p_{i,j-1}}{(\Delta r)^2}$$

$$+ \frac{h_i^3}{r_j}\frac{p_{i+1,j} + p_{i-1,j}}{(\Delta \theta)^2} - \left(\frac{3(h_{max} - h_{min})}{\beta}\right)\frac{h_i^2}{r_i}\frac{p_{i+1,j} - p_{i-1,j}}{\Delta \theta}$$

On rearranging

$$2\left(\frac{r_j h_i^3}{(\Delta r)^2} + \frac{h_i^3}{r_j(\Delta \theta)^2}\right)p_{i,j} = \left(\frac{6\eta U(h_{max} - h_{min})}{\beta}\right) + \left(\frac{h_i^3}{2(\Delta r)} + \frac{r_j h_i^3}{(\Delta r)^2}\right)p_{i,j+1}$$

$$+ \left(\frac{r_j h_i^3}{(\Delta r)^2} - \frac{h_i^3}{2(\Delta r)}\right)p_{i,j-1} + \left(\frac{h_i^3}{r_j(\Delta \theta)^2} - \left(\frac{3(h_{max} - h_{min})}{\beta}\right)\frac{h_i^2}{r_i \Delta \theta}\right)p_{i+1,j}$$

$$+ \left(\frac{h_i^3}{r_j(\Delta \theta)^2} + \left(\frac{3(h_{max} - h_{min})}{\beta}\right)\frac{h_i^2}{r_i \Delta \theta}\right)p_{i-1,j}$$

(4.2.30)

Equation 4.2.30 can be rearranged as

$$p_{i,j} = A_{i,j} \cdot p_{i,j+1} + B_{i,j} \cdot p_{i,j-1} + C_{i,j} \cdot p_{i+1,j} + D_{i,j} \cdot p_{i-1,j} + E \quad (4.2.31)$$

where,

$$A_{i,j} = \left(\frac{h_i^3}{2(\Delta r)} + \frac{r_j h_i^3}{(\Delta r)^2} \right)$$

$$B_{i,j} = \left(\frac{h_i^3}{2(\Delta r)} - \frac{r_j h_i^3}{(\Delta r)^2} \right)$$

$$C_{i,j} = \left(\frac{h_i^3}{r_j (\Delta \theta)^2} - \left(\frac{3(h_{max} - h_{min})}{\beta} \right) \frac{h_i^2}{r_i \Delta \theta} \right)$$

$$D_{i,j} = \left(\frac{h_i^3}{r_j (\Delta \theta)^2} + \left(\frac{3(h_{max} - h_{min})}{\beta} \right) \frac{h_i^2}{r_i \Delta \theta} \right)$$

Equation 4.2.31 can be solved for i = 1, N and j = 1, M with appropriate boundary conditions. Generally for i = 1 and i = N pressure at every j node (j = 1 to M) remain gauge pressure, which means p = 0. Similarly for j = 1 and j = M pressure at every i node (i = 1 to N) remain gauge pressure, which means p = 0. Using these pressure boundary conditions, Eq. 4.2.31 can be solved for i = 2, N-1 and j = 2, M-1 by writing sequence of equations, such as,

$$p_{2,2} = A_{2,2} \cdot p_{2,3} + B_{2,2} \cdot p_{2,1} + C_{2,2} \cdot p_{3,2} + D_{2,2} p_{1,2} + E$$

As most of values of nodal pressures $(p_{i,j})$ are unknown, therefore, iterative loop is employed to estimate the fluid pressure

$$p_{2,2}^{k+1} = A_{2,2} \cdot p_{2,3}^k + B_{2,2} \cdot p_{2,1}^{k+1} + C_{2,2} \cdot p_{3,2}^k + D_{2,2} p_{1,2}^{k+1} + E$$

Where k represents the iteration number. Process of iterations will be repeated till a convergence criterion is satisfied.

$$\frac{\left| \left(\sum_{i=1}^{N} \sum_{j=1}^{M} p_{i,j} \right)^{k+1} - \left(\sum_{i=1}^{N} \sum_{j=1}^{M} p_{i,j} \right)^{k} \right|}{\left| \left(\sum_{i=1}^{N} \sum_{j=1}^{M} p_{i,j} \right)^{k+1} \right|} \leq \varepsilon \qquad (4.2.32)$$

The value of ε is very small.

4.2.2 Tilting pad thrust bearing

Infinitely long pivoted slider has been described in section 3.3.2 of chapter 3. It has been mentioned that pivoted slider is better than fixed slider in providing flexibility related to the minimum film thickness. In other words, slider, which is pivoted, can take up an optimum inclination according to the external conditions. The slider can be pivoted about line (line pivot) and about point (spherical pivot). The simplest form of line pivot is shown in Fig. 4.2.2.1.

For a set of load and speed condition, pivot provides one equilibrium position. On changing load and/or speed, pressure distribution will change, and as a result there will be change in the film thickness to maintain ratio of film thickness constant.

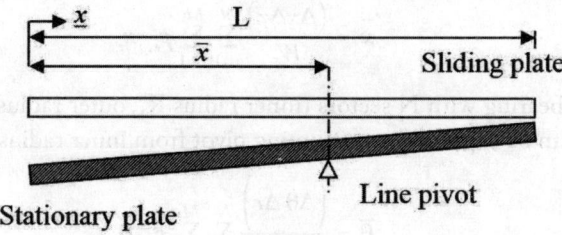

Fig. 4.9 Pivoted slider bearing

4.3 Load

In the previous section, pressure distribution was estimated using numerical method. The pressure developed in the lubricant film between the sliding surface (rotating disk) and the tilted pads (stationary disk) supports the external load applied to the sliding surface and thus prevents contact between the two surfaces. The load carrying capacity (LCC) for a given film profile can be obtained by integrating fluid pressure over the bearing pad area. The load capacity of pad (W_{pad})

$$W_{pad} = \int_0^\beta \int_{R_1}^{R_2} pr \, d\theta \, dr \qquad (4.3.1)$$

Equation 4.3.1 can be rewritten for numerical integration of pressure,

$$W_{pad} = \left(\Delta\theta \, \Delta r\right) \sum_{i=1}^{N} \sum_{j=1}^{M} p_{i,j} r_j \qquad (4.3.2)$$

4.4 Centre of Pressure

The centre of pressure defines the pivot position for tilting pad bearing. In the last chapter, the centre of pressure, hence the pivot position for infinitely long pivoted slider, has been determined by taking moment about the edge of bearing. It is expressed as

$$W \bar{x} = B \int_0^L px \, dx \qquad (3.3.13)$$

where, $\bar{x}$ = distance of edge from the centre of pressure, as depicted in Fig. 4.9. The location of the centre of pressure has been expressed in terms of inclination factor m '$(h_{max} - h_{min})/h_{min}$'

$$\bar{x} = \frac{2(3+m)(1+m)\ln(1+m) - m(6+5m)}{2m[(2+m)\ln(1+m) - 2m]} L \qquad (3.3.14)$$

In case of finite dimension bearings, $\bar{x}$ can be expressed by integrating product of fluid pressure and distance from leading edge over the area of pad and dividing the by applied force. For pivoted flat slider (Fig. 4.9), it can be expressed as:

$$\bar{x} = \frac{(\Delta x \Delta z)}{W} \sum_{i=1}^{N} \sum_{j=1}^{M} p_{i,j} x_i \qquad (4.4.1)$$

Similarly for thrust bearing with N sectors (inner radius R_1, outer radius R_2 and sector angle β) the centre of pressure can be expressed as (assuming pivot from inner radius to outer radius):

$$\bar{\theta} = \frac{(\Delta \theta \Delta r)}{W} \sum_{i=1}^{N} \sum_{j=1}^{M} p_{i,j} \theta_i \, r_j \qquad (4.4.2)$$

4.5 Friction

The estimation of frictional force, due to fluid shearing, is very important as it increases the oil temperature and as a consequence the lubricant viscosity is reduced. In large thrust bearings (consisting of a number of sector shaped pads distributed around a pitch–circle to form a bearing), to deal with excessive friction, space for grooves between pads is designed to enhance the cooling. The total friction (F) of the pad (as described in chapter 3) is determined by integrating the shear stress τ of the Newtonian viscous fluid in x and y directions.

$$F = \int_0^L \int_0^B \left(\pm \frac{dp}{dx} \cdot \frac{h}{2} + \frac{\eta u}{h} \right) .dx.dz \qquad (4.5.1)$$

Equation (4.5.1) is applicable for rectangular slider having length and width equal to L and B, respectively. In this equation ± sign indicates that the friction is different for sliding and stationary plates. To be conservative, only positive side is considered to estimate the friction force,

$$F = \int_0^L \int_0^B \left(\frac{dp}{dx} \cdot \frac{h}{2} + \frac{\eta u}{h} \right) .dx.dz \qquad (4.5.2)$$

To evaluate the friction force using numerical method, following expression can be used for flat slider,

$$F = (\Delta z) \sum_{i=2}^{N} \sum_{j=1}^{M} (p_{i,j} - p_{i-1,j}) \frac{h_i}{2} + \left(\eta U \Delta x \Delta z \right) \sum_{i=1}^{N} \sum_{j=1}^{M} \frac{1}{h_i} \qquad (4.5.3)$$

For annular thrust bearing with N sectors (inner radius R_1, outer radius R_2 and sector angle β) friction force can be expressed as:

$$F = N \int_0^\beta \int_{R_1}^{R_2} \left(\frac{dp}{rd\theta} \cdot \frac{h}{2} + \frac{\eta u}{h} \right) .rd\theta.dr \qquad (4.5.4)$$

Using numerical method, following equation can be used to evaluate the friction force

$$F = \left(N \frac{\Delta r}{\Delta \theta} \right) \sum_{i=1}^{N} \sum_{j=1}^{M} (p_{i,j} - p_{i-1,j}) \frac{h_i}{2} + (N \eta U \Delta \theta \Delta r) \sum_{i=1}^{N} \sum_{j=1}^{M} \frac{r_j}{h_i} \qquad (4.5.5)$$

Frequently Asked Questions

Q.1. How do we find clearance of a thrust pad bearing having leading and trailing film thicknesses of 0.04 and 0.02 mm, respectively?

Ans. Clearance of the thrust bearing can be calculated as $C = \dfrac{h_{min} + h_{max}}{2} = 0.03\,mm$

Q.2. A thrust pad bearing leading and trailing film thicknesses are h_l and h_t. What is the inclination factor of the thrust pad bearing?

Ans. The inclination factor (or ratio) K is defined as
$$K = \frac{h_l - h_t}{L}$$

Q.3. Leading and trailing film thicknesses of a thrust pad bearing are $h_l = 75$ micrometre and $h_t = 20$ micrometre, respectively. What is the inclination factor of the thrust pad bearing if length, L = 30 mm?

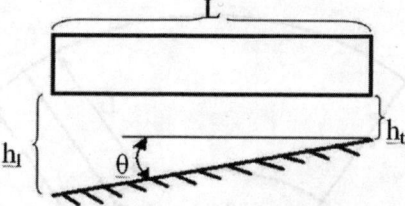

Fig. 4.10 Thrust Pad Bearing for Q3.

Ans. The angle of inclination is $\theta = \tan^{-1}\left(\dfrac{h_l - h_t}{L}\right)$ is 0.0018 radian (0.105 degree). It is worth noting that maintaining 0.105 degree is very difficult. To avoid this difficulty, tilting pad bearings (Fig. 4.9) are preferred over fixed taper thrust bearings.

Q.4. What is the most suitable location of the pivot in a pivoted thrust pad bearing?

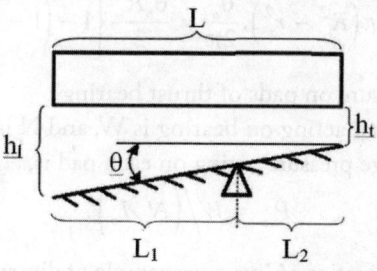

Fig. 4.11 Pivoted Thrust Pad Bearing for Q4.

Ans. Generally, the pivot is positioned at the distance L_1 from the leading edge. The ratio of $L_1:L_2$ is kept as 3:2. To account manufacturing and assembly errors, the L_1 can be kept in the range of 55–62% of L. If there is any possibility of variability in direction of sliding speed, then $L_1:L_2$ should be kept as 1:1.

Q.5. Depict the direction of oil flow on the thrust pad bearing.

Ans. The direction of oil flow has been depicted by series of arrows in Fig. 4.12.

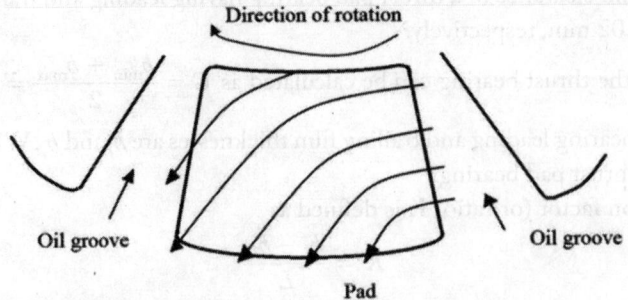

Fig. 4.12 Oil flow in a Thrust Pad Bearing for Q5.

Q.6. The geometry of the sector shape pad is shown in Fig. 4.13. How do we find the pad area?

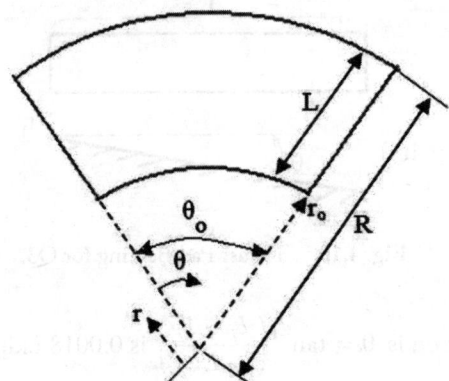

Fig. 4.13 Geometry of the sector shaped pad

Ans. The area of the pad is,

$$A_p = \pi(R^2 - r_o^2) \cdot \frac{\theta_o}{2\pi} = \frac{\theta_o R^2}{2}\left\{1 - \left(1 - \frac{L}{R}\right)^2\right\}$$

Q.7. How to find average pressure on pads of thrust bearing?

Ans. Let us assume that the load acting on bearing is W, and N is number of pads. Assuming A_p as pad of each area, average pressure acting on each pad is,

$$P_{av} = W/(N A_p).$$

Q.8. What is non-dimensionalization? Give one example of dimensional and one example of non-dimensional forms of Reynolds Equations.

Ans. Non-dimensionalization is the process of partial or full removal of units from an equation. Dimensional form of Reynolds equation is

$$\frac{\partial}{\partial x}\left(h^3 \frac{\partial P}{\partial x}\right) + \frac{\partial}{\partial z}\left(h^3 \frac{\partial P}{\partial z}\right) = 6\eta\left(U \frac{\partial h}{\partial x}\right)$$

Non-dimensional form of Reynolds equation is

$$\frac{\partial}{\partial \bar{x}}\left(\bar{h}^3 \frac{\partial \bar{p}}{\partial \bar{x}}\right) + \frac{X^2}{Z^2}\frac{\partial}{\partial \bar{z}}\left(\bar{h}^3 \frac{\partial \bar{p}}{\partial \bar{z}}\right) = \left(\frac{C}{X}\frac{\partial \bar{h}}{\partial \bar{x}}\right)$$

Q.9. Describe 'successive over-relaxation (SOR)' method. How this can be used in solving following equations, where i = 1, N and j = 1,M

$$P_{i,j} = a_i P_{i+1,j} + b_i \left(P_{i,j-1} + P_{i,j+1}\right) + c_i P_{i-1,j} - d_i.$$

Ans. SOR method is an efficient iterative solution method by which residual term is reduced in each iteration. To understand this let us rearrange the equation provided in the question,

$$P_{i,j} = P_{i,j} + w\left[\left(a_i P_{i+1,j} + b_i \left(P_{i,j-1} + P_{i,j+1}\right) + c_i P_{i-1,j} - d_i - P_{i,j}\right)\right]$$

Here, $\left[\left(a_i P_{i+1,j} + b_i \left(P_{i,j-1} + P_{i,j+1}\right) + c_i P_{i-1,j} - d_i - P_{i,j}\right)\right]$ is known as residual term. The term 'w' is termed as over-relaxation factor; chosen value of this factor is greater than 1.0. The objective is to reduce the residual term to zero by successive iterations.

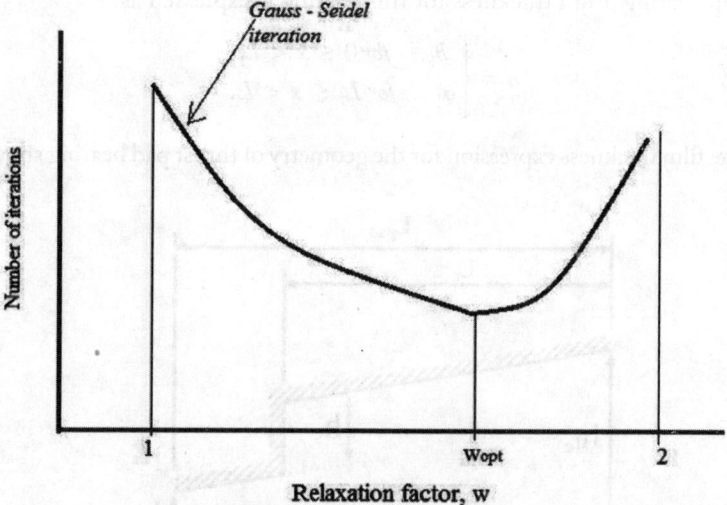

Fig. 4.14 Relaxation factor vs number of iterations

This factor 'w' always lies between $1 \leq w < 2$. When $w = 1$, the iteration scheme simply becomes the same as the Gauss–Seidel method, which has a much slower convergence rate. By trial and error, one can determine an optimum value of the relaxation factor w_{opt} for the fastest convergence. The optimum relaxation factor is a function of the mesh size. Normally the $w \approx 1.7$ is a good starting point for determining w_{opt}.

Q.10. What is the film thickness expression for the geometry of Rayleigh step bearing shown in Figure 4.15.

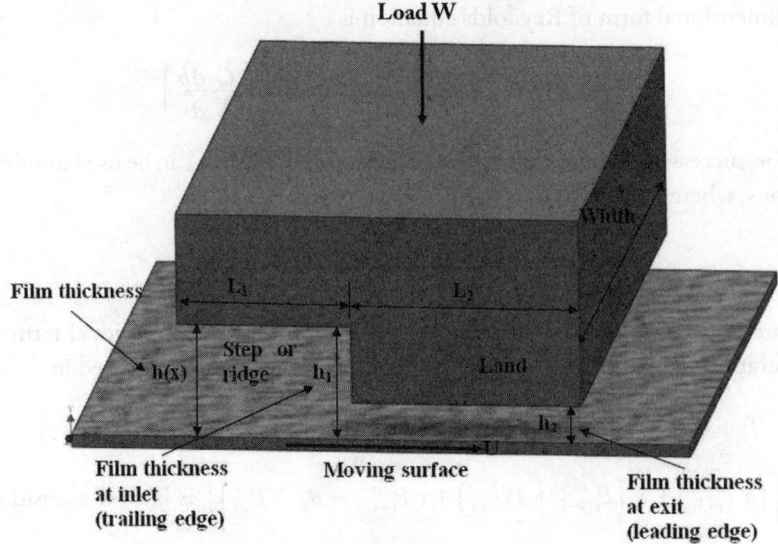

Fig. 4.15 Geometry of Rayleigh step bearing for Q10.

Ans. This is step bearing. Film thickness for this bearing is expressed as:

$$h = \begin{cases} h_1 & \text{for } 0 \leq x < L_1 \\ h_2 & \text{for } L_1 \leq x < L_2 \end{cases}$$

Q.11. What is the film thickness expression for the geometry of thrust pad bearing shown in Fig. 4.16.

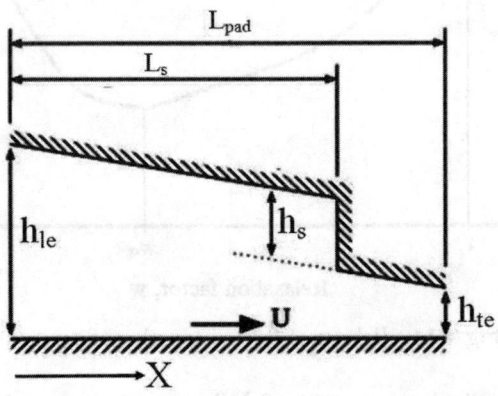

Fig. 4.16 Geometry of a thrust bearing for Q11.

Ans. Film thickness for this bearing is expressed as:

$$h = \begin{cases} h_{le} + \left(h_s - h_{le}\right)\dfrac{x}{L_s} & \text{for } 0 \leq x < L_s \\ \left(h_{le} - h_s\right) + \left(h_{te} - h_{le} + h_s\right)\dfrac{x}{L_{pad} - L_s} & \text{for } L_s \leq x \leq L_{pad} \end{cases}$$

Multiple Choice Questions

Q.1. What is the right representation of partial differential term $\frac{\partial}{\partial x}\left(h^3 \frac{\partial P}{\partial x}\right)$ in central difference scheme?

(a) $\dfrac{h_{i+1/2}^3 \cdot P_{i+1} + h_{i-1/2}^3 \cdot P_{i-1} - 2h_i^3 P_i}{(\Delta x)^2}$

(b) $\dfrac{2h_{i+1/2}^3 \cdot P_{i+1} + h_{i-1/2}^3 \cdot P_{i-1} - h_i^3 P_i}{(\Delta x)^2}$

(c) $\dfrac{h_{i+1/2}^3 \cdot P_{i+1} + h_{i-1/2}^3 \cdot P_{i-1} + 2h_i^3 P_i}{(\Delta x)^2}$

(d) $\dfrac{h_{i+1/2}^3 \cdot P_{i+1} - h_{i-1/2}^3 \cdot P_{i-1} + 2h_i^3 P_i}{(\Delta x)^2}$

Q.2. What is the right representation of partial differential term $\frac{\partial}{\partial z}\left(\frac{\partial P}{\partial z}\right)$ in central difference scheme?

(a) $\dfrac{P_{i+1} + P_{i-1} - 2P_i}{(\Delta z)^2}$

(b) $\dfrac{P_{i+1} + P_{i-1} + 2P_i}{(\Delta z)^2}$

(c) $\dfrac{P_{i+1} + P_{i-1} - P_i}{(\Delta z)^2}$

(d) $\dfrac{P_{i+1} + 2P_{i-1} - P_i}{(\Delta z)^2}$

Q.3. What is the right representation of partial differential term of $\frac{\partial h}{\partial x}$ in central difference scheme?

(a) $\dfrac{h_{i+1} - h_{i-1}}{2\Delta x}$

(b) $\dfrac{h_{i+1} - h_i}{2\Delta x}$

(c) $\dfrac{h_i - h_{i-1}}{2\Delta x}$

(d) $\dfrac{h_{i+1} - h_{i-1}}{\Delta x}$

Q.4. What is the appropriate finite difference operator for the partial derivative $\frac{\partial h}{\partial x}$?

(a) $\dfrac{h_{i+1} - h_{i-1}}{2\Delta x}$

(b) $\dfrac{h_{i+1} - h_i}{\Delta x}$

(c) $\dfrac{h_i - h_{i-1}}{\Delta x}$

(d) All of the above

Q.5. A fixed pad thrust bearing is shown in the following figure

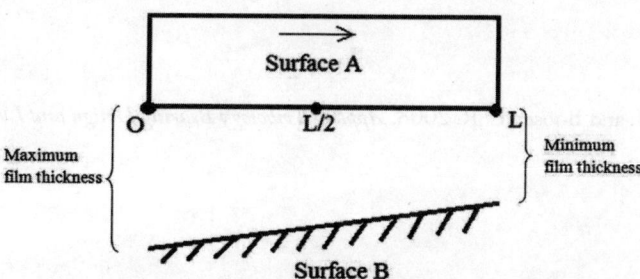

Fig. 4.17 Fixed pad thrust bearing for Q.5.

Which one of the following is true?
(a) Maximum film pressure occurs at the location of minimum film thickness.
(b) Maximum film pressure occurs somewhere in between L/2 and L.
(c) Maximum film pressure occurs somewhere in between 0 and L/2.
(d) Maximum film pressure occurs at the midpoint (L/2) location.

Q.6. What is the film thickness expression for the given geometry?

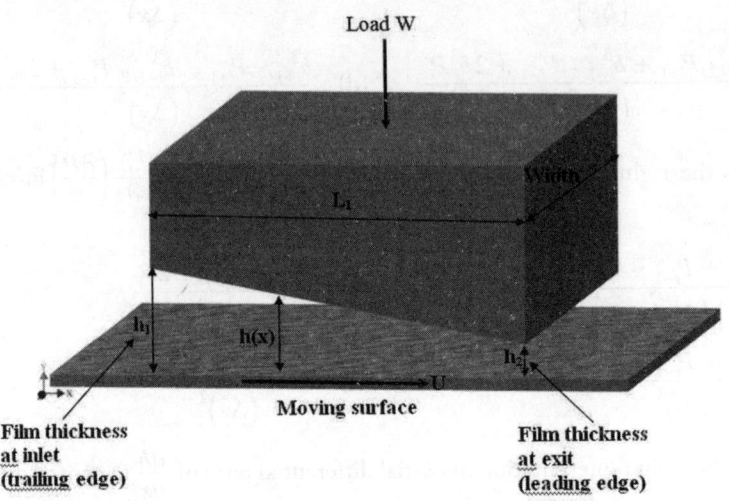

Fig. 4.18 Geometry of taper slider bearing for Q6.

(a) $h(x) = h_1 + (h_2 - h_1) \cdot \dfrac{x}{L}$
(b) $h(x) = h_1 + (h_2 + h_1) \cdot \dfrac{x}{L}$
(c) $h(x) = (h_2 - h_1) \cdot \dfrac{x}{L}$
(d) $h(x) = h_1 - (h_2 - h_1) \cdot \dfrac{x}{L}$

Answers

Q.1. (a)　　Q.2. (a)　　Q.3. (a)　　Q.4. (d)　　Q.5. (b)　　Q.6. (a)

References

- Khonsari, M. M. and Booser, E. R. 2008. *Applied Tribology Bearing Design and Lubrication*, John Wiley.

Chapter 5

Hydrostatic and Squeeze Film Lubrication

The importance of lubrication has been discussed in section 1.4 of this book. Various modes of lubrications have been introduced in section 1.5. Thick lubrication, governed by Reynolds' equation, has been classified into three categories: hydrodynamic, squeeze film, and hydrostatic, as discussed in chapter 3. The hydrodynamic lubrication has been discussed in detail in chapters 3 and 4. The hydrostatic and squeeze film lubrications are described in the present chapter.

5.1 Hydrostatic Lubrication

There are a number of tribo–pairs (i.e., trunnion bearing mechanisms, high precision machine tools, pumps) at interface of which heavy load needs to be supported. In such applications of lubrication, pressurized fluid (demonstrated in Fig. 5.1.1) is used to separate relatively moving (i.e., rotating, oscillating, sliding) surfaces. For this type of lubrication, the term hydrostatic is used to differentiate its operating mechanism from the self-acting hydrodynamic lubrication where speed and wedge effects are the essential requirements. The hydrostatic lubrication is suitable for extremely high load to be levitated at low (even zero) speed or at highly controlled precision position.

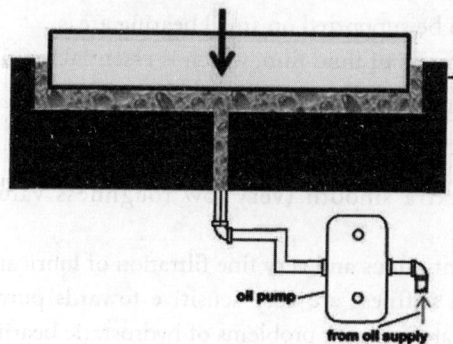

Fig. 5.1.1 Hydrostatic lubrication

5.1.1 Basic concept

This mechanism finds its application in large telescopes, radar tracking units, machine tools and gyroscopes. Similar to hydrostatic, there is a lubrication mechanism termed as aerostatic (here aero means air, nitrogen, helium or any other gas) which separates lightly loaded tribo–pairs operating in extreme temperature range varying from -200 °C to 2000 °C. Aerostatic bearings are also used where special problems of contamination of the system exist. Compared to hydrostatic bearing, aerostatic bearings have the following drawbacks:

- Stiffness of aerostatic bearings due to compressibility of gas is relatively lesser.
- Aerostatic bearings require very sophisticated manufacturing facilities.
- Aerostatic bearings sustain only 5-10% load compared to that of hydrostatic bearings.

As pressure is generated and supplied by external sources, hydrostatic as well as aerostatic are expensive mechanisms to separate two surfaces.

One of the most unique characteristics of hydrostatic bearing is its ability to self-adjusting separation between tribo–pair as per the applied load. For example, on reducing applied load corresponding separation between tribo–pair (film thickness) increases. Similarly on putting additional load on the moving surface, the film thickness decreases. But there is a limit on self-correction of hydrostatic bearings, which is decided by the maximum load for which bearing is designed. If the applied load is more than the design load, the tribo–surfaces do not get separated. To reduce this sensitivity, feedback control system is used, which increases the cost of overall system. Often concept of hybrid bearing, that is, 'hydrodynamic + hydrostatic' or 'aerodynamic + aerostatic' is used to achieve the best of both the mechanisms of fluid film lubrications.

5.1.2   Advantages and limitations

Advantages of hydrostatic bearings are:

- ☺ Low friction and zero wear; in hydrostatic bearings mechanical contact between solid surfaces is prohibited, even at start-up and at low speeds.
- ☺ Process fluids such as water, and liquid metals, which are poor lubricants, can be used in hydrostatic bearings.
- ☺ Very high loads can be supported on small bearing areas.
- ☺ Maintains high stiffness of fluid film, which is essential requirement for position control of the shaft.

Disadvantages of hydrostatic bearings are:

- ☹ Requirement of extra smooth (very low roughness values) surfaces increases its manufacturing cost
- ☹ Requirement of continuous and very fine filtration of lubricant
- ☹ Load capacity and stiffness are very sensitive towards pump performance. Therefore, problems of pump are inherent problems of hydrostatic bearings.
- ☹ High running cost

5.1.3   Viscous flow through rectangular slot

The hydrostatic bearings are meant to maintain a full lubricant film (no mechanical contact) between tribo–pair by employing externally pressurized lubricants to lift the rotating part. To fulfil this purpose, hydrostatic bearings require slots filled with pressurized lubricant. In this section an equation for the flow of viscous liquid through a rectangular slot has been derived. Let us consider a slot of width b, depth l, and thickness h, as shown in Fig. 5.1.3.1. To derive the relation between pressure difference $(P_1 - P_2)$ and lubricant flow rate, assume the width b is very large compared with the thickness h, so that the losses at the ends of the slot can be neglected.

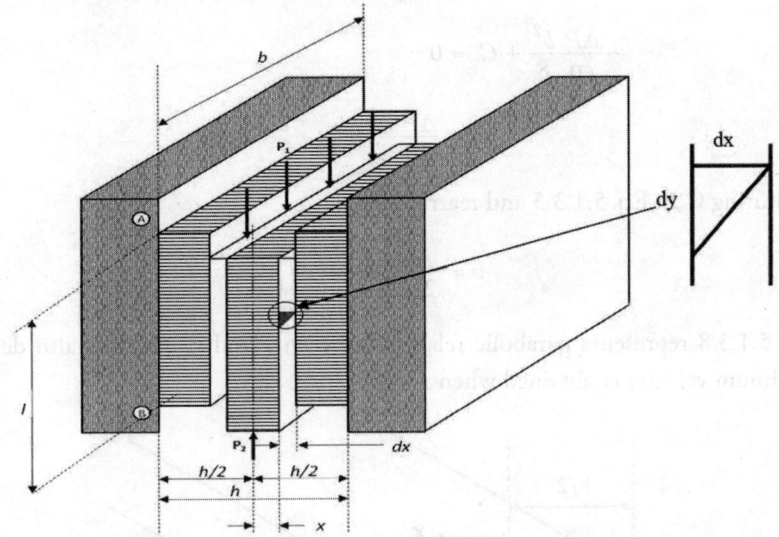

Fig. 5.1.3.1 Small element of oil film is extruded by pressure difference acting upon it

To derive the relation, consider a small rectangular liquid slab of thickness 2x, as shown in Fig. 5.1.3, being extruded down through the slot. The coordinate axis is in the centre of the slot, the force acting downward due to pressure difference is given by

$$F = 2xb\Delta P \qquad (5.1.3.1)$$

where, ΔP is the difference in pressure between points A and B.

If v is fluid velocity varying as a function of x, then resisting force upward (from Newton's law of viscosity) is given as:

$$F = \text{area} * \text{shear\_stress}$$
$$= (2lb) * \left(-\eta \frac{dv}{dx}\right) \qquad (5.1.3.2)$$

As slot is stationary, both of its walls will be stationary and fluid velocity will be the maximum at the centre. Here sign of $\frac{dv}{dx}$ will be negative which means as x increases v will decrease.

Using Eqs. 5.1.3.1 and 5.1.3.2 for static equilibrium, we get

$$2xb\Delta P = -2lb\eta \frac{dv}{dx} \qquad (5.1.3.3)$$

On rearranging,

$$-\frac{\Delta P}{l\eta} x \, dx = dv \qquad (5.1.3.4)$$

On integrating Eq. 5.1.3.4,

$$-\frac{\Delta P}{l\eta}\frac{x^2}{2} + C = v \qquad (5.1.3.5)$$

To find the value of constant C, use boundary conditions at $x = \pm h/2$, $v = 0$,

$$-\frac{\Delta P}{l\eta}\frac{h^2}{8} + C = 0 \qquad (5.1.3.6)$$

or

$$C = \frac{\Delta P}{8\eta}\frac{h^2}{l} \qquad (5.1.3.7)$$

On substituting C in Eq. 5.1.3.5 and rearranging,

$$v = \frac{\Delta P}{2\eta l}\left[\frac{h^2}{4} - x^2\right] \qquad (5.1.3.8)$$

Equation 5.1.3.8 represents parabolic relation between v and x, which is also depicted in Fig. 5.1.3.2. Maximum velocity is obtained when $x = 0$.

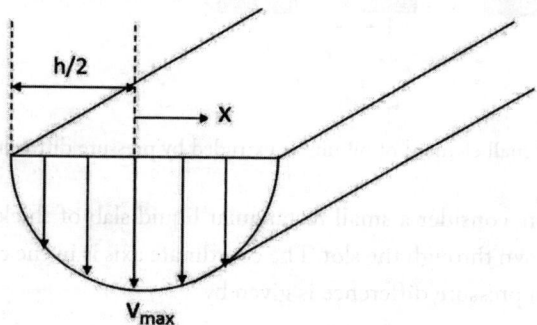

Fig. 5.1.3.2 Parabolic velocity profile across slot

$$v_{max} = \frac{\Delta P h^2}{8\eta l} \qquad (5.1.3.9)$$

Average height of the parabola is expressed as two-thirds of the maximum,

$$v_{avg} = \frac{2}{3}v_{max} = \frac{2\Delta P h^2}{24\eta l} = \frac{\Delta P h^2}{12\eta l} \qquad (5.1.3.10)$$

Volume of flow, Q, required to create the pressure difference in the slot is given by the average velocity multiplied by area of cross section

$$Q = v_{avg} * hb$$

$$= \frac{\Delta P h^2}{12\eta l}hb \qquad (5.1.3.11)$$

On rearranging

$$Q = \frac{\Delta P b h^3}{12\eta l} \qquad (5.1.3.12)$$

This equation provides an understanding that to pass viscous fluid through the rectangular slot causes pressure drop. To get an idea of such pressure drop on the load support imposed on hydrostatic bearing, it is necessary to understand 'type and configurations' of hydrostatic bearings.

5.1.4 Types and configurations

In the previous section, a need of lubricant slot to support the load has been expressed. With various arrangements of slots and bearing geometries, hydrostatic bearings can be designed to carry axial loads (thrust bearing), radial loads (journal bearings) and combined thrust and radial loads as illustrated in Fig. 5.1.4.1.

Figures 5.1.4.1(a) and (b) show configurations of thrust and journal bearings respectively. Figure 5.1.4.1(c) illustrates a hybrid (thrust and journal) bearing and Fig. 5.1.4.1(d) depicts a spherical bearing that finds its use when misalignment is of the concern.

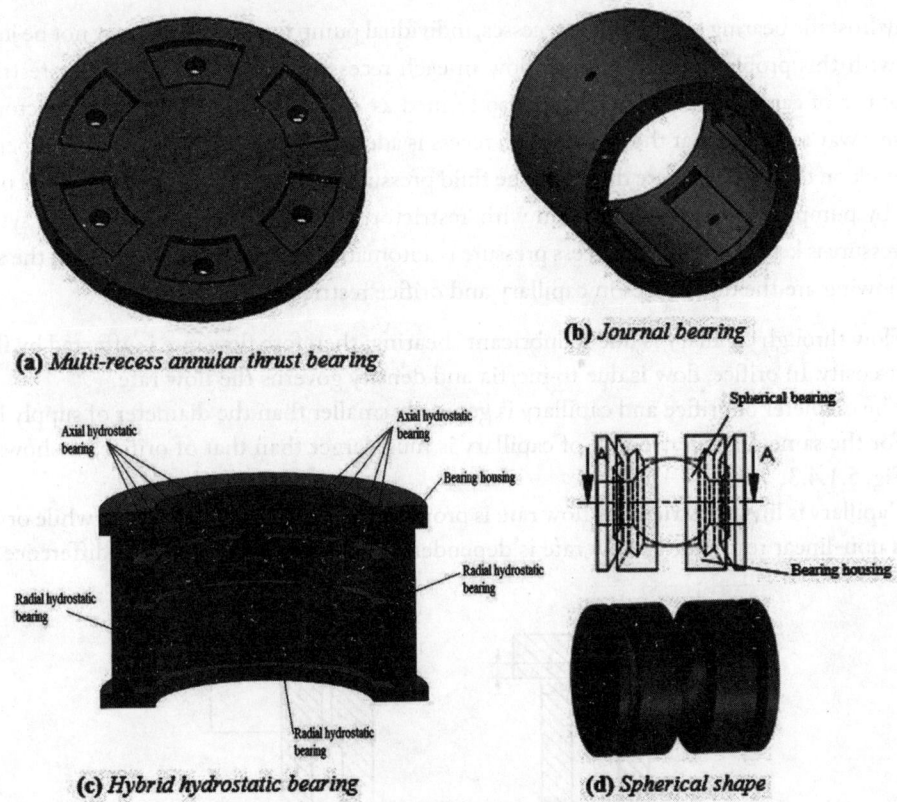

Fig. 5.1.4.1 Types of hydrostatic bearing

It is worth noting that in almost every design of hydrostatic bearings, shown in Fig. 5.1.4.1, multi-slots (also known as recesses, pockets) have been used. The multiple recesses (typically four to six in journal bearing) hydrostatic bearing can support asymmetric loads. Generally depth of recess is kept several times (i.e., 0.5 mm) greater than the film thickness. Pressurized oil in recess is delivered either directly by pump or through restrictor (i.e., orifice, capillary) as shown in Fig. 5.1.4.2.

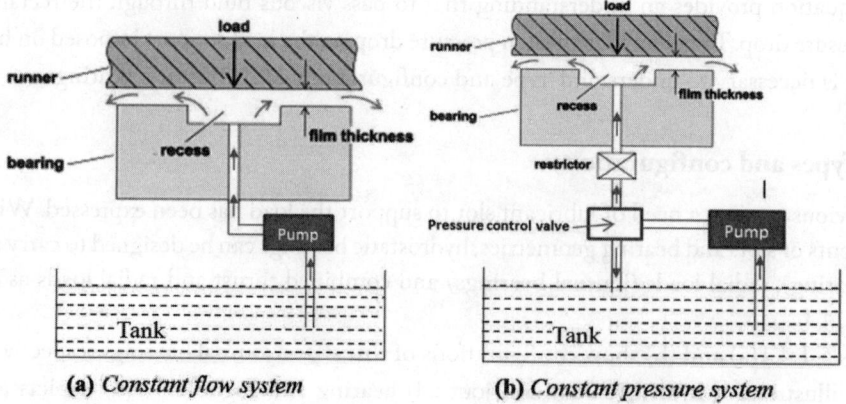

Fig. 5.1.4.2 Pressurized oil in recess directly by pump or through restrictor

In a hydrostatic bearing having many recesses, individual pump for each recess may not be justified. To deal with this problem, throttling the flow in each recess through capillary/orifice restrictor is used. The use of capillary/orifice, which is also termed as 'compensation device' or 'restrictor', is an inexpensive way to insure that the flow in each recess is adequately independent. Such compensation devices work on the fluid pressure drop and the fluid pressure in recess is lesser than the fluid pressure supplied by pump. The hydrostatic system with 'restrictor' is known as 'constant pressure system' as supply pressure is kept constant and recess pressure is automatically adjusted depending on the applied load. Following are the differences in capillary and orifice restrictors:

- Flow through capillary is due to lubricant shearing; therefore flow rate is affected by fluid viscosity. In orifice, flow is due to inertia and density governs the flow rate.
- The diameter of orifice and capillary is generally smaller than the diameter of supply line. For the same diameter, length of capillary is must larger than that of orifice, as shown in Fig. 5.1.4.3.
- Capillary is linear restrictor as flow rate is proportional to pressure difference, while orifice is non-linear restrictor as flow rate is dependent on square root of pressure difference.

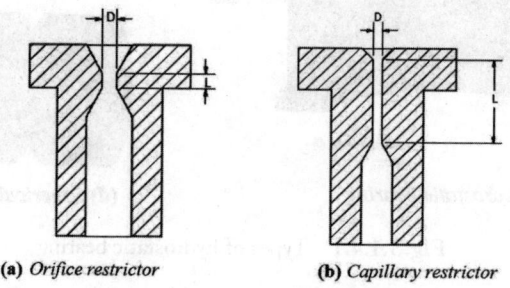

Fig. 5.1.4.3 Compensation devices

As per Fig. 5.1.4.3, compensation devices are connected between pump and recess, and pressure drop occurs when fluid flows through the compensation devices. Equations 5.1.4.1 and 5.1.4.2 relate the pressure with flow rate.

For an orifice
$$Q = A_o C_d \sqrt{\frac{2}{\rho}(P_s - P_r)} \qquad (5.1.4.1)$$

Here A_o is the cross section area of orifice and C_d (~0.8) is discharge coefficient.

For a capillary
$$Q = \frac{\pi d_c^4}{128 \eta L_c}(P_s - P_r) \qquad (5.1.4.2)$$

Here d_c and L_c are diameter and length of the capillary tube.

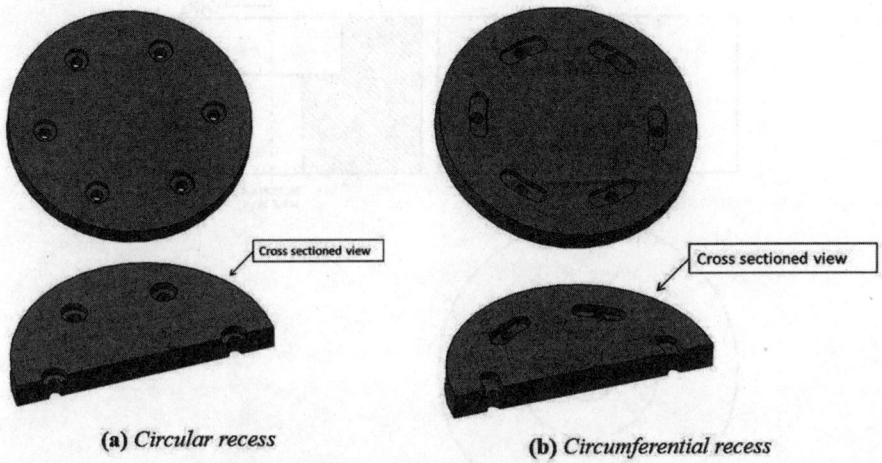

(a) *Circular recess*   (b) *Circumferential recess*

Fig. 5.1.4.4 Annular thrust bearings

Two examples of recesses for annular thrust bearings have been depicted in Fig. 5.1.4.4.

5.1.5 Circular step thrust bearing

Circular step bearings, commonly used in supporting large vertical shafts in turbo–generators, are the most common type of hydrostatic thrust bearings. Figure 5.1.5.1 shows a flat plate thrust bearing with a single lubricant supply at feed pressure P_S and recess pressure P_r. In the present heading load capacity, flow requirement, stiffness and power loss are described. Reduction in power loss, which depends on flow rate, supply pressure and speed of slider, is of special interest in the design of large hydrostatic thrust bearings. The stiffness of the hydrostatic bearing is of unique importance for centring of high precision milling machine, gyroscope, large arena movable seating areas, telescope bearing, etc.

To derive the pressure distribution for this bearing, refer to Eq. 3.2.3.1 of chapter 3,

$$\frac{\partial}{\partial x}\left(\frac{h^3}{\eta}\frac{\partial p}{\partial x}\right) + \frac{\partial}{\partial z}\left(\frac{h^3}{\eta}\frac{\partial p}{\partial z}\right) = 0 \qquad (3.2.31)$$

In cylindrical coordinates, this equation can be rewritten as:

$$\frac{d}{dr}\left(r\frac{h^3}{\eta}\frac{dp}{dr}\right) = 0 \qquad (5.1.5.1)$$

On integrating
$$r\frac{h^3}{\eta}\frac{dp}{dr} = C_1$$

On rearranging
$$dp = \frac{\eta}{rh^3} C_1 dr$$

On integrating
$$\int dp = \frac{\eta}{h^3} C_1 \int \frac{1}{r} dr + C_2$$

$$p = \frac{\eta}{h^3} C_1 \ln(r) + C_2 \quad (5.1.5.2)$$

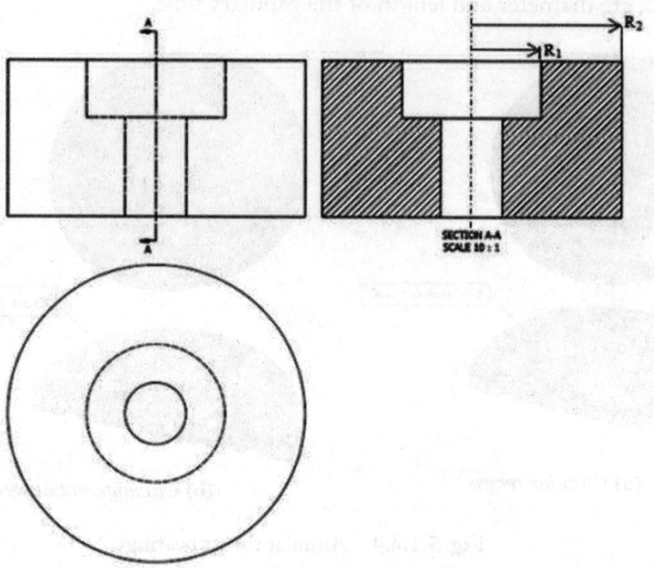

Fig. 5.1.5.1 Flat plate thrust bearing without a compensating element

Two boundary conditions ($p = P_r$ at $r = R_1$, and $p = 0$ at $r = R_2$) can be used to find constants C_1 and C_2 of Eq. 5.1.5.2. On using these boundary conditions,

$$P_r = \frac{\eta}{h^3} C_1 \ln(R_1) + C_2$$

$$0 = \frac{\eta}{h^3} C_1 \ln(R_2) + C_2$$

$$P_r = \frac{\eta}{h^3} C_1 \ln(R_1 / R_2)$$

$$C_1 = P_r \frac{h^3}{\eta} \ln(R_2 / R_1) \quad (5.1.5.3)$$

$$C_2 = -P_r \ln(R_2 / R_1) \ln(R_2) \quad (5.1.5.4)$$

$$p = \frac{\eta}{h^3} P_r \frac{h^3}{\eta} \ln(R_2 / R_1) \ln(r) - P_r \ln(R_2 / R_1) \ln(R_2)$$

$$p = P_r \ln(R_2 / R_1) \ln(r) - P_r \ln(R_2 / R_1) \ln(R_2)$$

$$p = P_r \frac{\ln(r/R_2)}{\ln(R_1/R_2)} \qquad (5.1.5.5)$$

The pressure profile predicted from Eq. 5.1.5.5 is shown in Fig. 5.1.5.2.

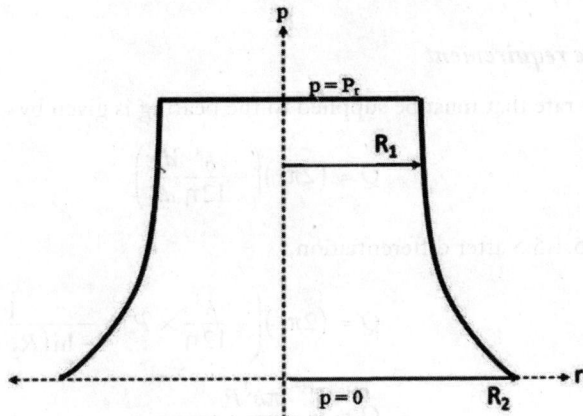

Fig. 5.1.5.2 Pressure distribution

5.1.5.1 *Load carrying capacity*

To determine the load carrying capacity let us integrate pressure over the bearing area and we get

$$W = \pi R_1^2 P_r + \int_{R_1}^{R_2} 2\pi r P \, dr$$

Using expression of P from Eq. 5.1.5.5

$$W = \pi R_1^2 P_r + \int_{R_1}^{R_2} 2\pi r P_r \frac{\ln(r/R_2)}{\ln(R_1/R_2)} dr$$

$$W = \pi R_1^2 P_r + \frac{2\pi P_r}{\ln(R_1/R_2)} \int_{R_1}^{R_2} r\left[\ln(r) - \ln(R_2)\right] dr$$

$$W = \pi R_1^2 P_r + \frac{2\pi P_r}{\ln(R_1/R_2)} \left[\left. \frac{r^2}{2} \ln(r) \right|_{R_1}^{R_2} - \int_{R_1}^{R_2} \frac{r^2}{2} \frac{1}{r} dr - \left. \frac{r^2}{2} \ln(R_2) \right|_{R_1}^{R_2} \right]$$

$$W = \pi R_1^2 P_r + \frac{2\pi P_r}{\ln(R_1/R_2)} \left[\frac{R_2^2}{2} \ln(R_2) - \frac{R_1^2}{2} \ln(R_1) - \frac{1}{4}\left(R_2^2 - R_1^2\right) - \frac{1}{2}\left(R_2^2 - R_1^2\right)\ln(R_2) \right]$$

$$W = \pi R_1^2 P_r + \frac{2\pi P_r}{\ln(R_1/R_2)} \left[\frac{R_1^2}{2}\left(\ln(R_2) - \ln(R_1)\right) - \frac{1}{4}\left(R_2^2 - R_1^2\right) \right]$$

$$W = \pi R_1^2 P_r + \frac{2\pi P_r}{\ln(R_1/R_2)} \left[\frac{R_1^2}{2}\left(\ln(R_2) - \ln(R_1)\right) \right] - \frac{2\pi P_s}{\ln(R_1/R_2)} \left[\frac{1}{4}\left(R_2^2 - R_1^2\right) \right]$$

$$W = \pi R_1^2 P_r - \pi R_1^2 P_r - \frac{\pi P_s}{\ln(R_1/R_2)}\left[\frac{1}{2}\left(R_2^2 - R_1^2\right)\right]$$

$$W = \frac{\pi P_r \left(R_2^2 - R_1^2\right)}{2\ln(R_2/R_1)} \tag{5.1.5.6}$$

5.1.5.2 Flow rate requirement

The volumetric flow rate that must be supplied to the bearing is given by

$$Q = (2\pi r)\left(-\frac{h^3}{12\eta}\frac{dp}{dr}\right) \tag{5.1.5.7}$$

Substituting Eq. 5.1.5.5 after differentiation,

$$Q = (2\pi r)\left(-\frac{h^3}{12\eta} \times P_r \times \frac{1}{-\ln(R_2/R_1)r}\right)$$

$$Q = \frac{\pi h^3 P_r}{6\eta \ln(R_2/R_1)} \tag{5.1.5.8}$$

P_r is common in Eqs. 5.1.5.6 and 5.1.5.8. On rearranging, flow rate and load carrying capacity can be related as:

$$W = \frac{3\eta Q(R_2^2 - R_1^2)}{h^3} \tag{5.1.5.9}$$

Using Eq. 5.1.5.9 it can be noted that η, R_2 and R_1 are constants for a particular bearing geometry, so the load carrying capacity (W) varies as a function of flow rate (Q) and film thickness (h). On maintaining flow rate constant, increase in load (W) decreases film thickness to maintain operating equilibrium. Hence bearing is self-compensating.

5.1.5.3 Bearing stiffness

Stiffness, an important parameter in hydrostatic bearing design, is defined as

$$K = -\frac{dW}{dh} \tag{5.1.5.10}$$

Here negative sign denotes that on decreasing load (W) film thickness (h) increases and vice-versa. On substituting expression of W from Eq. 5.1.5.9,

$$K = -\frac{d\left(\frac{3\eta Q(R_2^2 - R_1^2)}{h^3}\right)}{dh}$$

$$K = -\frac{(-3)}{h}\frac{3\eta Q(R_2^2 - R_1^2)}{h^3} = \frac{3}{h}W$$

$$K = \frac{3W}{h} \tag{5.1.5.11}$$

5.1.5.4 *Power loss*

The total power loss consists of viscous dissipation (H_v) and pumping loss (H_p). To evaluate viscous dissipation, it is essential to find friction torque due to shearing of oil film, which is given as

$$T = \int_A \tau r \, dA \tag{5.1.5.12}$$

$$T = \int_{R_1}^{R_2} \frac{\eta U}{h} r \, 2\pi r \, dr$$

or

$$T = \int_{R_1}^{R_2} 2\pi \frac{\eta \omega}{h} r^3 \, dr$$

or

$$T = \frac{\pi \eta \omega}{2h} \left(R_2^4 - R_1^4 \right) \tag{5.1.5.13}$$

$$H_v = T\omega = \frac{\pi \eta \omega^2}{2h} \left(R_2^4 - R_1^4 \right) \tag{5.1.5.14}$$

Pumping loss (H_p) depends on the fluid pressure and the flow rate. The fluid pressure supplied from pump (P_s) is generally greater than the recess pressure (P_r).

$$H_p = P_s Q = P_s \frac{\pi h^3 P_r}{6\eta \ln(R_2/R_1)} \tag{5.1.5.15}$$

5.1.6 Rectangular thrust bearing

There are several applications where rectangular thrust bearing is used. Unlike circular step bearings, there are pressure gradients in rectangular thrust bearings both in X- and Z-directions as shown in Fig. 5.1.6.1.

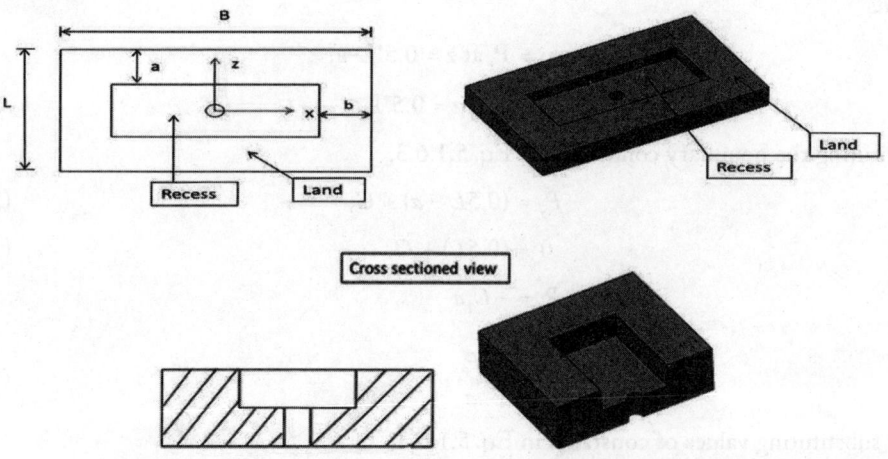

Fig. 5.1.6.1 Geometry of rectangular hydrostatic pad bearing

Considering an incompressible fluid, the Reynolds' equation for this type of bearing can be expressed as (refer to Eq. 3.2.31 of chapter 3),

$$\frac{\partial}{\partial x}\left(\frac{h^3}{\eta}\frac{\partial p}{\partial x}\right) + \frac{\partial}{\partial z}\left(\frac{h^3}{\eta}\frac{\partial p}{\partial z}\right) = 0 \qquad (3.2.31)$$

Considering a constant film thickness along X-and Z-axes (i.e., bearing and runner surfaces are parallel), the above equation is modified as

$$\frac{\partial^2 p}{\partial x^2} + \frac{\partial^2 p}{\partial z^2} = 0 \qquad (5.1.6.1)$$

To solve this equation, pressure boundary conditions are required. Pressure within recess remains constant and equals to recess pressure (P_r), which means at z = 0.5*L-a fluid pressure, p = P_r and at x = 0.5*B-b fluid pressure p = P_r. Pressure at outer edges of land will be atmospheric, which means at z = 0.5*L fluid pressure p = 0 and at x = 0.5*B fluid pressure p = 0. To solve Eq. 5.1.6.1 with boundary condition, numerical method (as described in section 4.2 of chapter 4) is required. However to get a feel of hydrostatic bearing rectangular recess, short/long bearing approximation can be made. For example if the length (L) of the lands in the bearing is much smaller than the width (B) of the lands, most of the fluid supplied to the bearing by the pump leaves by flowing from the recess over the lands in the direction of Z-axis. Such wide rectangular thrust bearings find their use in machine tool slide ways. It may be assumed that there is negligible change of pressure over the lands in the direction of X-axis at the ends of the recess. Thus Reynolds' equation reduces to

$$\frac{\partial^2 p}{\partial z^2} = 0 \qquad (5.1.6.2)$$

On integrating Eq. 5.1.5.2 twice,

$$p = C_1 z + C_2 \qquad (5.1.6.3)$$

where C_1 and C_2 are the constants of integration which can be evaluated by using the boundary conditions

$$p = P_r \text{ at } z = 0.5*L-a,$$

and
$$p = 0 \text{ at } z = 0.5*L.$$

Substituting the boundary conditions in Eq. 5.1.6.3,

$$P_r = (0.5L - a) + C_2 \qquad (5.1.6.4)$$

$$0 = (0.5L) + C_2 \qquad (5.1.6.5)$$

$\Rightarrow \qquad P_r = -C_1 a$

$\Rightarrow \qquad C_1 = -\dfrac{P_r}{a} \qquad (5.1.6.6)$

After substituting values of constants in Eq. 5.1.6.3,

$$p = \frac{P_r}{a}\left(\frac{L}{2} - z\right) \qquad (5.1.6.7)$$

The pressure gradient is linear as expressed in Eq. 5.1.6.7

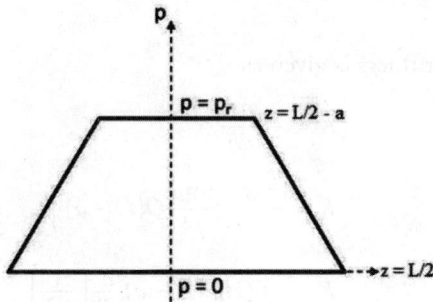

Fig. 5.1.6.2 Pressure distribution of rectangular hydrostatic pad bearing

5.1.6.1. Load carrying capacity

The load capacity of the bearing is given as

$$W = P_r.(L - 2a)(B - 2b) + 2(B - 2b) \int_{0.5L-a}^{0.5L} \frac{P_r}{a}(0.5L - z)dz \quad (5.1.6.8)$$

$$W = P_r.(L - 2a)(B - 2b) + 2(B - 2b)\left[\frac{P_r}{a}\left(\int_{0.5L-a}^{0.5L} 0.5L\,dz - \int_{0.5L-a}^{0.5L} z\,dz\right)\right]$$

$$W = P_r.(L - 2a)(B - 2b) + 2(B - 2b)\frac{P_r}{a}\left[0.5L(a) - 0.5 * \left\{(0.5L)^2 - (0.5L - a)^2\right\}\right]$$

$$W = P_r.(L - 2a)(B - 2b) + 2(B - 2b)\frac{P_r}{a}\left[0.5L(a) - 0.5 * \left\{La - a^2\right\}\right]$$

$$W = P_r.(L - 2a)(B - 2b) + (B - 2b)P_r.a$$

$$W = P_r.(L - a)(B - 2b)$$

$$W = P_r.(L - a)(B - 2b) \quad (5.1.6.9)$$

5.1.6.2 Flow rate requirement

The flow through wide rectangular thrust bearing, shown in Fig. 5.1.6.1, can be considered as one dimensional. The pressure in the recess, p_r is constant and linearly decreasing along the land of clearance h, in the Z-direction. The total (from two sides of land) volumetric flow rate, Q in Z-direction for uniform pressure along the Z-axis is given as

$$Q = 2 * width\left(-\frac{h^3}{12\eta}\frac{(P_r - 0)}{(0.5 * L - a - 0.5 * L)}\right)$$

$$\Rightarrow \quad Q = (B - 2b)\left(\frac{h^3}{6\eta}\frac{P_r}{a}\right) \quad (5.1.6.10)$$

The recess pressure, P_r is common in Eqs. 5.1.6.9 and 5.1.6.10. On rearranging, flow rate and load carrying capacity can be related as:

$$W = \frac{6\eta a}{h^3}Q.(L - a) \quad (5.1.6.11)$$

5.1.6.3 *Bearing stiffness*

As expressed in Eq. 5.1.5.10, stiffness is given as

$$K = -\frac{dW}{dh}$$

$$K = -\frac{d}{dh}\left(\frac{6\eta a}{h^3} Q(L - a)\right) \quad (5.1.6.12)$$

$$K = -6Q\eta a(L - a)\frac{d}{dh}\left(\frac{1}{h^3}\right)$$

$$K = -6Q\eta a(L - a)\left(\frac{-3}{h^4}\right)$$

$$K = 18Q\eta \frac{a(L - a)}{h^4} \quad (5.1.6.13)$$

5.1.6.4 *Power loss*

The total power loss consists of viscous dissipation (H_v) and pumping loss (H_p). However sliding is almost negligible and only pumping loss is calculated.

Pumping loss (H_p) depends on fluid pressure and flow rate.

$$H_p = P_s Q = P_s(B - 2b)\left(\frac{h^3}{6\eta}\frac{P_r}{a}\right) \quad (5.1.6.14)$$

5.1.7 Hydrostatic journal bearing

The basic mechanism of hydrostatic lubrication for journal bearing is the same as that for thrust bearing. Hydrostatic journal bearings are useful to avoid metal to metal contact under heavy static load conditions. The hydrostatic journal bearings can be used in synchronous condenser, rolling mills, etc.

The analysis of hydrostatic journal bearing is difficult as film thickness is not uniform (assumption of parallel surfaces does not work for journal bearings). To understand this let us consider hydrostatic lift mechanism shown in Fig. 5.1.7.1, in which 'e' is shown as eccentricity (separation between centres of journal and bearing) and 'h' is separation between surfaces of journal and bearing. It is easier to represent eccentricity 'e' in terms of eccentricity ratio=e/c. If the pressure and quantity of flow rate are in correct proportions, the journal (also termed as shaft), will be raised and supported on oil film. When the shaft has settled in the bearing and has made metal to metal contact, the eccentricity ratio ($\varepsilon = e/c$) is equal to 1.0. When the shaft and bearing are concentric, the eccentricity ratio is zero.

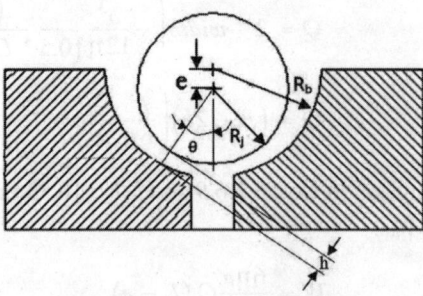

Fig. 5.1.7.1 Hydrostatic lift

In hydrostatic journal bearing, the film thickness h does not remain constant, but depends upon the angular position (θ) and eccentricity of journal in the bearing.

$$h = c(1 - \varepsilon \cos \theta) \quad (5.1.7.1)$$

The two main types of hydrostatic journal bearings are:

(a) Multi-pad bearings: A number of thrust pads with curved surfaces to form a cylindrical bearing surface.
(b) Multi-recess bearings

Basic difference between the operation of multi-pad and multi-recess bearings is the drainage grooves that exist between adjacent pads. The drainage grooves between pads are regions of low pressure and therefore contribute nothing to the bearing load capacity. One kind of multi-recess hydrostatic journal bearing has been shown in Fig. 5.1.7.2.

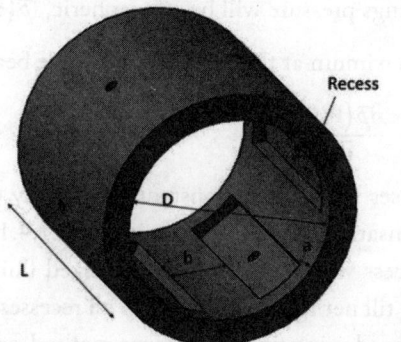

Fig. 5.1.7.2 Geometry of multi-recess hydrostatic journal bearing

A four recess hydrostatic journal bearing has been shown in Fig. 5.1.7.3. The lubricant is supplied from a constant pressure manifold at a pressure P_s, passing through restrictors in the clearance space. The pressures (P_{r1}, P_{r2}, P_{r3}, and P_{r4}) change from initial values to take up the net load on journal.

To derive the pressure distribution for this bearing, refer to Eq. 3.2.3.1 of chapter 3,

$$\frac{\partial}{\partial x}\left(\frac{h^3}{\eta}\frac{\partial p}{\partial x}\right) + \frac{\partial}{\partial z}\left(\frac{h^3}{\eta}\frac{\partial p}{\partial z}\right) = 0 \quad (3.2.31)$$

For an iso-viscous lubricant (as low viscosity lubricant is used in the hydrostatic bearings), Eq. 3.2.31 is given as

$$\frac{\partial}{\partial x}\left(h^3\frac{\partial p}{\partial x}\right) + \frac{\partial}{\partial z}\left(h^3\frac{\partial p}{\partial z}\right) = 0 \quad (5.1.7.2)$$

On non-dimensionalization using $\bar{p} = p/P_s ; \bar{h} = h/c; \theta = x/R; \bar{z} = z/(0.5L)$

Assuming bearing infinitely long, and replacing x with Rθ,

$$\frac{\partial}{\partial \theta}\left(\bar{h}^3\frac{\partial \bar{p}}{\partial \theta}\right) + \left(\frac{2R}{L}\right)^2 \frac{\partial}{\partial \bar{z}}\left(\bar{h}^3\frac{\partial \bar{p}}{\partial \bar{z}}\right) = 0 \quad (5.1.7.3)$$

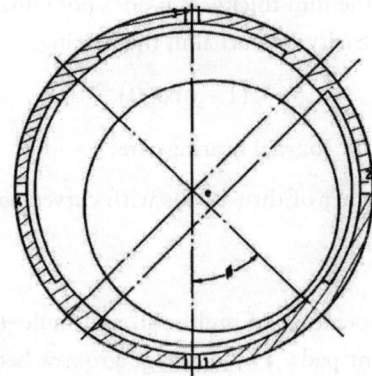

Fig. 5.1.7.3 Four recess hydrostatic journal bearing

As at both the ends of bearings pressure will be atmospheric, $\bar{p}(\theta, \pm 1) = 0$ (5.1.7.4)

Fluid pressure will be the maximum at the mid-length of the bearing,

$$\frac{\partial \bar{p}(\theta, 0)}{\partial \bar{z}} = 0 \qquad (5.1.7.5)$$

Assuming pressure in recesses will remain constant during any of iteration, the recess pressure depends on the type of compensation device, as given in Eq. 5.1.4.1 and Eq. 5.1.4.2. The flow rate passing through any of the recess will depend on the localized fluid film thickness and finding it requires a number of iterations till net force supported by all recesses equals to the applied load. The complete solutions can be obtained using finite difference method, appropriate boundary conditions, and Newton–Raphson iterations.

To get a feel of hydrostatic journal bearing, let us consider a typical hydrostatic bearing shown in Fig. 5.1.7.4. Such bearings are used in large alternators and large turbines. O_b is the centre of the bearing and O_J is the centre of the journal, which is displaced from O_b by a small eccentricity e. The film thickness given as $h = c(1 - \varepsilon \cos \theta)$

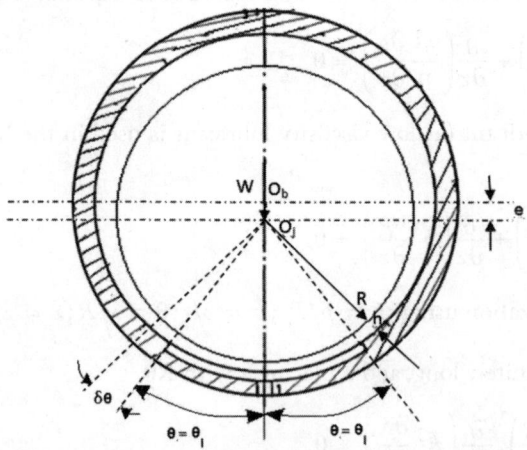

Fig. 5.1.7.4 A loaded hydrostatic journal bearing

Assuming the bearing is very long, L/D > 4, and the axial component of the flow can be neglected. The flow through an elemental slot is

$$Q = -\frac{Lh^3}{12\eta} \frac{dp}{Rd\theta} \quad (5.1.7.6)$$

Here L is bearing length. On substituting expression of film thickness, and rearranging, we get

$$dp = \frac{-12\eta QR}{Lc^3} \frac{d\theta}{(1 - \varepsilon\cos\theta)^3} \quad (5.1.7.5)$$

Integrating Eq. 5.1.7.5 with the boundary condition, p=0 at $\theta = \theta_1$.

$$p = \frac{-12\eta QR}{Lc^3} \left[\frac{\varepsilon\sin\theta(4 - \varepsilon^2 - 3\varepsilon\cos\theta)}{2(1-\varepsilon^2)(1-\varepsilon\cos\theta)} + \frac{2+\varepsilon^2}{(1-\varepsilon^2)^{5/2}} \tan^{-1}\left(\frac{1+\varepsilon}{\sqrt{1+\varepsilon^2}} \tan\frac{\theta}{2}\right) - K_1 \right] \quad (5.1.7.6)$$

where,

$$K_1 = \frac{\varepsilon\sin\theta_1(4 - \varepsilon^2 - 3\varepsilon\cos\theta_1)}{2(1-\varepsilon^2)(1-\varepsilon\cos\theta_1)^2} + \frac{2+\varepsilon^2}{(1-\varepsilon^2)^{5/2}} \tan^{-1}\left(\frac{1+\varepsilon}{\sqrt{1+\varepsilon^2}} \tan\frac{\theta_1}{2}\right) \quad (5.1.7.7)$$

5.1.7.1 Flow requirement

The value of Q from Eq. 5.1.7.6 can be determined using boundary condition $\theta = 0$, $p = p_r$, where p_r is the recess pressure. The expression for flow rate is given as

$$Q = \frac{Lc^3}{6\eta RK_1} \cdot p_r \quad (5.1.7.8)$$

5.1.7.2 Load carrying capacity

Load carrying capacity can be determined by integrating fluid pressure over supporting area, such as

$$W = 2\int_0^{\theta_1} L\, p\, r\cos\theta\, d\theta \quad (5.1.7.9)$$

For a 180° journal bearing ($\theta_1 = 90°$), the load capacity is given as

$$W = \frac{12\eta QR^2}{c^3} \left[\frac{2 + 3\varepsilon - \varepsilon^3}{(1-\varepsilon^2)^2} \right] \quad (5.1.7.10)$$

5.1.7.3 Bearing stiffness

The stiffness is defined as

$$K = -\frac{\partial W}{\partial e} \quad (5.1.7.3)$$

$$\Rightarrow \qquad K = -\frac{1}{c}\frac{\partial W}{\partial \varepsilon}$$

The stiffness of the bearing is dependent upon a number of factors such as:

(a) Ratio of recess pressure to supply pressure
(b) Type of compensator used
(c) Number of pads used

5.1.8 Energy losses

Hydrostatic bearing should be designed to minimize the energy losses. If the shaft rotates at angular speed ω, the power is required to overcome the viscous shear in the clearance space between the shaft and the bearing. In addition, pumping power is required to circulate the fluid. Therefore, to minimize energy loss, power loss in pumping the viscous liquid and viscous shearing should be minimized.

5.1.8.1 *System power loss*

The total power loss is a summation of the power loss due to friction and due to pumping losses:

$$H_{total} = H_v + H_P \qquad (5.1.8.1)$$

5.1.8.1.1 *Friction power loss*

The friction power lost in shearing lubricant layer is dissipated as heat in the lubricating oil. The mathematical model for this power loss depends on the geometry. For example, friction power loss in circular step thrust bearing, discussed in the section 5.1.5, is given as:

$$H_v = T\omega = \frac{\pi \eta \omega^2}{2h}\left(R_2^4 - R_1^4\right) \qquad (5.1.8.2)$$

In low speed applications, such as in machine tool slide ways, power loss is negligible compared to pumping losses.

5.1.8.1.2 *Pumping power loss*

If η_P denotes the efficiency of the pump then pumping power loss is given as:

$$H_P = \frac{P_S Q}{\eta_P} \qquad (5.1.8.3)$$

The flow rate, Q, depends on the geometry, piping and compensation device. For example, Q for circular step thrust bearing, expressed in Eq. 5.1.5.8, is

$$Q = \frac{\pi h^3 P_r}{6\eta \ln\left(R_2/R_1\right)}$$

On substituting this expression in Eq. 5.1.8.3

$$H_P = \frac{\pi h^3 P_S P_r}{6\eta_P \eta \ln\left(R_2/R_1\right)} \qquad (5.1.8.4)$$

Total power loss

$$H_{total} = \frac{\pi \eta \omega^2}{2h}(R_2^4 - R_1^4) + \frac{\pi h^3 P_S P_r}{6\eta_P \eta \ln(R_2/R_1)} \quad (5.1.8.5)$$

5.1.8.2 *Optimum design of circular step thrust hydrostatic bearing*

With respect to design of optimum hydrostatic bearings, pumping power loss is the biggest concern. If relative speed between the shaft and the bearing is high, then viscous friction loss also must be accounted. For optimization it is necessary to decide objective function (i.e., maximize load capacity, minimize power loss, maximize stiffness), design variables (i.e., film thickness, recess dimensions) and constraints. For example, due to space constraints the outer radius R_2 is fixed; however R_1 can be varied to minimize pumping power losses. Differentiating Eq. 5.1.8.5 with respect to R_1 and setting it equal to zero, we have

$$\frac{d}{dR_1}(H_{total}) = \frac{d}{dR_1}\left[\frac{\pi \eta \omega^2}{2h}(R_2^4 - R_1^4) + \frac{\pi h^3 P_S P_r}{6\eta_P \eta \ln(R_2/R_1)}\right]$$

$$0 = \frac{d}{dR_1}\left[\frac{\pi \eta \omega^2}{2h}(R_2^4 - R_1^4) + \frac{\pi h^3 P_S P_r}{6\eta_P \eta \ln(R_2/R_1)}\right] \quad (5.1.9.1)$$

Solution of Eq. 5.1.9.1 provides the value of R_1 to minimize the power loss.

Similarly differentiating Eq. 5.1.8.5 with respect to h and setting it equal to zero, we have

$$0 = \frac{d}{dh}\left[\frac{\pi \eta \omega^2}{2h}(R_2^4 - R_1^4) + \frac{\pi h^3 P_S P_r}{6\eta_P \eta \ln(R_2/R_1)}\right]$$

$$0 = \frac{d}{dh}\left[-\frac{\pi \eta \omega^2}{2h^2}(R_2^4 - R_1^4) + \frac{3\pi h^2 P_S P_r}{6\eta_P \eta \ln(R_2/R_1)}\right]$$

$$h^4 = \frac{\eta^2 \eta_P \omega^2}{P_S P_r}(R_2^4 - R_1^4)\ln(R_2/R_1) \quad (5.1.9.2)$$

5.2 Squeeze Film Lubrication

In section 3.1.2 of chapter 3, squeeze film lubrication has been introduced with note that fluctuation in load/velocity helps to develop positive fluid pressure. In the mechanism of squeeze lubrication, viscosity of lubricant plays very important role as viscous fluid cannot be squeezed out from the interface. The squeeze film effect can be quantified using Reynolds' equation (3.2.22), explained in chapter 3.

$$\frac{\partial}{\partial x}\left(\frac{h^3}{\eta}\frac{\partial p}{\partial x}\right) + \frac{\partial}{\partial z}\left(\frac{h^3}{\eta}\frac{\partial p}{\partial z}\right) = 12(V_h - V_0) \quad (3.2.33)$$

Right hand term of the above equation can be expressed as variation in film thickness with time ($\partial h/\partial t$). This means relative movement between tribo-surfaces in the normal (along film thickness) direction is the source of positive fluid pressure.

There are a number of applications of squeeze film lubrication. For example, in piston–pin and small end connecting rod bearings (shown in Fig. 5.2) squeeze film lubrication prevails. In such bearings, the time required to produce a film of certain thickness is much greater than the duration the force is actually applied for. Similarly, squeeze film lubrication occurs in human body joints (i.e., shoulder, hip, knee, ankle) that employ articular cartilage as bearing and synovial (non-Newtonian shear thinning) fluid as the lubricant.

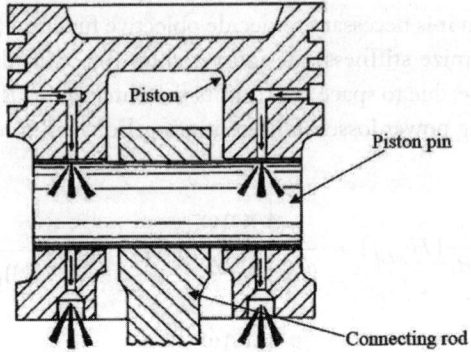

Fig. 5.2 Squeeze film engine bearings

The squeeze film lubrication also occurs in piston ring assembly of the IC engine. The piston ring assembly provides a mechanism to seal the combustion gases passing from cylinder to crankcase during the power and compression strokes. But piston cannot be designed to have a tight fit as friction would increase and lead to tremendous amount of wear in a very short span of time. Therefore piston rings are designed with suitable profile so that the hydrodynamic lubrication prevails between the ring and the cylinder walls during major portion of the stroke, and squeeze lubrications prevails when the piston velocity reaches zero at the top/bottom dead centre. In other words, piston sliding velocity drops towards the start and end of its stroke, hydrodynamic action diminishes and only squeeze film lubrication save the piston rings.

5.2.1 Basic concept

It is interesting to note that dynamic/pulsating/reciprocating loads may develop positive lubricant film to support the applied load. This load carrying mechanism comes from the fact that a viscous lubricant takes time to squeeze out from tribo–surfaces and if within that time, load reverses its direction, then the tribo–surfaces will never meet. In other words, if load/velocity varies periodically, then a periodic variation in lubricant film thickness occurs and surfaces may not contact at all. A sketch of squeeze action due to variable speeds has been shown in Fig. 5.2.1.

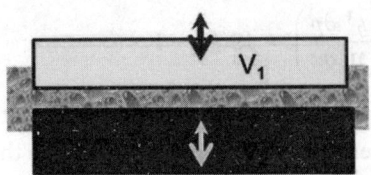

Fig. 5.2.1 Squeeze lubrication

One of negative effect of squeeze film lubrication occurs in tires of automobiles and road in rainy seasons, where due to squeeze film action water film builds-up between tires and road. When this happens, traction is lost and steering/braking becomes uncontrollable.

In the remaining of the present chapter, analysis of squeeze film action has been presented. One of the interesting examples of squeeze film lubrication is dynamically loaded journal bearings used in aircraft and automotive engines. Under dynamic conditions, these bearings safely carry greater loads than unidirectional loads.

5.2.2 Squeeze action between circular flat plates

Let us consider two circular flat plates of radius R approaching each other, or a single circular plate of radius R approaching a flat (very large radius) surface, as shown in Fig. 5.2.2.1. The space between both the plates is filled with a viscous liquid which is being displaced radially outward by the relative motion of the plates. Flow rate of an incompressible viscous fluid through a gap (between parallel plates) of h finite slot is given by Eq. 5.1.3.12 expressed in previous heading,

$$Q = \frac{bh^3}{12\eta} \frac{\Delta P}{l} \qquad (5.2.2.1)$$

Applying this equation to the present case, where fluid pressure is a function of radius and $b = 2\pi r$, equation becomes

$$Q = -\frac{2\pi r h^3}{12\eta} \frac{dp}{dr} \qquad (5.2.2.2)$$

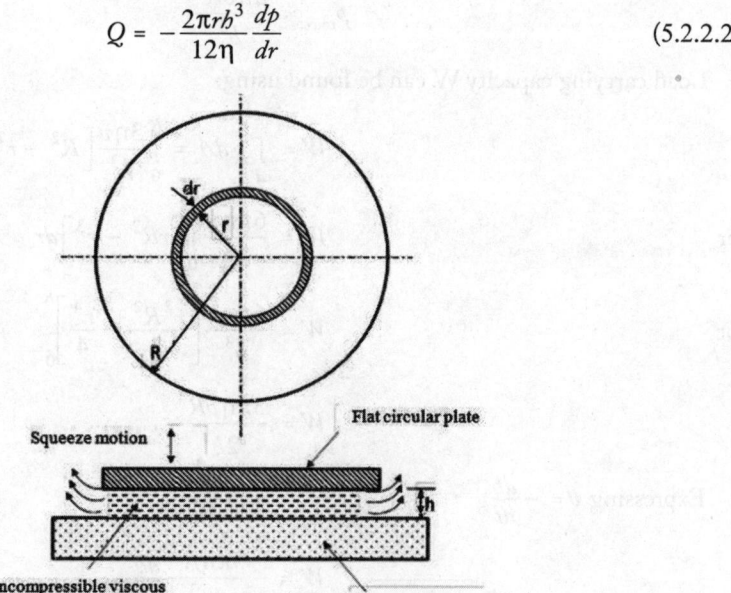

Fig. 5.2.2.1 Representation between squeeze film between disk and flat plate

Under squeezing action (when two plates are approaching each other), volume of fluid forced out through the gap h can be expressed as:

$$Q = \pi r^2 v \qquad (5.2.2.3)$$

where v is the velocity of fluid getting squeeze out. On equating Eqs. 5.2.2.2 and 5.2.2.3,

$$\pi r^2 v = -\frac{2\pi r h^3}{12\eta}\frac{dp}{dr} \qquad (5.2.2.4)$$

On rearranging
$$dp = -\frac{6\eta v r \cdot dr}{h^3} \qquad (5.2.2.5)$$

On integrating
$$p = -\frac{3\eta v r^2}{h^3} + C \qquad (5.2.2.6)$$

The constant of integration in Eq. 5.2.2.6 can be evaluated using the atmospheric pressure boundary condition (p = 0) at r = R, which gives

$$C = \frac{3\eta v R^2}{h^3} \qquad (5.2.2.7)$$

So fluid pressure at any point on the circular plate is given as

$$p = \frac{3\eta v}{h^3}\left[R^2 - r^2\right] \qquad (5.2.2.8)$$

The pressure distribution is parabolic with maximum pressure at the centre i.e., r = 0. The maximum pressure is given as

$$p_{max} = \frac{3\eta v R^2}{h^3} \qquad (5.2.2.9)$$

Load carrying capacity W can be found using:

$$W = \int_A p\, dA = \int_0^R \frac{3\eta v}{h^3}\left[R^2 - r^2\right] 2\pi r\, dr$$

or
$$W = \frac{6\pi\eta v}{h^3}\int_0^R \left[rR^2 - r^3\right] dr$$

or
$$W = \frac{6\pi\eta v}{h^3}\left[\frac{r^2 R^2}{2} - \frac{r^4}{4}\right]_0^R$$

$$W = \frac{3\pi\eta v R^4}{2h^3} \qquad (5.2.2.10)$$

Expressing $v = -\dfrac{dh}{dt}$

$$W = -\frac{3\pi\eta R^4}{2h^3}\frac{dh}{dt} \qquad (5.2.2.11)$$

Integrating with appropriate limits to find the time of approach,

$$\int_{t_1}^{t_2} dt = -\frac{3\pi\eta R^4}{2W}\int_{h_1}^{h_2}\frac{dh}{h^3} \qquad (5.2.2.12)$$

$$\Delta t = \frac{3\pi\eta R^4}{4W}\left[\frac{1}{h_2^2} - \frac{1}{h_1^2}\right] \quad (5.2.2.13)$$

where $\Delta t = t_2 - t_1$, is called as the time of approach for the film gap to reduce from an initial value h_1 to final value h_2. As can been seen from Eq. 5.2.2.13 that as $\Delta t \to \infty$ then $h \to 0$. Hypothetically it can be said that to completely squeeze out the fluid it takes infinite amount of time.

5.2.3 Squeeze action between rectangular plates

In the previous section, pressure distribution, load capacity and time to approach have been derived for cylindrical plate. There is possibility of using rectangular plates (i.e., crosshead-type mechanisms), as depicted in Fig. 5.2.3.1, to generate squeeze film lubrication. In this figure breadth b is very large compared to length l, which means pressure gradient in Z-direction (along breadth direction) can be neglected.

Fig. 5.2.3.1 Squeeze lubrication under variable load

From Eq. 5.1.3.12,

$$Q = -\frac{bh^3}{12\eta}\frac{dp}{dx} \quad (5.2.3.1)$$

Under squeezing action (when two plates approach each other), volume of fluid would be forced out through the gap h. The volume flow rate can be expressed as:

$$q = (bx)v \quad (5.2.3.2)$$

Equating Eqs. 5.2.3.1 and 5.2.3.2

$$vbx = -\frac{bh^3}{12\eta}\frac{dp}{dx} \quad (5.2.3.3)$$

on rearranging

$$dp = -\frac{12\eta}{h^3}vx\,dx$$

Integrating the above equation to express p algebraically

$$p = -\frac{12\eta}{h^3}v\frac{x^2}{2} + C$$

Evaluating the constant of integration by using the boundary condition p = 0 at x = l/2, we get

$$p = \frac{6\eta v}{h^3}\left[\frac{l^2}{4} - x^2\right] \quad (5.2.3.4)$$

Equation 5.2.3.4 represents parabolic pressure distribution. The maximum value of pressure occurs at x = 0 and is given by

$$p_{max} = \frac{3\eta v l^2}{2h^3} \quad (5.2.3.5)$$

Load carrying capacity W can be calculated using:

$$W = \int_A p \, dA = \int_{-1/2}^{1/2} \frac{6\eta v}{h^3}\left[\frac{l^2}{4} - x^2\right] b \, dx$$

Or

$$W = \frac{6\eta v b}{h^3}\left[\frac{l^2}{4}\left(\frac{l}{2} + \frac{l}{2}\right) - \frac{1}{3}\left(\frac{l^3}{8} + \frac{l^3}{8}\right)\right]$$

Or

$$W = \frac{6\eta v b}{h^3} \frac{l^3}{6}$$

$$W = \frac{b\eta v l^3}{h^3} \quad (5.2.3.6)$$

Expressing $v = -\frac{dh}{dt}$

$$W = -\frac{b\eta l^3}{h^3} \frac{dh}{dt} \quad (5.2.3.7)$$

Integrating with appropriate limits to find the time of approach,

$$\int_{t_1}^{t_2} dt = -\frac{b\eta l^3}{W} \int_{h_1}^{h_2} \frac{dh}{h^3}$$

The time of approach is given as

$$\Delta t = \frac{b\eta l^3}{2W}\left[\frac{1}{h_2^2} - \frac{1}{h_1^2}\right] \quad (5.2.3.8)$$

where $\Delta t = t_2 - t_1$ is the time in seconds for the film thickness to reduce from an initial value of h_1 to final value of h_2 with a constant force W acting on the rectangular plates.

5.2.4 Squeeze action under variable and alternating loads

In Fig. 5.2.4.1 load variation with time has been shown by double headed arrow. Due to variation in applied load, film thickness varies ($\partial h/\partial t$) with time. The connecting rod and crankshaft bearings used in reciprocating compressors and internal combustion engines are subjected to fluctuating load. The load diagram for such bearings has been shown in Fig. 5.2.4.2.

It is interesting to note that crank angle (theta) varies with time and, therefore, loads vary with time. Variation in gas pressure for one of the typical engine has been shown in Fig. 5.2.4.3.

To consider the squeeze action under fluctuating load, let us first consider application of squeeze action in journal bearings.

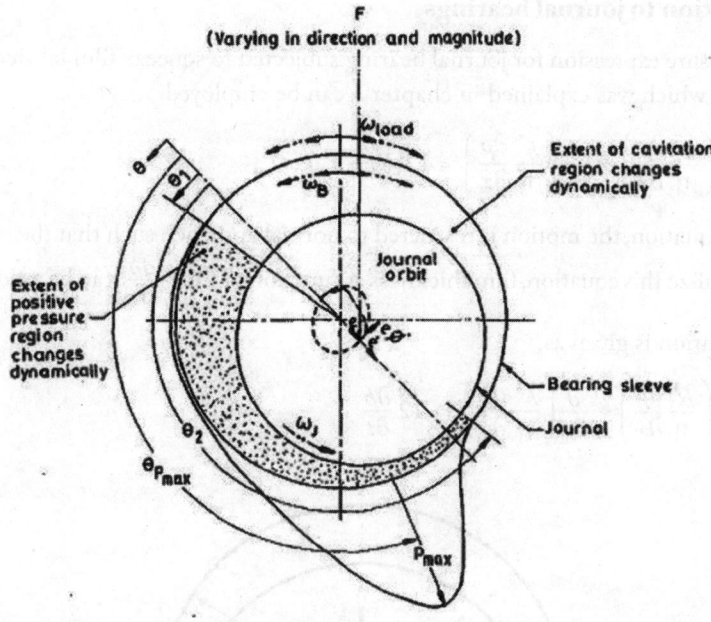

Fig. 5.2.4.1 Representation for calculating oil–cushion effect in a slipper type of bearing

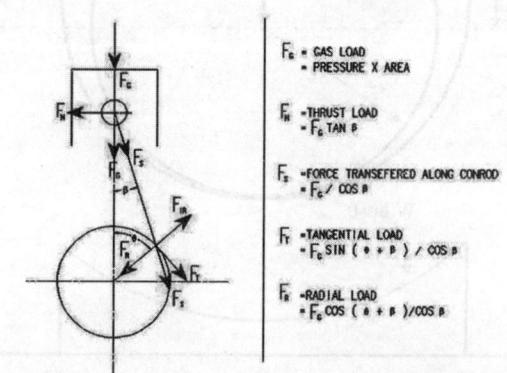

Fig. 5.2.4.2 Load as a function of angles (time)

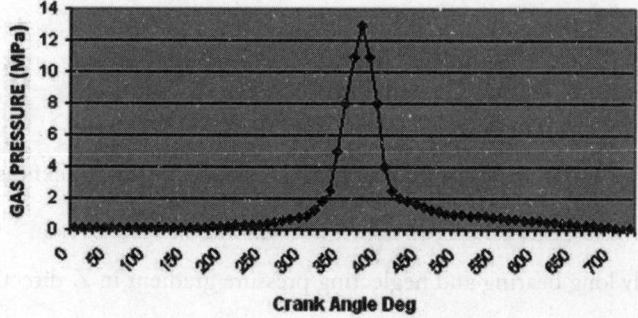

Fig. 5.2.4.3 Variation in gas pressure as a function of crank angle (time)

5.2.5 Application to journal bearings

To derive the pressure expression for journal bearing subjected to squeeze film lubrication, Reynolds' equation (3.2.22) which was explained in chapter 3 can be employed.

$$\frac{\partial}{\partial x}\left(\frac{h^3}{\eta}\frac{\partial p}{\partial x}\right) + \frac{\partial}{\partial z}\left(\frac{h^3}{\eta}\frac{\partial p}{\partial z}\right) = 12(V_h - V_0) \tag{3.2.33}$$

As per above equation, the motion is restricted to normal approach such that the sliding velocities are zero. To generalize this equation, film thickness as function of time $\frac{\partial h}{\partial t}$ can be used. The Reynolds' with this modification is given as,

$$\frac{\partial}{\partial x}\left(\frac{h^3}{\eta}\frac{\partial p}{\partial x}\right) + \frac{\partial}{\partial z}\left(\frac{h^3}{\eta}\frac{\partial p}{\partial z}\right) = 12\frac{\partial h}{\partial t} \tag{5.2.5.1}$$

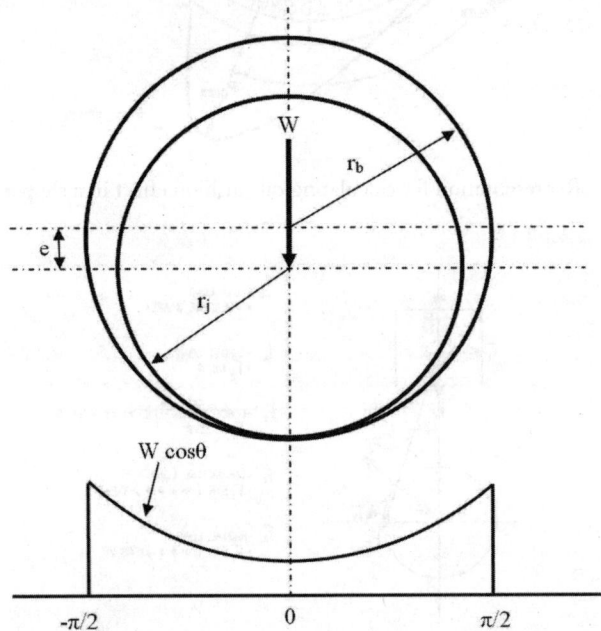

Fig. 5.2.5.1 Journal bearing with squeeze film action

For journal bearing, $x = R\theta$ can be used,

$$\frac{1}{R^2}\frac{\partial}{\partial \theta}\left(\frac{h^3}{\eta}\frac{\partial p}{\partial \theta}\right) + \frac{\partial}{\partial z}\left(\frac{h^3}{\eta}\frac{\partial p}{\partial z}\right) = 12\frac{\partial h}{\partial t} \tag{5.2.5.2}$$

As per Fig. 5.2.5.1, if θ starts with minimum film thickness, the film thickness can be defined as

$$h = c(1 - \varepsilon \cos \theta) \tag{5.2.5.3}$$

Assuming infinitely long bearing and neglecting pressure gradient in Z-direction,

$$\frac{\partial}{\partial \theta}\left(h^3 \frac{\partial p}{\partial \theta}\right) = 12\eta R^2 \frac{\partial h}{\partial t} \tag{5.2.5.4}$$

Substituting Eq. 5.2.5.3,

$$\frac{\partial}{\partial \theta}\left(h^3 \frac{\partial p}{\partial \theta}\right) = 12\eta cR^2 \frac{\partial(1-\varepsilon\cos\theta)}{\partial t}$$

$$\frac{\partial}{\partial \theta}\left(h^3 \frac{\partial p}{\partial \theta}\right) = 12\eta cR^2\left(-\varepsilon\sin\theta\frac{d\theta}{dt} - \dot{\varepsilon}\cos\theta\right) \quad (5.2.5.5)$$

As it is assumed that the motion is restricted to normal approach, which means rate of change of θ will be zero,

$$\Rightarrow \quad \frac{\partial}{\partial \theta}\left(h^3 \frac{\partial p}{\partial \theta}\right) = 12\eta cR^2\left(-\dot{\varepsilon}\cos\theta\right) \quad (5.2.5.6)$$

Integrating Eq. 5.2.5.6 with respect to θ, we get

$$\frac{\partial p}{\partial \theta} = \frac{-12\eta cR^2\dot{\varepsilon}\sin\theta}{h^3} + C_1 \quad (5.2.5.7)$$

To determine the integration constant C_1, maximum pressure at $\theta = 0$ can be used, which means $C_1 = 0$.

$$\frac{\partial p}{\partial \theta} = \frac{-12\eta cR^2\dot{\varepsilon}\sin\theta}{h^3} \quad (5.2.5.8)$$

Substituting the value of film thickness from Eq. 5.2.5.3 in Eq. 5.2.5.8, and rearranging

$$\frac{\partial p}{\partial \theta} = -12\eta\frac{R^2}{c^2}\frac{\dot{\varepsilon}\sin\theta}{(1-\varepsilon\cos\theta)^3} \quad (5.2.5.9)$$

Integrating Eq. 5.2.5.9,

$$p = -12\eta\frac{R^2}{c^2}\dot{\varepsilon}\int\frac{\sin\theta}{(1-\varepsilon\cos\theta)^3}d\theta + C_2$$

$$p = -6\eta\frac{R^2}{c^2}\frac{\dot{\varepsilon}}{\varepsilon}\frac{1}{(1-\varepsilon\cos\theta)^2} + C_2 \quad (5.2.5.10)$$

The integration constant C_2 can be obtained by using boundary condition $p = 0$ at $\theta = \pm 0.5\pi$,

$$0 = -6\eta\frac{R^2}{c^2}\frac{\dot{\varepsilon}}{\varepsilon}\frac{1}{(1-0)^2} + C_2$$

$$C_2 = 6\eta\frac{R^2}{c^2}\frac{\dot{\varepsilon}}{\varepsilon} \quad (5.2.5.11)$$

On substituting Eq. 5.2.5.11 in Eq. 5.2.5.10,

$$p = 6\eta\frac{R^2}{c^2}\frac{\dot{\varepsilon}}{\varepsilon}\left[1 - \frac{1}{(1-\varepsilon\cos\theta)^2}\right] \quad (5.2.5.12)$$

The normal load carrying capacity W is determined by integrating pressure,

$$W = L \int_{-0.5\pi}^{0.5\pi} p R \cos\theta \, d\theta \tag{5.2.5.13}$$

On substituting expression of pressure from Eq. 5.2.5.12 and assuming symmetric condition,

$$W = 2L \int_{0}^{0.5\pi} 6\eta \frac{R^2}{c^2} \frac{\dot{\varepsilon}}{\varepsilon} \left[1 - \frac{1}{(1 - \varepsilon \cos\theta)^2} \right] R \cos\theta \, d\theta$$

On rearranging,

$$W = 12\eta L \frac{R^3}{c^2} \frac{\dot{\varepsilon}}{\varepsilon} \int_{0}^{0.5\pi} \left[1 - \frac{1}{(1 - \varepsilon \cos\theta)^2} \right] \cos\theta \, d\theta$$

Using Sommerfeld substitution, [Booker, 1965]

$$W = 12\eta L \frac{R^3}{c^2} \dot{\varepsilon} \frac{\pi}{(1 - \varepsilon^2)^{1.5}} \tag{5.2.5.14}$$

The time of approach can be calculated by rearranging Eq. 5.2.5.14,

$$W = 12\eta L \frac{R^3}{c^2} \frac{\pi}{(1 - \varepsilon^2)^{1.5}} \frac{d\varepsilon}{dt}$$

$$\Rightarrow \quad \frac{W}{12\eta\pi L} \frac{c^2}{R^3} dt = \frac{d\varepsilon}{(1 - \varepsilon^2)^{1.5}}$$

On integrating,

$$\frac{W}{12\eta\pi L} \frac{c^2}{R^3} \int_{t_1}^{t_2} dt = \int_{\varepsilon_1}^{\varepsilon_2} \frac{d\varepsilon}{(1 - \varepsilon^2)^{1.5}}$$

$$\Delta t = t_2 - t_1 = \frac{12\eta\pi L R^3}{Wc^2} \left[\frac{\varepsilon_2}{(1 - \varepsilon_2^2)^{0.5}} - \frac{\varepsilon_1}{(1 - \varepsilon_1^2)^{0.5}} \right]$$

$$\tag{5.2.5.15}$$

From Fig. 5.2.5.1, it can be observed that film thickness is minimum when $\theta = 0$. Hence

$$h_{min} = c(1 - \varepsilon) \tag{5.2.5.16}$$

Example: A piston pin (Fig. 5.2) of I.C. engine is lubricated with a lubricating oil having operating viscosity $\eta = 0.03$ Pa.s. The pin diameter, length, and radial clearance are D = 0.020 m, L = 0.038 m, and c = 25.4 µm, respectively. Assume an effective pressure of 689475 Pa is applied on top of the piston (diameter is 0.1016 m) during the power stroke. Assuming negligible side leakage, find

(a) the time required for the film thickness at the top of the bushing to reduce from 1.27×10^{-5} m to 2.54×10^{-6} m.

(b) the squeeze velocity at the onset of the power stroke.

Solution: The load imposed on the bearing during the power stroke is
$$W = P_{eff} A_p = P_{eff}\left(\pi D_p^2 / 4\right) = (689475)\pi(0.381)^2 / (4) = 78606 \text{ N}$$

The minimum film thickness as per Eq. 5.2.5.16 is $h_{min} = c(1-\varepsilon)$. Eccentricity ratio corresponding two given minimum film thicknesses:

For $h_{min} = 0.0000127$ m $\varepsilon_1 = 0.5$.
For $h_{min} = 0.00000254$ m $\varepsilon_2 = 0.9$.

From Eq. 5.2.5.15, we have

(a)
$$\Delta t = t_2 - t_1 = \frac{12\pi\eta L R^3}{Wc^2}\left[\frac{\varepsilon_2}{\left(1-\varepsilon_2^2\right)^{0.5}} - \frac{\varepsilon_1}{\left(1-\varepsilon_1^2\right)^{0.5}}\right]$$

$$\Delta t = \frac{12\pi(0.03)(0.038)(0.01)^3}{(78606)(25.4*10^{-6})^2}[2.0647 - 0.5774]$$

$$= 0.0013 \text{ s}$$

(b) To evaluate onset velocity, we can use Eq. 5.2.5.14
$$W = 12\eta L \frac{R^3}{c^2}\dot{\varepsilon}\frac{\pi}{\left(1-\varepsilon^2\right)^{1.5}}$$

Rearranging $\quad \dfrac{Wc^2}{12\pi\eta L R^3}\left(1-\varepsilon^2\right)^{1.5} = \dot{\varepsilon}$

On multiplying with c on both the sides

$$V_s = c\dot{\varepsilon} = \frac{W\left(1-\varepsilon_1^2\right)^{3/2}}{12\pi\eta L}\left(\frac{c}{R}\right)^3 = 0.0195 \, m/s$$

5.3 Engine Bearing Lubrication

An engine bearing operates on 'hydrodynamic plus squeeze' film lubrication mechanism. The following Reynolds' equation is used to determine the pressure distribution.

$$\frac{1}{R^2}\frac{\partial}{\partial\theta}\left(\frac{h^3}{\eta}\frac{\partial p}{\partial\theta}\right) + \frac{\partial}{\partial z}\left(\frac{h^3}{\eta}\frac{\partial p}{\partial z}\right) = 6(U_1 + U_2)\frac{\partial h}{\partial x} + 12\frac{\partial h}{\partial t} \quad (5.3.1)$$

Solving this equation is tedious. It is interesting to note that length to diameter ratio in the range of 0.4–0.6 is normally preferred for engine journal bearing. The primary reason for this is that any lesser length results in excessive loss of oil pressure from the ends of bearing, while any increase in length may cause misalignment problem leading to edge loading. The upper bound (L/D=0.6) of preferred range is often selected for bearing having complete circumferential groove midway, while the lower bound (L/D=0.4) is adopted for bearings with a feeding hole. Hirani et al. [Hirani et al., 1999] provided analytical method to evaluate engine bearing performance.

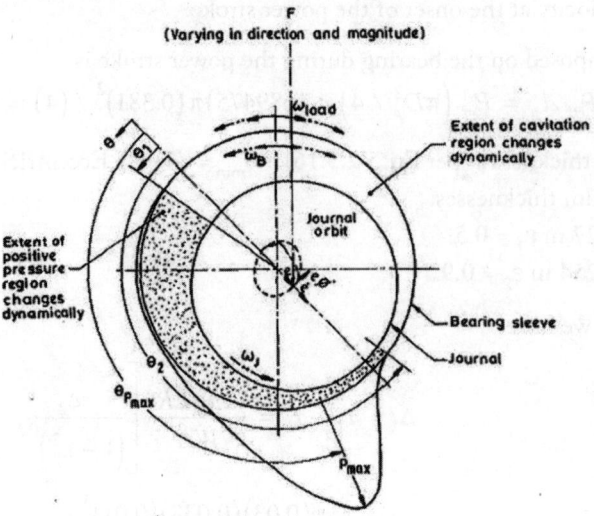

Fig. 5.3.1 Schematic of engine bearing

It is important to note that fluid pressure is generated due to relative speed (hydrodynamic lubrication) and change in journal position with time (squeeze lubrication) and can be given by [Hirani et al., 1999]:

$$P = \frac{-6\eta \left(L/c_r\right)^2 \left(1/4 - \bar{z}^2\right)\{\dot{\varepsilon}\cos\theta + \varepsilon(\dot{\phi} - \bar{\omega})\sin\theta\}}{H^3\left(1 + \frac{\left(1/4 - \bar{z}^2\right)}{B_1}\right)} \quad (5.3.2)$$

where $B_1 = \dfrac{H(1+H)}{(L/R)^2}$, $H = 1 + \varepsilon\cos\theta$, $\bar{z} = z/L$, c_r is the radial clearance, $\dot{\varepsilon}$ is rate of change of eccentricity ratio, $\dot{\phi}$ is rate of change of attitude angle and $\bar{\omega}$ is the average angular velocity between the journal and bearing relative to load line.

For a stationary journal bearing arrangement, in which the journal rotating at an angular velocity of (ω_j) and load rotating at a constant magnitude and constant angular velocity (ω_1), the average angular velocity is given by following equation

$$\bar{\omega} = \frac{\omega_j}{2}\left(1 - \frac{2\omega_1}{\omega_j}\right) \quad (5.3.3)$$

For a connecting rod bearing where the polar load diagram is relative to the connecting rod axis, the average angular velocity is given by Eq. 5.3.4

$$\bar{\omega} = \frac{\omega_j}{2}\left(1 - \frac{2\omega_1}{\omega_j} + \frac{\omega_b}{\omega_j}\right) \quad (5.3.4)$$

where ω_b is the angular velocity of the connecting rod bearing, and it is given as:

$$\omega_b = \omega_j \left(1 - \frac{\text{length of crank}}{\text{length of connecting rod}}\right) \tag{5.3.5}$$

The detailed description of calculation of average velocity is given in Stolarski [Stolarski, 1990]. The suggested minimum radial clearance by Stolarski [Stolarski, 1990] is 0.000375 D to avoid bearing overheating. Constantinescu et al. [Constantinescu et al., 1985] presented clearance as function of D and N, such as:

$$c_r = 0.7N(1000D)^{1.5} 10^{-8} \text{ mm for car and tractor engines (D < 130mm)} \tag{5.3.6}$$

$$c_r = 0.000018 N^{0.25} (1000D)^{5/4} \text{ mm for large industrial reciprocating engines.} \tag{5.3.7}$$

It is worth noting that suggested radial clearance (c_r) increases with increase in rotational speed. Further, if calculated value of c_r from Eq. 5.3.6/ 5.3.7 is lesser than 0.000375 D, then c_r = 0.000375 D.

To determine the pressure distribution using Eq. 3.5.2, there is need to find $\dot{\varepsilon}$, $\dot{\phi}$, ε, and the range of positive pressure (lower limit = θ_1, upper limit = θ_2). The values of θ_1 and θ_2 can be expressed in terms of $\dot{\varepsilon}$, $\dot{\phi}$ and ε using the following equation:

$$\left.\begin{array}{l}\theta_1 = \tan^{-1}\left(\dfrac{\dot{\varepsilon}}{\varepsilon(\overline{\omega} - \dot{\phi})}\right) \\ \theta_2 = \theta_1 + K\pi\end{array}\right\} \tag{5.3.8}$$

where $K = \begin{cases} +1 & \text{for} \quad \varepsilon(\overline{\omega} - \dot{\phi}) > 0 \\ -1 & \text{for} \quad \varepsilon(\overline{\omega} - \dot{\phi}) < 0 \end{cases}$.

In Eq. 5.3.8 Gumbal (half Sommerfeld) boundary condition has been used.

To evaluate $\dot{\varepsilon}$, $\dot{\phi}$ and ε, there is a need to solve inverse problem. For that purpose, forces along the line of centre and perpendicular to line of centres are expressed as:

$$F^\varepsilon = 6\eta \left(\frac{R}{c_r}\right)^2 LR\left[\dot{\varepsilon}I_1 + \varepsilon(\dot{\phi} - \overline{\omega})I_2\right] \tag{5.3.9}$$

$$F^\phi = 6\eta \left(\frac{R}{c_r}\right)^2 LR\left[\dot{\varepsilon}I_2 + \varepsilon(\dot{\phi} - \overline{\omega})I_3\right] \tag{5.3.10}$$

where;

$I_1 = \int_{\theta_1}^{\theta_2} B_2 \cos^2\theta \, d\theta$, $I_2 = \int_{\theta_1}^{\theta_2} B_2 \cos\theta \sin\theta \, d\theta$, $I_3 = \int_{\theta_1}^{\theta_2} B_2 \sin^2\theta \, d\theta$,

$B_2 = \left[1 - \dfrac{B_1}{\sqrt{B_1 + \frac{1}{4}}} \log\left(\dfrac{\sqrt{B_1 + \frac{1}{4}} + \frac{1}{2}}{\sqrt{B_1 + \frac{1}{4}} - \frac{1}{2}}\right)\right]\left(\dfrac{1+H}{H^2}\right)$, Integral I_1, I_2 and I_3 are evaluated using

Weddle's numerical integral formula (as expressed in Eq. 3.4.29 of chapter 3). On inverting Eqs. 5.3.9 and 5.3.10,

$$\begin{Bmatrix} \dot{\varepsilon} \\ \varepsilon(\dot{\phi} - \overline{\omega}) \end{Bmatrix} = \frac{F(c_r/R)^2}{6\eta LR} \begin{Bmatrix} M^\varepsilon \\ M^\phi \end{Bmatrix} \qquad (5.3.11)$$

where; $M^\varepsilon = \dfrac{I_3 \cos\varphi + I_2 \sin\varphi}{I_1 I_3 - I_2^2}$ and $M^\varphi = \dfrac{I_1 \sin\varphi + I_2 \cos\varphi}{I_2^2 - I_1 I_3}$

The solution of Eq. 5.3.11 requires forces at number of time steps and eccentricity (ε) can be expressed in terms of rate of change of eccentricity ratio and time step. After evaluating and $\dot{\phi}$ and $\dot{\varepsilon}$ at every time step, the maximum pressure at each time steps can be determined using Eq. 5.3.12.

$$P_{max} = \frac{-1.5\eta(L/c_r)^2 (\dot{\varepsilon}\cos\theta + \varepsilon(\dot{\phi} - \overline{\omega})\sin\theta)}{H_{P\,max}^3 \left(1 + \dfrac{(L/D)^2}{H_{P\,max}(1 + H_{P\,max})}\right)} \qquad (5.3.12)$$

where; $H_{P\,max} = 1 + \varepsilon \cos\theta_{P\,max}$, $\theta_{P\,max} = \theta_{f3} + \dfrac{(\theta_{f4} - \theta_{f3})}{1 - \Theta}$,

$\theta_{f3} = \theta_{f1} + \varepsilon \Delta\theta_f$, $\theta_{f4} = \theta_{f2} - (1 - \varepsilon)\Delta\theta_f$

$\Delta\theta_f = 0.5(\theta_{f2} - \theta_{f1})$, $\theta_{f1} = \theta_1 + K\pi/2$, $\theta_{f2} = K\pi$, if $K\theta_1 < 0$ then $\theta_{f2} = \theta_2$

$\Theta = \dfrac{G|_{\theta_{f4}}}{G|_{\theta_{f3}}}$, $G|_\theta = \dfrac{\left[\varepsilon(\overline{\omega} - \dot{\phi})(3\varepsilon + \cos\theta - 2\varepsilon\cos^2\theta) + \dot{\varepsilon}\sin\theta(1 - 2\varepsilon\cos\theta)\right]}{(1 + \varepsilon\cos\theta)}$.

5.3.1 Oil flow

The oil flow required to maintain the lubricant oil film and limit the temperature rise consists of:

- Hydrodynamic flow, Q_H, caused by shaft rotation relative to bearing and the resulting film pressure. Flow Q_H in a dynamically loaded journal bearing is presented as Hirani [Hirani,1998]:

$$Q_H = c_r LR \left[\varepsilon(\dot{\phi} - \overline{\omega})\cos\theta_1 - \dot{\varepsilon}\sin\theta_1\right] \qquad (5.3.13)$$

- Feed flow, Q_p, which is pumped in the bearing through groove under supply pressure higher than atmospheric pressure. The feed pressure flow (Q_p) for different groove arrangements has been described in section 3.4.2.1 (oil hole: Eq. 3.4.11; circumferential groove: Eq. 3.4.12; partial groove Eq. 3.4.13) of chapter 3. As per Martin and Lee [Martin and Lee, 1983] algebraic summation of hydrodynamic and feed flow overestimated the total flow, therefore they provided empirical equation such as;

$$Q = Q_H + Q_p - 0.3\sqrt{Q_H \times Q_p} \qquad (5.3.14)$$

5.3.2 Power loss

There are two sources of power loss. First is power due to hydrodynamic action (shearing) and second is due to squeeze action.

Total power loss,

$$P_{loss} = P_{shear} + P_{squeeze} \quad (5.3.15)$$

$$P_{shear} = \frac{\eta R^3 L (\omega_j - \omega_B)^2}{c_r} \left[\int_{\theta_1}^{\theta_2} \frac{d\theta}{(1 + \varepsilon \cos \theta)} + (1 - \varepsilon) \int_{\theta_2}^{\theta_2 + \pi} \frac{d\theta}{(1 + \varepsilon \cos \theta)^2} \right] + \frac{c_r (\omega_j + \omega_B)}{2} \varepsilon F^{\varphi}$$

$$\quad (5.3.16)$$

$$P_{squeeze} = F c_r \left(\dot{\varepsilon} \cos \varphi - \varepsilon (\dot{\varphi} - \omega_{av}) \sin \varphi \right) \quad (5.3.17)$$

5.3.3 Temperature rise

Effective temperature rise [Hirani et al., 1998] is provided as:

[fluid density][Q][Heat Capacity of Oil][Effective Temperature Rise] = $\varepsilon[P_{loss}]$

Or,

$$\Delta t = \frac{\varepsilon P_{loss}}{\rho_0 C_0 Q} \quad (5.3.18)$$

Effective Temperature is: $\quad T_{eff} = T_{in} + \Delta t \quad (5.3.19)$

Maximum Temperature: $\quad T_{max} = T_{eff} + \Delta t \quad (5.3.20)$

5.3.4 Design procedure

The journal diameter, angular velocities of journal and bearing, and load cycles (load vs. shaft rotation) are generally known prior to the bearing design stage, because they invariably form a part of the overall system requirements. For similar reason the grade of oil, used for other engine components, is also generally imposed, leaving only diametral clearance and bearing length to be decided. The proposed design methodology is illustrated in Fig. 5.3.2. Here NDATA is number of data for one force cycle and EPS is the convergent criterion on eccentricity ratio, attitude angle and oil viscosity. The bearing design can be summarised by the following steps:

Step 1: Input NDATA, EPS, F_x, F_y, D, L, c_r, ω_b, ω_j, ω_L:

(a) In 4-stroke IC engine, the complete load cycle contains two rotations of crankshaft. If load data are available for at each degree, the number of data's (NDATA) becomes 721 (including the first and last data).

(b) Calculate the time interval by $[\Delta t = 4\pi/(\text{NDATA}-1)(\omega_j)]$.

(c) Calculate the value of c_r using Eq. 5.3.3 for D < 130 mm. If D > 130 mm, use Eq. 5.3.4. To calculate this, c_r needs to be equal or greater than 0.000375 D.

Step 2: Make initial guess for ε_1, ϕ_1, $\dot{\varepsilon}_1$, $\dot{\phi}_1$

(a) The value of ε_1 is calculated using the short bearing static load Eq. 3.3.52:

$$F = \sqrt{F_x^2 + F_y^2} = \frac{\eta U L^3 \varepsilon_1}{4c_r^2\left(1-\varepsilon_1^2\right)^2}\left[\pi^2\left(1-\varepsilon_1^2\right)+16\varepsilon_1^2\right]^{1/2},$$

where F_x is loading horizontal direction and F_y loading in vertical direction. On the basis of calculated eccentricity ratio, the attitude angle is evaluated by using $\varphi_1 = \tan^{-1}\left(\dfrac{\sqrt{1-\varepsilon_1^2}}{\varepsilon_1}\right)$.

(b) The initial value of $\dot{\varepsilon}_1$ and $\dot{\varphi}_1$ is assumed to be zero.

Step 3: Evaluate θ_1, θ_2

The values of θ_1, θ_2 are calculated using Eq. 5.3.8. Using the values of θ_1 and θ_2, the new values of $\dot{\varepsilon}_1$ and $\dot{\varphi}_1$ are evaluated using Eq. 5.3.11.

Step 4: If the newly found $\dot{\varepsilon}_1$ and $\dot{\varphi}_1$ are not equal to the initially assumed values $\dot{\varepsilon}_1 = 0$ and $\dot{\varphi}_1 = 0$, step 2 and 3 will be repeated until required accuracy (0.01%) in journal velocity is obtained.

Step 5: Calculate ε_{i+1} and φ_{i+1} at the next time step using $\varepsilon_{i+1} = \varepsilon_i + \Delta t \dot{\varepsilon}_i$ and $\varphi_{i+1} = \varphi_i + \Delta t \dot{\varphi}_i$.

Step 6: Evaluate the feed pressure flow (Q_p) for different groove arrangements, as has been described in section 3.4.2.1 (oil hole: Eq. 3.4.8; circumferential groove: Eq. 3.4.9; partial groove Eq. 3.4.10) of chapter 3. Calculate the hydrodynamic flow (Q_H) using Eq. 5.3.10. The complete oil flow is evaluated using Eq. 5.3.11.

Step 7: Determine the power loss using Eq. 5.3.15 by summing up the power loss due to shear (Eq. 5.3.16) and squeeze (Eq. 5.3.17), respectively.

Step 8: Estimate the effective and maximum temperature using Eqs. 5.3.19 and 5.3.20. Determine the effective viscosity using the Walther equation (Eq. 3.1.2.3).

Step 9: Repeat the steps 4–8, until the last time step (t = NDATA) is reached

Step 10: Check the convergence using $\left|\dfrac{\varphi_{NDATA} - \varphi_I}{\varphi_I}\right|$ <EPS, $\left|\dfrac{\varepsilon_{NDATA} - \varepsilon_I}{\varepsilon_I}\right|$ <EPS and $\left|\dfrac{\eta_{NDATA} - \eta_I}{\eta_I}\right|$

<EPS. The value of EPS is subjective, as it depends on the required accuracy and execution time. In the present study EPS is kept as 0.01%. If the convergence is satisfied, then move to step 1, otherwise assign $\varepsilon_I = \varepsilon_{NDATA}, \varphi_I = \varphi_{NDATA}, \dot{\varepsilon}_I = \dot{\varepsilon}_{NDATA}$ and $\dot{\varphi}_I = \dot{\varphi}_{NDATA}$

Step 11: Repeat steps 2–10 until convergence.

Step 12: Determine the maximum pressure (P_{max}) and eccentricity ratio (ε) in complete cycle. Calculate minimum film thickness h_{min}. For safe design, (h_{min}) should be greater than permissible limit. This limit depends on the surface finish of journal and sleeve, thermal distortion, etc. It is difficult to quote a precise value of h_{min} at which bearing damage might occur. After carrying out a survey with various bearing manufacturers, Booker [Booker, 1979] provided some guidance on danger levels for film thickness, which is listed in Table 5.3.4.1. If h_{min} is lower than the prescribed limit, then change input parameters (increase oil viscosity, increase bearing length, etc.) and repeat the step 2 onwards.

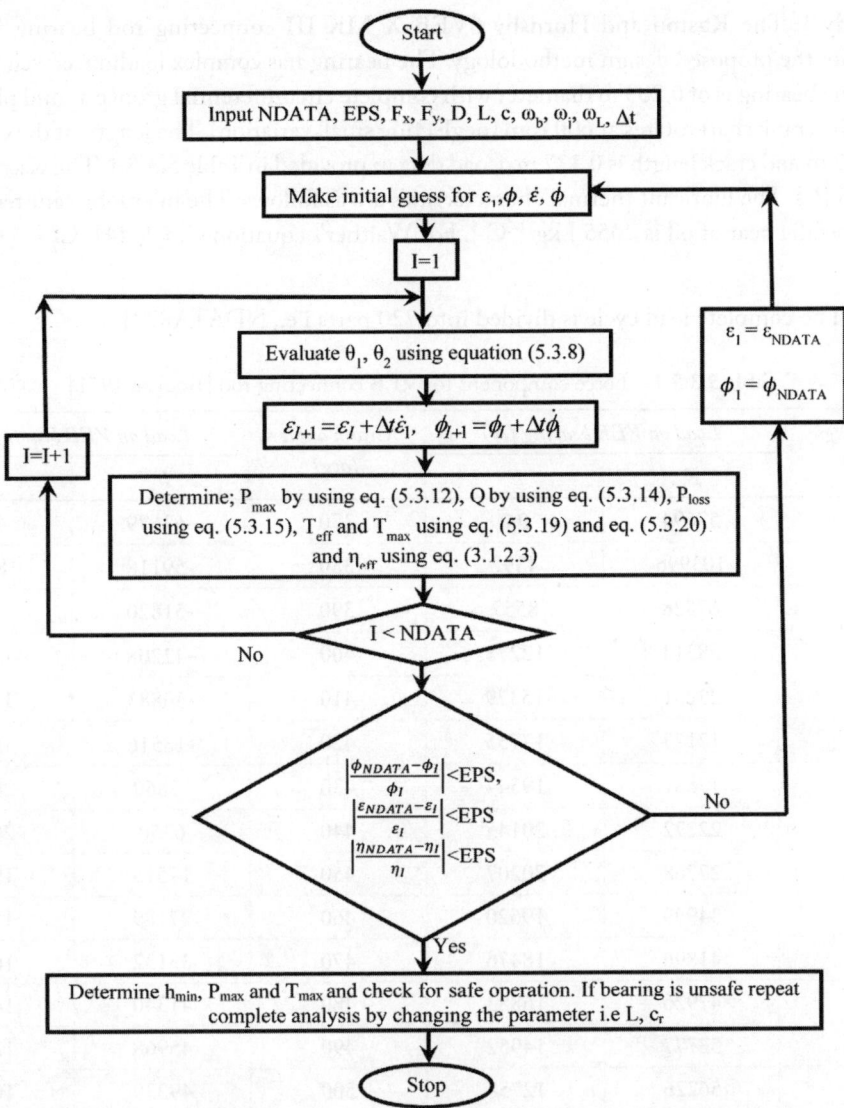

Fig.5.3.2 Computational scheme for bearing analysis using pressure model

Table 5.3.4.1 Guidance on dangerous levels of film thickness Booker [Booker, 1979]

Type of Engine	D (m)	h_{min} (μm)
Automotive (Gasoline)	0.05	1
Automotive (Diesel)	0.075–0.1	1.75
Industrial (Diesel)	0.25	2.5

5.3.5 Case studies

To illustrate the application of the suggested methodology, following two case studies have been detailed.

Case Study 1: The Ruston and Hornsby 6VEB-X MK III connecting rod bearing is used to demonstrate the proposed design methodology. The bearing has complex loading as well as angular motion. The bearing is of 0.203 m diameter with complete circumferential groove at mid plane of the bearing. The crank shaft rotates at 600 rpm (neglecting small variation). The length of the connecting rod is 0.782 m and crack length is 0.127 m. Load data in provided in Table 5.3.5.1. The viscosity of the oil is 0.015 Pas. The lubricant thermal characteristics are as follows: The inlet lubricant temperature is 70 °C, specific heat of oil is 2055 J kg^{-1} °C^{-1}. For Walther's equation $C_1 = 9.441$, $C_2 = 3.665$.

Solution:

Step 1(a): The complete load cycle is divided into 720 parts i.e., NDATA=721

Table 5.3.5.1 Force component for VEB connecting rod [Booker, 1971]

Crank angle (deg)	Load on VEB bearing (N)		Crank angle (deg)	Load on VEB bearing (N)	
	F_x	F_y		F_x	F_y
0	57694	0.0	370	-63679	4392
10	103996	4392	380	-59118	8557
20	67528	8557	390	-51820	12273
30	38314	12273	400	-42208	15379
40	23251	15379	410	-30883	17755
50	17177	17755	420	-18516	19344
60	17581	19344	430	-5860	20145
70	22272	20145	440	6350	20207
80	27768	20207	450	17515	19620
90	34999	19620	460	27189	18476
100	41896	18476	470	35132	16883
110	47926	16883	480	41340	14952
120	52777	14952	490	45968	12758
130	56226	12758	500	49239	10386
140	58718	10386	510	51442	7881
150	59207	7880	520	52799	5295
160	57961	5295	530	53533	2656
170	56849	2656	540	53778	0.0
180	55247	0.0	550	53533	-2656
190	54090	-2656	560	52799	-5295
200	53177	-5295	570	51442	-7881
210	51442	-7880	580	49239	-10386
220	49239	-10386	590	46525	-12758

Contd.

Contd.

Crank angle (deg)	Load on VEB bearing (N)		Crank angle (deg)	Load on VEB bearing (N)	
	F_x	F_y		F_x	F_y
230	44647	-12758	600	42475	-14952
240	41340	-14952	610	36646	-16883
250	35133	-16883	620	29637	-18476
260	27189	-18476	630	20915	-19620
270	17515	-19620	640	11637	-20207
280	6350	-20207	650	2042	-20145
290	-5860	-20145	660	-5202	-19344
300	-18516	-19344	670	-10724	-17755
310	-30883	-17755	680	-11939	-15379
320	-42208	-15379	690	-5135	-12273
330	-51820	-12273	700	-8575	-8557
340	-59118	-8557	710	25899	-4392
350	-63679	-4392	720	57694	0.0
360	-65215	0.0			

Step 2(a): The value of ε_1 for F_x = 57694 N and F_y = 0 N is 0.88 and attitude angle φ_1 for ε_1 = 0.88 is 28.3°.

Step 2(b): The initial values of $\dot{\varepsilon}_1$ and $\dot{\varphi}_1$ are assumed to be zero.

Step 3: The values of θ_1, θ_2 are evaluated using Eq. 5.3.8 and found to be 0 and π, respectively. Using the values of θ_1 and θ_2, the new values of $\dot{\varepsilon}_1$ and $\dot{\varphi}_1$ are found to be 28.7 and -153, respectively.

Step 4: The newly found values of $\dot{\varepsilon}_1$ and $\dot{\varphi}_1$ are different than initial assumed values, the step 2 and step 3 are repeated until required accuracy (0.01%) in $\dot{\varepsilon}_1$ and $\dot{\varphi}_1$ is obtained. After three iterations the converged values are $\dot{\varepsilon}_1 = -3.32$ and $\dot{\varphi}_1 = -89$.

Step 5: At next time step $\varepsilon_{i+1} = \varepsilon_i + \Delta t \dot{\varepsilon}_i$ and $\varphi_{i+1} = \varphi_i + \Delta t \dot{\varphi}_i$ are used and found ε_2 = 0.879 and φ_2 = 26.52°.

Step 6: The value of Q_p for circumferential groove, Q_H and Q are found to be 1.75 x 10^{-5} m³/s, 3.74 x 10^{-5} m³/s and 4.72 x 10^{-5} m³/s, respectively.

Step 7: The power loss is estimated to be 15.94W.

Step 8: The effective and maximum temperature is estimated to be 348.7 K and 354.3 K respectively. For the estimated temperatures and using the coefficient value C_1 and C_2, the effective viscosity using the Walther equation is calculated to be 0.0175 Pas.

Step 9: Repeat the steps 4–8, until last time step (t=NDATA) is reached.

Step 10: The convergence using $\left|\frac{\phi_{NDATA} - \phi_I}{\phi_I}\right| < 0.0001$, $\left|\frac{\varepsilon_{NDATA} - \varepsilon_I}{\varepsilon_I}\right| < 0.0001$ and $\left|\frac{\eta_{NDATA} - \eta_I}{\eta_I}\right|$ <0.0001 is checked. In the first cycle iteration the following results are obtained:

$$\varepsilon_1 = 0.61, \varphi_1 = -31, \dot{\varepsilon}_1 = 46.33, \dot{\phi}_1 = 180.83 \text{ and } \mu_1 = 0.012.$$

Step 11: Steps 2–10 are repeated till required convergence is achieved.

Step 12: Finally, the values of ε_{max} and P_{max} are estimated to be 0.9588 and 33.8 MPa. The value of h_{min} for eccentricity ratio 0.9588 is estimated to be 3.14μm. By comparing the value of h_{min} given in Table 5.3.4.1, the estimated value is higher than the dangerous level. So the bearing is safe.

Case Study 2: This case study contains the design of crankshaft main bearing. The journal of diameter 0.072 m rotates at 5000 rpm. The bearing has circumferential groove of 160° extend and lubricant supply pressure is 3.5 x 10⁵ N/m². The horizontal and vertical forces acting on the bearing are provided in Table 5.3.4.3. The lubricant properties are: specific heat (C_0) = 2083 J kg⁻¹ °C⁻¹, viscosity at 40 °C and 100°C are 96.871 cS and 11.659 cS, respectively. The diametral clearance and bearing length are to be determined using proposed design methodology. This bearing has 160° extend of circumferential groove.

Table 5.3.4.2 Force component for main bearing Paranjpe and Goenka. [Paranjpe and Goenka, 1990]

Crank angle (deg)	Load on crankshaft bearing (N)		Crank angle (deg)	Load on crankshaft bearing (N)	
	F_x	F_y		F_x	F_y
0	3396	-4062	370	-3707	-6218
10	7003	-7750	380	-2554	-6451
20	8287	-7565	390	-228	-4506
30	7476	-4726	400	1926	-3206
40	7169	-2926	410	3940	-2529
50	7379	-2060	420	5367	-1559
60	7890	-1909	430	6649	-840
70	8434	-2307	440	8056	-1018
80	8739	-3198	450	9046	-1271
90	8586	-4640	460	9554	-1547
100	7894	-6531	470	9559	-1823
110	6899	-8611	480	9104	-2068
120	2195	-16901	490	8240	-2285
130	-1783	-24879	500	7037	-2520

Contd.

Contd.

Crank angle (deg)	Load on crankshaft bearing (N)		Crank angle (deg)	Load on crankshaft bearing (N)	
	F_x	F_y		F_x	F_y
140	-1967	-25561	510	5599	-2755
150	-17	-20288	520	4011	-2979
160	655	-16297	530	2329	-3217
170	478	-13672	540	787	-3132
180	-509	-11663	550	-734	-3059
190	-1711	-10449	560	-2412	-3284
200	-2699	-10360	570	-4029	-3441
210	-3508	-10859	580	-5520	-3516
220	-4019	-11806	590	-6833	-3466
230	-4258	-12725	600	-7901	-3257
240	-865	-19057	610	-8634	-2923
250	3074	-24329	620	-8961	-2485
260	3433	-22971	630	-8872	-1934
270	504	-16973	640	-8385	-1298
280	-1556	-12316	650	-7561	-608
290	-2733	-8884	660	-6489	-57
300	-2745	-5530	670	-5273	601
310	-2280	-2877	680	-3993	929
320	-2119	-1591	690	-2683	929
330	-1975	-810	700	-1336	562
340	-1914	-520	710	-13	44
350	-1827	-566	720	3396	-4062
360	-3031	-3331			

Solution: As the bearing has a 160° extend of circumferential groove, bearing length (L) is calculated by considering the value of length to diameter ratio (L/D) between 0.4–0.6. In the present case L/D is kept as 0.5. For diameter (D) = 0.072 m length (L) is calculated to be 0.036 m. Since the bearing diameter (D) < 130 mm, the clearance is calculated using Eq. 5.3.3. The value of C is calculated to be 0.0214 mm. The lower bound for the clearance is given by 0.000375 D i.e., 0.027 mm. Since the lower bound value is greater than the calculated value from Eq. 5.3.3, the clearance (c_r) value is 0.027 mm. The estimated values of important parameters are listed in Table 5.3.5.3.

Table 5.3.5.3 Values of different parameters

Parameters	Values
Max. Pressure (MPa)	85.7
Min. film thickness (μm)	2.47
Average flow (cc/s)	12.9
Power loss (watts)	573.9

Frequently Asked Questions

Q.1. Why is external source of pressure required for hydrostatic lubrication?

Ans. Generally hydrostatic lubrication regime is required where the load is very high and the relative velocity between the rubbing surfaces is comparatively low. In such situations fluid pressure cannot be generated by hydrodynamic action, and external source to pressurize the lubricant is required to separate the rubbing surfaces. The external pressure should be high enough so that the lubricant is able to bear the load of the rubbing surfaces.

Q.2. What happens if the external source of pressure fails to generate sufficient pressure in hydrostatic lubrication regime?

Ans. There are two ways to deal with insufficient pressure in hydrostatic lubrication mechanism. First to incorporate the control system so that based on the load, external pressure is adjusted to avoid mechanical contact between tribo–surfaces. Second method is to hybridize the hydrostatic mechanism with hydrodynamic mechanism. In such a case hydrostatic mechanism bears static load, while the hydrodynamic mechanism deals with dynamic load on hybrid bearing.

Q.3. Where does the hydrostatic bearing find its application?

Ans. Hydrostatic bearing finds its application where the load is very high and the relative velocity between the rubbing surfaces is comparatively low such as in large telescopes, radar tracking units, machine tools and gyroscopes.

Q.4. Why hydrostatic lubrication should not be used in automotive axles since friction losses are very less in hydrostatic lubrication?

Ans. Friction, due to viscous shearing of lubricant, is low in hydrostatic lubrication if relative speed is very low. In automotive axles, if relative speed is low then hydrostatic lubrication will be a good choice, but generation of external pressure through pump in hydrostatic lubrication requires complex and sophisticated systems which consume huge amount of energy to generate hydrostatic lift. Hence hydrostatic lubrication finds its use in static applications such as telescopes and observatory domes.

Q.5. Provide some examples of physical structures where hydrostatic lubrication mechanism has been used.

Ans. Magellan telescope (Fig. 5.4.1) of the observatory of Carnegie Institution of Washington, which includes a mirror of 6.5 m diameter, uses hydrostatic lubrication supporting mechanism. This system utilizes 18 hydrostatic bearing pads to support the weight and allows the operator to gently rotate the telescope.

Fig. 5.4.1 Magellan telescope (http://www.gmto.org)

Second example is Mile High stadium in Denver, Colorado (Fig. 5.4.2) where the entire seating section of the football stadium can be moved for conversion into a baseball field using water lubricated hydrostatic bearing. The system utilizes multiple rubber pad bearings 1.22 m in diameter to lift and move nearly 20 million-Kg (4500 ton) load.

Fig. 5.4.2 Mile High stadium in Denver

Q.6. What is the basic difference between the operating mechanism of hydrodynamic and hydrostatic lubrication?

Ans. Although both hydrodynamic and hydrostatic lubrication utilize thick fluid film separating the surfaces, the term hydrostatic is introduced to differentiate its operating mechanism (flowing pressurized fluid from external source to the interface) from the self-acting hydrodynamic lubrication where speed and wedge effects are the main requirements. In other words, in hydrostatic lubrication even at zero relative speed adequate film thickness can be developed to separate the bearing surfaces.

Q.7. Can air or gas be used as a lubricant in hydrostatic bearing?

Ans. Yes, air or gas can be used as a lubricant in hydrostatic bearing where contamination of the system or sub system by the lubricant is an issue and operating temperatures are high. Viscosity of gases increase with temperature and that is advantageous.

Q.8. What is a recess in a hydrostatic bearing? What is its purpose?

Ans. Each hydrostatic bearing is equipped with a recess (pocket) whose depth is several times greater than the film thickness. With help of recess, appropriate support area with relatively high fluid pressure is possible. In other words, a recess in hydrostatic bearing improves its load capacity and stiffness.

Q.9. What is the use of compensating element in a hydrostatic bearing?

Ans. In hydrostatic bearing, lubricant flows from supply pump into a pressure control valve and then passes through the compensating element and then into the bearing recess. Capillary and orifice restrictors are the most commonly used compensating elements. With use of compensating elements, lubricant can be supplied to two or more bearing recesses with a single pump. The pressure in the recess is not equal to the supply pressure from the pump, and the recess pressure may vary based on operating condition. In other words, due to restrictor and recess, hydrostatic bearing gets compensating (auto adjusting) characteristics.

Q.10. What is the significance of oil hydraulic lift in journal bearings?

Ans. Oil hydraulic lift is useful for eliminating high starting friction and reducing journal bearing wear during start/stop cycles. The following diagram shows a schematic representation of hydrostatic lift.

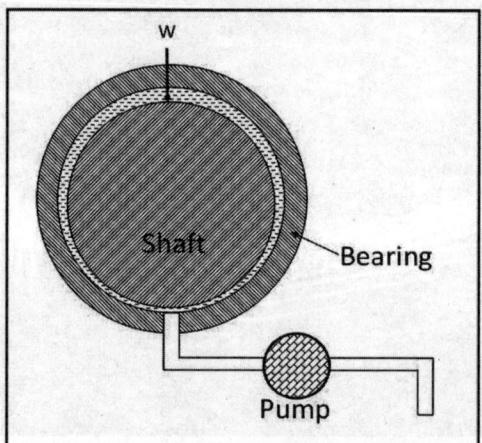

Fig. 5.4.3 Schematic representation of hydrostatic lift

Q.11. What are the important design objectives of hydrostatic bearings?

Ans. Important design objectives related to hydrostatic bearings are load carrying capacity, bearing stiffness, flow rate and friction losses.

Q.12. How can we represent the compensating elements such as capillary/orifice compensators, pressure relief valve and pump with a help of a schematic diagram to understand the functioning of multi recess hydrostatic bearing?

Ans. The diagram represents three systems each having a different load W_1, W_2 and W_3 and supported by different pressures P_1, P_2 and P_3 with the help of compensating elements (capillary/orifice restrictors). The diagram also represents the flow of used lubricant to the oil sump. Each recess is having a different film thickness h_1, h_2 and h_3. If the applied load W increases, the corresponding film thickness h decreases and simultaneously the flow through the capillary or orifice also decreases. In case the load is removed from any of the bearing/recess, the flow of lubricant through the feed line would increase.

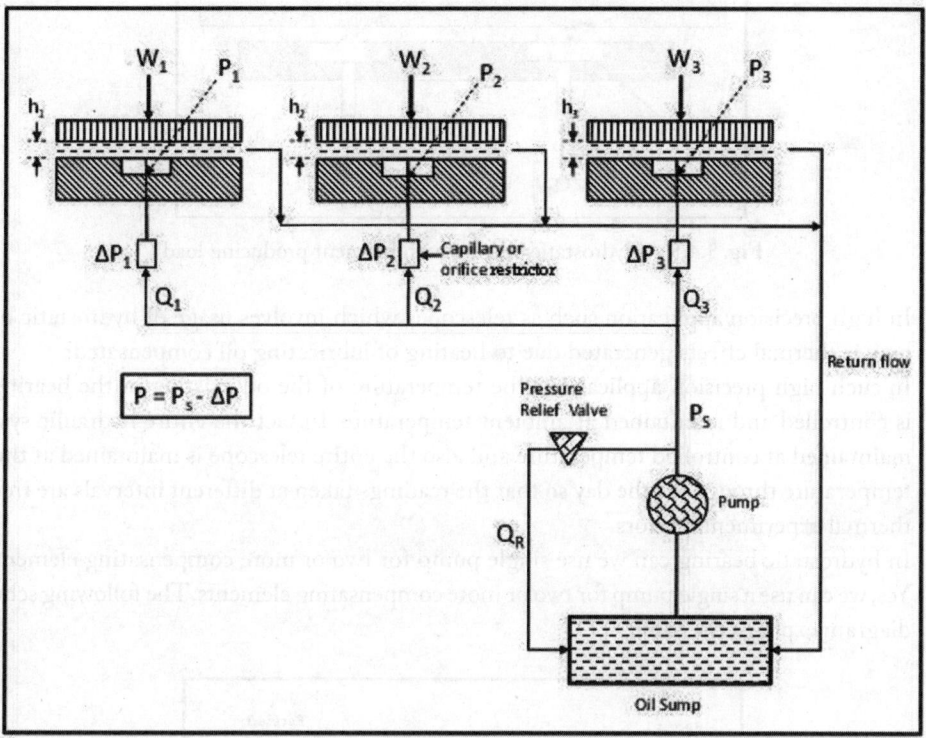

Fig. 5.4.4 Schematic representation of multi recess system with orifice/capillary compensation

Q.13. Keeping in view all the advantages of hydrostatic bearings i.e., low friction, high stiffness and strong viscous damping can hydrostatic bearings be used in machine tools?

Ans. Hydrostatic bearings have been used in commercial lathe machines as well as grinding wheels which can produce a rotational accuracy of the spindle of the order of 0.05 microns. In case if the spindle is impacted the oil film in hydrostatic bearings restores the equilibrium in a short span of time (roughly around 0.002 sec). Hence hydrostatic bearings find extensive usage in machine tools.

Q.14. What is hydrostatic shimming?

Ans. The term shimming is used to indicate levelling in assembly. . The process of fine tuning the film thickness in hydrostatic bearing after assembly is called hydrostatic shimming. The following schematic diagram shows a hydrostatic bearing with moment producing load W.

The flow to each recess Q_1, Q_2, Q_3 and Q_4 would be adjusted until necessary film thickness h_1, h_2, h_3 and h_4 have been attained.

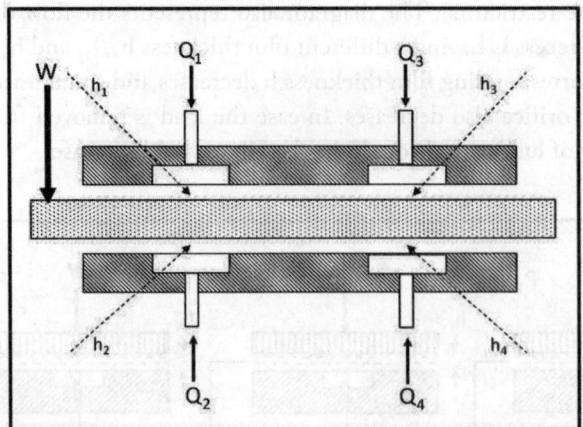

Fig. 5.4.5 Hydrostatic bearing with moment producing load

Q.15. In high precision application such as telescopes which involves usage of hydrostatic bearing how is thermal effects generated due to heating of lubricating oil compensated?

Ans. In such high precision applications the temperature of the oil existing in the bearing pads is controlled and maintained at ambient temperature. In fact the entire hydraulic system is maintained at controlled temperature and also the entire telescope is maintained at the same temperature throughout the day so that the readings taken at different intervals are free from thermal experimental errors.

Q.16. In hydrostatic bearing can we use single pump for two or more compensating elements?

Ans. Yes, we can use a single pump for two or more compensating elements. The following schematic diagram explains the same.

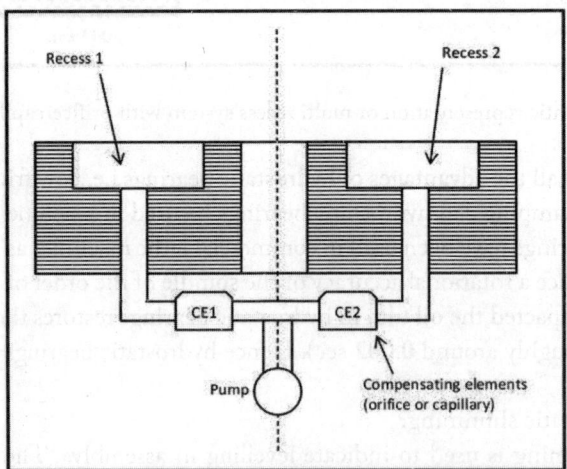

Fig. 5.4.6 Hydrostatic bearing with multiple compensating elements where one pump supplies lubricant to several bearings

Lubricant will flow from the supply pump into pressure control valve and then through the restrictor into the bearing recess. The pressure in the recess is not equal to the supply pressure. Capillary and orifice restrictors are the two most commonly used compensating elements.

Q.17. On what principle, is the load carrying capacity of the squeeze film lubrication based?

Ans. The load carrying phenomenon in squeeze film lubrication arises from the fact that a viscous lubricant cannot be instantaneously squeezed out from between two surfaces that are approaching each other. As a result it takes a finite length of time for the fluid to be squeezed out from the sides. Hence the fluid offers a certain amount of cushioning effect or load-carrying capacity even in the absence of sliding motion or physical wedge effect.

Q.18. Out of the two compensating elements i.e., orifice and capillary, which one of them offers better stiffness?

Ans. Orifice compensated bearing has better stiffness than capillary but the stiffness offered by both these compensating elements is much lesser than the constant flow system. The following diagram shows the comparison among the stiffness factor for various compensating devices.

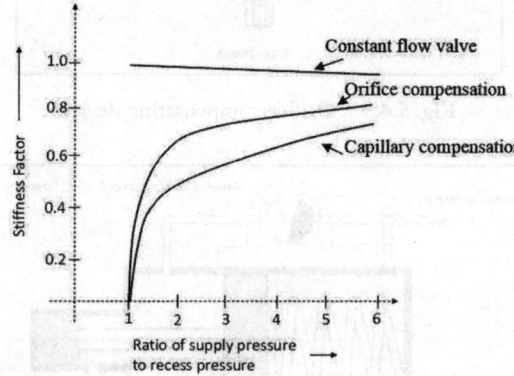

Fig. 5.4.7 Comparison of stiffness factor for constant flow rate, orifice and capillary compensated bearings

Q.19. How can we represent the different orifice compensating elements with the help of a schematic diagram in order to have a basic idea about their functioning?

Ans. The following schematic diagrams represent the commonly used compensating elements such as capillary, orifice and constant flow compensating elements.

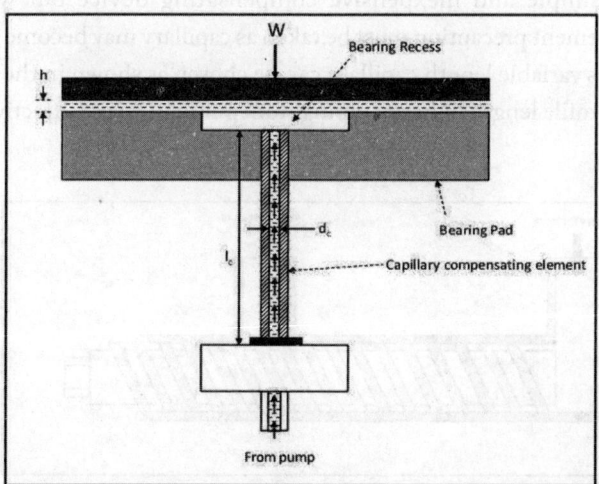

Fig. 5.4.8 Capillary compensating device

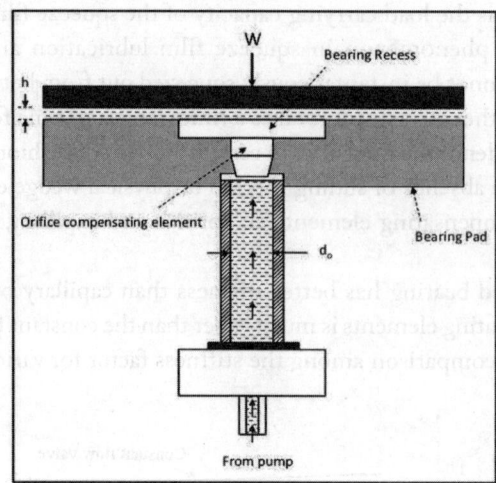

Fig. 5.4.9 Orifice compensating device

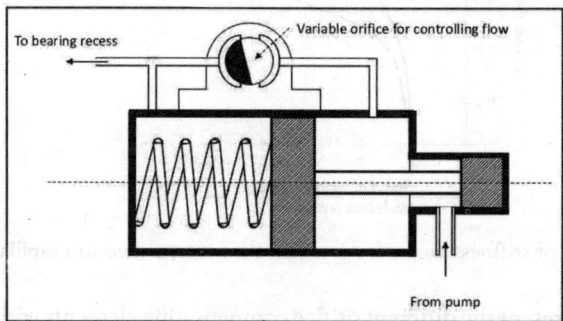

Fig. 5.4.10 Constant flow valve compensating device

Q. 20. What is adjustable length capillary compensating device?

Ans. Capillary is a simple and inexpensive compensating device but while choosing this compensating element precaution must be taken as capillary may become excessively long. For such applications variable length capillary can be chosen as shown in the following figure. By varying thread profile length between piping from pump to recess, effective length of capillary can be adjusted.

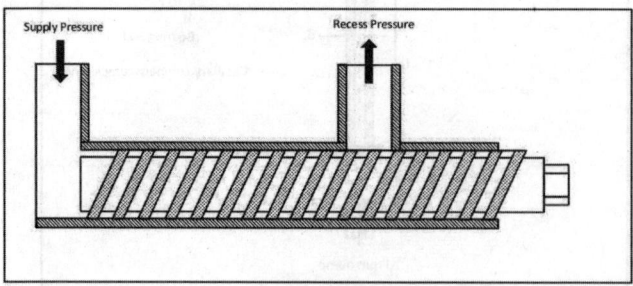

Fig. 5.4.11 Adjustable length capillary compensating device

Q.21. What kind of precautions may be taken while using orifice compensating device in a bearing?

Ans. Orifice is a temperature sensitive device hence for applications where precise control is needed orifice compensation without appropriate control on temperature MUST NOT be used. Orifice is also affected by the dirt and contamination from the lubricant and this combined with pressure surges may cause damage to the orifice plate and hydrostatic bearing.

Q.22. Is there any equation that governs the flow through an orifice compensating device?

Ans. The flow through the orifice is governed by the following equation:

$$Q = \pi R^2 C_D \sqrt{\frac{2}{\rho}(P_S - P_R)}$$

where
R = orifice radius.
C_D = coefficient of discharge which is the function of Reynolds' number R_e.
P_S = supply pressure.
P_R = recess pressure.
ρ = density of lubricant.

Q.23. What are the general guidelines for design of compensated bearings?

Ans. Some of the important points regarding the design of compensated bearings are as follows:

1. Once the dimensions and the load has been defined, compute the recess pressure.
2. Choose the compensating elements for maximum stiffness (load bearing capacity).
3. Determine the supply pressure and calculate the flow rate with the help of equations. Select a pump of suitable capacity.
4. With the help of flow rate information, select compensating device such as capillary. Select a capillary with appropriate diameter and compute the capillary length. A capillary compensating device with adjustable length can also be selected.
5. Compute the Reynolds' number to determine that the flow is laminar.

Q.24. In case of squeeze film lubrication, when two circular disks approach each other what can be interpreted if the computed time of approach for the film gap to reduce from h_1 to h_2 comes out to be infinity when h_2 reaches 0?

Ans. When the computed time of approach comes out to be infinity when h_2 is zero it means that it would take infinite amount of time to squeeze out all the fluid. However, it is hypothetical and detailed analysis is required.

Q.25. For a journal bearing what is the practical significance of assuming L/D ratio equal to ∞?

Ans. The assumption of L/D ratio equal to ∞ is made to signify that there is no side leakage. Assumption of infinitely long bearing is made to make the problem one dimensional that is the term $\frac{\partial}{\partial z}$ vanishes from the equation. The solutions from this assumption would be sufficiently accurate when L/D ratio is greater than 2. But in many practical applications L/D ratio would be less than 2, hence for such cases numerical analysis is required.

Q.26. What kind of lubricant is present in knee joint and how does squeeze film action takes place within a knee joint?

Ans. Synovial fluid is the natural lubricant which is present in the knee joint. It involves a long chain polymer known as Sodium hyaluronate which exhibits a remarkable degree of shear thinning. Impact between the femur and tibia is cushioned by the body synovial fluid during exercises demonstrating squeeze film action.

Q.26. How can we rank the compensating elements on the basis of different parameters such as initial cost, useful life, serviceability, etc.?

Ans. The following compensating elements such as capillary, orifice and constant flow has been ranked in the table 5.4.1 on the basis of various parameters.

Q.27. Is there any minimum diameter limitation for capillary compensating device?

Ans. Although very small diameters are available and even hypodermic needle serves quite well as capillary tubing for hydrostatic bearings. Generally diameters less than 50 micrometers should not be used because of tendency to clog.

Q.28. Compare compensating elements used for hydrostatic bearings.

Ans. Following table provides the required comparison.

Table 5.4.1 Ranking of compensating elements [Rippel, 1963]

Parameters		Compensating Element		
		Capillary	Orifice	Constant-flow valve
1.	Initial Cost	2	1	3
2.	Cost to fabricate and install	2	3	1
3.	Space Required	2	1	3
4.	Reliability	1	2	3
5.	Useful life	1	2	3
6.	Commercial availability	2	3	1
7.	Tendency to log	1	2	3
8.	Serviceability	2	1	3
9.	Adjustability	3	2	1

Note: Rank 1 denotes the best or the most desirable element for that particular parameter.

Multiple Choice Questions

Q.1. The synovial fluid present in the knee joint is a
(a) Shear thickening
(b) Shear thinning
(c) Newtonian fluid
(d) Thermal thickening

Q.2. In case of squeeze film lubrication the time of approach between two circular plates (radius R) for the film gap to reduce from initial value h_1 to final value h_2 with a constant force W acting on the plates is given as?

(a) $t = \dfrac{3\pi\eta R^4}{4W}\left[\dfrac{1}{h_1^2}\right]$

(b) $t = \dfrac{3\pi\eta R^4}{4W}\left[\dfrac{1}{h_2^2}\right]$

(c) $t = \dfrac{3\pi\eta R^4}{4W}\left[\dfrac{1}{h_2^2} - \dfrac{1}{h_1^2}\right]$

(d) $t = \dfrac{3\pi\eta R^4}{4W}\left[\dfrac{1}{h_1^2} - \dfrac{1}{h_2^2}\right]$

Q.3. In the case of parabolic pressure distribution, the average height of the parabola is
(a) 2/3 of the maximum height
(b) 1/4 of the maximum height
(c) 1/2 of the maximum height
(d) 1/3 of the maximum height

Q.4. In the case of squeeze film lubrication the time of approach between two rectangular plates for the film gap to reduce from initial value h_1 to final value h_2 with a constant force W acting on the plates is given as:

(a) $t = \dfrac{b\eta l^3}{2W}\left[\dfrac{1}{h_1^2}\right]$

(b) $t = \dfrac{b\eta l^3}{2W}\left[\dfrac{1}{h_2^2}\right]$

(c) $t = \dfrac{b\eta l^3}{2W}\left[\dfrac{1}{h_2^3} - \dfrac{1}{h_1^3}\right]$

(d) $t = \dfrac{b\eta l^3}{2W}\left[\dfrac{1}{h_2^2} - \dfrac{1}{h_1^2}\right]$

Q.5. Which of the following are the characteristics of hydrostatic bearing?
(a) Extremely low friction
(b) High load carrying capacity at low speeds
(c) High positional accuracy in high speed
(d) All of the above

Q.6. The concept of squeeze film lubrication when applied to journal bearing (radius r_b), the time of approach is given as:

(a) $\Delta t = \dfrac{12\pi\eta r_b^3}{Wc^2}\left(\dfrac{\varepsilon_2}{(1-\varepsilon_2^2)^{1/2}} - \dfrac{\varepsilon_1}{(1-\varepsilon_1^2)^{1/2}}\right)$

(b) $\Delta t = \dfrac{12\pi\eta r_b^3}{Wc^2}\left(\dfrac{\varepsilon_1}{(1-\varepsilon_1^2)^{1/2}}\right)$

(c) $\Delta t = \dfrac{12\pi\eta r_b^3}{Wc^2}\left(\dfrac{\varepsilon_2}{(1-\varepsilon_2^2)^{1/2}}\right)$

(d) $\Delta t = \dfrac{12\pi\eta r_b^3}{Wc^2}$

Q.7. Between the top dead centre (TDC) and bottom dead centre (BDC), which of the following lubrication regime is mostly present between the piston rings and cylinder walls during the normal operation of the IC engine?
    (a) Hydrostatic lubrication
    (b) Hydrodynamic lubrication
    (c) Squeeze film lubrication
    (d) Boundary lubrication

Q.8. At the top dead centre (TDC) and bottom dead centre (BDC) when the piston velocity is almost zero which of the following lubrication regime is present between the piston rings and cylinder walls?
    (a) Hydrostatic lubrication
    (b) Hydrodynamic lubrication
    (c) Squeeze film lubrication
    (d) Boundary lubrication

Q.9. Reynolds' equation that governs the pressure distribution in finite length journal bearing assuming V_s as the squeeze velocity, is given as:

(a) $\dfrac{\partial}{\partial x}\left(\dfrac{h^3}{\eta}\dfrac{\partial P}{\partial x}\right) + \dfrac{\partial}{\partial z}\left(\dfrac{h^3}{\eta}\dfrac{\partial P}{\partial z}\right) = 6U\dfrac{\partial h}{\partial x} - 12V_s$

(b) $\dfrac{\partial}{\partial x}\left(\dfrac{h^3}{\eta}\dfrac{\partial P}{\partial x}\right) = 6U\dfrac{\partial h}{\partial x} - 12V_s$

(c) $\dfrac{\partial}{\partial y}\left(\dfrac{h^3}{\eta}\dfrac{\partial P}{\partial y}\right) = 6U\dfrac{\partial h}{\partial x} - 12V_s$

(d) $\dfrac{\partial}{\partial x}\left(\dfrac{h^3}{\eta}\dfrac{\partial P}{\partial x}\right) + \dfrac{\partial}{\partial z}\left(\dfrac{h^3}{\eta}\dfrac{\partial P}{\partial z}\right) = 6U\dfrac{\partial h}{\partial x}$

Q.10. Hydrostatic bearing finds its application in
    (a) Large radio antennas
    (b) Large telescopes
    (c) Observatory domes
    (d) All of the above.

Q.11. For a circular step hydrostatic thrust bearing having R_1 and R_2 are the inner and outer radii of the bearing and P_s is the supply pressure the load bearing capacity is given as

(a) $W = \dfrac{\pi P_s\left(R_2^2 - R_1^2\right)}{2}$

(b) $W = \dfrac{\pi\left(R_2^2 - R_1^2\right)P_s \ln(R_2/R_1)}{2}$

(c) $W = \dfrac{\pi R_2^2 P_s \ln(R_2/R_1)}{2\left[1 - \left(R_1/R_2\right)^2\right]}$

(d) $W = \dfrac{\pi P_s\left(R_2^2 - R_1^2\right)}{2\ln(R_2/R_1)}$

Q.12. The provision of recess in a hydrostatic bearing
    (a) Increases load bearing capacity
    (b) Increases bearing stiffness
    (c) Decreases load bearing capacity
    (d) Both (a) and (b).

Q.13. Stiffness of hydrostatic bearing
    (a) Decreases as film thickness increases
    (b) Increases as film thickness increases
    (c) Remains unchanged
    (d) None of these

Q.14. The load carrying capacity in squeeze film lubrication arises from the fact that
(a) Fluid offers a certain amount of cushioning effect
(b) Lubricant is instantaneously squeezed out.
(c) Viscous lubricant cannot be instantaneously squeezed out from between two surfaces that are approaching each other.
(d) Both (a) and (c)

Q.15. Out of the following compensating elements for hydrostatic bearing which one offers the maximum bearing stiffness?
(a) Orifice compensating element
(b) Capillary compensating element
(c) Constant flow system
(d) Both (a) and (b)

Q.16. Which of the following problems are associated with orifice compensating device?
(a) Temperature sensitiveness
(b) Dirt contamination
(c) Contamination from lubricant
(d) All of the above

Q.17. Which of the following equations correctly represents the flow of an orifice compensating device for a hydrostatic bearing given that C_D is the discharge coefficient, P_S is the supply pressure and P_R is the recess pressure?

(a) $Q = \pi R C_D \sqrt{\frac{2}{\rho}(P_S - P_R)}$
(b) $Q = \pi R^2 C_D \sqrt{\frac{2}{\rho}(P_R - P_S)}$
(c) $Q = \pi R^2 C_D \sqrt{\frac{2}{\rho}(P_S - P_R)}$
(d) $Q = \pi R^2 C_D \sqrt{(P_S - P_R)}$

Q.18. The operational feasibility of the hydrostatic bearing is determined using:
(a) Initial cost, operating cost
(b) Load carrying capacity
(c) Bearing and system reliability
(d) All of the above

Answers

Q.1. (b)	Q.2. (c)	Q.3. (a)	Q.4. (d)	Q.5. (d)	Q.6. (a)
Q.7. (b)	Q.8. (c)	Q.9. (a)	Q.10. (d)	Q.11. (d)	Q.12. (d)
Q.13. (a)	Q.14. (d)	Q.15. (c)	Q.16. (d)	Q.17. (c)	Q.18. (d)

References

1. Fuller, Dudley D. 1984. *Theory and Practice of Lubrication for Engineers*. Second Edition. John Wiley and Sons.
2. Hamrock, Bernard J. 2004. *Fundamentals of Fluid Film Lubrication*. Second Edition. Marcel Dekker Inc.
3. Hays, D. F. Dec 1961. 'Squeeze Films: A Finite Journal Bearing with Fluctuating Load.' *Transactions of American Society of Mechanical Engineers.*

4. Rippel, H. C. 1963. *Cast Bronze Hydrostatic Bearing Design Manual.* Second Edition. Cast Bronze Bearing Institute, Inc. Cleveland.
5. Hirani, H. Athre, K. and Biswas, S. 1999. 'Dynamically loaded finite length journal bearings: analytical method of solution'. *Journal of Tribology.* 121 (4):844–852.
6. Stolarski, T. A. 1990. *Tribology in machine design.* Heinemann Newnes.
7. Constantinescu, Nica, Pascovici, Ceptureance and Nedelcu. 1985. *Sliding Bearing.* Translated from the Rumanian by Al. Nica. Allerton Press Inc. New York. , P.91
8. Martin F. A. and Lee C. S. 1983. Feed Pressure Flow in Plain Journal Bearing, *Trans. ASLE.* 26:381–392.
9. Booker, J. F. 1979. Design of Dynamically Loaded Journal Bearing. *Fundamentals of Design of Fluid Film Bearing.* ASME Publication. 31–44.
10. Booker, J F., A Table of the Journal Bearing Integral, Transactions of ASME, Journal of Basic Engineering, Vol. 87, pp 533-535.

Chapter 6

Elasto–Hydrodynamic Lubrication

In previous chapters, tribo–pairs were considered as rigid solids (i.e., no deformation under load), but tribo–elements carrying heavy load with low geometrical conformity are likely to deform elastically. The elastic deformation is order of film thickness. Lubrication of such elastically deforming tribo–pairs is classified as elasto–hydrodynamic lubrication (EHL). EHL occurs in rolling element bearings, gears, O-rings, human synovial joints and cam-follower contact.

6.1 Principles and Applications

The EHL mechanism is advantageous as it provides the lowest friction among all lubrication mechanisms. In addition, EHL reduces the chances of mechanical wear as lubricant film thickness increases due to: (1) local elastic deformation of surfaces, and (2) piezo–viscous increase in lubricant viscosity. Viscosity–pressure relation has been defined in section 3.1.2.3 of chapter 3. It is interesting to note that calculation of minimum film thickness (also termed as central film thickness) and the maximum pressure are two main design parameters in EHL contacts.

One of the notable points of EHL compared to hydrodynamic (covered in chapters 3 and 4), hydrostatic and squeeze lubrications (covered in chapter 5) is the negligible effect of load on the minimum film thickness $h_{min} \alpha \ W^{-0.075}$. This effect is illustrated in Fig. 6.1.1. In this figure, it has been assumed that 'component 1' is rigid while 'component 2' is elastic. On increasing load from W to 1.5W, the spread of elastic deformation of 'component 2' increases without noticeable reduction in the minimum film thickness.

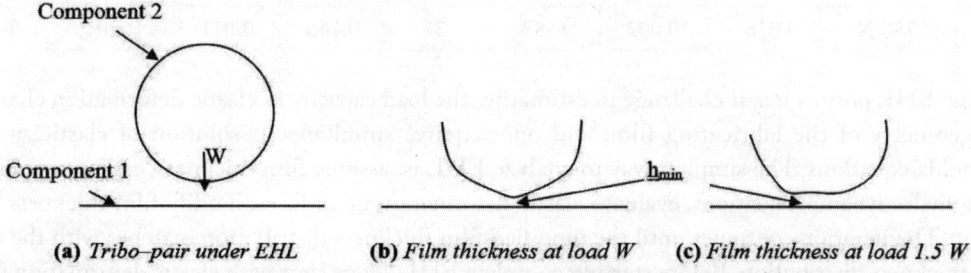

(a) *Tribo–pair under EHL*   (b) *Film thickness at load W*   (c) *Film thickness at load 1.5 W*

Fig. 6.1.1   Effect of load on film thickness in elasto–hydrodynamic lubrication

It is also important to note that due to elastic properties of material, geometry of approaching bodies ('component 2' in Fig. 6.1.1) changes and as a result film thickness expression and corresponding fluid pressure distribution change. It can be said that the maximum pressure generated in the lubricant at the interface of tribo–pair depends on the materials of contacting bodies. To understand this two bearing materials, 'acrylic' having low elastic limit and 'brass' having comparatively higher elastic limit were selected. The reason behind choosing the acrylic material was to observe EHL at comparatively low load. Bearings (inner radius = 20.175 mm, length = 40 mm, radial clearance = 0.1 mm) fabricated from these materials are shown in Fig. 6.1.2.

Fig. 6.1.2 Brass and Acrylic bearings

Various results obtained on torque, maximum pressure, temperature rise, and coefficient of friction are tabulated in Table 6.1.1. These results prove that for same speed and load conditions the values of maximum pressure and coefficient of friction are different for different materials. It is worth noting that increasing load from 600 N to 750 N, the value of maximum pressure in Acrylic bearing remains almost same. The results of Table 6.1.1 encourage studying EHL.

Table 6.1.1 Torque, friction coefficient, temperature and maximum pressure values for different combination of speed and load for acrylic and brass bearing

Combination	Acrylic Bearing				Brass Bearing			
	Torque N-m	Fric. Coeff.	P_{max} kPa	Temp diff °C	Torque N-m	Fric. Coeff.	P_{max} kPa	Temp diff °C
500 rpm-300 N	0.064	0.011	596	2	0.178	0.0296	696	3
500 rpm-600 N	0.050	0.004	1393	3	0.126	0.0105	1441	5
500 rpm-750 N	0.036	0.002	1387	3	0.166	0.011	1620	5

The EHL poses a major challenge in estimating the load capacity as elastic deformation changes the geometry of the lubricating film; and one requires simultaneous solution of elasticity and Reynolds' equations. The simplest way to analyze EHL is: assume film thickness, estimate pressure using hydrodynamic equations, evaluate elastic deformation of surfaces, modify film thickness and iterate. The iteration continues until the modified film thickness distribution matches with the new film thickness distribution. Before starting complete EHL, let us start with elastic deformation. The study of the deformation of solids was performed by Hertz in 1882, description of which is provided in the next section.

6.2 Hertz Theory

Hertz developed a theory to calculate the elastic deformation and contact pressure between the two non-conforming surfaces. Here non-conforming contact means contacting bodies have convex–convex profiles. Examples of such contacts are gears, cam follower, ball–bearing, etc. When non-conforming bodies, subject to load are brought in contact; whole load is supported by smaller area of body and high compressive stress induces. Hertz provided expressions to estimate the compression and stress induced in the objects.

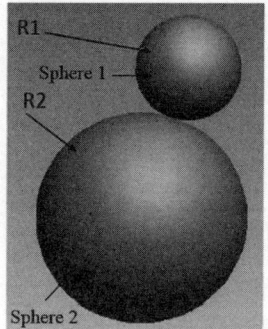

Fig. 6.2.1 Contact between two solid spheres

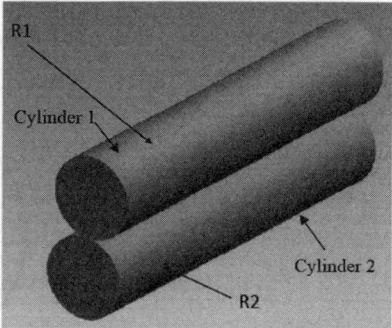

Fig. 6.2.2 Contact between two solid cylinders

Consider sphere on sphere (Fig. 6.2.1) and cylinder on cylinder (Fig. 6.2.2) in mechanical contact as these two examples simulate the major engineering components (i.e., involute gears, rolling element bearings). On compressing spheres with force F, as shown in Fig. 6.2.3(b), elastic deformation of sphere occurs. On constraining spheres in X- and Y-directions, and loading in Z-direction, elastic deformation of spheres occurs in the Z-direction. Under these conditions, distance between centres of spheres reduces by distance 'd' as shown in Fig. 6.2.3(b). The deformation, 'd', is very small due to non-conformity between convex–convex contact. In the present chapter, it is assumed that this deformation is perfectly elastic. To quantify this deformation, material and geometric properties must be considered. In Fig. 6.2.3(b), E_1 is the Young's modulus, v_1 is a Poisson's ratio and R_1 is the radius of sphere 1. Similarly, E_2 is the Young's modulus, v_2 is a Poisson's ratio and R_2 is the radius of sphere 2. F is compressive load on contacting solids.

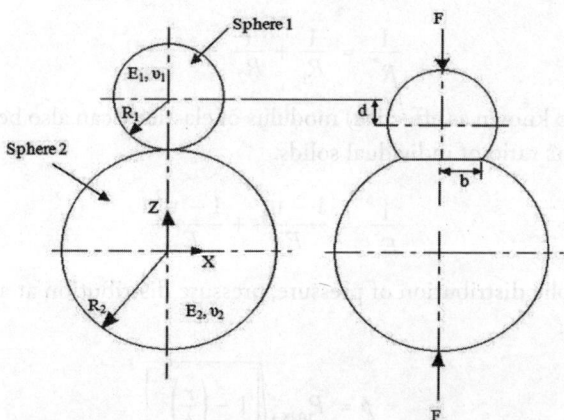

Fig. 6.2.3 Elastic deformation of two contacting solid spheres

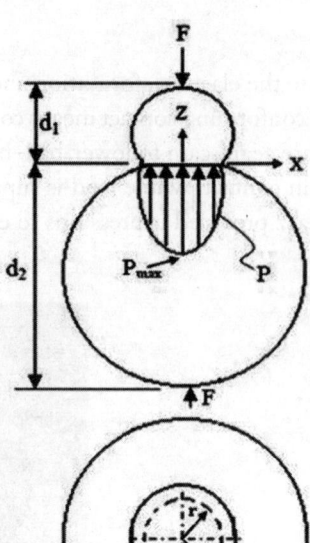

Fig. 6.2.4 Elastic deformation between two solid spheres

In Fig. 6.2.3(b) and Fig. 6.2.4, contact width of sphere, after deformation, is 2b. This dimension 2b is very small compared to diameter of sphere 1 (2b << d_1) as well as diameter of sphere 2 (2b << d_2). Here d_1 and d_2, as shown in Fig. 6.2.4, are diameters of spheres. The pressure generated at the contact interface follows a parabolic distribution and the maximum pressure ranges 0.2–3 GPa depending on the applied load and material properties.

The deformation also occurs in the perpendicular direction. In the case of spherical contact, contact patch is a circle as shown in Fig. 6.2.4. In the case of cylindrical contact the contact patch will be elliptical.

To estimate the maximum pressure, p_{max} as depicted in Fig. 6.2.4, concepts of 'effective radius' and 'equivalent elastic modulus' are used. The effective radii of curvature is defined by the radii of curvature of the two solids as,

$$\frac{1}{R^*} = \frac{1}{R_1} + \frac{1}{R_2} \tag{6.2.1}$$

The equivalent (also known as effective) modulus of elasticity can also be defined by the modulus of elasticity and poison's ratio of individual solids.

$$\frac{1}{E^*} = \frac{1-v_1^2}{E_1} + \frac{1-v_2^2}{E_2} \tag{6.2.2}$$

Considering parabolic distribution of pressure, pressure distribution at any radius r (in the range of 0–b) is expressed as:

$$p = P_{max}\sqrt{\left[1-\left(\frac{r}{b}\right)^2\right]} \tag{6.2.3}$$

Pressure will be zero at the boundaries (periphery of circle, shown in Fig. 6.2.4) and the maximum at the centre. Following expressions may be used to find the maximum fluid pressure:

$$F = \int_0^b \int_0^{2\pi} p\, r d\theta\, dr \tag{6.2.4}$$

On substituting expression of pressure from Eq. 6.2.3

$$F = 2\pi \int_0^b P_{max} \sqrt{\left[1 - \left(\frac{r}{b}\right)^2\right]} r dr$$

$$F = \frac{2\pi P_{max}}{b} \int_0^b \sqrt{\left[b^2 - r^2\right]} r dr$$

On assuming, $b^2 - r^2 = t^2$

$$F = \frac{2\pi P_{max}}{b} \int_b^0 t(-t dt)$$

$$F = \frac{2\pi P_{max}}{b} \frac{b^3}{3}$$

$$P_{max} = \frac{3}{2} \frac{F}{\pi b^2} \tag{6.2.5}$$

In case of cylindrical contact (as shown in Fig. 6.2.2) such as in rolling element bearing, there will be elliptical contact and pressure distribution may be provided as:

$$p = P_{max} \sqrt{\left[1 - \left(\frac{x}{b}\right)^2 - \left(\frac{y}{b}\right)^2\right]} \tag{6.2.6}$$

Assuming length of cylinder is much larger than its diameter, pressure variation in y direction may be neglected, and

$$p = P_{max} \sqrt{\left[1 - \left(\frac{x}{b}\right)^2\right]} \tag{6.2.7}$$

To evaluate the maximum value of pressure, following force equilibrium may be used,

$$F = 2L \int_0^b P_{max} \sqrt{\left[1 - \left(\frac{x}{b}\right)^2\right]} dx \tag{6.2.8}$$

Let $x = b \sin \theta$,

$$F = 2LP_{max} \int_0^{\pi/2} b(\cos^2 \theta d\theta)$$

$$F = \frac{\pi}{2} bLP_{max} \tag{6.2.9}$$

Therefore,

$$P_{max} = \frac{2}{\pi} \frac{F}{bL} \tag{6.2.10}$$

The maximum value of pressure at the interface of spherical and cylindrical contacts can be estimated using Eq. 6.2.5 and Eq. 6.2.10, respectively. However, without finding contact width '2b', the maximum pressure cannot be determined.

To determine the elastic deformation and corresponding contact width, expression based on theory of elasticity can be used. For spherical contact deformation δ_1 and δ_2, as shown in Fig. 6.2.5, can be expressed:

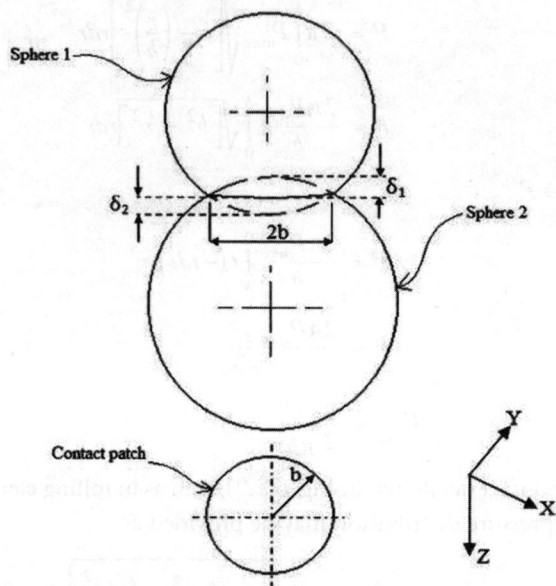

Fig. 6.2.5 Elastic deformation of Spheres

$$\delta_1 = \frac{(1-v_1^2)F}{2\pi E_1 r} \qquad (6.2.11)$$

$$\delta_2 = \frac{(1-v_2^2)F}{2\pi E_2 r} \qquad (6.2.12)$$

On substituting expression of F from Eq. 6.2.4 and expression of pressure from Eq. 6.2.3

$$\delta_1(r,\theta) = \frac{(1-v_1^2)}{2\pi E_1} \int_0^b \int_0^{2\pi} \frac{P_{max}\sqrt{1-(r/b)^2}}{r} r d\theta\, dr$$

$$\delta_1 = \frac{(1-v_1^2)}{2\pi E_1} 2\pi \int_0^b \frac{P_{max}\sqrt{1-(r/b)^2}}{r} r dr$$

$$\delta_1 = \frac{(1-v_1^2)P_{max}}{E_1} \int_0^b \sqrt{1-(r/b)^2}\, dr$$

On assuming, $r = b \sin\theta$

$$\delta_1 = \frac{(1-v_1^2)P_{max}}{E_1} \int_0^{\pi/2} \cos\theta \left(b\cos\theta \, d\theta\right)$$

$$\delta_1 = \frac{b(1-v_1^2)P_{max}}{2E_1} \int_0^{\pi/2} (\cos 2\theta + 1) \, d\theta$$

$$\delta_1 = \frac{b(1-v_1^2)P_{max}}{2E_1} \left[\frac{\sin 2\theta}{2} + \theta\right]_0^{\pi/2}$$

$$\delta_1 = \frac{b(1-v_1^2)P_{max}}{2E_1} \cdot \pi/2$$

Similarly
$$\delta_2 = \frac{b(1-v_2^2)P_{max}}{2E_2} \cdot \pi/2$$

Total deflection

$$\delta_1 + \delta_2 = \left(\frac{(1-v_1^2)}{E_1} + \frac{(1-v_2^2)}{E_2}\right)\frac{b\pi P_{max}}{4} \qquad (6.2.13)$$

Substituting $P_{max} = \frac{3F}{2\pi b^2}$ from Eq. 6.2.5,

In terms of force F, $\delta_1 + \delta_2 = \left(\frac{(1-v_1^2)}{E_1} + \frac{(1-v_2^2)}{E_2}\right)\frac{3F}{8b} \qquad (6.2.14)$

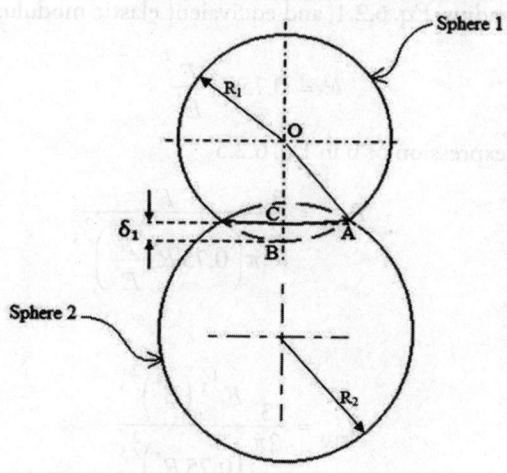

Fig. 6.2.6 Schematic of elastic deformation

From Fig. 6.2.6, elastic deformation (δ_1) can be expressed as

$$\delta_1 = OB - OC \qquad (6.2.15)$$

Or, $$\delta_1 = R_1 - \sqrt{OA^2 - AC^2}$$

Or, $$\delta_1 = R_1 - \sqrt{R_1^2 - b^2}$$

Or, $$\delta_1 = R_1\left(1 - \sqrt{1 - \left(\frac{b}{R_1}\right)^2}\right)$$

Or, $$\delta_1 = R_1\left(1 - \left(1 - \frac{1}{2}\left(\frac{b}{R_1}\right)^2 + negligible\_term\right)\right)$$

$$\Rightarrow \delta_1 = \frac{b^2}{2R_1}$$

Similarly $$\delta_2 = \frac{b^2}{2R_2}$$

Total deflection $$\delta_1 + \delta_2 = \frac{b^2}{2R_1} + \frac{b^2}{2R_2} \qquad (6.2.16)$$

On equating Eq. 6.2.14 to Eq. 6.2.16, and rearranging it

$$b^3 = 0.75\left(\frac{1-v_1^2}{E_1} + \frac{1-v_2^2}{E_2}\right)\frac{F}{(1/R_1 + 1/R_2)}$$

In terms of effective radius, Eq. 6.2.1, and equivalent elastic modulus, Eq. 6.2.2

$$b^3 = 0.75 R^* \frac{F}{E^*} \qquad (6.2.17)$$

On substituting this expression of b in Eq. 6.2.5

$$P_{max} = \frac{3}{2}\frac{F}{\pi\left(0.75 R^* \frac{F}{E^*}\right)^{2/3}}$$

On rearranging it

$$P_{max} = \frac{3}{2\pi}\frac{F^{1/3}(E^*)^{2/3}}{(0.75 R^*)^{2/3}} \qquad (6.2.18)$$

Similar equation can be derived for cylindrical contacts.

Example: A ball thrust bearing (as shown in Fig. 6.2.7) with 7 balls is loaded with 700 N across its races through the balls. Diameter of spherical balls is 10 mm. Assume load is equally shared by all balls, and race is a flat surface ($R_2 = \infty$). Determine the size of contact patch on the race. Assume Poisson's ratio = 0.28 and E = 207 GPa for ball and race materials.

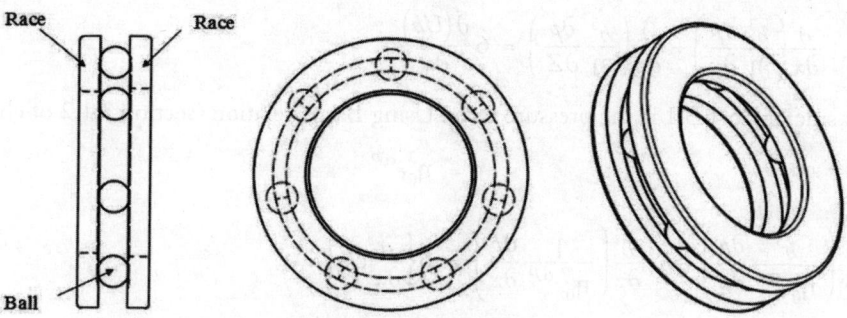

Fig. 6.2.7 Thrust bearing with races and balls

Answer: The size of contact patch, 2*b can be calculated using Eq. 6.2.17

$$b^3 = 0.75 R^* \frac{F}{E^*} \qquad (6.2.17)$$

Here
$$F = \frac{Load}{No\ of\ balls} = \frac{700}{7} = 100\ N; \frac{1}{R^*} = \frac{1}{0.005} + \frac{1}{\infty}$$

$\Rightarrow \qquad R^* = 0.005\ m$

$$\frac{1}{E^*} = \frac{1-v_1^2}{E_1} + \frac{1-v_2^2}{E_2}$$

$\Rightarrow \qquad E^* = \frac{E}{2(1-v^2)} = 1.123 * 10^{11}\ N/m^2$

On substituting these values, size of contact patch

$2 * (0.75 * 0.005 * 100/1.123e11)^{1/3} = 2.99 * 10^{-4} m$

Therefore, contact patch is 299 microns.

To realize the value of maximum pressure, use Eq. 6.2.18

$$P_{max} = \frac{3}{2\pi} \frac{F^{1/3} (E^*)^{2/3}}{(0.75 R^*)^{2/3}} \qquad (6.2.18)$$

On substituting the values, P_{max} = 2.1372 GPa

The value of maximum pressure is very high, which surely increases the lubricant viscosity. As the aim of the present chapter is to explore elasto–hydrodynamic lubrication, there is a need to incorporate pressure–viscosity relation (section 3.1.2 of chapter 3) in the Reynolds' equation.

6.3 Pressure–Viscosity Term in Reynolds' Equation

The Reynolds' equation has been discussed in section 3.1.3 of chapter 3. For hydrodynamic lubrication, this equation (Eq. 3.2.38) can be expressed as:

$$\frac{\partial}{\partial x}\left(\frac{h^3}{\eta}\frac{\partial p}{\partial x}\right) + \frac{\partial}{\partial z}\left(\frac{h^3}{\eta}\frac{\partial p}{\partial Z}\right) = 6\frac{\partial(Uh)}{\partial x} \tag{6.3.1}$$

The left side of Eq. 6.3.1 is the pressure term. Using Barus' relation (section 3.1.2 of chapter 3)

$$\eta = \eta_0 e^{\alpha P}$$

$$\frac{\partial}{\partial x}\left(\frac{h^3}{\eta_0 e^{\alpha P}}\frac{\partial p}{\partial x}\right) + h^3 \frac{\partial}{\partial z}\left(\frac{1}{\eta_0 e^{\alpha P}}\frac{\partial p}{\partial z}\right) = 6\left\{\frac{\partial}{\partial x}Uh\right\} \tag{6.3.2}$$

Or, $\quad \frac{\partial}{\partial x}\left(h^3 e^{-\alpha P}\frac{\partial p}{\partial x}\right) + h^3 \frac{\partial}{\partial z}\left(e^{-\alpha P}\frac{\partial p}{\partial z}\right) = 6\eta_0 U\left\{\frac{\partial}{\partial x}h\right\}$

Let us assume $\quad q = \frac{1-e^{-\alpha P}}{\alpha} \Rightarrow \frac{\partial q}{\partial x} = e^{-\alpha p}\frac{\partial p}{\partial x} \tag{6.3.3}$

On replacing P with q, $\frac{\partial}{\partial x}\left(h^3\frac{\partial q}{\partial x}\right) + h^3\frac{\partial}{\partial z}\left(\frac{\partial q}{\partial z}\right) = 6\eta_0 U\left\{\frac{\partial}{\partial x}h\right\} \tag{6.3.4}$

Here it is assumed that h does not depend on z and U does not depend on x. To get a feel of pressure–viscosity relation, a computer program was made for the journal bearing shown in Fig. 6.3.1. The following dimensions of the bearing are considered: bearing radius of 25.3 mm, axial length (L) = 25 mm, radial clearance (C) = 43 μm, journal velocity (1 m/s) and oil viscosity (10 mPa.s). From this figure it can be noted that the contacting pair comprises concave and convex surfaces. The effective radius of this contacting pair is given by $1/R^* = (1/R_1 - 1/R_2)$. Figure 6.3.2(a) provides pressure profile obtained from solving Eq. 6.3.1, while Fig. 6.3.2(b) depicts the profile obtained by solving Eq. 6.3.4 for eccentricity ratio 0.89. The fluid pressures along angular coordinate at the mid-plane of axial length have been shown in Fig. 6.3.2(c). The estimated load for these different cases is 15.79N. From these figures it can be inferred that there is hardly any difference in the maximum values of pressure. This clearly indicates that the variation in viscosity at low pressure is negligible.

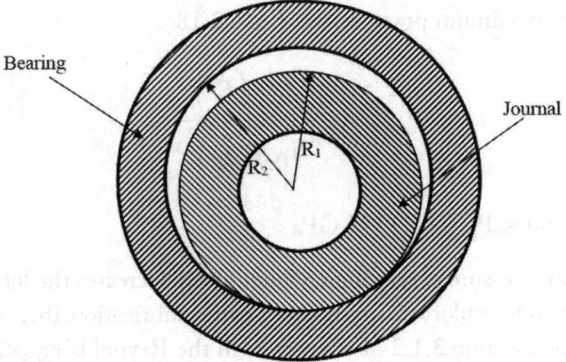

Fig. 6.3.1 Journal bearing

To observe the EHL effect, viscosity was increased by twenty times (200 mPa.s). The pressure distributions for highly viscous oil are shown in Fig. 6.3.3. On accounting the pressure–viscosity term, the maximum pressure value increases from 44.83 MPa to 59.43 MPa.

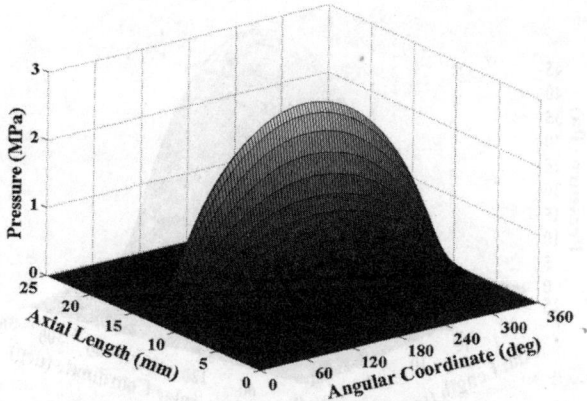

(a) Without pressure viscosity relation

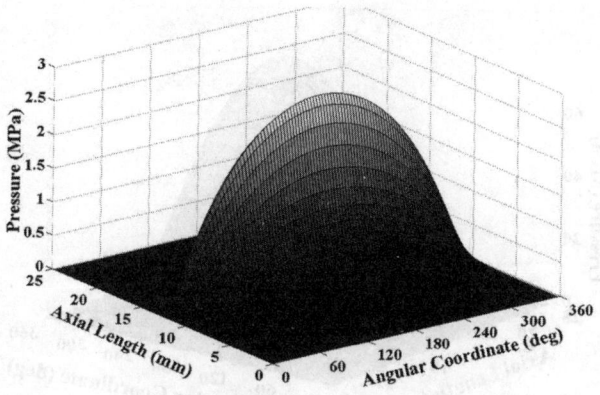

(b) With pressure viscosity relation

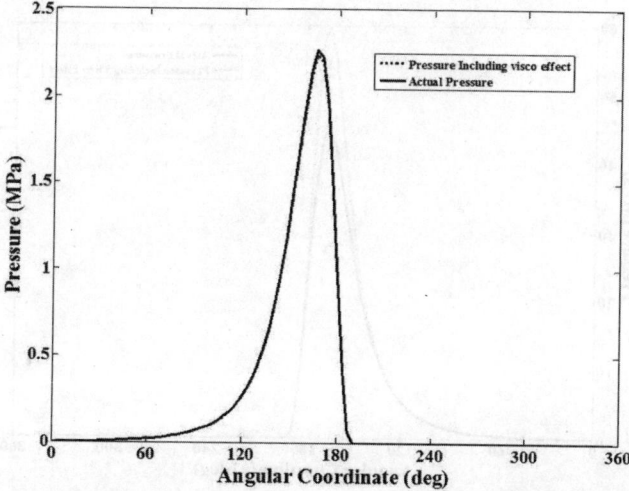

(c) Pressure comparison along the mid plane of axial length for angular coordinate

Fig. 6.3.2 The effect of pressure–viscosity relation

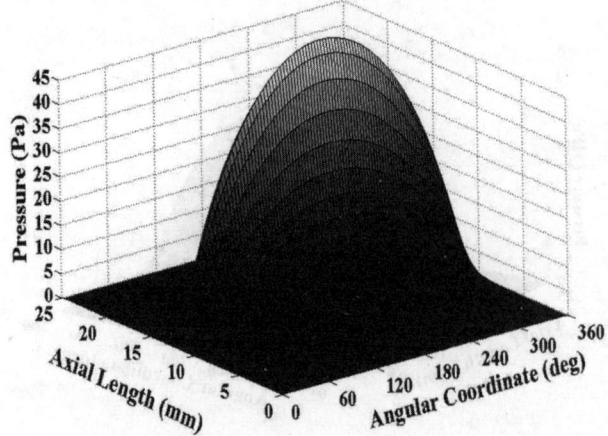

(a) Without pressure viscosity relation

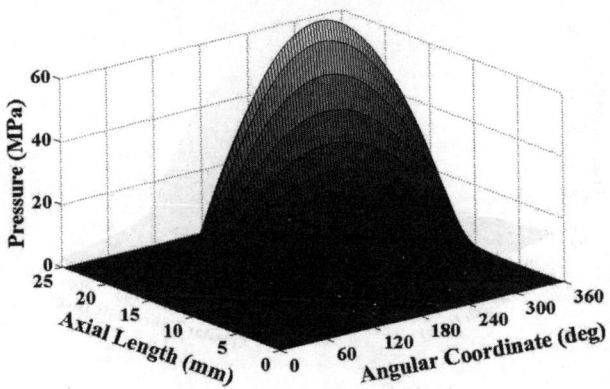

(b) With pressure viscosity relation

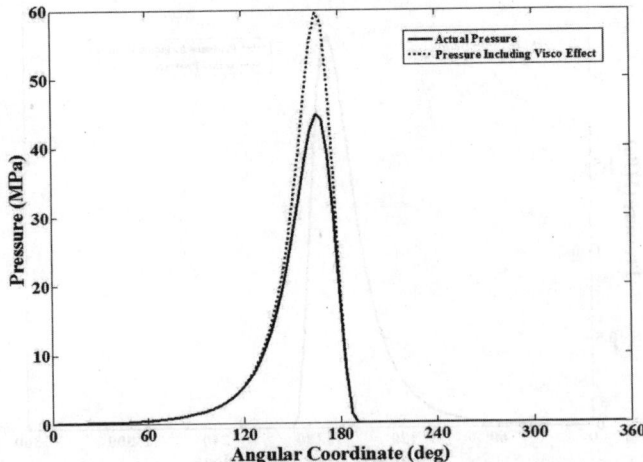

(c) Pressure comparison along the mid plane of axial length for angular coordinate

Fig. 6.3.3 The effect of pressure–viscosity relation for thicker oils

Elasto–Hydrodynamic Lubrication   273

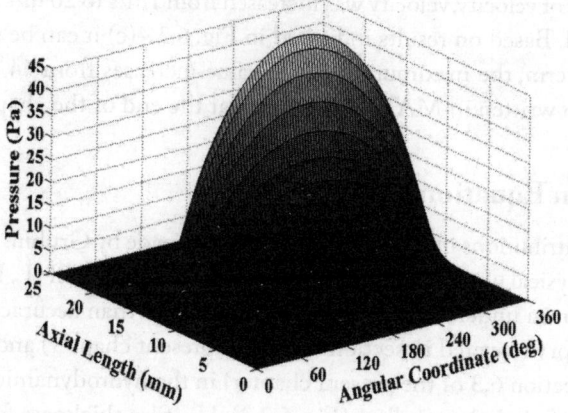

(a) Without pressure viscosity relation

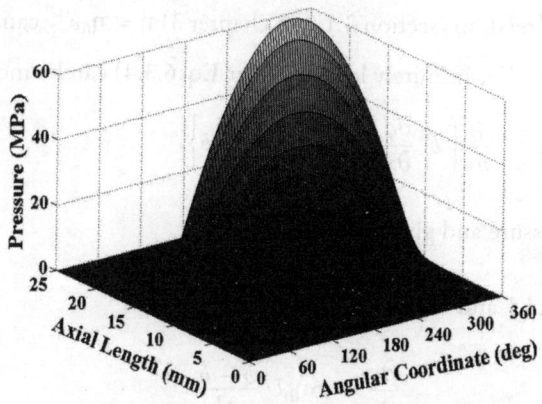

(b) With pressure viscosity relation

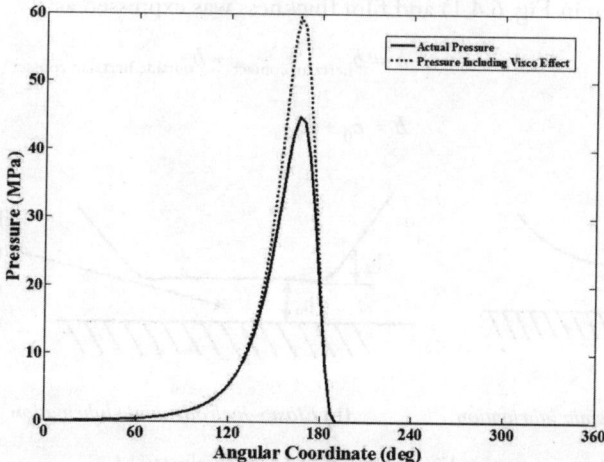

(c) Pressure comparison along the mid plane of axial length for angular coordinate

Fig. 6.3.4 The effect of pressure–viscosity relation on increasing relative velocity

To observe the effect of velocity, velocity was increased from 1m/s to 20 m/s and the obtained results are plotted in Fig. 6.3.4. Based on results indicated in Fig. 6.3.4(c) it can be said that on accounting the pressure–viscosity term, the maximum pressure value increases from 44.83 MPa to 59.43 MPa. The computer program written in MATLAB is given at the end of the chapter.

6.4 Ertel–Grubin Equation

One of the valuable contributions in the theory of EHL was made by Grubin. He adopted a simplified approach to provide physical understanding and practical solution for EHL. In other words, Grubin's solution is important from understanding point of view rather than accuracy point of view. Grubin introduced Hertz's theory (covered in section 6.2 of the present chapter) and piezo–viscous effect of lubricant (covered in section 6.3 of the present chapter) in the hydrodynamic lubrication (covered in chapters 3 and 4) for infinitely long rollers (Fig. 6.2.2). His film thickness formula is nearly true for highly loaded line contacts. It appears that Grubin adopted the approach that was first studied by Ertel. Therefore, the work is referred to as Ertel–Grubin work.

As per Grubin, Barus' relation (section 3.1.2 of chapter 3) $\eta = \eta_0 e^{\alpha P}$ can be used to represent the viscosity thickening effect. For infinitely long cylinder Eq. 6.3.4) can be modified as,

$$\frac{\partial}{\partial x}\left(h^3 \frac{\partial q}{\partial x}\right) = 6\eta_0 U \left\{\frac{\partial}{\partial x} h\right\} \tag{6.4.1}$$

Here q is reduced pressure and given as $q = \dfrac{1 - e^{-\alpha P}}{\alpha}$

On integrating Eq. 6.4.1 and using $\left.\dfrac{\partial q}{\partial x}\right|_{h=\bar{h}} = 0$

$$\frac{\partial q}{\partial x} = 6\eta_0 U \frac{h - \bar{h}}{h^3} \tag{6.4.2}$$

To provide expression of film thickness h, constant film thickness within Hertzian contact area was assumed (as shown in Fig. 6.4.1) and film thickness was expressed as:

$$h = h_{\text{hertzian contact}} + h_{\text{outside hertzian contact}}$$

$$h = h_0 + h_s \tag{6.4.3}$$

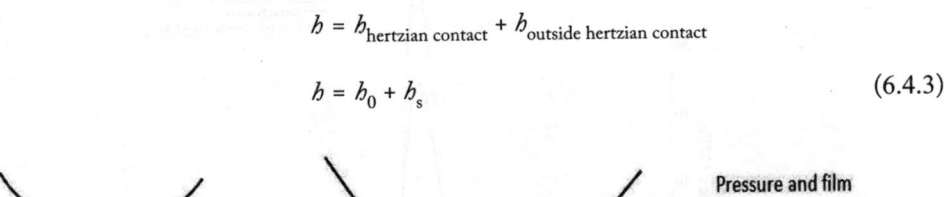

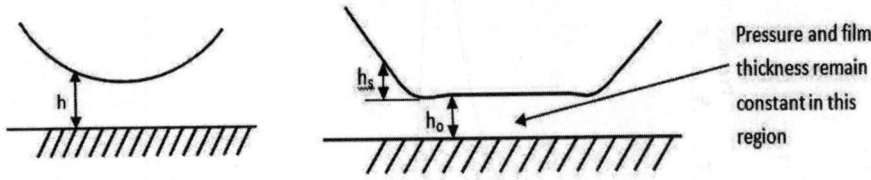

(a) *Hydrodynamic lubrication*   (b) *Elasto–hydrodynamic lubrication*

Fig. 6.4.1 Fluid film thickness

It is worth noting that as per Grubin the reduced pressure and film thickness remain constant in the elastically deformed region. On substituting Eq. 6.4.3 in Eq. 6.4.2,

$$\frac{\partial q}{\partial x} = 6\eta_0 U \frac{h_0 + h_s - h_0}{(h_0 + h_s)^3}$$

$$\Rightarrow \quad \frac{\partial q}{\partial x} = 6\eta_0 U \frac{h_s}{(h_0 + h_s)^3} \quad (6.4.4)$$

The expression of h_s can be written as given in Cameron [Cameron, 1996],

$$h_s = \left(\frac{1-v_1^2}{E_1} + \frac{1-v_2^2}{E_2}\right) p_{max} \, b \left\{ \frac{x}{b}\sqrt{\frac{x^2}{b^2} - 1} - \log_e\left[\frac{x}{b} + \sqrt{\frac{x^2}{b^2} - 1}\right] \right\} \quad (6.4.5)$$

Here b is half of the contact width. Equation 6.4.5 is applicable for $x \geq b$. Expressing Eq. 6.4.5) in terms of 'equivalent elastic modulus' as defined in Eq. 6.2.2),

$$h_s = \frac{p_{max} \, b}{E^*}\left\{ \frac{x}{b}\sqrt{\frac{x^2}{b^2} - 1} - \log_e\left[\frac{x}{b} + \sqrt{\frac{x^2}{b^2} - 1}\right] \right\} \quad (6.4.6)$$

Expressing p_{max} in terms of force using Eq. 6.2.10), $p_{max} = \frac{2}{\pi}\frac{F}{bL}$

$$h_s = \frac{2F}{\pi L E^*}\left\{ \frac{x}{b}\sqrt{\frac{x^2}{b^2} - 1} - \log_e\left[\frac{x}{b} + \sqrt{\frac{x^2}{b^2} - 1}\right] \right\} \quad (6.4.7)$$

To free Eq. 6.4.7 from units, it can be non-dimensionalized such as

$$\overline{h}_s = \frac{L\pi E^* h_s}{F} = 2\left\{ \frac{x}{b}\sqrt{\frac{x^2}{b^2} - 1} - \log_e\left[\frac{x}{b} + \sqrt{\frac{x^2}{b^2} - 1}\right] \right\} \quad (6.4.8)$$

Evaluating Eq. 6.4.8 is a bit difficult and therefore from practical point of view it is approximated as [Cameron, 1966]

$$\overline{h}_s = 4.23\left(\frac{x}{b} - 1\right)^{1.55} \quad (6.4.9)$$

Film thicknesses h and h_0 can also be written in non-dimensional form, such as

$$\overline{h} = \frac{L\pi E^* h}{F}$$

$$\overline{h}_0 = \frac{L\pi E^* h_0}{F} \quad (6.4.10)$$

On substituting these expressions in Eq. 6.4.4,

$$\frac{\partial q}{\partial x} = \frac{6\eta_0 U}{\left(F/L\pi E^*\right)^2} \frac{\overline{h}_s}{(\overline{h}_0 + \overline{h}_s)^3}$$

On non-dimensionalizing, x, $\bar{x} = x/b$

$$\frac{\partial q}{\partial \bar{x}} = \frac{6\eta_0 U b}{\left(F/L\pi E^*\right)^2} \frac{\bar{h}_s}{\left(\bar{h}_0 + \bar{h}_s\right)^3} \qquad (6.4.11)$$

On non-dimensionalizing,

$$q = \bar{q}\frac{6\eta_0 U b}{\left(F/L\pi E^*\right)^2} \qquad (6.4.12)$$

$$\frac{\partial \bar{q}}{\partial \bar{x}} = \frac{\bar{h}_s}{\left(\bar{h}_0 + \bar{h}_s\right)^3} \qquad (6.4.13)$$

On integrating

$$\bar{q} = \int_{-\infty}^{-1} \frac{\bar{h}_s}{\left(\bar{h}_0 + \bar{h}_s\right)^3} d\bar{x} \qquad (6.4.14)$$

Equation 6.4.14 can only be solved numerically. However, it can be shown that at $x = -b$ [Cameron, 1966],

$$\bar{q} = 0.0986 \left(\bar{h}_0\right)^{-1.375}$$

$$q = 1/\alpha \qquad (6.4.15)$$

Using Eq. 6.4.12 and Eq. 6.4.15

$$\frac{1}{\alpha} = 0.0986 \left(\bar{h}_0\right)^{-1.375} \frac{6\eta_0 U b}{\left(F/L\pi E^*\right)^2}$$

On rearranging

$$\bar{h}_0 = \left[0.5916 \frac{\eta_0 \alpha U b}{\left(F/L\pi E^*\right)^2}\right]^{8/11} \qquad (6.4.16)$$

For cylindrical contacts $b = \left(\frac{4FR^*}{L\pi E^*}\right)^{1/2}$

$$\Rightarrow \qquad \bar{h}_0 = \left[0.5916 \frac{\eta_0 \alpha U \left(\frac{4FR^*}{L\pi E^*}\right)^{1/2}}{\left(F/L\pi E^*\right)^2}\right]^{8/11}$$

$$\Rightarrow \qquad \bar{h}_0 = \left[1.1832 \frac{\eta_0 \alpha U \left(R^*\right)^{1/2}}{\left(F/L\pi E^*\right)^{3/2}}\right]^{8/11} \qquad (6.4.17)$$

Using Eqs. 6.4.10 and 6.4.17 dimensional minimum film thickness can be expressed as:

$$h_0 = \bar{h}_0 \left(\frac{F}{L\pi E^*} \right) = \left[1.1832 \frac{\eta_0 \alpha U (R^*)^{1/2}}{(F/L\pi E^*)^{3/2}} \right]^{8/11} \left(\frac{F}{L\pi E^*} \right)$$

$$\Rightarrow h_0 = \left[1.1832 \eta_0 \alpha U (R^*)^{1/2} \right]^{8/11} \left(\frac{L\pi E^*}{F} \right)^{1/11}$$

$$\Rightarrow h_0 = 1.13 \left[\eta_0 \alpha U \right]^{8/11} (R^*)^{4/11} \left(\frac{L\pi E^*}{F} \right)^{1/11}$$

This equation can be represented in terms of important parameters (film thickness parameter, material parameter, velocity parameter, load parameter). To do that dividing both the sides by R^* and rearranging

$$\frac{h_0}{R^*} = 1.13 \left(\frac{\eta_0 \alpha U}{R^*} \right)^{8/11} \left(\frac{R^* L\pi E^*}{F} \right)^{1/11} \quad (6.4.18)$$

Here film thickness parameter = $\dfrac{h_0}{R^*}$,

Material velocity parameter = $\dfrac{\eta_0 \alpha U}{R^*}$,

Load parameter = $\dfrac{F}{R^* L\pi E^*}$

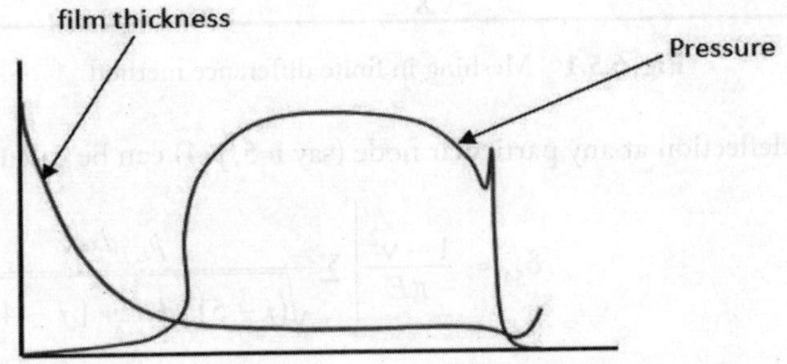

Fig. 6.4.2 Film thickness and pressure distribution in EHL

The Eq. 6.4.18 provides the film thickness for a line contact problem. This equation indicates that viscosity, velocity and piezo–viscous contacts are equally important, and the effects of applied force and effective modulus are only 12.5% of afore-mentioned variables. It is important to note that Grubin's formula for finding EHL film thickness is simple, but not accurate. The reason is a number of assumptions, such as constant film thickness and parabolic pressure distribution in deformed regions. The actual variation in film thickness corresponding to the pressure distribution is shown in Fig. 6.4.2. A sudden peak pressure and sharp decrease in film thickness near outlet, as shown in Fig. 6.4.2, are characteristics of EHL of non-confirming contact. This also indicates that central film thickness is always greater than minimum film thickness. In Grubin's equation central film thickness was the same as the minimum film thickness.

6.5 Numerical Method for Determining Oil Film Thickness in Elasto–Hydrodynamic Lubrication

In order to consider the effect of EHL, it is necessary to consider the change in the surface geometry, change in film thickness, and change in deflection of surfaces. In previous section, infinite long (cylindrical) geometry was considered. Analysis of spherical contacts (i.e., ball bearings) is relativity more difficult. To evaluate the deformation due to a distributed normal pressure on the surface following expression can be used,

$$\delta = \frac{1-v^2}{\pi E} \iint_A \frac{p(x,z)\,dxdz}{\sqrt{x^2+z^2}} \tag{6.5.1}$$

To incorporate elastic deflection in finite difference scheme, contact surface can be developed such as shown in Fig. 6.5.1. In this figure, $P_{i,j}$ is the fluid pressure at one particular node.

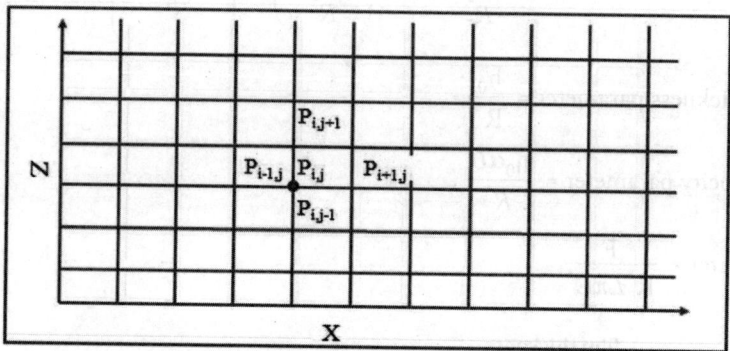

Fig. 6.5.1 Meshing in finite difference method

Using Eq. 6.5.1, deflection at any particular node (say i=5, j=4) can be calculated as:

$$\delta_{54} = \frac{1-v^2}{\pi E}\left[\sum \frac{p_{i,j}\,dxdz}{\sqrt{(i-5)^2\,dx^2 + (j-4)^2\,dz^2}}\right]$$

Expressing in non-dimensional terms:

$$x = \bar{x}X,\; y = \bar{y}Y,\; p_{i,j} = \frac{6\eta U X^2}{C^3}\bar{p}_{i,j}$$

$$\delta_{54} = \frac{1-\nu^2}{\pi E} \frac{6\eta U X^2}{C^3} \left[\sum \frac{\bar{p}_{i,j} X Z \, d\bar{x} \, d\bar{z}}{\sqrt{(i-5)^2 X^2 d\bar{x}^2 + (j-4)^2 Z^2 d\bar{z}^2}} \right]$$

$$\delta_{54} = \frac{1-\nu^2}{\pi E} \frac{6\eta U X^2}{C^3} \left[\sum \frac{\bar{p}_{i,j} X \, d\bar{x} \, d\bar{z}}{\sqrt{(i-5)^2 \left(X/Z\right)^2 d\bar{x}^2 + (j-4)^2 d\bar{z}^2}} \right]$$

Assuming $d\bar{x} = d\bar{z}$

$$\delta_{53} = \frac{1-\nu^2}{\pi E} \frac{6\eta U X^3}{C^3} \left[\sum \frac{\bar{p}_{i,j} d\bar{x}}{\sqrt{(i-5)^2 \left(X/Z\right)^2 + (j-4)^2}} \right]$$

$$\bar{\delta}_{53} = \frac{1-\nu^2}{\pi E C} \left[\frac{6\eta U X^3}{C^3} \sum \frac{\bar{p}_{i,j} d\bar{x}}{\sqrt{(i-5)^2 \left(X/Z\right)^2 + (j-3)^2}} \right] \quad (6.5.2)$$

Applying finite difference method,

$$\frac{\partial}{\partial \bar{x}}\left(\bar{h}^3 \frac{\partial \bar{p}}{\partial \bar{x}}\right) + \frac{x^2}{z^2}\frac{\partial}{\partial \bar{z}}\left(\bar{h}^3 \frac{\partial \bar{p}}{\partial \bar{z}}\right) = \frac{C}{X}\frac{\partial \bar{h}}{\partial \bar{x}} \quad (6.5.3)$$

$$\frac{\partial}{\partial \bar{x}}\left(\bar{h}^3 \frac{\partial \bar{p}}{\partial \bar{x}}\right) = 3\bar{h}_{i,j}^2 \frac{\bar{h}_{i+1,j} - \bar{h}_{i-1,j}}{2\Delta\bar{x}} \frac{\bar{p}_{i+1,j} - \bar{p}_{i-1,j}}{2\Delta\bar{x}} + \bar{h}_{i,j}^3 \frac{\bar{p}_{i+1,j} + \bar{p}_{i-1,j} - 2\bar{p}_{i,j}}{(\Delta\bar{x})^2}$$

$$\frac{\partial}{\partial \bar{z}}\left(\bar{h}^3 \frac{\partial \bar{p}}{\partial \bar{z}}\right) = 3\bar{h}_{i,j}^2 \frac{\bar{h}_{i,j+1} - \bar{h}_{i,j-1}}{2\Delta\bar{z}} \frac{\bar{p}_{i,j+1} - \bar{p}_{i,j-1}}{2\Delta\bar{z}} + \bar{h}_{i,j}^3 \frac{\bar{p}_{i,j+1} + \bar{p}_{i,j-1} - 2\bar{p}_{i,j}}{(\Delta\bar{z})^2}$$

$$\frac{\partial \bar{h}}{\partial \bar{x}} = \frac{\bar{h}_{i+1,j} - \bar{h}_{i-1,j}}{2\Delta\bar{x}}$$

On substituting these expressions in Eq. 6.5.3 and rearranging

$$\bar{p}_{i+1,j}\left(3\bar{h}_{i,j}^2 \frac{\bar{h}_{i+1,j} - \bar{h}_{i-1,j}}{4\Delta\bar{x}^2} + \frac{\bar{h}_{i,j}^3}{\Delta\bar{x}^2}\right) - \bar{p}_{i-1,j}\left(3\bar{h}_{i,j}^2 \frac{\bar{h}_{i+1,j} - \bar{h}_{i-1,j}}{4\Delta\bar{x}^2} - \frac{\bar{h}_{i,j}^3}{\Delta\bar{x}^2}\right) -$$

$$2\bar{h}_{i,j}^3 \bar{p}_{i,j}\left(\frac{1}{(\Delta\bar{x})^2} + \frac{1}{(\Delta\bar{z})^2}\left(\frac{X}{Z}\right)^2\right) + \left(\frac{X}{Z}\right)^2\left[\bar{p}_{i,j+1}\left(3\bar{h}_{i,j}^3 \frac{\bar{h}_{i,j+1} - \bar{h}_{i,j-1}}{4\Delta\bar{z}^2} + \frac{\bar{h}_{i,j}^3}{(\Delta\bar{z})^2}\right) - \right.$$

$$\left. \bar{p}_{i,j-1}\left(3\bar{h}_{i,j}^3 \frac{\bar{h}_{i,j+1} - \bar{h}_{i,j-1}}{4\Delta\bar{z}^2} - \frac{\bar{h}_{i,j}^3}{(\Delta\bar{z})^2}\right)\right] = \frac{C}{X}\frac{\bar{h}_{i+1,j} - \bar{h}_{i-1,j}}{2\Delta\bar{x}}$$

On rearranging

$$\bar{P}_{i+1,j}\left(\frac{3}{4}\left(\bar{h}_{i+1,j}-\bar{h}_{i-1,j}\right)+\bar{h}_{i,j}\right)-\bar{P}_{i-1,j}\left(\frac{3}{4}\left(\bar{h}_{i+1,j}-\bar{h}_{i-1,j}\right)-\bar{h}_{i,j}\right)-2\bar{h}_{i,j}\bar{P}_{i,j}\left(1+\left(\frac{X}{Z}\right)^{2}\right)$$

$$+\left(\frac{X}{Z}\right)^{2}\left[\bar{P}_{i,j+1}\left(\frac{3}{4}\left(\bar{h}_{i+1,j}-\bar{h}_{i-1,j}\right)+\bar{h}_{i,j}\right)-\bar{P}_{i,j+1}\left(\frac{3}{4}\left(\bar{h}_{i,j-1}-\bar{h}_{i,j-1}\right)-\bar{h}_{i,j}\right)\right]$$

$$=\frac{C}{X}\frac{\Delta\bar{x}}{2}\frac{\left(\bar{h}_{i+1,j}-\bar{h}_{i-1,j}\right)}{\bar{h}_{i,j}^{2}}$$

$$\bar{P}_{i,j}=\bar{P}_{i+1,j}\frac{\left(\frac{3}{4}\left(\bar{h}_{i+1,j}-\bar{h}_{i-1,j}\right)+\bar{h}_{i,j}\right)}{2\bar{h}_{i,j}\left(1+\left(\frac{X}{Z}\right)^{2}\right)}+\bar{P}_{i-1,j}\frac{\left(\bar{h}_{i,j}-\frac{3}{4}\left(\bar{h}_{i+1,j}-\bar{h}_{i-1,j}\right)\right)}{2\bar{h}_{i,j}\left(1+\left(\frac{X}{Z}\right)^{2}\right)}+\frac{C}{X}\frac{\Delta\bar{x}}{4}\frac{\left(\bar{h}_{i-1,j}-\bar{h}_{i+1,j}\right)}{\bar{h}_{i,j}^{3}\left(1+\left(\frac{X}{Z}\right)^{2}\right)}$$

$$+\left(\frac{X}{Z}\right)^{2}\left[\bar{P}_{i+1,j}\frac{\left(\frac{3}{4}\left(\bar{h}_{i+1,j}-\bar{h}_{i-1,j}\right)+\bar{h}_{i,j}\right)}{2\bar{h}_{i,j}\left(1+\left(\frac{X}{Z}\right)^{2}\right)}+\bar{P}_{i-1,j}\frac{\left(\bar{h}_{i,j}-\frac{3}{4}\left(\bar{h}_{i+1,j}-\bar{h}_{i-1,j}\right)\right)}{2\bar{h}_{i,j}\left(1+\left(\frac{X}{Z}\right)^{2}\right)}\right] \quad (6.5.4)$$

To illustrate the procedure, journal bearing described in section 6.3 is considered. It is assumed that the bearing is made of gun metal (v_1 = 0.29 and E_1 = 90 GPa) and material of journal is steel (v_2 = 0.30 and E_2 = 205 GPa). To compare the pressure results with that of presented in Fig. 6.3.4 same viscosity (10 mPa.s) and same velocity (20 m/s) have been considered. The deflection of the bearing due to finally convergent fluid pressure has been shown in Fig. 6.5.2.

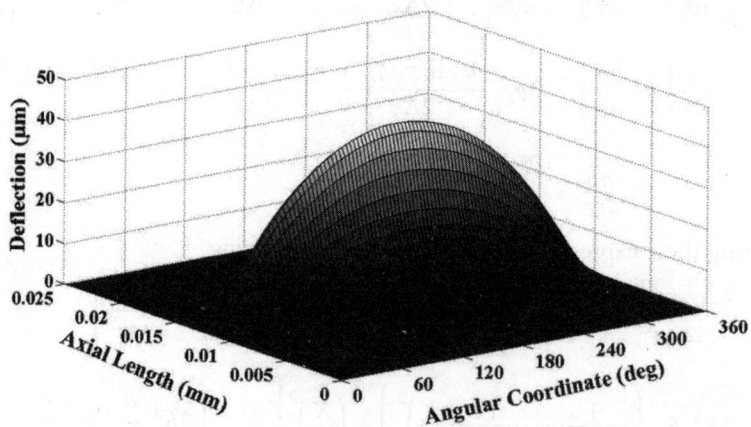

Fig. 6.5.2 Deflection curve in elasto–hydrodynamic lubrication

Since the contact of the journal and bearing surfaces is confined to a line contact, the maximum deflection is observed at the location of maximum pressure. On comparing pressure distributions of Figs. 6.3.4 and 6.5.2, it can be said that due to elastic deformation, the maximum value of pressure decreases and relatively uniform distribution of pressure occurs.

6.6 Rolling Element Bearings

In headings 6.1–6.5 of this chapter, the focus has been to understand the EHL mechanism and to develop the expression/methodology to quantify the mechanism. In this and the next section, EHL is applied to specific machine elements.

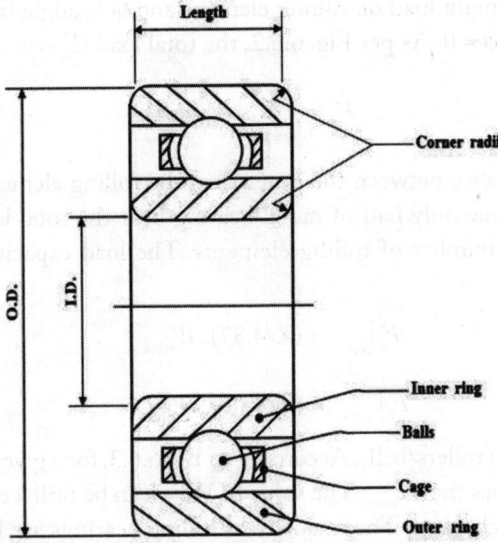

Fig. 6.6.1 Four main components of rolling element bearing

Every rotating machine has some rolling elements as rotation is always easier than linear motion. In rolling element bearings, friction and wear are very low. Another important and noteworthy point is standardization, which motivated mass production and easy replacements of rolling element bearings. These bearings have very good surface finish and bearings are subject to elasto-hydrodynamic lubrication. There are four main components: inner ring, rolling element, cage and outer ring, as shown in Fig. 6.6.1. A cage is used to separate the rolling elements. In absence of cage, the rolling elements start colliding with each other and reduce the rolling velocity. The corners are always rounded to avoid concentration of stress on the corners.

To bear the applied load, the bearings are classified in three major categories: radial (shown in Fig. 6.6.2), axial (shown in Fig. 6.2.7) and combined load (i.e., taper roller) bearings.

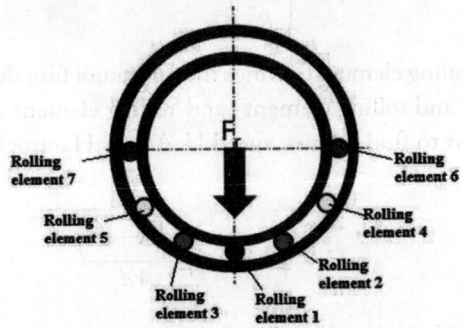

Fig. 6.6.2 Rolling element arrangement in the bearing

In axial (thrust) rolling bearing (as shown in Fig. 6.2.7) load on rolling elements is nearly uniformly distributed. However, in radial bearing (Fig. 6.6.2), the load on each rolling element varies from zero to maximum. The equation of load on the rolling element is expressed as:

$$W_\psi = W_{max} (\cos \psi)^n \tag{6.6.1}$$

Here W_{max} is the maximum load on rolling element and ψ is angle from load line, which means load will be maximum at $\psi = 0$. As per Fig. 6.6.2, the total load (F_r) can be expressed as follows.

$$F_r = \sum_{\psi=-\psi_1}^{\psi=\psi_1} W_\psi \cos \psi \tag{6.6.2}$$

In the absence of clearance between the inner ring, the rolling element and the outer ring, $\psi_1 = 90°$. This means at any time only half of radial bearing bear the total load. The load capacity will increase with increase in number of rolling elements. The load capacity (F_r) can be expressed as (Hamrock et al., 2004)

$$F_r\big|_{ball} = (Z/4.37) . W_{max}$$

$$F_r\big|_{roller} = (Z/4.06) . W_{max} \tag{6.6.3}$$

Here Z is the number of rollers/balls. According to Eq. 6.6.3, for a given F_r, increase in the number of rolling elements decreases the W_{max}. The value of W_{max} can be utilized to find the minimum film thickness under EHL mechanism. The majority of rolling bearings are lubricated by grease (semi-solid) and liquid lubricants. The shields/seals are used to retain the grease within the bearings. The liquid lubrication requires some sort of circulating system.

To determine the film thickness numerical method (as explained in session 6.5) or curve fit method can be used. In case of rolling bearing, there is a possibility of curvature in X- as well as Z-direction. Therefore, instead of using Eq. 6.2.1 for effective radius, the following equations are used:

$$\frac{1}{R_x^*} = \frac{1}{R_{1x}} + \frac{1}{R_{2x}} \tag{6.6.4}$$

$$\frac{1}{R_z^*} = \frac{1}{R_{1z}} + \frac{1}{R_{2z}} \tag{6.6.5}$$

Based on these two effective radii, a new dimensionless 'ellipticity' parameter is defined as,

$$k_e = \left(\frac{R_z^*}{R_x^*}\right)^{2/\pi} \tag{6.6.6}$$

It is worth noting that in rolling element bearings the minimum film thicknesses must be calculated at the interface of 'inner ring and rolling element', and 'rolling element and outer ring'.

Special attention is needed to find relative speed U. As per Hamrock et al (2004) relative speed is given by:

$$U = |\omega_i - \omega_o| \frac{(d_e^2 - d^2)}{4 d_e} \tag{6.6.7}$$

where ω_i and ω_o are the angular velocities of inner and outer rings, respectively. The diameter of rolling element is expressed as d. The pitch diameter, d_e, is given by

$$d_e = \frac{d_o + d_i}{2} \tag{6.6.8}$$

Non-dimensional film thickness [Hamrock et al., 2004]

$$H_{min} = \frac{h_0}{R_x^*} = 3.63 \frac{\left(\eta_0 U/E'R_x\right)^{0.63} (\alpha E')^{0.49}}{\left(F/E'R_x^2\right)^{0.073}} \left(1 - e^{-0.68 k_e}\right) \tag{6.6.9}$$

where $E' = \dfrac{2}{\left(1-v_1^2\right)/E_1 + \left(1-v_2^2\right)/E_2} = 2E^*$

In case of roller bearing k_e is ∞ and Eq. 6.6.9) reduces to

$$H_{min} = \frac{h_0}{R_x^*} = 3.63 \frac{\left(\eta_0 U/E'R_x\right)^{0.63} (\alpha E')^{0.49}}{\left(W_{max}/E'R_x^2\right)^{0.073}} \tag{6.6.10}$$

Example: A roller bearing having rollers of diameter 16mm and outer race diameter in contact of roller is 96 mm. Assume the length of the roller is equal to its diameter. There are total nine rollers bearing the load of 10 kN. Outer ring is fixed to the stationary housing, while inner ring rotates at the angular speed of 524 rad/s. The inner ring, the outer ring and the rollers are made of steel (E=205 GPa, v=0.3). Viscosity and piezo–viscous coefficient are 10 mPa.s and $2*10^{-8}$ m²/N, respectively. Find the minimum film thickness at inner and outer rings.

Solution: The maximum load on each roller is given by Eq. 6.6.3

$$F_r\big|_{roller} = (Z/4.06) \cdot W_{max}$$

$$\Rightarrow \quad W_{max} = \frac{4.06}{Z} F_r\big|_{roller}$$

According to the data given in the present example

$$W_{max} = 4.06*10000/9 = 4511 \text{ N} \tag{6.6.11}$$

$$E^* = \frac{2}{\left(1-v_1^2\right)/E_1 + \left(1-v_2^2\right)/E_2} = \frac{1}{\left(1-v^2\right)/E} = \frac{E}{1-v^2} = 225.2747 \, GPa \tag{6.6.12}$$

Radius of roller, $R_1 = 0.008$ m
Radius of outer ring contacting roller, $R_{2,o} = -0.048$ m (negative sign represents concave surface)
Radius of inner ring contacting roller, $R_{2,i} = 0.048$ m-0.016 (assuming zero clearance between the inner ring and the roller).

$$\frac{1}{R_{x,i}^*} = \frac{1}{0.008} + \frac{1}{0.032}$$

$$\Rightarrow \quad R_{x,i}^* = 0.0064 \tag{6.6.13}$$

$$\frac{1}{R_{x,0}^*} = \frac{1}{0.008} - \frac{1}{0.048}$$

$$\Rightarrow \qquad R^*_{x,0} = 0.0096 \qquad (6.6.14)$$

Finding relative velocity using Eq. 6.6.7

$$U = |\omega_i - \omega_o| \frac{\left(\frac{(d_i + d_o)^2}{4} - d^2\right)}{2(d_i + d_o)}$$

$$U = 10.061 \text{ m/s} \qquad (6.6.15)$$

Substituting the values from Eqs. 6.6.11, 6.6.12, 6.6.13 and 6.6.15 in Eq. 6.6.10 to find minimum film thickness at inner ring,

$$\frac{h_0|_i}{R^*_{x,i}} = 3.63 \frac{\left(\eta_0 U / E' R_{x,i}\right)^{0.63} (\alpha E')^{0.49}}{\left(W_{max} / E' R^2_{x,i}\right)^{0.073}}$$

$$\Rightarrow \qquad \frac{h_0|_i}{R^*_{x,i}} = 156.13 * 10^{-6}$$

$$\Rightarrow \qquad h_0|_i = 1 \mu m$$

Similarly minimum film thickness at the outer ring,

$$\frac{h_0|_o}{R^*_{x,0}} = 3.63 \frac{\left(\eta_0 U / E' R_{x,o}\right)^{0.63} (\alpha E')^{0.49}}{\left(W_{max} / E' R^2_{x,0}\right)^{0.073}}$$

$$\Rightarrow \qquad \frac{h_0|_o}{R^*_{x,0}} = 128.31 * 10^{-6}$$

$$\Rightarrow \qquad h_0|_o = 1.232 \mu m$$

6.7 EHL of Gear–Teeth Contact

Involute gear pairs (Fig. 6.7.1) are non-conforming contact with pure rolling at the pitch point. However other than pitch point, there will be sliding at each of contact points. The sliding motion causes a friction at bare gear surfaces in gears. The friction is not desirable, as gears are required to transmit power. To reduce friction, lubrication is a must. In short, the torque in the gear pair is transmitted through the EHL contacts.

In order to understand the EHL mechanism at the interface of contacting gear teeth, it is necessary to describe involute geometry and its kinematics. Figure 6.7.2 depicts the involute tooth of gear/pinion, enclosed between addendum and base circles. Two (A and B) points are considered on spur gear involute profile at radii r_A and r_B from the centre (O) of the base circle. The tangents drawn on base circle through points intersect the base circle at points C and D, respectively. From $\triangle AOC$ and $\triangle BOD$

$$OC = r_A \cos \varphi_A = r_b$$

and
$$OD = r_B \cos \varphi_B = r_b \qquad (6.7.1)$$

From Eq. 6.7.1 it can be inferred that the pressure angle at any point on involute profile can be determined if the distance of that from centre is known. From Eq. 6.7.1 it is also indicated that at base radius pressure angle is zero.

Fig. 6.7.1 Gear pair showing contact between two pairs of teeth

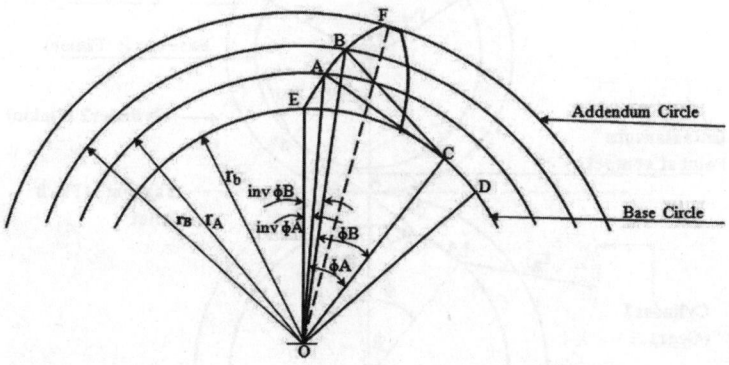

Fig. 6.7.2 Involute profile of gear/pinion tooth (phi A and phi B, inv phi A and inv phi B, addendum circle)

Gear and pinion in contact at the pitch point A have been shown in Fig. 6.7.3. The contact length is along $a^* b^*$, which is a common tangent to the base circles of base radii r_{bg} and r_{bp}. Figure 6.7.3 indicates that contact begins at 'a' and ends at 'b'. Due to involute gear profile local radii of pinion and gear at the contact points change and as a consequence sliding speed vary at contact points. To understand this, let us consider Fig. 6.7.4.

286 *Fundamentals of Engineering Tribology with Applications*

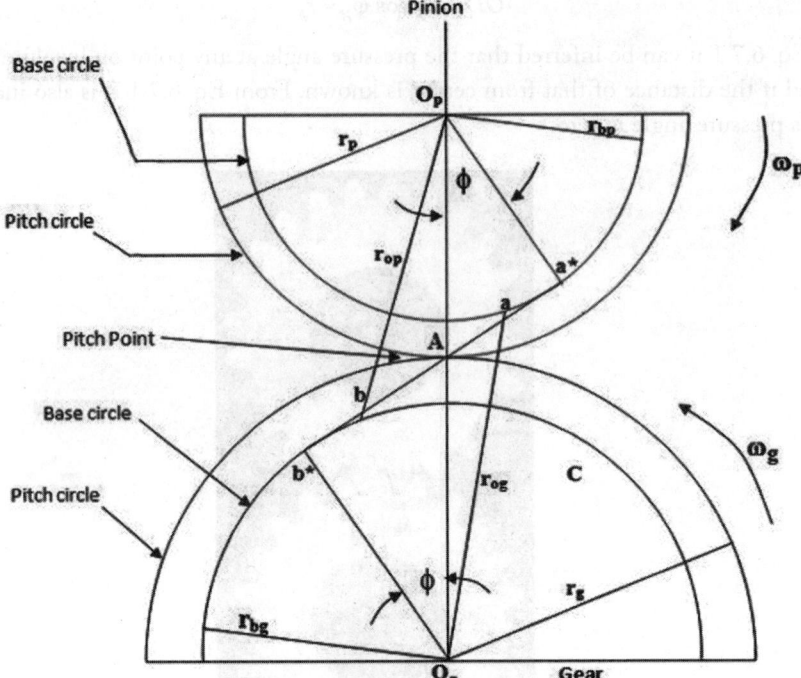

Fig. 6.7.3 Gear and pinion in mesh

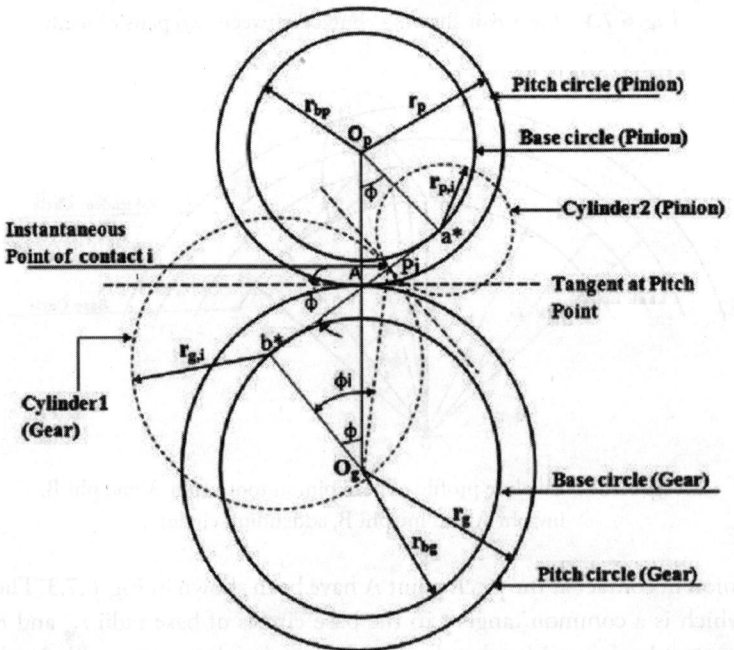

Fig. 6.7.4 Cylindrical contact analogy for gear and pinion tooth interaction

In Fig. 6.7.4 pinion and gear tooth interaction has been treated as contact between cylinders. The radii of the cylinder vary with rotation of gear–pair. Let us consider P_i is the instantaneous point of contact between gear and pinion teeth. This means radius of cylinder 1 is a^*P_i and radius of cylinder 2 is b^*P_i.

$$r_{p,i} = a^* P_i = r_p \cos \varphi \tan \varphi_{pi}$$

$$r_{g,i} = b^* P_i = r_g \cos \varphi \tan \varphi_{gi} \qquad (6.7.2)$$

$$\frac{1}{R_i^*} = \frac{1}{r_{p,i}} + \frac{1}{r_{g,i}} \qquad (6.7.3)$$

In addition, the normal load at the contact point changes. As a result the minimum film thickness gets modified to account the unsteady speed and load conditions. The minimum lubricant film thickness is given by,

$$\frac{h_{min}}{R_i^*} = 1.714 \left(\frac{W_N}{LE'R_i^*}\right)^{-0.128} \left(\frac{\eta_o(V_1+V_2)}{E'R_i^*}\right)^{0.694} (\alpha E')^{0.568} \qquad (6.7.4)$$

h_{min} is the minimum separation between the gear pairs in micron level. R_i^* is the radius of curvature of the interface, W_N is normal load on gear teeth and L is face width.

$$E' = \frac{2}{(1-v_1^2)/E_1 + (1-v_2^2)/E_2} \qquad (6.7.5)$$

Example: A pinion having number of teeth 20, face width 30 mm and module 5mm meshes with gear of 30 teeth. The pressure angle at pitch point comes out to be 20°. The angular speed of pinion is 300 rad/s. Other elastic and lubricant properties are: η_0 = 90 mPa.s, α = 2.0*10^{-8}, E = 205GPa, Normal load at pitch point = 2500 N, and v = 0.3. Find the minimum film thickness.

Solution: Equation 6.7.4 can be used to determine the minimum film thickness

$$\frac{h_{min}}{R_i^*} = 1.714 \left(\frac{W_N}{LE'R_i^*}\right)^{-0.128} \left(\frac{\eta_o(V_1+V_2)}{E'R_i^*}\right)^{0.694} (\alpha E')^{0.568} \qquad (6.7.6)$$

From Eq. 6.7.5 $\qquad E' = \dfrac{1}{(1-v^2)/E} = \dfrac{E}{1-v^2} = 225.2747\, GPa \qquad (6.7.7)$

Pitch radius of pinion r_p = number of teeth*module/2

$\Rightarrow \qquad\qquad\qquad\qquad r_p = 0.05$

Pitch radius of gear r_g = number of teeth*module/2

$\Rightarrow \qquad\qquad\qquad\qquad r_g = 0.075$

From Eq. 6.7.3 $\qquad\qquad \dfrac{1}{R_i^*} = \dfrac{1}{r_{p,i}} + \dfrac{1}{r_{g,i}}$

At pitch point $\qquad\qquad \dfrac{1}{R_i^*} = \left(\dfrac{1}{r_p} + \dfrac{1}{r_g}\right)\dfrac{1}{\sin \varphi}$

$$\Rightarrow \quad R_i^* = \frac{r_p r_g \sin\varphi}{(r_p + r_g)} = 0.0103\, m \quad (6.7.8)$$

Angular speed of gear ω_g = gear ratio*angular speed of pinion = 300/1.5 = 200 rad/s (6.7.9)
Substituting Eqs. 6.7.7–6.7.9 in Eq. 6.7.6

$$\frac{h_{min}}{0.0103} = 1.714 \left(\frac{2500}{0.03 * 225.2747 * 10^9 * 0.0103}\right)^{-0.128} \left(\frac{0.09*(200*0.075 + 300*0.05)\cos(20°)}{2*225.2747*10^9 * 0.0103}\right)^{0.694}$$

$$(2.2* 10^{-8} * 225.2747 * 10^9)^{0.568}$$

$h_{mn} = 1.714 * 0.0103 * (3.5914 * 10^{-5})^{-0.128} (5.4673 * 10^{-10})^{0.694} (4.956 * 10^3)^{0.568}$

$h_{min} = 3\, \mu m$

Frequently Asked Questions

Q.1. In EHL effective Young's modulus is represented in three different format
$\left(\frac{1}{E^*} = \frac{1-v_1^2}{E_1} + \frac{1-v_2^2}{E_2}\right)$, $\frac{1}{E^*} = \frac{1}{2}\left(\frac{1-v_1^2}{E_1} + \frac{1-v_2^2}{E_2}\right)$ and $\frac{1}{E^*} = \frac{1}{\pi}\left(\frac{1-v_1^2}{E_1} + \frac{1-v_2^2}{E_2}\right)$.
What are the reasons for such different representations?

Ans. In deriving the equation for elastic deflection one expression $\left(\frac{1-v_1^2}{E_1} + \frac{1-v_2^2}{E_2}\right)$ is used. As this expression is long, therefore the expression is replaced with $\frac{1}{E^*}$. Expressing effective Young's modulus E^* in this manner $\left(\frac{1}{E^*} = \frac{1-v_1^2}{E_1} + \frac{1-v_2^2}{E_2}\right)$ provides value of E^* greater than E_1 and E_2, which difficult to justify as the value of effective Young's modulus must be lesser than the values E_1 and E_2. Therefore expression $\frac{1}{E^*} = \frac{1}{2}\left(\frac{1-v_1^2}{E_1} + \frac{1-v_2^2}{E_2}\right)$ is more common compared to $\left(\frac{1}{E^*} = \frac{1-v_1^2}{E_1} + \frac{1-v_2^2}{E_2}\right)$. Similarly expression $\frac{1}{E^*} = \frac{1}{\pi}\left(\frac{1-v_1^2}{E_1} + \frac{1-v_2^2}{E_2}\right)$ is more common than $\left(\frac{1}{E^*} = \frac{1-v_1^2}{E_1} + \frac{1-v_2^2}{E_2}\right)$ as it simplifies the terms (πE^* as E^*) provided in Eqs. 6.4.8 and 6.4.10.

Q.2. Which contact will have less conformity in contact and why?
(a) Ball and inner ring
(b) Ball and outer ring
(c) Shaft and inner ring
(d) Housing and outer ring

Ans. The ball and inner race will have less conformity in contact between them. This is due to the geometry of the contacting bodies; the ball is convex and so also is the inner race. It is well known that there is least conformity between convex–convex shapes. This can be explained by estimating effective radius. Assume a roller bearing with roller radius of $R_1 = 0.008$ m, radius of the outer ring contacting the roller, $R_{2,o} = -0.048$ m (negative sign represents concave surface) and radius of the inner ring contacting the roller, $R_{2,i} = 0.032$. The effective radii can be estimated using following equation.

$$\frac{1}{R^*_{x,i}} = \frac{1}{0.008} + \frac{1}{0.032}$$

$$\Rightarrow R^*_{x,i} = 0.0064 \quad (6.6.13)$$

$$\frac{1}{R^*_{x,o}} = \frac{1}{0.008} - \frac{1}{0.048}$$

$$\Rightarrow R^*_{x,o} = 0.0096$$

Since $R^*_{x,i} = 0.0064$ is smaller than $R^*_{x,o} = 0.0096$, which means conformity in convex–convex is lower than conformity in convex–concave bodies.

Q.3. How is the relative speed calculated in the gear pair contact?

Ans. The surface contact velocity is expressed as (refer to Fig. 6.7.4):

$$V = \frac{V_1 + V_2}{2} = \frac{\omega_1 \cdot r_{p1} \cdot \sin\phi + \omega_2 \cdot r_{p2} \cdot \sin\phi}{2};$$

Since $V_1 = V_2$, either one of them could be used.

Q.4. What are the differences between hydrodynamic and elasto–hydrodynamic lubrications?

Ans. In hydrodynamic lubrication, relative motion and wedge formation between the contacting surfaces is used to pump lubricant in convergent region to separate the contacting surfaces. In elasto–hydrodynamic lubrication nonconforming surfaces or higher load conditions elastically deform the bodies. Such deformation helps to raise the fluid pressure, which in turn increases the lubricant viscosity considerably. In short, relatively high fluid pressure, elastic deformation of contacting surfaces and viscosity thickening make elasto–hydrynamic lubrication different than hydrodynamic lubrication.

Q.5. What is the difference between 'soft EHL' and 'Hard EHL'?

Ans. In 'soft EHL', lubricant viscosity does not get affected by fluid pressure. The 'soft EHL' occurs when at least one of the materials has a low modulus such as rubber. This means only elastic deformation of material is considered in the 'soft EHL'. The pressure-viscosity coefficient is important in 'hard EHL' as the viscosity may increase by 10 times inside the contact. The 'hard EHL' occurs between materials with a high elastic modulus.

Q.6. Why applied load in EHL has little effect on the minimum film thickness?

Ans. In EHL, increase in load increases the spread of elastic deformation, which means increase in the contact area rather than decrease in the minimum film thickness. In other words, due to elastic properties of materials in contact deformation spreads over the area and effect a little to the minimum film thickness.

Q.7. Why does viscosity of lubricant increase with increase in fluid pressure?

Ans. The liquid lubricants are normally incompressible and an increase in pressure doesn't really bring the molecules significantly closer together to increase viscosity significantly. However, the liquid lubricant under extreme pressure often experiences an increase in viscosity as molecules come very close and liquids turn into semi-solids.

Q.8. Which kind of lubricant is preferred for EHL?

Ans. A lubricant having high but stable viscosity and higher value of pressure viscosity (piezo–viscous) coefficient is preferred over lubricant having low viscosity and lower value of pressure viscosity coefficient.

Q.9. Is there any possibility of squeeze film action in elasto–hydrodynamic lubrication mechanism?

Ans. Yes, squeeze film action and the elasto–hydrodynamic lubrication can occur simultaneously particularly in dynamic conditions. But solving transient EHL is very difficult and should be used only when contribution of squeeze film action is significant.

Q.10. Provide an example of contacting pair that operate in 'soft EHL' regime.

Ans. In 'soft EHL' regime, fluid pressures are relatively low, but the elastic deflections of one or both contacting bodies are order of film thickness. In 'soft EHL' the fluid pressure is not high enough to increase the lubricant viscosity. Typical examples of 'soft EHL' are rubber–seals, wind–screen wipers and wet tyres.

References

Cameron, A. 1966. *The Principles of Lubrication*. Longmans.

Hamrock, B. J. Schmid, S. R. and Jacobson, B. O. 2004. *Fundamentals of Fluid Film Lubrication*. Second edition. Marcel Dek.

Harris, T. A. 2001. *Rolling Bearing Analysis*. Wiley.

Program Listing in MATLAB for Figure 6.3.2–6.3.4

```
clear all

% For estimating the pressure-visco elastic effect, following are variables
    are required as input:
% L-Length of bearing (m)
% R- Radius of the bearing (m)
% Clc- Clearance (m)
% w = (2*pi*N/60)- angular velocity (N is the rotational speed) (rad)
% eta- Viscosity of lubricant (Pas)
% N - Number of division in angular direction
% M - Number of Division in Axial Direction
% Fapp = Applied load (N)
% V- Velocity of Journal
```

```
% variables to be given as Inputs:
N=100;
M=100;
K=zeros(N,M);

eta=0.010;
L=0.025;
R = 0.0253;
Clc=0.000043;

deltet = 2*pi/N;
delz = L/M;
ITER = 1000000;
V=20;
w=V/R;
al=1e-8;

for I=1:N
tet(I)=deltet*(I);
zz(I)=delz*(I);
p(N+1,I)=0;
end

eps=0.89

    for I=1:N+1
        for J=1:M+1
p(I,J)=0;
        end
    end

for K = 1:ITER
 K=K+1;
sumij=0;
for I=2:N

tet(I)=deltet*(I-1);
h(I)=(1+eps.*cos(tet(I)));
A=((-3/2.*eps*sin(tet(I)))./(h(I)*deltet))+(1/deltet^2);
B=((3/2.*eps*sin(tet(I)))./(h(I)*deltet))+(1/deltet^2);
C=(R^2/(delz)^2);
D=2*(1/deltet^2 +R^2/((delz)^2));
E=eps*sin(tet(I))/h(I).^3;
  for J=2:M
p(I,J)=C/D*p(I,J+1)+C/D*p(I,J-1)+E/D+A/D*p(I+1,J)+B/D*p(I-1,J);

if p(I,J)<0
    p(I,J)=0;
else
```

```
            p(I,J)=p(I,J);
    end

sumij=sumij+p(I,J);
    end
    end
sum(K+1)=sumij;
percentage = abs(sum(K+1)-sum(K))/abs(sum(K+1));
if percentage < 1e-6

        break
        end
end


for I=1:N
for J=1:M
Pr(I,J)=p(I,J).*(6*eta*w*(R/Clc)^2);
Q(I,J) = -log((1-abs(al*Pr(I,J))))/al;
end
end


for i=1:N
      tett(i)=deltet*(i-1)*360/2/pi;
end

for j=1:M
      z(j)=delz*(j-1);
end

figure
surf(tett,z,Pr/1e6) % Plot the surface plot of Pressure
xlabel('Angular Coordinate (deg)','FontName','Times New Roman','fontsize',16);
ylabel('Axial Length (mm)','FontName','Times New Roman','fontsize',16);
zlabel('Pressure (Pa)','FontName','Times New Roman','fontsize',16);


figure

plot(tett,Pr(:,26)/1e6,'b')
hold on
plot(tett,Q(:,26)/1e6,'r')
xlabel('Angular Coordinate (deg)','FontName','Times New Roman','fontsize',16);
ylabel('Pressure (MPa)','FontName','Times New Roman','fontsize',16);

for i=2:N-1
      for j=2:M
```

```
                F(i,j)=dblquad(@(t,z)Pr(i,j)*R*sin(t),tett(i),tett(i+1),0,delz);
            end
end

for i=2:N-1
    for j=2:M
        FF(i,j)=-dblquad(@(t,z)Pr(i,j)*R*cos(t),tett(i),tett(i+1),0,delz);
        end
end

Fx=0;

for i=2:N-1
    for j=2:M
        Fx=Fx+(F(i,j));
    end
end
Fy=0;
for i=2:N-1
    for j=2:M
        Fy=Fy+(FF(i,j));
    end
end
fi=atand(Fx/Fy);
W=abs((Fy*cosd(fi)+Fx*sind(fi)))
```

Program listing in MATLAB for Figure 6.5.2 – 6.5.3

```
% To estimate pressure-viscoelastic effect, following inputs are required
% L=Length of bearing (m)
% R=Radius of the bearing (m)
% Clc=Clearance (m)
% w = (2*pi*N/60)- angular velocity (N is the rotational speed) (rad)
% eta=Viscosity of lubricant (Pas)
% N=Number of division in angular direction
% M=Number of Division in Axial Direction
% Fapp=Applied load (N)
% V=Velocity of Journal (m/s)
% al=pressure Coefficient
% p= Non-dimensional pressure
% Pr= Pressure (Pa)
% Q= Estimated Pressure using pressure viscosity relation

%%Input
clear all
N=50;
M=50;
K=zeros(N,M);
```

```
eta=0.010;
L=0.025;
R = 0.0253;
Clc=0.000043;
mu1=0.3;
E1=91*10^9;
mu2=0.29;
E2=205*10^9;
EE=((1-mu1^2)/E1+(1-mu2^2)/E2)^-1;
deltet = 2*pi/N;
delz = L/M;
ITER = 10000;
V=20;
w=V/R;
al=1e-8;
eps=0.89;

%%Initialization
p=zeros(N+1,M+1);
Del=zeros(N,M);
Fx=0;
Fy=0;

%%Iteration Block
for H=1:1
for K = 1:ITER
 K=K+1;
sumij=0;
for I=2:N
    for J=2:M
tet(I)=deltet*(I-1);
h0(I,J)=(1+eps.*cos(tet(I)));
h(I,J)=h0(I,J)+Del(I,J)./Clc;
A=((-3/2.*eps*sin(tet(I)))./(h(I,J)*deltet))+(1/deltet^2);
B=((3/2.*eps*sin(tet(I)))./(h(I,J)*deltet))+(1/deltet^2);
C=(R^2/(delz)^2);
D=2*(1/deltet^2 +R^2/(delz)^2));
E=eps*sin(tet(I))/h(I,J).^3;
p(I,J)=C/D*p(I,J+1)+C/D*p(I,J-1)+E/D+A/D*p(I+1,J)+B/D*p(I-1,J);   %EQ 4.2.31

if p(I,J)<0
            p(I,J)=0;
else
            p(I,J)=p(I,J);
end

sumij=sumij+p(I,J);
end
```

```
        end
sum(K+1)=sumij;
percentage = abs(sum(K+1)-sum(K))/abs(sum(K+1));
if percentage < 1e-6
          break
              end
end

for I=1:N
for J=1:M
Pr(I,J)=p(I,J).*(6*eta*w*(R/Clc)^2);
Q(I,J) = -log((1-abs(al*Pr(I,J))))/al;
end
end




%% Output Block
for i=1:N
     tett(i)=deltet*(i-1)*360/2/pi;
end

for j=1:M
     z(j)=delz*(j-1);
end

for i=2:N-1
    for j=2:M-1
          F(i,j)=dblquad(@(t,z)Pr(i,j)*R*sin(t),tett(i),tett(i+1),0,delz);
   FF(i,j)=-dblquad(@(t,z)Pr(i,j)*R*cos(t),tett(i),tett(i+1),0,delz);
     end
end

for i=2:N-1
    for j=2:M-1
          Fx=Fx+(F(i,j));
        Fy=Fy+(FF(i,j));
     end
end

fi=atand(Fx/Fy);
W=abs((Fy*cosd(fi)+Fx*sind(fi)));

%% Iteration Block

AA=(1/(pi*EE));

for I=2:N
```

```
        for J=2:M
            del=0;
            for i=2:N-1
                for j=2:M-1
                    if I==i && J==j
                        j=j+1;
                    else
                        del=del+Pr(I,J)*R*deltet*delz/sqrt((I-i)^2*(R*deltet)^2+(J-j)^2*(delz)^2);
                    end
                end
            end
            Del(I,J)=Del(I,J)+AA*del;

        end
end
```

Chapter 7

Gas (Air) Lubricated Bearings

7.1 Introduction

An air bearing is a category of bearings, in which compressed air is used as lubricant. Major advantage of air lubricant is very low viscosity due to which power loss in air bearing is negligible and heat generated due to lubricant shearing compared to liquid-lubricated bearings is insignificant. Additional advantage of the air bearings is no requirement of oil lubrication. This means 'contamination free operation', 'no limit on operating temperature', and 'no requirement of complex sealing' if air/gas bearing technology is employed.

Table 7.1 Viscosity of gases [Booser, 1984]

Gas	Chemical	molecular weight	Viscosity (msec/m² × 10⁴)	
			27° C	127° C
Air	$N_2H_7O_2$	28.96	18.5	23.0
Ammonia	NH_2	17.03	15.76	16.35
Argon	Ar	39.94	22.7	28.9
Carbon dioxide	CO_2	44.01	14.9	19.4
Fluorine	F_2	38.00	24.0	30.0
Freon 12	CF_2Cl_2	120.92	12.4	15.3
Freon 11	CF_2Cl_3	137.39	11.5	13.7
Helium	He	4.003	19.81	24.02
Hydrogen	H_2	2.016	8.96	10.8
Krypton	Kr	83.8	25.6	33.1
Nitrogen	N_2	28.02	17.9	22.1
Oxygen	O_2	32.00	20.7	25.8

Viscosities of the common gases at two temperatures have been listed in Table 7.1. From this table it is clear that viscosity of gases increases with increase in temperature. This means, in case of gases, lubrication provides better results at higher temperature (due to thermal thickening effect) compared to that liquid lubricants that exhibit thermal thinning behaviour (i.e., decrease in viscosity with increase in temperature). This difference in the behaviour of viscosities of gases and liquids with temperature occurs due to the dominant role of momentum transfer in gases, while significance of intermolecular forces in liquids. In other words, there are two effects: intermolecular forces and momentum transfer, which influence the viscosity of any fluid. Viscosity of liquid is predominantly governed by intermolecular forces. As intermolecular forces decrease with increase in temperature, liquid viscosity drops with increase in temperature.

There are a number of applications of air bearings, such as high precision instruments (i.e., coordinate measuring machines), high speed rotating machines, etc. However, there are some disadvantages of air/gas bearings such as 'low load carrying capacity', 'poor damping', and 'expensive fabrication'. The difficulty in manufacturing and assembly of air bearings arises due to the functional requirement of very small gap (as depicted in Fig. 7.1.1) between tribo–pairs and that too without any mechanical contact anywhere along the length/circumference at any instance of bearing operation. This means manufacturing with very close dimensional and geometric tolerances is a must for gaining the advantages from air/bearing.

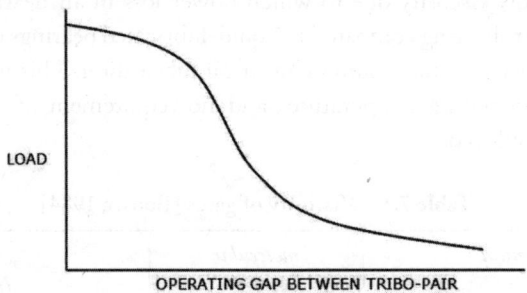

Fig. 7.1.1 General load clearance characteristics of air bearing [Neale, 2001]

There are two main mechanisms, aerostatic and aerodynamic, to support the applied load on the air bearings. The basic mechanism is similar to that of hydrostatic (discussed in chapter 5) and hydrodynamic (discussed in chapters 3 and 4). In other words, aerodynamic bearing is analogous to the hydrodynamic bearings; and aerostatic bearings are analogous to the hydrostatic bearings except that the fluid is compressible. For ideal gas, pressure (P) and volume (V) of the gas are related by:

$$pV = nRT$$

Or
$$p = \rho RT \qquad (7.1)$$

Here T is absolute temperature expressed in Kelvin, R is the universal gas constant and ρ is the density. A gas obeys a polytropic relation, which means $P \propto \rho^n$ where n is the polytropic gas expansion exponent. For an isothermal flow n = 1 and for adiabatic flow n = 1.4. In the present chapter n = 1 has been assumed.

An aerodynamic bearing mechanism has been sketched in Fig. 7.1.2. In this mechanism, rotating cylindrical element rotates smoothly at relatively high speed in a fixed cylindrical component (shown in Fig. 7.1.2 by section lines bounded by two circles) and generates aerodynamic pressure within the bearing clearance.

The main difference between aerostatic and aerodynamic bearings is that the aerostatic bearings require supply of pressurized air for their operation, while the gas pressure is developed at the surfaces of aerodynamic bearings by action of high relative speed. Due to this difference in operating mechanism, the aerodynamic bearings should be used for light rotors operating with very high speeds, while aerostatic bearings should be used for somewhat higher loads without any constraint on its minimum speed. It is worth noting that even for aerodynamic bearings it may be necessary to supply externally pressurized gas (as shown in Fig. 7.1.3) to prevent mechanical contact at the start and end of rotation.

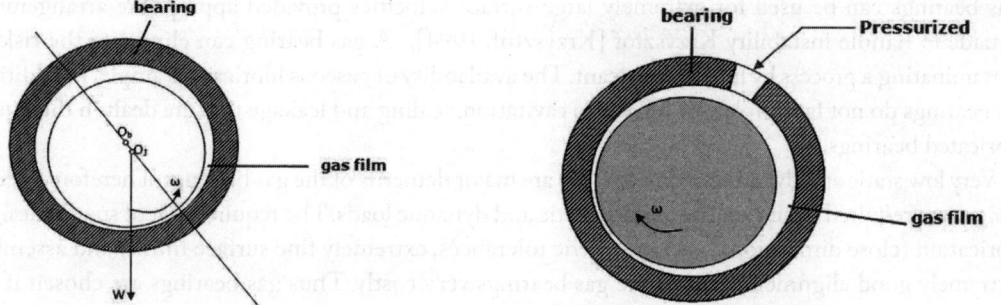

Fig. 7.1.2 Cylindrical aerodynamic bearing

Fig.: 7.1.3 Cylindrical aerodynamic with pressurised air supply arrangement

The aerostatic bearings provide better dynamic (stiffness and damping) characteristics compared to the aerodynamic bearings. An annular orifice configuration of aerostatic bearing has been shown in Fig. 7.1.4.

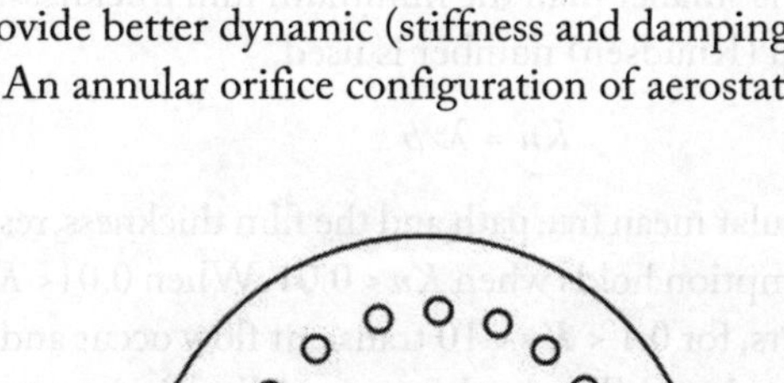

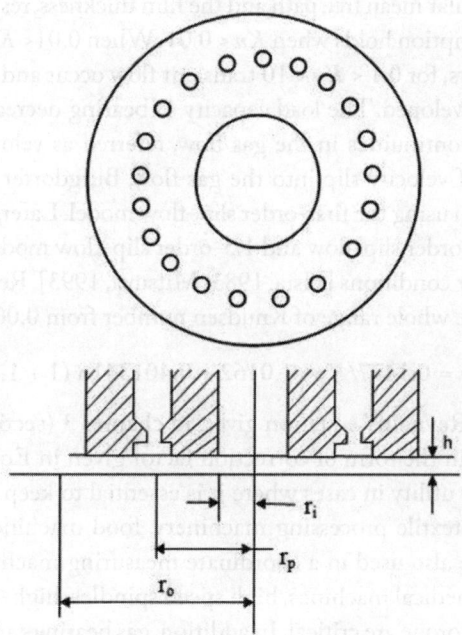

Fig. 7.1.4 Annular pocketed aerostatic bearing

Both the aerostatic and aerodynamic air bearings can be used to support axial and/or radial loads. In other words, aerostatic and aerodynamic bearings are of two types: (a) radial bearings that support loads perpendicular to shaft axis, and (b) thrust bearings that support axial loads.

7.2 Merits, Demerits and Applications

Gas lubricated bearings have numerous advantages over bearings lubricated with liquid and solid lubricants. A gas bearing is virtually frictionless (due to inherently low viscosity) and silent. Generally gases are chemically stable over a wide range of temperature. Additionally, viscosity of gas increases with increase in operating temperature, which minimizes the thermal issue in gas-lubricated films. Gas bearings can be used for extremely large surface velocities provided appropriate arrangement is made to handle instability Krzysztof [Krzysztof, 1994]. A gas bearing can eliminate the risk of contaminating a process by liquid- lubricant. The availability of gaseous lubricant is ample. In addition gas bearings do not face problems related to cavitation, sealing and leakage that are dealt in the liquid lubricated bearings.

Very low static and dynamic characteristics are major demerits of the gas bearings. Therefore, special designs are required to sustain the applied static and dynamic loads. The requirement of special design, fabrication (close dimensional and geometric tolerances, extremely fine surface finish) and assembly (extremely good alignment) makes the gas bearings very costly. Thus gas bearings are chosen if no other lubricant (liquid/solid) could possibly be used for the application.

It is often very difficult to analyze the performance of gas bearings. The continuity equation (Eq. 3.2.13), required to derive the Reynolds' equation (Eq. 3.2.29), becomes invalid if the mean free path between the gas molecules is smaller than the minimum film thickness. To simplify the computation, following non-dimensional (Knudsen) number is used,

$$Kn = \lambda/h \qquad (7.2.1)$$

where λ and h are the molecular mean free path and the film thickness, respectively. As per Gross [Gross, 1980], the continuum assumption holds when $Kn < 0.01$. When $0.01 < Kn < 0.1$, slip flow (between the surfaces of tribo–pair) occurs, for $0.1 < Kn < 10$ transient flow occur and for $Kn > 10$, molecular flow is thought to become fully developed. The load capacity of bearing decreases with increase in Knudsen number as this creates discontinuities in the gas flow, referred as velocity slip [Maxwell, 1867]. By accounting for the effect of velocity slip into the gas flow, Burgdorfer [Burgdorfer, 1959] derived a modified Reynolds' equation using the first-order slip–flow model. Later, two other modified Reynolds' equations based on second-order slip-flow and 1.5-order slip-flow models were respectively proposed for higher Knudsen number conditions [Hsia, 1983; Mitsuya, 1993]. Recently, Huang and Hu [2008] proposed a factor Q_p for the whole range of Knudsen number from 0.009 to 88 as given in Eq. 7.2.2).

$$Qp = 0.1477/K_n + 1.0162 + 0.40134 \ln(1 + 1.477 K_n) \qquad (7.2.2)$$

In the present chapter, Reynolds' equation given in chapter 3 (section 3.3.2) will be modified to incorporate the slip factor in the form of correction factor given in Eq. 7.2.2).

Gas bearings finds their utility in cases where it is essential to keep environment free from liquid contamination, such as in textile processing machinery, food machinery, flying head in read/write devices. These bearings are also used in a coordinate measuring machines, precision machine tools, rotary tables, rheometers, medical machines, high speed spindles, high-speed dental drills, etc., where precision and constancy of torque are critical. In addition, gas bearings are used in gas-cycle machinery, where the cycle gas is used in the bearings.

As the distance of nearest approach between the bearing surfaces is very small, special precautions must be taken in manufacturing gas bearing. Well designed and manufactured air bearings are able to float relatively moving surfaces on a pressurized film of air without any physical contact. This means only the shearing of the molecules contributes to the friction. The static and dynamic coefficients of friction at startup are identical and there is no stick–slip effect. This minimizes lost motion and results are more repeatable.

7.3 Aerodynamic Bearings

Principle of aerodynamic bearing function is the same as that of hydrodynamic bearing, which was explained in chapter 3 (section 3.3.2), except compressibility of gas. To account the compressibility, the Reynolds' equation is written as:

$$\frac{\partial}{\partial x}\left(\frac{\rho h^3}{\eta}\frac{\partial p}{\partial x}\right) + \frac{\partial}{\partial z}\left(\frac{\rho h^3}{\eta}\frac{\partial p}{\partial z}\right) = 6U\frac{\partial(\rho h)}{\partial x} + 12\frac{\partial(\rho h)}{\partial t} \qquad (7.3.1)$$

Gas lubricating films are nearly isothermal because the ability of the bearing materials to conduct away heat is generally greater than the heat generating capacity of the gas film. Thus, the flow can be assumed isothermal and Eq. 7.1 can be used to express density in terms of pressure.

On substituting expression of ρ from Eq. 7.1 and rearranging

$$\frac{\partial}{\partial x}\left(\frac{p h^3}{\eta}\frac{\partial p}{\partial x}\right) + \frac{\partial}{\partial z}\left(\frac{p h^3}{\eta}\frac{\partial p}{\partial z}\right) = 6U\frac{\partial(p h)}{\partial x} + 12\frac{\partial(p h)}{\partial t} \qquad (7.3.2)$$

For static conditions,

$$\frac{\partial}{\partial x}\left(\frac{p h^3}{\eta}\frac{\partial p}{\partial x}\right) + \frac{\partial}{\partial z}\left(\frac{p h^3}{\eta}\frac{\partial p}{\partial z}\right) = 6U\frac{\partial(p h)}{\partial x} \qquad (7.3.3)$$

7.3.1 Pad bearings

To solve the Eq. 7.3.3, geometry of the bearing is required. One of typical radial gas bearing is shown in the Fig. 7.3.1. In this figure, there are five pads, which are supported on pivots so that the pads may tilt about the pivot and change the geometry of lubricant film. The major advantage of these bearings over full cylindrical bearings is the absence of cross stiffness [Rangwala, 2006], which means free from half speed whirl instability problem.

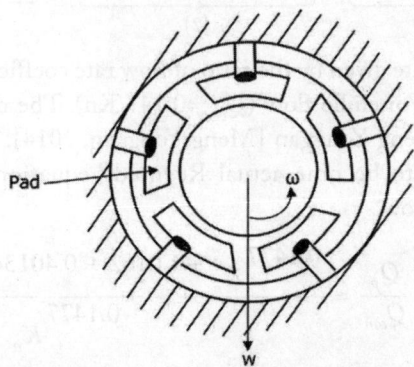

Fig. 7.3.1 Tilting pad journal bearing

To analyze the performance of tilting pad bearing, each pad can be analyzed separately and then vector summation of force, supported by each pad, can be used to find the final load capacity of pad journal bearing. For simplicity, the curvature of pad may be neglected and pad may be represented as shown in Fig. 7.3.2. In this figure U = ω.R.

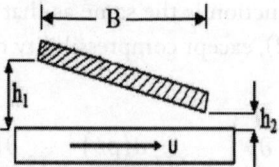

Fig. 7.3.2 Infinitely long plane slider

Non-dimensionalizing Eq. 7.3.3 by substituting

$$P = p/p_a; H = h/h_{min};$$

$$X = x/B\ ;\ Z = z/L$$

$$\frac{\partial}{\partial X}\left(PH^3 \frac{\partial P}{\partial X}\right) + \left(\frac{B}{L}\right)^2 \frac{\partial}{\partial Z}\left(PH^3 \frac{\partial P}{\partial Z}\right) = \frac{6\eta UB}{p_a h_{min}^2} \frac{\partial(PH)}{\partial X} \quad (7.3.4)$$

In Eq. 7.3.4, it is assumed that the gas viscosity (η) is somewhat insensitive to change in the fluid pressure, and the operating temperature is virtually constant; therefore the gas viscosity is constant.

On substituting $\Lambda = \dfrac{6\eta UB}{p_a h_{min}^2}$ as bearing (compressibility) number

$$\frac{\partial}{\partial X}\left(PH^3 \frac{\partial P}{\partial X}\right) + \left(\frac{B}{L}\right)^2 \frac{\partial}{\partial Z}\left(PH^3 \frac{\partial P}{\partial Z}\right) = \Lambda \frac{\partial(PH)}{\partial X} \quad (7.3.5)$$

To account slip (as discussed in section 7.2), a correction factor Q [Meng Yonggan, 2014] needs to be incorporated in Eq. 7.3.5,

$$\underbrace{\frac{\partial}{\partial X}\left(QPH^3 \frac{\partial P}{\partial X}\right)}_{(1)} + \underbrace{\left(\frac{B}{L}\right)^2 \frac{\partial}{\partial Z}\left(QPH^3 \frac{\partial P}{\partial Z}\right)}_{(2)} = \underbrace{\Lambda \frac{\partial(PH)}{\partial X}}_{(3)} \quad (7.3.6)$$

where Q is the relative flow rate given by the ratio of flow rate coefficient of Poiseuille flow (Q_p) and flow rate of the continuum Poiseuille flow (Q_{con}=0.147/Kn). The detailed discussion on the flow rate has been provided by Meng Yonggan [Meng Yonggan, 2014]. The above modified Reynolds' equation. (Eq. 7.3.6) tends to become actual Reynolds' equation (Eq. 7.3.5) when ($K_n \to 0$) ($Q_n \to 1$), which is shown below.

$$Q = \frac{Q_p}{Q_{con}} = \frac{0.1477/K_n + 1.0162 + 0.40134\ln(1+1.477K_n)}{0.1477/K_n}$$

$$= 1 + 6.88K_n + 2.7K_n \ln(1+1.477K_n) = 1|_{Kn \to 0}$$

Assuming $q = QH^3$ and substituting $P\dfrac{\partial P}{\partial X} = \dfrac{1}{2}\dfrac{\partial P^2}{\partial X}$, Eq. 7.3.6 modifies as:

$$\frac{1}{2}\frac{\partial}{\partial X}\left(q\frac{\partial P^2}{\partial X}\right) + \frac{1}{2}\left(\frac{B}{L}\right)^2 \frac{\partial}{\partial Z}\left(q\frac{\partial P^2}{\partial Z}\right) = \Lambda\frac{\partial(PH)}{\partial X}$$

$$\Rightarrow \underbrace{\frac{\partial}{\partial X}\left(q\frac{\partial P^2}{\partial X}\right)}_{(1)} + \underbrace{\left(\frac{B}{L}\right)^2 \frac{\partial}{\partial Z}\left(q\frac{\partial P^2}{\partial Z}\right)}_{(2)} = \underbrace{2\Lambda\frac{\partial(PH)}{\partial X}}_{(3)} \qquad (7.3.7)$$

Equation (7.3.7) can be discretized as discussed in section 4.2. The discretized form of the three terms of Eq. 7.3.7 is given below:

(1)
$$\frac{\partial}{\partial X}\left(q\frac{\partial P^2}{\partial X}\right) = \frac{\partial q}{\partial X}\frac{\partial P^2}{\partial X} + q\frac{\partial^2 P^2}{\partial X^2}$$

$$\Rightarrow \left(\frac{q_{i+1,j} - q_{i-1,j}}{2\Delta X}\right)\left(\frac{P_{i+1,j}^2 - P_{i-1,j}^2}{2\Delta X}\right) + q_{i,j}\left(\frac{P_{i+1,j}^2 - 2P_{i,j}^2 + P_{i-1,j}^2}{\Delta X^2}\right)$$

$$\Rightarrow \left(\frac{q_{i+1,j} - q_{i-1,j} + 4q_{i,j}}{4\Delta X^2}\right)P_{i+1,j}^2 - \left(\frac{q_{i+1,j} - q_{i-1,j} - 4q_{i,j}}{4\Delta X^2}\right)P_{i-1,j}^2 - q_{i,j}\left(\frac{2P_{i,j}^2}{\Delta X^2}\right) \qquad (7.3.8(a))$$

(2)
$$\left(\frac{B}{L}\right)^2 \frac{\partial}{\partial Z}\left(q\frac{\partial P^2}{\partial Z}\right) = \left(\frac{B}{L}\right)^2 \left(\frac{\partial q}{\partial Z}\frac{\partial P^2}{\partial Z} + q\frac{\partial^2 P^2}{\partial Z^2}\right)$$

$$\Rightarrow \left(\frac{B}{L}\right)^2 \left[\left(\frac{q_{i,j+1} - q_{i,j-1}}{2\Delta Z}\right)\left(\frac{P_{i,j+1}^2 - P_{i,j-1}^2}{2\Delta Z}\right) + q_{i,j}\left(\frac{P_{i,j+1}^2 - 2P_{i,j}^2 + P_{i,j-1}^2}{\Delta Z^2}\right)\right]$$

$$\Rightarrow \left(\frac{B}{L}\right)^2 \left[\left(\frac{q_{i,j+1} - q_{i,j-1} + 4q_{i,j}}{4\Delta Z^2}\right)P_{i,j+1}^2 - \left(\frac{q_{i,j+1} - q_{i,j-1} - 4q_{i,j}}{4\Delta Z^2}\right)P_{i,j-1}^2 - q_{i,j}\left(\frac{2P_{i,j}^2}{\Delta Z^2}\right)\right] \qquad (7.3.8(b))$$

(3)
$$2\Lambda\frac{\partial(PH)}{\partial X} = 2\Lambda\left[H\frac{\partial P}{\partial X} + P\frac{\partial H}{\partial X}\right]$$

$$\Rightarrow \Lambda\left[H_{i,j}\left(\frac{P_{i+1,j} - P_{i-1,j}}{\Delta X}\right) + P_{i,j}\left(\frac{H_{i+1,j} - H_{i-1,j}}{\Delta X}\right)\right] \qquad (7.3.8(c))$$

Substituting the Eqs. 7.3.8 in Eq. 7.3.7, we get

$$\left(\frac{q_{i+1,j} - q_{i-1,j} + 4q_{i,j}}{4\Delta X^2}\right)P_{i+1,j}^2 - \left(\frac{q_{i+1,j} - q_{i-1,j} - 4q_{i,j}}{4\Delta X^2}\right)P_{i-1,j}^2 - \Lambda H_{i,j}\left(\frac{P_{i+1,j} - P_{i-1,j}}{\Delta X}\right)$$

$$\left(\frac{B}{L}\right)^2 \left[\left(\frac{q_{i,j+1} - q_{i,j-1} + 4q_{i,j}}{4\Delta Z^2}\right)P_{i,j+1}^2 - \left(\frac{q_{i,j+1} - q_{i,j-1} - 4q_{i,j}}{4\Delta Z^2}\right)P_{i,j-1}^2\right]$$

$$= \Lambda \left[P_{i,j} \left(\frac{H_{i+1,j} - H_{i-1,j}}{\Delta X} \right) \right] + q_{i,j} \left(\frac{2P_{i,j}^2}{\Delta Z^2} \right) + q_{i,j} \left(\frac{2P_{i,j}^2}{\Delta X^2} \right)$$

It is worth noticing that, in the above equation power of $P_{i,j}$ term is two, while in a liquid lubricated bearing the power of term $P_{i,j}$ is one. Such non-linearity increases the complexity in gas bearing. One way to solve the above equation is to express it in following manner:

$$P_{i,j}^{(k+1)} = \frac{\left[\left(\frac{q_{i+1,j} - q_{i-1,j} + 4q_{i,j}}{4\Delta X^2} \right) P_{i+1,j}^{2(k)} - \left(\frac{q_{i+1,j} - q_{i-1,j} - 4q_{i,j}}{4\Delta X^2} \right) P_{i-1,j}^{2(k)} - \Lambda H_{i,j} \left(\frac{P_{i+1,j}^{(k)} - P_{i-1,j}^{(k)}}{\Delta X} \right) + \left(\frac{B}{L} \right)^2 \left[\left(\frac{q_{i,j+1} - q_{i,j-1} + 4q_{i,j}}{4\Delta Z^2} \right) P_{i,j+1}^{2(k)} - \left(\frac{q_{i,j+1} - q_{i,j-1} - 4q_{i,j}}{4\Delta Z^2} \right) P_{i,j-1}^{2(k)} \right] \right]}{\Lambda \left[\left(\frac{H_{i+1,j} - H_{i-1,j}}{\Delta X} \right) \right] + q_{i,j} \left(\frac{2P_{i,j}^{(k)}}{\Delta Z^2} \right) + q_{i,j} \left(\frac{2P_{i,j}^{(k)}}{\Delta X^2} \right)}$$

(7.3.9)

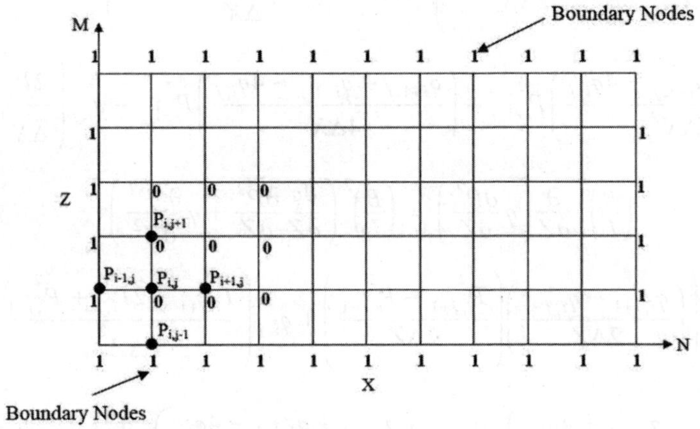

Fig 7.3.3 Discretization of pad surface

Equation 7.3.9 can be solved for pad bearing by discretizing it (i = 1, N and j = 1, M) and using appropriate boundary conditions as shown in Fig. 7.3.3. At the beginning (iteration = 0), values of nodal pressures ($P_{i,j}$), other than boundary nodes, are unknown and it is assumed that value of those nodal pressures is zero at the beginning. For example, substituting the value of $P^{(0)}$ for i=1 and j=1, $P_{1,1}^{(0)} = 0; P_{0,1}^{(0)} = 1; P_{2,1}^{(0)} = 0; P_{1,0}^{(0)} = 1; P_{1,2}^{(0)} = 0$ (as shown in Fig. 7.3.3) in Eq. 7.3.9, we obtain:

$$P_{1,1}^{(1)} = \frac{\left[\left(\frac{q_{2,1} - q_{0,1} + 4q_{1,1}}{4\Delta X^2} \right) P_{2,1}^{2(0)} - \left(\frac{q_{2,1} - q_{0,1} - 4q_{1,1}}{4\Delta X^2} \right) P_{0,1}^{2(0)} - \Lambda H_{1,1} \left(\frac{P_{2,1}^{(0)} - P_{0,1}^{(0)}}{\Delta X} \right) + \left(\frac{B}{L} \right)^2 \left[\left(\frac{q_{1,2} - q_{1,0} + 4q_{1,1}}{4\Delta Z^2} \right) P_{1,2}^{2(0)} - \left(\frac{q_{1,2} - q_{1,0} - 4q_{1,1}}{4\Delta Z^2} \right) P_{1,0}^{2(0)} \right] \right]}{\Lambda \left[\left(\frac{H_{2,1} - H_{0,1}}{\Delta X} \right) \right] + q_{1,1} \left(\frac{2P_{1,1}^{(0)}}{\Delta Z^2} \right) + q_{1,1} \left(\frac{2P_{1,1}^{(0)}}{\Delta X^2} \right)}$$

$$\Rightarrow \quad P_{1,1}^{(1)} = \frac{\left[-\left(\dfrac{q_{2,1}-q_{0,1}-4q_{1,1}}{4\Delta X^2}\right) - \left(\dfrac{B}{L}\right)\left[\left(\dfrac{q_{1,2}-q_{1,0}-4q_{1,1}}{4\Delta Z^2}\right)\right] + \left(\dfrac{\Lambda H_{1,1}}{\Delta X}\right)\right]}{\Lambda\left[\left(\dfrac{H_{2,1}-H_{0,1}}{\Delta X}\right)\right]}$$

In the next iteration $P_{i,j}^{(0)}$ is replaced by $P_{i,j}^{(1)}$ and value of $P_{i,j}^{(2)}$ is calculated.

$$P_{1,1}^{(2)} = \frac{\left[\left(\dfrac{q_{2,1}-q_{0,1}+4q_{1,1}}{4\Delta X^2}\right)P_{2,1}^{2(1)} - \left(\dfrac{q_{2,1}-q_{0,1}-4q_{1,1}}{4\Delta X^2}\right) - \Lambda H_{1,1}\left(\dfrac{P_{2,1}^{(1)}-1}{\Delta X}\right) + \left(\dfrac{B}{L}\right)^2\left[\left(\dfrac{q_{2,1}-q_{0,1}+4q_{1,1}}{4\Delta Z^2}\right)P_{1,2}^{2(1)} - \left(\dfrac{q_{2,1}-q_{0,1}-4q_{1,1}}{4\Delta Z^2}\right)\right]\right]}{\Lambda\left[\left(\dfrac{H_{2,1}-H_{0,1}}{\Delta X}\right)\right] + q_{1,1}\left(\dfrac{2P_{1,1}^{(1)}}{\Delta Z^2}\right) + q_{1,1}\left(\dfrac{2P_{1,1}^{(1)}}{\Delta X^2}\right)}$$

The process of iterations will be repeated till the following convergence criterion is met.

$$\frac{\left|\left(\sum_{i=1}^{n}\sum_{j=1}^{m}P_{i,j}\right)^{k+1} - \left(\sum_{i=1}^{n}\sum_{j=1}^{m}P_{i,j}\right)^{k}\right|}{\left|\left(\sum_{i=1}^{n}\sum_{j=1}^{m}P_{i,j}\right)^{k+1}\right|} \leq \varepsilon \qquad (7.3.10)$$

The load carrying capacity, for a given film profile, can be obtained by integrating fluid pressure (estimated by multiplying p_a with non-dimensional fluid pressure obtained from Eq. 7.3.9 i.e., $p = Pp_a$) over the bearing pad area. The load capacity of pad (W_{pad})

$$W_{pad} = \int_{0}^{B}\int_{0}^{L}(p - p_a)\,dx\,dz \qquad (7.3.11)$$

The Eq. 7.3.11 can be rewritten for numerical integration of pressure,

$$W_{pad} = \left(\Delta x\,\Delta z\right)\sum_{i=1}^{N}\sum_{j=1}^{M}\left(p_{i,j} - p_a\right) \qquad (7.3.12)$$

If spread of pad (B, as shown in Fig. 7.3.1) is much smaller than its length (L), then it can be treated as infinitely long plane fixed sliders (discussed in section 3.3.1), as shown in Fig. 7.3.2. The flow of gas in and out of the sides of the bearing (side flow) is insignificant. For this geometry Eq. 7.3.7 reduces to:

$$\frac{\partial}{\partial X}\left(q\frac{\partial P^2}{\partial X}\right) = 2\Lambda\frac{\partial(PH)}{\partial X} \qquad (7.3.13)$$

Here $H = \dfrac{h}{h_1}, n = \dfrac{h_1}{h_2}, X = \dfrac{x}{B}$ and $H = n-(n-1)X$. The discretized form of Eq. 7.3.13 can be obtained from Eq. 7.3.9 by removing terms with variable with 'Z'. The final discretized equation is given by Eq. 7.3.14.

$$P_i = \frac{\left[\left(\dfrac{q_{i+1} - q_{i-1} + 4q_i}{4\Delta X^2}\right)P_{i+1}^2 - \left(\dfrac{q_{i+1} - q_{i-1} - 4q_i}{4\Delta X^2}\right)P_{i-1}^2 - \Lambda H_i \left(\dfrac{P_{i+1} - P_{i-1}}{\Delta X}\right)\right]}{\Lambda\left[\left(\dfrac{H_{i+1} - H_{i-1}}{\Delta X}\right)\right] + q_i \left(\dfrac{2P_i}{\Delta X^2}\right)} \tag{7.3.14}$$

The load carrying capacity for the obtained pressure can be calculated by integrating the pressure values along the bearing length i.e.,

$$W_{pad} = L \int_0^B (p - p_a) dx \tag{7.3.15}$$

Equation 7.3.15 can be rewritten for numerical integration of pressure,

$$W_{pad} = L(\Delta x) \sum_{i=1}^{N} (p_i - p_a) \tag{7.3.16}$$

7.3.2 Cylindrical bearings

A pad bearing can be a full cylinder journal bearing. The equations described in section 7.3.1 for the pad bearing can be used for a journal bearing by substituting $P = p/p_a; H = h/C$

$\theta = x/R; Z = z/L; \Lambda = \dfrac{6\eta\omega}{p_a}\left(\dfrac{R}{C}\right)^2$ and $U = \omega R$ in Eq. 7.3.7 will be;

$$\frac{\partial}{\partial \theta}\left(q \frac{\partial P^2}{\partial \theta}\right) + \left(\frac{R}{L}\right)^2 \frac{\partial}{\partial Z}\left(q \frac{\partial P^2}{\partial Z}\right) = 2\Lambda \frac{\partial (PH)}{\partial \theta} \tag{7.3.17}$$

$$P_{i,j} = \frac{\left[\left(\dfrac{q_{i+1,j} - q_{i-1,j} + 4q_{i,j}}{4\Delta\theta^2}\right)P_{i+1,j}^2 - \left(\dfrac{q_{i+1,j} - q_{i-1,j} - 4q_{i,j}}{4\Delta\theta^2}\right)P_{i-1,j}^2 - \Lambda H_{i,j}\left(\dfrac{P_{i+1,j} - P_{i-1,j}}{\Delta\theta}\right)\right] + \left(\dfrac{R}{L}\right)^2\left[\left(\dfrac{q_{i,j+1} - q_{i,j-1} + 4q_{i,j}}{4\Delta Z^2}\right)P_{i,j+1}^2 - \left(\dfrac{q_{i,j+1} - q_{i,j-1} - 4q_{i,j}}{4\Delta Z^2}\right)P_{i,j-1}^2\right]}{\Lambda\left[\left(\dfrac{H_{i+1,j} - H_{i-1,j}}{\Delta\theta}\right)\right] + q_{i,j}\left(\dfrac{2P_{i,j}}{\Delta Z^2}\right) + q_{i,j}\left(\dfrac{2P_{i,j}}{\Delta\theta^2}\right)}$$

The dimensional load carrying capacity for the journal bearing can be obtained by integrating fluid pressure over the full surface of the journal. The load capacity ($W_{journal}$) is given by;

$$W_{journal} = \int_0^{2\pi}\int_0^L (p - p_a) R \, d\theta \, dz \tag{7.3.18}$$

Equation 7.3.18 can be rewritten for numerical integration of pressure,

$$W_{journal} = (R \Delta\theta \Delta z) \sum_{i=1}^{N} \sum_{j=1}^{M} (p_{i,j} - p_a) \tag{7.3.19}$$

This similar treatment can be used if curvature of pad is to be accounted. In section 7.3.1 curvature of pad was neglected. Considering the curvature of the pad, the load capacity of pad (W_{pad})

$$W_{pad} = \int_{\theta_1}^{\theta_2} \int_{0}^{L} (p - p_a) \cdot R \cdot d\theta \cdot dz \qquad (7.3.20)$$

where θ_1 is the starting and θ_2 is the ending subtanded angle of the bearing pad. In other words ($\theta_2 - \theta_1$) is the angular length of the bearing pad. The Eq. 7.3.20 can be rewritten for numerical integration of pressure,

$$W_{pad} = \left(R \Delta \theta \Delta z \right) \sum_{i=1}^{N} \sum_{j=1}^{M} \left(p_{i,j} - p_a \right) \qquad (7.3.21)$$

7.3.3 Magnetic recording discs with flying head

In section 7.2 of this chapter, it was mentioned that slip flow occurs if Knudsen number (Eq. 7.2.1) is greater than 0.01 and modification in the Reynolds' equation is required. The Knudsen number depends on the mean free path among molecules. The typical values of mean free path (λ) at atmospheric conditions for Hydrogen, Helium, Air and Neon are 0.1125 µm, 0.186 µm, 0.064 µm and 0.132 µm respectively.

Air is used as lubricant in magnetic recording discs, which rotates at speeds of up to 10,000 rpm and separation of read/write head is very low (~50 nm). This means Knudsen number for magnetic recording disc application is $K_n = \lambda/h = 0.064/0.050 = 1.28$; which is far greater than 0.01 and slip will occur.

The flying heads, required in the magnetic recording devices is shown in Fig. 7.3.4. The magnetic recording head can be schematically represented as shown in Fig. 7.3.5, which is similar to the slider, discussed in section 7.3.1.1.

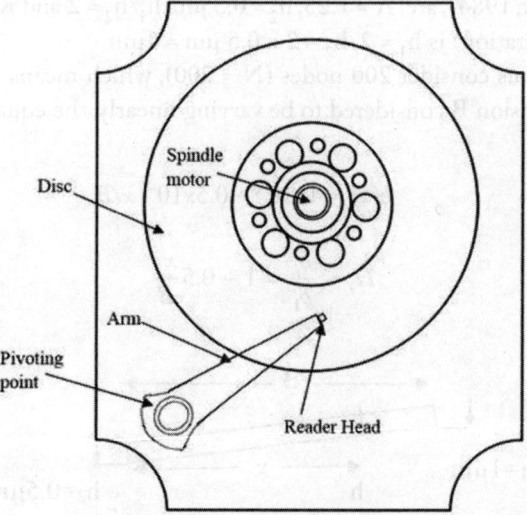

Fig. 7.3.4 Magnetic recording device

308 *Fundamentals of Engineering Tribology with Applications*

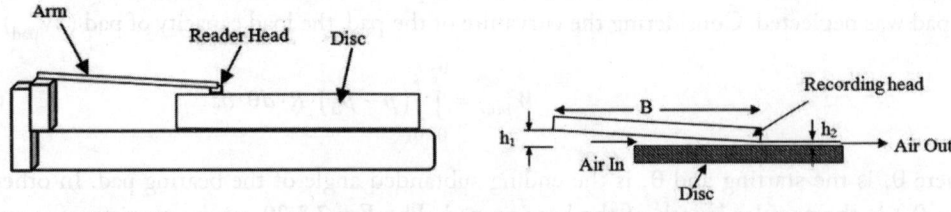

(a) *Arrangement of arm, read/write head and disc* (b) *Infinitely long plane slider*

Fig. 7.3.5 Pad bearing arrangement for hard disc drive

In the magnetic recording device, the flying head must maintain its separation gap without crashing into the rotating disk. Analysis of such gas bearings requires modelling aspects of gas film lubrication. To model gas flow, following assumptions have been made:

- Air in and out of the sides of the bearing is negligible (as shown in Fig. 7.3.5)
- The flow is isothermal

For calculating the pressure along the length of the recoding disc, the discretized pressure equation, Eq. 7.3.14, for infinitely long plane slider can be used.

$$P_i = \frac{\left[\left(\frac{q_{i+1} - q_{i-1} + 4q_i}{4\Delta X^2}\right)P_{i+1}^2 - \left(\frac{q_{i+1} - q_{i-1} - 4q_i}{4\Delta X^2}\right)P_{i-1}^2 - \Lambda H_i\left(\frac{P_{i+1} - P_{i-1}}{\Delta X}\right)\right]}{\Lambda\left[\left(\frac{H_{i+1} - H_{i-1}}{\Delta X}\right)\right] + q_{i,j}\left(\frac{2P_i}{\Delta X^2}\right)} \qquad (7.3.22)$$

To illustrate the pressure distribution in an infinite long plane slider gas bearing, an example provided by Fuller [Fuller, 1984] has been considered. The parameters of the slider gas bearing, provided by Fuller [Fuller, 1984], are: $\Lambda = 1.25$, $h_2 = 0.5$ μm, $h_1/h_2 = 2$ and $K_n = 0.01, 0.005$ and 0.167. The maximum gap (separation) is $h_1 = 2$, $h_2 = 2 \times 0.5$ μm = 1μm.

To use Eq. 7.3.22, let us consider 200 nodes (N = 200), which means $i = 1,200$. If the variation of the 'h' along the dimension B considered to be varying linearly, the equation of 'h' from Fig. 7.3.6 is derived to be:

$$h(x) = 1\times10^{-6} - 0.5\times10^{-6}\, x/B$$

$$H_i = \frac{h_i}{h_1} = 1 - 0.5\frac{x_i}{B}$$

$h_1 = 1$μm $h_2 = 0.5$μm

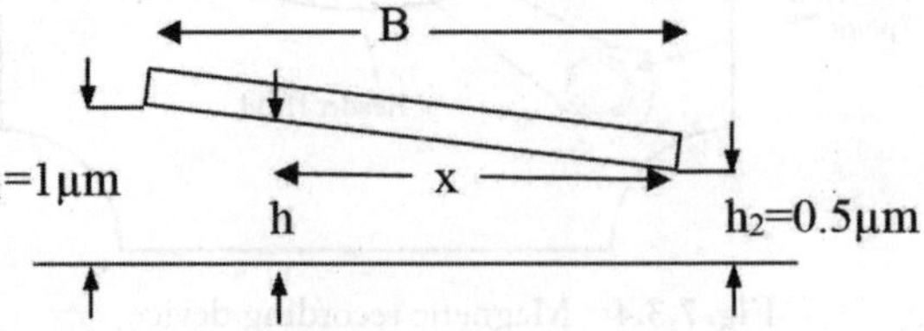

Fig. 7.3.6 Long plane slider

To find q, we need to calculate Q, which is given as:

$$Q = \frac{Q_p}{Q_{con}}$$

$$= \frac{0.1477/K_n + 1.0162 + 0.40134\ln(1 + 1.477K_n)}{0.1477/K_n}$$

$$= 1 + 6.88K_n + 2.7K_n \ln(1 + 1.477K_n)$$

$$q_i = QH_i^3$$

The pressure results obtained from Eq. 7.3.22 are plotted in Fig. 7.3.7. The values obtained are having reasonable correlation with results presented by Fuller [Fuller, 1984]. From this figure it can also be inferred that the pressure increases with reduction in K_n value. In other words, load carrying capacity decreases with increase in Knudsen number. The computer program in MATLAB is given at the end of this chapter.

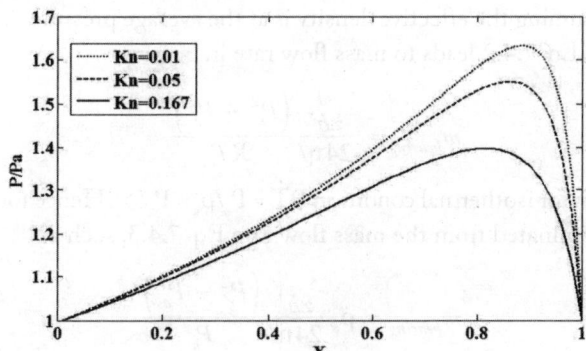

Fig. 7.3.7 Non-dimensional pressure along non-dimensional length

7.4 Aerostatic Bearings

The term static with aero in aerostatic bearings indicates load support even at zero speed. Aerostatic bearing operates with an external pressurized air or gas supply. Gas is fed through a restrictor and escapes through the gap between the bearing lands. This means, the aerostatic bearings require external compressor powerful enough to generate sufficient pressures to separate the tribo–surfaces at zero relative speed and eliminate any problem of mechanical wear.

The mass flow rate of the bearing ($m_{bearing}$) is derived by assuming the flow of air between the parallel plates. The plates are stationary while the fluid moves with velocity 'u'. The volumetric flow rate of fluid has been discussed in the section 3.2 of chapter 3. The difference between the flow rate of liquid and air is that, flow of liquid is represented in volume flow rate while the flow of air is represented by mass flow rate. The mass flow rate (m) is calculated by multiplying density (ρ) of the fluid with volumetric flow rate (Q): i.e., m = ρQ.

The flow due to the pressure gradient (Poiseuille flow) has been given in Eq. 3.2.4 in chapter 3. The modified flow with zero plate velocity (U=0) is presented as Eq. 7.4.1

$$u = \frac{1}{2\eta}\frac{dp}{dx}\left(y^2 - yh\right) \quad (7.4.1)$$

The mass flow through an annulus of dimension 'z' and depth 'h' is expressed as

$$m_{bearing} = z\rho\int_0^h u\,dy$$

$$\Rightarrow \quad m_{bearing} = z\rho\int_0^h \frac{1}{2\eta}\frac{dp}{dx}\left(y^2 - yh\right) dy$$

$$\Rightarrow \quad m_{bearing} = z\rho\frac{1}{2\eta}\frac{dp}{dx}\left|\frac{y^3}{3} - \frac{y^2 h}{2}\right|_0^h$$

$$m_{bearing} = \rho z \frac{1}{12\eta}\frac{dp}{dx} h^3 \quad (7.4.2)$$

For isothermal gas flow through a thin parallel film of length 'l' (l can be the circumferential distance or the axial distance between the pockets), the pressure gradient is $dp/dx = (P_r - P_a)/l$; where P_a is the atmospheric pressure and P_r is the restrictor pressure. Incompressible flow requires a constant density, hence for simplicity assuming the effective density is at the average pressure of the film. By using Eq. 7.1 $\rho = (P_r + P_a)/(2RT)$, Eq. 7.4.2 leads to mass flow rate in bearing:

$$m_{bearing} = \frac{zh^3}{24\eta l}\frac{\left(P_r^2 - P_a^2\right)}{\Re T} \quad (7.4.3)$$

According to gas law, for isothermal condition $RT = P_s/\rho_s = P_a/\rho_a$. Hence for isothermal condition, RT can therefore be eliminated from the mass flow rate Eq. 7.4.3, such as

$$m_{bearing} = \rho_a \frac{zh^3}{24\eta l}\frac{\left(P_r^2 - P_a^2\right)}{P_a} \quad (7.4.4)$$

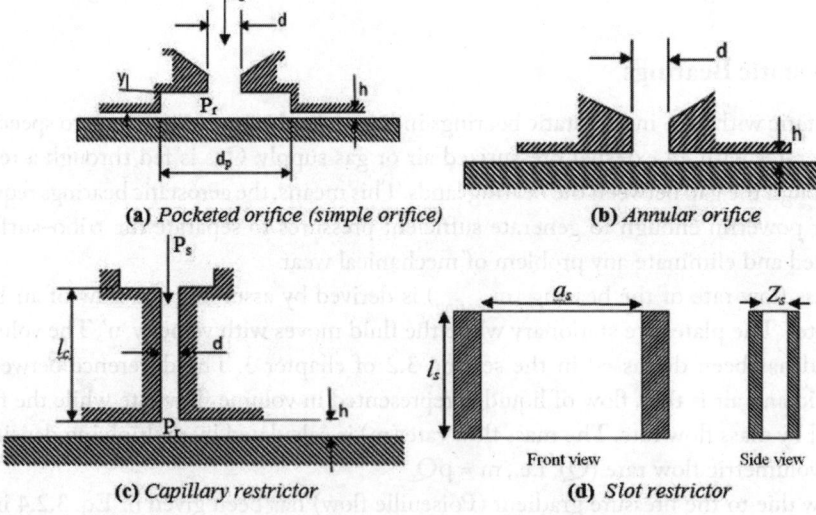

(a) Pocketed orifice (simple orifice)   (b) Annular orifice

(c) Capillary restrictor   (d) Slot restrictor

Fig. 7.4.1 Various types of restrictors [Neale, 2001]

7.4.1 Flow through restrictors

To supply the pressurized gas at the interface of tribo–surfaces, four types of restrictors, as shown in Fig. 7.4.1, can be used. The restrictors (orifices, capillaries or slot) generally have lesser area to create resistance in the flow path of the compressed gas from compressor to the bearing interface. It is interesting to note that flow rate through restrictors is very important parameter as it is related to the fluid pressure. Finding the relationship between the pressure and flow is therefore essential, which can be derived using mass flow continuity, $m_{restrictor} = m_{bearing}$. If there are several restrictors in a bearing, the sum of the restrictor flow must be equal to the total flow out of the bearing.

7.4.1.1 Flow through orifice restrictors

An orifice is a hole of short length to diameter ratio as shown in Fig. 7.4.1(a–b). There are two types of orifice restrictors, as shown in Fig. 7.4.1(a) and Fig. 7.4.1(b). The difference between these two types is the presence/absence of pocket (pocket is like recess in hydrostatic bearing) at the end of orifice. Providing a pocket after the orifice–restrictor ensures a constant cross-section of the gas flow and makes it independent of the operating gap. On the other hand, in case of annular orifice (Fig. 7.4.1 b) the cross-section changes with the film thickness. The depth of pocket must be designed appropriately. In the case of a gas contained in a large deep pocket, hammer blow self-excited oscillations [Brian Rowe, 2012] can occur, which may be quite destructive. To prevent this, the pocket must be shallow (about 4–7 times the film thickness). To derive the equation for mass flow, assumption of 'pressure at the entry to the orifice equal to the manifold supply pressure' and 'constant value of pressure in the bearing pockets' can be made. The gas flow through an orifice is given by [Brain Rowe, 2012]

$$m_{bearing} = C_d A_0 P_s \left(\frac{2}{\Re T_s} \frac{k}{k-1} \left[\left(\frac{P_r}{P_s}\right)^{2/k} - \left(\frac{P_r}{P_s}\right)^{k+1/k} \right] \right)^{1/2} \quad (7.4.5)$$

For air, exponent k can be approximated as $C_p/C_v = 1.4$ and the value of C_d can be considered as 0.96 [Grewal, 1979]. According to Eq. 7.4.5, the flow rate through restrictor is a function of pressure ratio ($K_s = P_r/P_s$). At certain pressure ratio, mass flow rate with respect to change in pressure ratio becomes zero and that condition is termed as chocked flow condition. To find the condition for chocking, the rate of change of mass flow rate with respect to K_s is equated to zero.

$$\frac{dm}{dK_s} = \frac{1}{2} C_d A_0 P_s \left(\frac{2}{\Re T_s} \frac{k}{k-1} \left[(K_s)^{2/k} - (K_s)^{k+1/k} \right] \right)^{-1/2} \left(\frac{2}{k}(K_s)^{\frac{2-k}{k}} - \frac{k+1}{k}(K_s)^{\frac{1}{k}} \right)$$

$$0 = \left(\frac{2}{k}(K_s)^{\frac{2-k}{k}} - \frac{k+1}{k}(K_s)^{\frac{1}{k}} \right)$$

$$\frac{2}{k}(K_s)^{\frac{2-k}{k}} = \frac{k+1}{k}(K_s)^{\frac{1}{k}}$$

$$K_s = \left(\frac{2}{k-1}\right)^{\frac{k}{k+1}} \quad (7.4.6)$$

For chocked flow Szeri [Szeri, 1980] provided the following equation:

$$m_{restrictor} = C_d A_0 P_s \left(\left(\frac{k}{\Re T_s}\right)^{1/2} \left(\frac{2}{k+1}\right)^{\frac{k+1}{2(k-1)}} \right) \quad (7.4.7)$$

The condition for chocking can be found using value of k. For example k for air, nitrogen and oxygen is 1.4, and the value of K_s for these gases is 0.526. In other words, when the pressure ratio ($K_s = P_r/P_s$) of air through an orifice is $\leq .526$, chocked flow occurs. The mass flow rate through the orifice (Eq. 7.4.7) must be equal to mass flow through the bearing (Eq. 7.4.6) i.e.:

$$m_{restrictor} = m_{bearing}$$

$$\Rightarrow \quad m_{restrictor} - m_{bearing} = 0$$

$$\rho_a \frac{zh^3}{24\eta l} \frac{(P_r^2 - P_a^2)}{P_a} - C_d A_0 P_s \left(\frac{2}{\Re T_s} \frac{k}{k-1} \left[\left(\frac{P_r}{P_s}\right)^{2/k} - \left(\frac{P_r}{P_s}\right)^{k+1/k} \right] \right)^{1/2} = 0 \quad (7.4.8\ a)$$

For chocked flow

$$\rho_a \frac{zh^3}{24\eta l} \frac{(P_r^2 - P_a^2)}{P_a} - C_d A_0 P_s \left(\left(\frac{k}{\Re T_s}\right)^{1/2} \left(\frac{2}{k+1}\right)^{\frac{k+1}{2(k-1)}} \right) = 0 \quad (7.4.8\ b)$$

The value of the restrictor pressure (P_r) is estimated by solving. Eq. 7.4.8(a) for non-chocked flow and Eq. 7.4.8(b) for chocked flow.

7.4.1.2 Flow through slot restrictors

The rate of isothermal viscous flow through a slot (as shown in Fig. 7.4.1 d) of length width l_s, width a_s and film thickness Z_s is

$$m_{restrictor} = \rho_a \frac{(P_s^2 - P_r^2)}{P_a} \frac{a_s Z_s^3}{24\eta l_s} \quad (7.4.9)$$

Using

$$m_{restrictor} - m_{bearing} = 0$$

$$\rho_a \frac{(P_s^2 - P_r^2)}{P_a} \frac{a_s Z_s^3}{24\eta l_s} - \rho_a \frac{zh^3}{24\eta l} \frac{(P_r^2 - P_a^2)}{P_a} = 0$$

or

$$(P_s^2 - P_r^2) \frac{a_s Z_s^3}{l_s} = \frac{zh^3}{l} (P_r^2 - P_a^2)$$

or

$$\frac{a_s Z_s^3}{l_s} P_s^2 + \frac{zh^3}{l} P_a^2 = P_r^2 \left(\frac{zh^3}{l} + \frac{a_s Z_s^3}{l_s} \right)$$

or

$$P_r = \sqrt{\frac{\left(\frac{a_s Z_s^3}{l_s} P_s^2 + \frac{zh^3}{l} P_a^2\right)}{\left(\frac{zh^3}{l} + \frac{a_s Z_s^3}{l_s}\right)}} \quad (7.4.10)$$

The value of the restrictor pressure (Pr) is estimated by solving Eq. 7.4.10.

7.4.1.3 Flow through a capillary

For the capillary tube restrictor (Fig. 7.4.1 c) the depth of the pocket (l_c) must be much larger compared to the diameter of pocket (d), i.e., l_c/d should be greater than 20. Due to the extremely small bore diameter of the capillary tube, required for gas bearing, its usage is least preferred.

The mass flow rate through the capillary (shown in Fig. 7.4.1 c) for isothermal viscous condition is given by

$$m = \rho_a \frac{(P_s^2 - P_r^2)}{P_a} \frac{\pi d^4}{128\eta l_c} \quad (7.4.11)$$

Using $m_{restrictor} - m_{bearing} = 0$

$$(P_s^2 - P_r^2)\frac{\pi d^4}{128\eta l_c} - \frac{zh^3}{24\eta l}(P_r^2 - P_a^2) = 0$$

or

$$\frac{(P_s^2 - P_r^2)}{128 l_c}\pi d^4 = \frac{zh^3}{24l}(P_r^2 - P_a^2)$$

or

$$\frac{\pi d^4}{l_c}P_s^2 + \frac{\pi d^4}{l_c}P_a^2 = P_r^2 \frac{16}{3}\left(\frac{zh^3}{l} + \frac{\pi d^4}{l_c}\right)$$

or

$$P_r = \sqrt{\frac{\frac{\pi d^4}{l_c}P_s^2 + \frac{\pi d^4}{l_c}P_a^2}{\frac{16}{3}\left(\frac{zh^3}{l} + \frac{\pi d^4}{l_c}\right)}}$$

The value of the restrictor pressure (Pr) is estimated by solving Eq. 7.4.12.

The aerostatic bearings can be classified based on the load carrying capacity as (i) Radial aerostatic bearings and (ii) Thrust aerostatic bearing. The detailed discussion of each bearing is provided in the next section.

7.4.2 Radial aerostatic bearings

An externally pressurized radial aerostatic bearing with multiple supply holes is shown in Fig. 7.4.2. Compressed air/gas at a constant supply pressure is fed into the bearing through a number of orifices/capillary/slot restrictors. In a radial aerostatic bearing, two or more number of restrictors in axial as well as circumferential direction are provided.

A radial aerostatic bearing with eight restrictors is shown in Fig. 7.4.2. In this figure there are two rows of restrictors in axial direction. Each row of restrictors contains four restrictors in the circumferential direction.

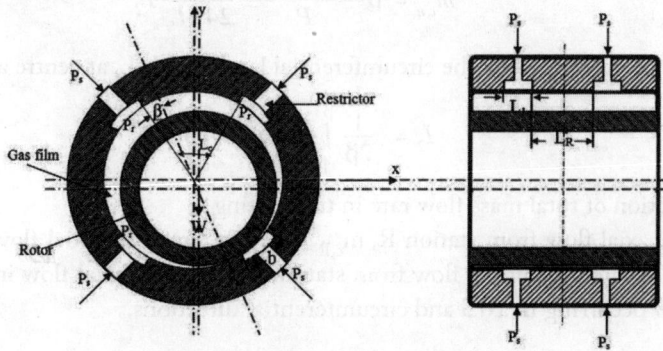

Fig. 7.4.2 Radial aerostatic bearing

In Fig. 7.4.2, 'b' is length of the restrictor in circumferential direction, L_R is the minimum axial distance between the restrictors, minimum circumferential land length between two restrictors L_c and L_1 is the length of the restrictor in axial direction. The air is supplied with pressure P_s at the entry of the restrictors. Pressure at the exit of restrictor (P_r) is generally lesser than P_s and the reduction depends on the type and geometry of the restrictor. Finally the air is leaked out of the bearings at ends at atmospheric pressure (P_a). The supply and atmospheric pressures are known but the restrictor pressure at each station must be evaluated, as discussed in the previous section, that is, 7.4.1.

It is to be noted that the air flows out from restrictors in circumferential as well as axial direction. Hence the total air flow in the bearing will be summation of the air flowing out from the circumferential and axial directions of each pocket. A step by step procedure to calculate the pressure in aerostatic bearing having 'm' and 'n' numbers of restrictors in axial and angular directions respectively is given below.

Step 1: Determination of axial mass flow rate for the mth restrictor in the bearing

The total mass flow rate entering in axial direction is given by [Panday 1980]:

$$m_{a,m} = \rho_a \frac{\left(P_{r,m}^2 - P_a^2\right)}{P_a} \frac{RC^3}{24\eta L_R} \int_{\theta_1}^{\theta_2} (1 + \varepsilon \cos\theta)^3 \, d\theta$$

$$m_{a,m} = \rho_a \frac{\left(P_{r,m}^2 - P_a^2\right)}{P_a} \frac{RC^3}{24\eta L_R} I_1 \tag{7.4.13}$$

where θ_1 and θ_2 is the starting and end points of the restrictor in the circumferential coordinate,

$$I_1 = \int_{\theta_1}^{\theta_2} (1 + \varepsilon \cos\theta)^3 \, d\theta$$

Step 2: Determination of circumferential mass flow rate for the nth restrictor in the bearing

In calculating the circumferential mass flow rate, the value of film thickness (h) is taken as the average of the cube of clearance (this reduces the error at higher eccentricities). Thus the circumferential inflow from pocket n^{th} is [Panday 1980]:

$$m_{c,n} = \rho_a \frac{\left(P_r^2 - P_a^2\right)}{P_a} \frac{L_1 C^3}{24\eta L_c} \frac{1}{2\beta} \int_{-\beta}^{\beta} (1 + \varepsilon \cos\theta)^3 \, d\theta$$

$$m_{c,n} = \rho \frac{\left(P_{r,n}^2 - P_a^2\right)}{P_a} \frac{L_1 C^3}{24\eta L_c} I_2 \tag{7.4.14}$$

where β is half-angle subtended by the circumferential land length L, at centre and

$$I_2 = \frac{1}{2\beta} \int_{-\beta}^{\beta} (1 + \varepsilon \cos\theta)^3 \, d\theta$$

Step 3: Determination of total mass flow rate in the bearing

Let $m_{a,R}$ be the axial flow from station R, $m_{c,R \to S}$ be the circumferential flow from station R to S and $m_{c,S \to R}$ be the circumferential flow from station S to R. The total flow in the bearing is the summation of flow occurring in axial and circumferential directions,

$$m_{bearing} = m_{a,R} + \sum_{n=1}^{l} m_{c,R \to S} + \sum_{n=1}^{l} m_{c,S \to R} \qquad (7.4.15)$$

$$m_{bearing} = \rho_a \frac{\left(P_{r,R}^2 - P_a^2\right)}{P_a} \frac{RC^3}{24\eta L_R} I_1 + \sum_{n=1}^{l} \rho_a \frac{\left(P_{c,R}^2 - P_a^2\right)}{P_a} \frac{L_1 C^3}{24\eta L_c} I_2 + \sum_{n=1}^{l} \rho_a \frac{\left(P_{c,S}^2 - P_a^2\right)}{P_a} \frac{L_1 C^3}{24\eta L_c} I_2 \qquad (7.4.16)$$

Step 4: Determination of mass flow rate through the restrictor

Mass flow rate through the ith orifice restrictor in a radial aerostatic bearing (shown in Fig. 7.4.2) can be estimated using Eq. 7.4.5 / 7.4.7. However in the case of slot or capillary restrictors, the mass flow rate equation is replaced by Eqs. 7.4.9 and 7.4.11, respectively

Step 5: Determination of restrictor pressure (P_r)

The restrictor pressure (P_r) is estimated by equating the mass flow rate through the bearing (explained in step 3) and mass flow through the restrictor (explained in step 4) and then solving that algebraic equation.

Step 6: Solving the Reynolds' equation for Aerostatic condition

The Reynolds' equation for aerodynamic bearing has been given in Eq. 7.3.17 in section 7.3.2. This equation can be used for aerostatic bearing by equating the RHS of the equation equal to zero. The modified Reynolds' equation for estimating the pressure variation in the aerostatic bearing is given in Eq. 7.4.17.

$$\frac{\partial}{\partial \theta}\left(q \frac{\partial P^2}{\partial \theta}\right) + \left(\frac{R}{L}\right)^2 \frac{\partial}{\partial Z}\left(q \frac{\partial P^2}{\partial Z}\right) = 0 \qquad (7.4.17)$$

The solution of the Eq. 7.4.17 is obtained by discretizing the equation, as discussed in section 7.3.1 and substituting $\Lambda = 0$. The descretized form is given by the Eq. 7.4.18.

$$P_{i,j} = \frac{\left[\left(\frac{q_{i+1,j} - q_{i-1,j} + 4q_{i,j}}{4\Delta\theta^2}\right) P_{i+1,j}^2 - \left(\frac{q_{i+1,j} - q_{i-1,j} - 4q_{i,j}}{4\Delta\theta^2}\right) P_{i-1,j}^2 + \left(\frac{R}{L}\right)^2 \left[\left(\frac{q_{i,j+1} - q_{i,j-1} + 4q_{i,j}}{4\Delta Z^2}\right) P_{i,j+1}^2 - \left(\frac{q_{i,j+1} - q_{i,j-1} - 4q_{i,j}}{4\Delta Z^2}\right) P_{i,j-1}^2\right]\right]}{q_{i,j}\left(\frac{2P_{i,j}}{\Delta Z^2}\right) + q_{i,j}\left(\frac{2P_{i,j}}{\Delta\theta^2}\right)} \qquad (7.4.18)$$

Step 7: Iterations for calculation of pressure distribution

The Eq. 7.4.18 can be solved for radial aerostatic bearing by discretizing it (i = 1, N and j = 1, M). At beginning (iteration = 0)

(i) The nodes denoting the restrictor location is substituted with restricted pressure value, estimated in step 5 (as shown in Fig. 7.4.3).
(ii) The boundary nodes are assigned the ambient pressure value
(iii) All other nodes are equated to zero

The process of iterations will be repeated till the convergence criterion, (Eq. 7.3.10), is satisfied.

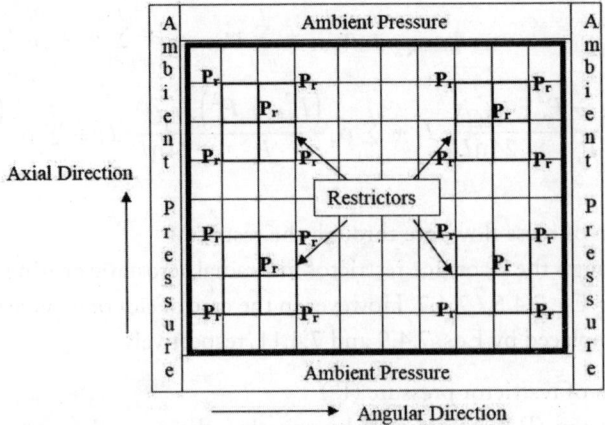

Fig. 7.4.3   Pressure distribution

Step 8: Estimation of load carrying capacity

The dimensional load carrying capacity for the journal bearing can be obtained by integrating fluid pressure over the full surface of the journal. The load capacity ($W_{journal}$) is given by

$$W_{journal} = \int_0^{2\pi}\int_0^L (p - p_a) R\, d\theta\, dz \qquad (7.4.19)$$

Equation 7.3.19 can be rewritten for numerical integration of pressure,

$$W_{journal} = \left(R\, \Delta\theta\, \Delta z\right) \sum_{i=1}^{N} \sum_{j=1}^{M} \left(p_{i,j} - p_a\right)$$

7.4.3   Thrust aerostatic bearings

The procedure for estimating the load carrying capacity of an aerostatic and hydrostatic bearing is similar, but only difference is the restrictor pressure in the case of hydrostatic bearing is calculated based on volumetric flow rate of liquid while in the case of aerostatic bearing, mass flow rate of the air is considered.

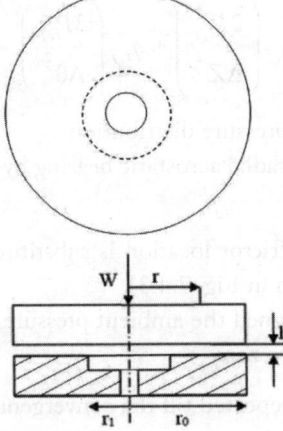

Fig. 7.4.4   Circular thrust gas bearing

The volumetric flow rate of liquid supplied in a circular thrust bearing (shown in Fig. 7.4.4) was discussed in section 5.1.5 of chapter 5. In the case of aerostatic bearings, conservation of mass is considered instead of conservation of volume. The mass flow rate of the bearing is estimated by multiplying the density (ρ) of the gas to the volumetric flow rate (Q) and equation is presented below:

$$m = \rho Q = -\rho \frac{h^3}{12\eta} \frac{dP}{dr} 2\pi r \qquad (7.4.20)$$

By mass continuity condition:

$$m_{restricotr} = m_{bearing}$$

$$\Rightarrow \qquad \rho_r Q_r = \rho Q$$

$$\Rightarrow \qquad \rho_r = \frac{\rho Q}{Q_r} \qquad (7.4.21)$$

From continuity consideration and for isothermal condition, the flow through restrictor must be equal to flow in the bearing

$$P/\rho = P_r/\rho_r = cons$$

$$\Rightarrow \qquad \rho_r = \frac{P_r \rho}{P} \qquad (7.4.22)$$

Substituitng Eq. 7.4.21 in Eq. 7.4.22

$$\frac{\rho Q}{Q_r} = \frac{P_r \rho}{P}$$

$$\Rightarrow \qquad P\left(-\frac{h^3}{12\eta}\frac{dP}{dr}2\pi r\right) = P_r Q_r$$

$$\Rightarrow \qquad PdP = -P_r Q_r \frac{12\eta}{h^3}\left(\frac{1}{2\pi r}dr\right)$$

Integrating on both sides:

$$\int_{P_r}^{P_a} PdP = -P_r Q_r \frac{12\eta}{h^3}\left(\int_{r_1}^{r_0}\frac{1}{2\pi r}dr\right)$$

$$\Rightarrow \qquad \frac{P_a^2 - P_r^2}{2} = -P_r Q_r \frac{12\eta}{2\pi h^3}\ln\frac{r_0}{r_1}$$

$$\Rightarrow \qquad Q_r = \frac{P_r^2 - P_a^2}{P_r}\frac{\pi h^3}{12\eta \ln\frac{r_0}{r_1}}$$

The mass flow rate of flow through the bearing is

$$m_r = \rho_a \frac{\pi h^3}{12\eta \ln\left(\frac{r_0}{r_1}\right)}\frac{\left(P_r^2 - P_a^2\right)}{P_r} \qquad (7.4.23)$$

For continuity consideration flow through the orifice should be equal to the bearing outflow. Flow through a single orifice as a restrictor is given in Eq. 7.4.24

$$P_a \frac{\pi h^3}{12\eta \ln\left(\frac{r_0}{r_1}\right)} \frac{\left(P_r^2 - P_a^2\right)}{P_r} - C_d A_0 P_s \left(\frac{2}{\Re T_s} \frac{k}{k-1}\left[(K_s)^{2/k} - (K_s)^{k+1/k}\right]\right)^{1/2} = 0 \qquad (7.4.24\text{ a})$$

For chocked flow:

$$P_a \frac{\pi h^3}{12\eta \ln\left(\frac{r_0}{r_1}\right)} \frac{\left(P_r^2 - P_a^2\right)}{P_r} - C_d A_0 P_s \left(\left(\frac{k}{\Re T_s}\right)^{1/2}\left(\frac{2}{k+1}\right)^{\frac{k+1}{2(k-1)}}\right) = 0 \qquad (7.4.24\text{ b})$$

Equation (7.4.24) can be solved numerically by Newton-Raphson method in order to obtain the pressure at the downstream of the orifice for different values of restrictor parameters (area of the restrictor, supply pressure, surface are of the thrust bearing etc.).

The load bearing capacity for the aerostatic thrust bearing shown in Fig. 7.4.4 can be estimated by the same procedure followed for hydrostatic bearing, discussed in section 5.1.5 chapter 5. For convenience, the equations are re-presented:

$$W = P_r \pi r_i^2 - P_a \pi r_0^2 + \int_{r_i}^{r_0} 2\pi r P dr \qquad (7.4.25)$$

If it is assumed that the pressure drops linearly from r_i to r_o, $P = P_r - P_a$, $r_i < r < r_0$

$$P = P_a - (P_a - P_r)\left(\frac{r_0 - r}{r_0 - r_i}\right)$$

$$W = \pi(P_r - P_a)r_i^2 + \int_{r_i}^{r_0} 2\pi r(P - P_a) dr$$

$$\Rightarrow \qquad W = \pi r_i^2(P_r - P_a) + 2\pi \int_{r_i}^{r_0} r\left(-(P_a - P_r)\left(\frac{r_0 - r}{r_0 - r_i}\right)\right) dr$$

$$W = \pi(P_r - P_a)r_i^2 - \frac{2\pi(P_a - P_r)}{(r_0 - r_i)}\left|\left(\frac{r^2}{2}r_0 - \frac{r^3}{3}\right)\right|_{r_i}^{r_0}$$

$$W = \pi(P_r - P_a)(r_i^2) - \pi\frac{(P_a - P_r)}{3(r_0 - r_i)}\left[3r_0^3 - 2r_0^3 - 3r_0 r_i^2 + 2r_i^3\right]$$

$$W = \frac{\pi(P_r - P_a)}{3(r_0 - r_i)}\left[3(r_i^2)(r_0 - r_i) + \left[r_0^3 - 3r_0 r_i^2 + 2r_i^3\right]\right]$$

$$W = \frac{\pi(P_r - P_a)}{3(r_0 - r_i)}\left[3r_0 r_i^2 - 3r_i^3 + r_0^3 - 3r_0 r_i^2 + 2r_i^3\right]$$

$$W = \frac{\pi(P_r - P_a)}{3(r_0 - r_i)}\left[-r_i^3 + r_0^3\right]$$

$$W = \frac{\pi}{3}(P_r - P_a)\left(\frac{r_0^3 - r_i^3}{r_0 - r_i}\right) \qquad (7.4.26)$$

Frequently Asked Questions

Q.1. List down the advantages of air lubricated bearings over liquid lubricated bearings.

Ans. The advantages of air bearing over liquid lubricated bearings are:
- Very low viscosity of air lubricant due to which power loss in air bearing is negligible and heat generated due to lubricant shearing, compared to liquid-lubricated bearings is insignificant.
- No requirement of oil lubrication. This means 'contamination free operation' and 'no limit on operating temperature' from lubricant point of view.
- Gas bearings do not face problems related to cavitation, sealing and leakage that are dealt in the liquid lubricated bearings.

Q.2. Why does load carrying capacity of air lubricated bearing improve (compared to liquid lubricated bearings) with increase in operating temperature?

Ans. The load carrying capacity of any bearing increases in proportion to the viscosity of the lubricant used. The viscosity of gases increases with increase in temperature due to the thickening effect. This means, gas–lubrication provides better results at higher temperature compared to that of liquid–lubrication, which exhibits thermal thinning behaviour (i.e., decrease in viscosity with increase in temperature). This difference in the behaviour of viscosities of gases and liquids with change in temperature occurs due to the dominant role of momentum transfer in gases.

Q.3. What are the disadvantages of air bearing?

Ans. Air bearings have low load carrying capacity, poor damping, and expensive fabrication. The difficulty in manufacturing and assembly of air bearings arises due to the functional requirement of very small gap between tribo–pairs and that too without any mechanical contact anywhere along the length/circumference at any instance of bearing operation. This means manufacturing with very close dimensional and geometric tolerances is a must for gaining the advantages from air bearing.

Q.4. What are the differences between aerostatic and aerodynamic bearings?

Ans. The main difference between aerostatic and aerodynamic bearings is that the aerostatic bearings require supply of pressurized air to levitate the imposed load, while the gas pressure is developed at the surfaces of aerodynamic bearings by action of high relative speed. Due to this difference in operating mechanism, the aerodynamic bearings should be used for light rotors operating with very high speeds, while aerostatic bearings should be used for somewhat higher loads without any constraint on its minimum speed. It is worth noting that even for aerodynamic bearings it may be necessary to supply externally pressurized gas to prevent mechanical contact at the start and end of rotation. The aerostatic bearings provide better dynamic (stiffness and damping) characteristics compared to the aerodynamic bearings.

Q.5. How are pad bearing advantages over full cylindrical bearing?

Ans. The major advantage of pad bearing over full cylindrical bearings is the absence of cross stiffness, which means free from half speed whirl instability problem.

Q.6. What is the purpose of providing pocket with orifice restrictor in the aerostatic bearings?

Ans. Providing a pocket after the orifice–restrictor ensures a constant cross-section of the gas flow and makes it independent of the operating gap.

Q.7. Write the expression for Reynolds' equation for polytropic flow in Cartesian coordinate.

Ans.
$$\frac{\partial}{\partial x}\left(\frac{p^{1/n} h^3}{\eta}\frac{\partial p}{\partial x}\right) + \frac{\partial}{\partial z}\left(\frac{p^{1/n} h^3}{\eta}\frac{\partial p}{\partial z}\right) = 6U\frac{\partial\left(p^{1/n} h\right)}{\partial x} + 12\frac{\partial\left(p^{1/n} h\right)}{\partial t}$$

where n is the polytropic gas expansion exponent. For isothermal conditions n=1.

Q.8. Which types of restrictors are most commonly used in the aerostatic bearings?

Ans. Four types of restrictors are used in the aerostatic bearing
- Pocketed orifice
- Annular orifice
- Capillary restrictor
- slot restrictor

Q.9. How hammer blow self-excited oscillations can be avoided in aerostatic bearing?

Ans. Hammer blow self-excited oscillations occur due to the large deeper pocket. To prevent this, the pocket must be shallow (about 4-7 times the film thickness).

Q.10. Explain the Knudsen number. How is the load carrying capacity of the bearing affected by Knudsen number?

Ans. The continuity equation becomes invalid if the value of minimum film thickness (h) approaches to value of the mean free path (λ) between the gas molecules. To use Reynolds' equation even in such cases, a non-dimensional number, Knudsen number is used. This number is expressed in following manner:

$$Kn = \lambda/h$$

The continuum assumption holds when $Kn < 0.01$. When $0.01 < Kn < 0.1$, slip occurs, for $0.1 < Kn < 10$ transient flow occur and for $Kn > 10$, molecular flow is thought to become fully developed.

The load capacity of bearing decreases with increase in Knudsen number as this creates discontinuities in the gas flow.

Q.11. Specify the assumptions which can be made (without significant loss in accuracy) to evaluate the performance of flying head of magnetic recording discs?

Ans. Assumptions considered for flying head of magnetic recording discs are:
- Due to negligible width of the arm of the magnetic recording device compared to its length, arm can be considered as infinitely long slider.
- Air in and out of the sides of the bearing is negligible (as shown in Fig. 7.3.5)
- The flow is isothermal

Q.12. List ten applications of air bearings.

Ans. Applications of air bearings are:
1. In textile processing machinery,
2. Food machinery,
3. Flying head in read/write devices
4. Coordinate measuring machines
5. Precision machine tools
6. Rotary tables
7. Rheometers
8. Medical machines

9. High speed spindles
10. High-speed dental drills
11. Gas–cycle machinery

Q.13. Why derivation carried out for flow rate in hydrostatic bearing cannot be used for aerostatic bearing?

Ans. In hydrostatic bearing, due to incompressible nature of liquid, conservation of volume flow holds good. In the case of aerostatic bearing, air being compressible, the conservation of volume cannot be used. It is necessary to account the variation in the air density due to change in gas-pressure. In other words, in the case of aerostatic bearing the flow of air is represented by mass flow rate, which can be obtained by multiplying density (ρ) of the fluid with volumetric flow rate (Q): i.e., m=ρQ.

Q.14. Explain how restrictor pressure is estimated in an aerostatic bearing?

Ans. The restrictor pressure is determined by the computation of flow rate. In an aerostatic bearing flow rate through restrictors is related to the fluid pressure. Finding the relationship between the pressure and flow is therefore essential, which can be derived using mass flow continuity $m_{restrictor} = m_{bearing}$. If there are several restrictors in a bearing, the sum of the restrictor flows must be equal to the total flow out of the bearing.

Q.15. Explain the procedure for calculation of chocked flow in orifice restrictors.

Ans. Chocked flow occurs when the rate of change of mass flow rate with respect to K_s is equated to zero.

$$\frac{dm}{dK_s} = \frac{1}{2}C_d A_0 P_s \left(\frac{2}{\Re T_s} \frac{k}{k-1} \left[(K_s)^{2/k} - (K_s)^{k+1/k} \right] \right)^{-1/2} \left(\frac{2}{k}(K_s)^{\frac{2-k}{k}} - \frac{k+1}{k}(K_s)^{\frac{1}{k}} \right)$$

$$0 = \left(\frac{2}{k}(K_s)^{\frac{2-k}{k}} - \frac{k+1}{k}(K_s)^{\frac{1}{k}} \right)$$

$$\frac{2}{k}(K_s)^{\frac{2-k}{k}} = \frac{k+1}{k}(K_s)^{\frac{1}{k}}$$

$$K_s = \left(\frac{2}{k-1} \right)^{\frac{k}{k+1}}$$

Multiple Choice Questions

Q.1. The viscosity of the air
(a) Increases with temperature
(b) Decreases with temperature
(c) Not effected with temperature
(d) Depends on the temperature at which the air bearing is operated

Q.2. Choose the right statement
(a) Air bearing is designed only for radial load
(b) Air bearing is designed only for Axial load
(c) Air bearing can be designed for both axial and radial loads
(d) Air bearing cannot deal with axial or radial load

Q.3. Gas lubricated bearings provide:
(a) Higher static load carrying capacity than liquid lubricated bearing.
(b) Higher dynamic load carrying capacity than liquid lubricated bearings.
(c) Load capacity equal to load carrying capacity of liquid lubricated bearings.
(d) Lesser static and dynamic load carrying capacities compared to liquid lubricated bearings.

Q.4. Thermal thickening of gases at higher temperature is due to.
(a) Momentum transfer
(b) intermolecular forces
(c) Both intermolecular forces and Momentum transfer
(d) Sometime it is due to momentum transfer and sometimes it is due to intermolecular forces.

Q.5. Air bearings have
(a) Good damping
(b) Higher Load capacity
(c) Low friction
(d) Low fabrication

Q.6. Which one of the following is correct for isothermal gases:
(a) $p = \rho^2 RT$
(b) $p = \rho^{1.4} RT$
(c) $p = \rho RT$
(d) $p\rho = nRT$

Q.7. Aerodynamic bearing must be used for
(a) Low mass rotor
(b) Very small gap between tribo–pairs
(c) High speed
(d) All of the above

Q.8. The Reynolds' equation can be used to evaluate the pressure in gas bearings:
(a) Only if the mean free path is lesser than 0.01 times of the minimum film thickness
(b) Only if the mean free path is lesser than 0.1 times of the minimum film thickness
(c) Only if the mean free path is equal to the minimum film thickness
(d) By incorporating necessary modification factor(s)

Q.9. Load carrying capacity of gas bearings increases with:
(a) increase in K_n
(b) Decrease with K_n
(c) Decrease with K_n but below certain value of K_n load capacity does not depend on K_n
(d) Independent of K_n

Q.10. Correct formula for coefficient of Poiseuille flow in terms of Knudsen number is
(a) $Q_p = 0.1477 + 1.0162 + 0.40134 \ln (1 + 1.477 K_n)$
(b) $Q_p = 0.1477/K_n$
(c) $Q_p = 0.1477/K_n + 1.0162 + 0.40134 \ln (1.477 K_n)$
(d) $Q_p = 0.1477 K_n$

Q.11. Select the most appropriate sentence. Gas bearings are used
(a) where precision in shaft position and consistency of torque are critical
(b) where precision in shaft position is critical
(c) where consistency of torque is critical
(d) where there is need to support the shaft, irrespective of any requirement on shaft position and friction torque.

Q.12. In the case of zero relative velocity, which one of following will be followed to derive the restrictor pressure for the aerostatic bearing:
(a) Conservation of mass
(b) Conservation of volume
(c) Conservation of momentum
(d) All the above

Q.13. In aerostatic bearing, hammer blow self-excited oscillations occurs when
  (a) Restrictor have large deep pocket and larger film thickness
  (b) Restrictor have Shallow pocket and larger thickness
  (c) Restrictor have large deep pocket and smaller film thickness
  (d) None of the above

Q.14. The condition for chocked flow in orifice restrictors is expressed by:
  (a) $K_s = \left(\dfrac{2}{k-1}\right)^{\frac{k}{k+1}}$
  (b) $K_s = \left(\dfrac{2}{k+1}\right)^{\frac{k}{k+1}}$
  (c) $K_s = \left(\dfrac{2}{k-1}\right)^{\frac{k}{k-1}}$
  (d) None of the above

Q.15. For a capillary restrictors, the ratio of depth of the pocket to the diameter of pocket must be
  (a) Greater than 20
  (b) Lesser than 10
  (c) Lesser than 20 but greater than 10
  (d) In the range of 12–18

Answers

Q.1. (a)   Q.2. (c)   Q.3. (d)   Q.4. (a)   Q.5. (c)   Q.6. (c)
Q.7. (d)   Q.8. (d)   Q.9. (c)   Q.10. (c) Q.11. (a) Q.12. (a)
Q.13. (c) Q.14. (a) Q.15. (a)

References

Booser, E. R. 1984. *Handbook of Lubrication, Theory and Practice of Tribology.* Vol II. Theory and Design, CRC Press.

Brunner, R. K. and Harker, J. M. 1959. 'A Gas Film Lubrication Study Part III: Experimental Investigation of Pivoted Slider Bearings.' *IBM Journal of Research and Development.* Vol. 8: pp. 260–274.

Burgdorfer, A. 1959. 'The Influence of the Molecular Mean Free Path on the Performance of Hydrodynamic Gas Lubricated Bearings.' *ASME J. Basic Eng.* Vol. 81: pp. 94–100.

Ding-Wen Yang. Cheng-Hsien Chen. Yuan Kang. Ren-Ming Hwang. Shyh-ShyongShyr. August 2009. 'Influence of orifices on stability of rotor–aerostatic bearing system.' *Tribology International.* Volume 42. Issue 8: Pages 1206–1219.

Grewal S. S. 'An investigation of externally pressurised orifice-compensated air journal bearings with particular reference to misalignment and inner orifice variations.' *CNAA PhD thesis.* Liverpool Polytechnic.

Gross, W. 1980. *Fluid Film Lubrication.* Wiley Interscience.

Hsia, Y. T. and Domoto, G. A. 1983. 'An Experimental Investigation of Molecular Rarefaction Effects in Gas Lubricated Bearings at Ultra-low Clearances.' *ASME J. Lubr. Technol.* Vol. 105: pp. 120–130.

Huang, P. Niu, R. J. and Hu, H. H. 2008. 'A New Numerical Method to Solve Modified Reynolds' Equation for Magnetic Head/Disk Working in Ultra Thin Gas Films,' *Sci. China, Ser. E: Technol. Sci.* Vol. 51. No. 4: pp. 337–480.

Krzysztof, C. March 1994. 'Stability of high stiffness gas journal bearings.' *Wear.* Volume 172. Issue 2: , Pages 175–183.

Maxwell, J. 1879. 'On Stress in Rarefied Gases Arising from In-equalities of Temperature.' *Philos. Trans. R. Soc. London.* Vol. 170: , pp. 231–261.

MengYonggang. 2014. 'Physics and Chemistry of Micro-Nanotribology- Chapter 6 Gas Lubrication in Nano-Gap.' *ASTM International.*, pp.96–115.

Mitsuya, Y. 1993. 'Modified Reynolds' Equation for Ultra-thin Film Gas Lubrication Using 1.5-order Slip–Flow Model and Considering Surface Accommodation Coefficient.' *ASME J. Tbol.* Vol. 115: pp. 289–294.

Neale, M. J. 2001. *The Tribology Handbook*. Second Edition. Butterworth Heinemann Press.

Pande, S. S. and Somasundaram, S. 1981. 'Analysis of an aerostatic journal bearing with a position-sensing restrictor.' *Wear*, 70: 141–154.

Rangwala, A. S. 2006. 'Reciprocating Machinery Dynamics.' *New Age International Publisher*. New Delhi.

Szeri, A. Z. 1980. 'Tribology.' *Friction, Lubrication and Wear*. Hemisphere. New York.

Program Listing in MATLAB for Figure 7.3.7

```
%To estimate the non-dimensional pressure vs non-dimensional length for
    different Knudsen number.
% N and M =Number of division of along length
%Kn= Knudsen number
%lamda=6*eta*w*(R/C)2/Pa
%eta=viscosity of oil (Pas)
%w=angular velocity (rpm)
%R=Radius (m)
%C=Clearance (m)
%Pa=Atmospheric pressure (N/mm^2)
%h1 and h2 = film thickness (m)
%P=nodal pressure
%h0= minimum film thickness (m)


clear all
N=100;
Kn=0.167;
delx = 1/N;
lamda=1.25*2;
h2=0.5e-6;
h1=h2*2;
P(1)=1;
P(N)=1;
for I=2:N-1
P(I)=0;
end

for I=1:N
q(I)=0;
h0(I)=(h1-h2*delx*I)/h2;
Q=1+6.88*Kn+2.7*Kn*log10(1+1.477*Kn);   %eqn 7.2.2
q(I)=Q.*h0(I).^3;                                        % eqn in page 302
x(I)=delx*I;
end
```

```
    ITER=10000;     % more number of division more number of Iterations

    for K = 1:ITER
      sumij=0 ;
    for I=2:N-1
      A=q(I).*(P(I+1)+P(I-1))./delx.^2;
      B=q(I).*((P(I+1)-P(I-1))/2/delx)^2;
      C=lamda*h0(I).*(P(I+1)-P(I-1))/delx;
      D=lamda*(h0(I+1)-h0(I-1))/delx;
      E=((q(I+1)-q(I))/delx).*((P(I+1)-P(I-1))/2/delx);
      F=2*q(I)/delx.^2;
    P(I)=(A+B-C)./(D-E+F);
    sumij=sumij+P(I);

    if P(I)<0
        P(I)=0;
    else
        P(I)=P(I);
    end
    end
    sum(K+1)=sumij;
    percentage = abs(sum(K+1)-sum(K))/abs(sum(K+1));
    if percentage < 1e-6
        break
        end
    end
    hold on
    plot(x,P,'r')
    xlabel('X','FontName','Times New Roman','fontsize',16);
    ylabel('P/Pa','FontName','Times New Roman','fontsize',16);
```

Chapter 8

Mixed Lubrication

Many machine components like gears, cams, heavily loaded sliding bearings, etc., operate under mixed lubrication regime, which occurs when the applied load is shared between the full-film lubrication and asperity contact regions. In the asperity contact regions, only a small fraction of the total area supports the major fraction of load, and as a result very high compressive stresses are induced at the contacting asperities. To predict the real contact area, finding the stresses at asperity contacts and relating the same to the friction and wear is very difficult task. The performance of these tribo–systems is limited due to the damages caused by wear, scuffing, etc. The key feature of mixed lubrication is the presence of asperity interactions at several locations along with separation of surfaces by thin lubricant film at numerous other locations, their individual proportions being dependent upon the operating conditions. Due to the presence of asperity contacts and very thin lubricant film, the surface roughness assumes significance in determination of the tribological performance in mixed lubrication. This chapter covers the current understanding of the mixed lubrication.

8.1 Introduction

In many applications like gears, cams, valves, the operating conditions warrant the operation to lie in mixed lubrication regime. In addition, under very heavy loads, slow relative speed, insufficient surface area and scarcity of lubricant, the formation of thick film necessary for hydrodynamic lubrication becomes difficult and interacting surfaces contact each other at several locations and tribo–pairs operate in *mixed lubrication*. This lubrication regime has been depicted as zone 2 on a Stribeck curve described in Fig. 1.7.13 (chapter 1).

The sketch shown in Fig. 8.1 depicts a model of mixed lubrication regime, in which the applied load is shared by the fluid film and the asperity contacts. Two distinct regions: first, where fluid film separates the tribo–pairs and second where the asperities are in contact are shown in the Fig. 8.1. The asperity contacts under relative speed results in wear. The worn out particles may also scratch the contacting surfaces and increase wear rate.

Many researchers [Lu and Khonsari, 2005; Khonsari and Booser, 2006; Moshkovich et al., 2011] have proposed different methods of identifying the operative lubricating regime. However, the most commonly used method is to specify the specific film thickness (Λ) [Khonsari and Booser, 2006]. This term has been described in section 1.5.2 and expressed by equation 1.4 (chapter 1). As per Khonsari and Booser [Khonsari and Booser, 2006] the specific film thickness (Λ) should be in the range of 3–10 for tribo–pairs to operate in the full film lubrication. In cases of misalignment, thermal and elastic deformation of the shaft, Khonsari and Booser [Khonsari and Booser, 2006] raised this value from 10 to 20.

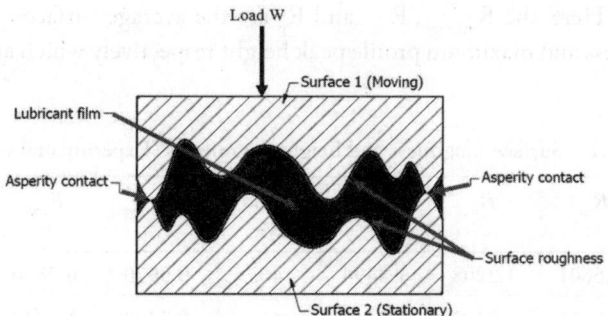

Fig. 8.1 Model of mixed lubrication

To demonstrate the conditions under which mixed lubrication occurs, consider example 3 of chapter 3. In that example, the journal bearing with the following data was considered: Dynamic viscosity of lubricant, η = 20 mPa.s; journal diameter, D = 50 mm; bearing length, L = 50 mm; journal speed, N = 1440 rpm; load, W = 1000 N and radial clearance, c = 0.00005 m. The eccentricity ratio for the bearing was estimated to be 0.36. This results in a value of minimum film thickness (h_{min}=0.64*c) equals to 32 μm. If the surface roughness of the journal and bearing are given as 0.25 μm and 0.50 μm, respectively, then the specific film thickness as per Eq. 1.4 of chapter 1 is 57.2433. This indicates that the journal bearing operates under full film (hydrodynamic) lubrication regime. However, if for the same bearing dimensions, the load is increased ten times (10,000 N) the previous value and speed is reduced to half (720 rpm), of the previous value, then the value of (ho/c) is 0.08 (obtained from Fig. 3.5.2, corresponding to the Sommerfeld Number of 0.015). This change in the operating conditions reduces the minimum film thickness to 4 μm. The specific film thickness corresponding to this minimum film thickness of 4 μm is determined to be 7.15. Since the specific film thickness value falls in the range of 3–10, there is a possibility of full film lubrication, but there is also a probability of mixed lubrication regime.

If for the same example, different surface roughness of the journal and bearing are considered (say, for example $R_{rms,journal}$ = 1.6 μm, $R_{rms,bearing}$ = 1.6 μm for journal and bearing respectively) and the bearing operates in the same conditions with the minimum film thickness of 4 μm) then, the specific film thickness is determined to be 1.77, which is lesser than 3. This indicates that the bearing operative regime will be mixed lubrication.

This indicates a strong dependence of the mixed lubrication on the surface roughness of contacting surfaces, but it is difficult to ascertain the lubrication regime with any certainty using the concept of specific film thickness. The specific film thickness, even though is a very important parameter to identify the regime of operation of a journal bearing, but it does not conclusively establish the operative lubrication regime. Sharma et al. [Sharma et al., 2009] suggested specific film thickness in the range 1.5–3 for mixed lubrication regime and specific film thickness >10 for hydrodynamic operative regime. Khonsari and Booser [Khonsari and Booser, 2006] suggested specific film thickness < 3 for mixed lubrication regime and specific film thickness in the range 3–10 for hydrodynamic operative regime. Since different researchers have used a different value of this parameter, experimental verification of the operative regime of bearing is essential.

In order to understand the significance and limitations of the specific film thickness parameter in the determination of lubrication regime, the roughness of several bearing surfaces, which were finished by final surface grinding process, was measured using the stylus method and the observations are listed in table 8.1. Here, the $R_{average}$, R_{rms} and R_p are the average surface roughness, root mean square surface roughness and maximum profile peak height respectively which are described in detail in the next section.

Table 8.1 Surface roughness and height parameters (Experimental values)

Serial No.	$R_{average}$	R_{rms}	R_p	R_p/R_{rms}	Serial No.	$R_{average}$	R_{rms}	R_p	R_p/R_{rms}
1.	0.5376	0.6601	1.2208	1.8494	40.	0.5970	0.7830	2.2322	2.8508
2.	0.5248	0.6561	1.2102	1.8445	41.	0.7498	1.0849	2.4988	2.3033
3.	0.3642	0.4613	1.1955	2.5916	42.	0.7387	1.0543	2.6760	2.5382
4.	0.3700	0.4443	0.9126	2.0540	43.	0.6850	0.9847	2.4651	2.5034
5.	0.2435	0.2984	0.6769	2.2684	44.	0.5972	0.7705	1.9639	2.5489
6.	0.3009	0.3637	0.8841	2.4308	45.	0.5514	0.7024	1.7410	2.4786
7.	0.2378	0.2899	0.6939	2.3936	46.	0.5752	0.7261	2.0214	2.7839
8.	0.3037	0.3710	0.9751	2.6283	47.	0.5656	0.7301	1.9761	2.7066
9.	0.5041	0.6265	1.1127	1.7761	48.	0.6458	0.8403	2.1174	2.5198
10.	0.2597	0.3276	0.6617	2.0198	49.	0.6025	0.7749	1.9771	2.5514
11.	0.3061	0.4135	0.6903	1.6694	50.	0.6166	0.7832	2.0780	2.6532
12.	0.2448	0.3298	0.6167	1.8699	51.	0.6421	0.8287	2.2744	2.7445
13.	0.3017	0.3740	1.0029	2.6816	52.	0.5859	0.7571	2.0667	2.7298
14.	0.3030	0.3783	1.1563	3.0566	53.	0.5779	0.7281	1.8502	2.5411
15.	0.3373	0.4341	1.0415	2.3992	54.	0.5866	0.7471	1.7757	2.3768
16.	0.2880	0.3477	0.8278	2.3808	55.	0.5678	0.7136	2.0018	2.8052
17.	**0.3945**	**0.6698**	**2.6182**	**3.9089**	56.	0.6290	0.8011	2.0481	2.5566
18.	0.8034	1.0023	2.6626	2.6565	57.	0.6246	0.8234	2.2315	2.7101
19.	0.6945	0.8845	2.3229	2.6262	58.	0.5864	0.7637	2.0557	2.6918
20.	0.6676	0.8465	2.2172	2.6193	59.	0.6576	0.8383	1.9533	2.3301
21.	0.5171	0.6653	1.8061	2.7147	60.	0.6300	0.8088	2.1430	2.6496
22.	0.6411	0.8148	2.0146	2.4725	61.	0.6145	0.7796	2.0658	2.6498
23.	0.6193	0.7905	2.1388	2.7056	62.	0.5342	0.7123	1.8090	2.5397
24.	0.6305	0.8154	2.3547	2.8878	63.	0.6773	0.8665	2.0569	2.3738
25.	0.6247	0.7984	2.0968	2.6263	64.	0.6353	0.8116	2.0554	2.5325
26.	0.6079	0.7850	1.9173	2.4424	65.	0.6961	0.8847	2.0693	2.3390
27.	0.6616	0.8549	2.3859	2.7909	66.	0.7874	1.3442	3.0172	2.2446
28.	0.6658	0.8467	1.9014	2.2457	67.	0.7361	0.9312	2.2384	2.4038
29.	0.7525	0.9613	2.3942	2.4906	68.	0.6323	0.8037	2.2032	2.7413

Contd.

Contd.

Serial No.	$R_{average}$	R_{rms}	R_p	$R_p R_{rms}$	Serial No.	$R_{average}$	R_{rms}	R_p	$R_p R_{rms}$
30.	**0.7300**	**0.9488**	**3.0199**	**3.1829**	69.	0.6240	0.7929	1.9960	2.5173
31.	0.7822	0.9988	2.6663	2.6695	70.	0.6502	0.8295	2.2307	2.6892
32.	0.7569	0.9802	2.6867	2.7410	71.	0.6295	0.7993	2.2900	2.8650
33.	0.5715	0.7225	1.8989	2.6282	72.	0.6420	0.8217	2.1432	2.6083
34.	0.5816	0.7313	1.8287	2.5006	73.	0.7829	0.9786	2.4578	2.5115
35.	0.5881	0.7425	1.9366	2.6082	74.	0.8059	1.0039	2.3131	2.3041
36.	0.5962	0.7728	2.0006	2.5888	75.	0.7879	0.9871	2.2949	2.3249
37.	0.5837	0.7501	2.0228	2.6967	76.	0.7485	0.9480	2.4153	2.5478
38.	0.6086	0.7656	1.8349	2.3967	77.	0.8034	1.0023	2.6626	2.6565
39.	0.6283	0.7969	1.9776	2.4816					

It is observed from Table 8.1 that, on an average, the highest asperity peaks are about 2.5 times the R_{rms} surface roughness values, with few even going as high as 3.18 and 3.90. In such a situation even a specific film thickness greater than 3 [Khonsari and Booser, 2006] will not be sufficient enough to ensure full film lubrication and the asperity interaction will take place and mixed lubrication condition will exist. This situation exists when no other uncertainties (like shaft misalignment, etc.,) are considered. If all the uncertainties are considered, then it will further require a greater value of specific film thickness ensuring full film lubrication. Therefore, in the present book, specific film thickness greater than 5 has been considered as requirement for full film lubrication. This is substantiated by the fact that there is a variation in surface roughness obtained by different machining/finishing processes and due to these errors in manufacturing the height of highest asperity peaks is increased. Some manufacturing errors result in the highest asperity peaks, which may be even more than 2.5 times the R_{rms} surface roughness. As a representative example, one such measurement is depicted in Fig. 8.2 corresponding to the surface roughness given at Serial No. 17 in Table 8.1.

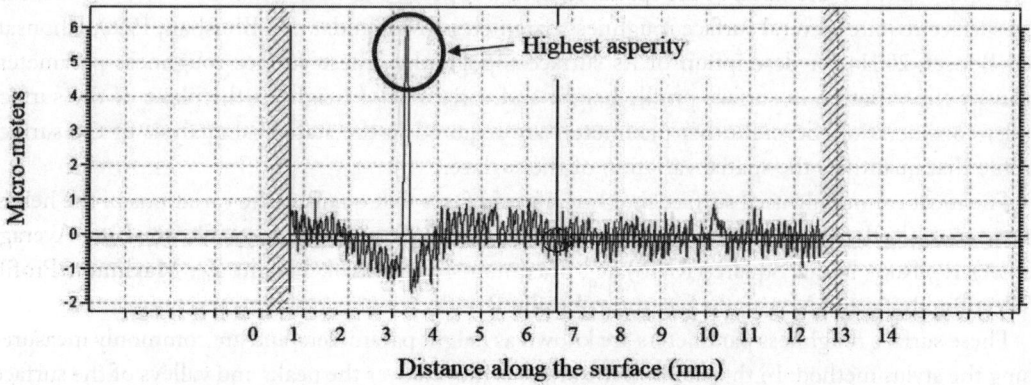

Fig. 8.2 Surface roughness profile of bearing (corresponding to data at Serial No. 17)

The surfaces, obtained by different methods of manufacturing, contain irregularities or deviations from the desired geometric form. Figure 8.3 gives the general range of the surface roughness values obtained by using different production methods but the actual surface roughness may contain asperities that may be higher than that given in Fig. 8.3.

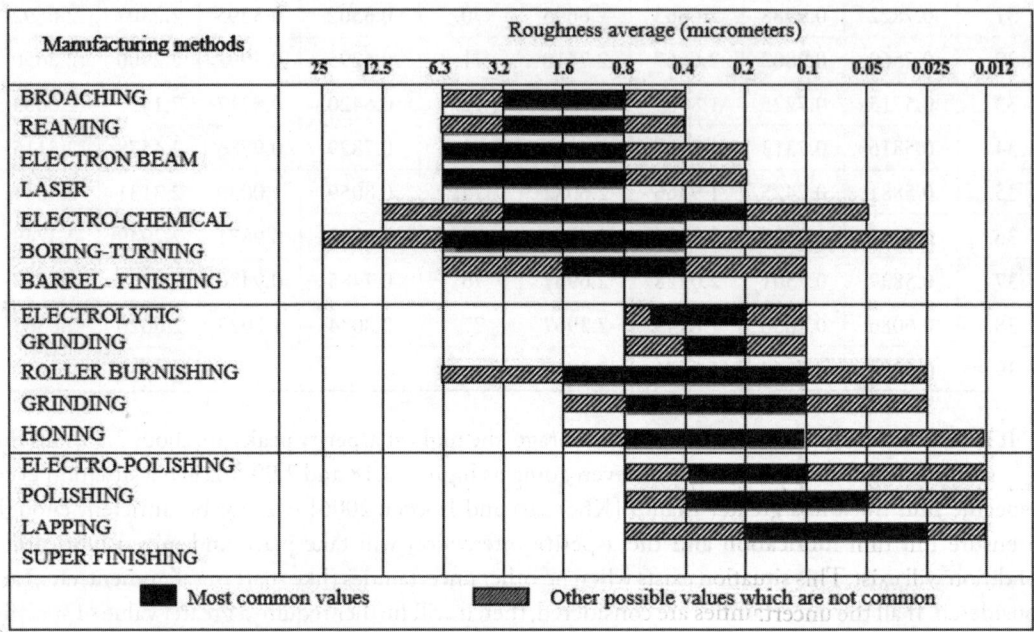

Fig. 8.3 Surface roughness produced by common production methods [ANSI B46.1-1985]

Therefore, each surface needs to be investigated separately and the general range of the surface roughness as represented in Fig. 8.3 is to be used as a general guideline only.

8.2 Surface Topography

The topography represents all of the spatial structure of peaks and valleys that exist on a surface. The real surface requires several surface roughness parameters [Majumdar and Bhushan, 1990; Khonsari and Booser, 2001] for description of its surface topography. These surface roughness parameters quantify the variation in surface profile height and variations of height in the plane of the surface under consideration. Several other parameters are required in the statistical analysis of the surface profile that quantifies the spatial variation of the surface.

The most commonly used surface roughness parameters that quantify the variations in the height of the surface relative to a reference plane are: Arithmetic Average (AA) or Centre Line Average (CLA) R_a, Root Mean Square (RMS) R_q, Maximum Profile Peak Height Z_p, Maximum Profile Valley Depth Z_v, and Maximum Height of Profile R_z.

These surface roughness parameters are known as height parameters, and are commonly measured using the stylus method. In this method the stylus is moved over the peaks and valleys of the surface. The vertical motion of the stylus is converted into an electrical signal by the help of a transducer and is analyzed to determine the roughness parameters, as shown in Fig. 8.4.

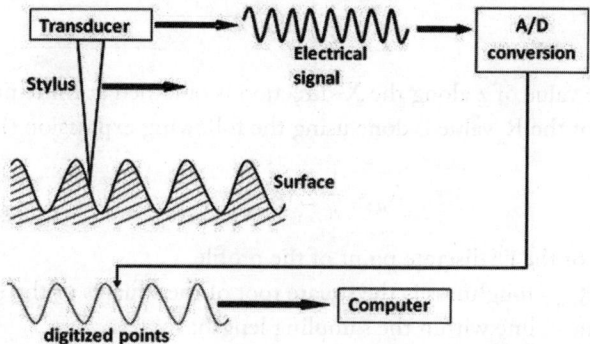

Fig. 8.4 Stylus method of surface roughness measurement

The stylus traverses through the surface for a distance equal to traversing length, in which the data at the beginning and end of the stylus motion is ignored to reduce the error at the starting and stopping. This means more sampling data is collected and data collected at the start and end of stylus are rejected. To determine statistically reliable surface roughness parameters, averaging of roughness data over five sampling lengths is performed, as shown in Fig. 8.5.

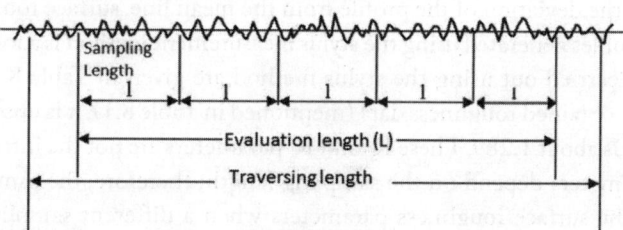

Fig. 8.5 Standard method to estimate roughness parameters

The average roughness ($R_{average}$ or R_a) is the arithmetic mean deviation of the surface height from mean line through the profile. The mean line is defined as the one having equal areas of the profile above and below it, as shown in Fig. 8.6.

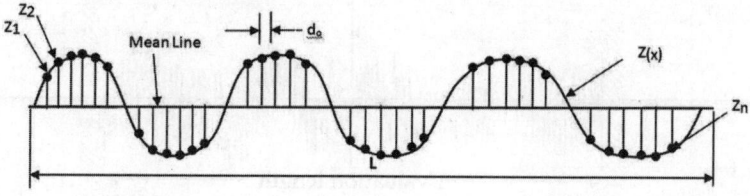

Fig. 8.6 Discretization of surface to find $R_{average}$ or R_a

$$R_a = CLA = AA = \frac{1}{L}\int_0^L |z(x) - m|\,dx \qquad (8.1)$$

(where CLA and AA are abbreviations for Centre Line Average and Arithmetic Average, respectively), and

$$m = \frac{1}{L}\int_0^L z(x)\,dx \qquad (8.2)$$

However, since the value of **z** along the **X**-direction is obtained at finite number of locations, the actual quantification of the R_a value is done using the following expression (in discrete form):

$$R_a = \frac{1}{n}\sum_{i=1}^{n}(z_i - m) \qquad (8.3)$$

where z_i is the height of the ith discrete point of the profile.

The RMS (R_q or R_{rms}) roughness is the square root of the squares of the individual deviations of the profile from the mean line within the sampling length:

$$R_q \text{ or } R_{rms} = \sqrt{\frac{1}{L}\left(\int_0^L [z(x)-m]^2\,dx\right)} \qquad (8.4)$$

However, since the value of z along the **X**-direction is obtained at finite number of locations, the actual quantification of the R_q value is done using the following expression (in discrete form):

$$R_q^2 = \frac{1}{n}\sum_{i=1}^{n}(z_i - m)^2 \qquad (8.5)$$

With increase in the deviation of the profile from the mean line, surface roughness increases. The surface roughness profiles generated using the stylus measurement method is shown in Fig. 8.7. Several such measurements carried out using the stylus method are given in Table 8.1. On the analysis of these experimentally obtained roughness data (mentioned in Table 8.1), it is observed that the average ratio of $R_{rms}/R_{average}$ is about 1.289. These R_q and R_a parameters are not the intrinsic properties of the profile as these parameters depend on the sampling length; therefore, the same surface will exhibit different values of the surface roughness parameters when a different sampling length is used for measurement or when an instrument of a different resolution is used for measurement. According to Thomas and Sayles [Thomas and Sayles, 1978], the values of parameter increase as the square root of the sampling length. If these considerations are not accounted for or mentioned in the results, proper characterization of the surface will not be achieved.

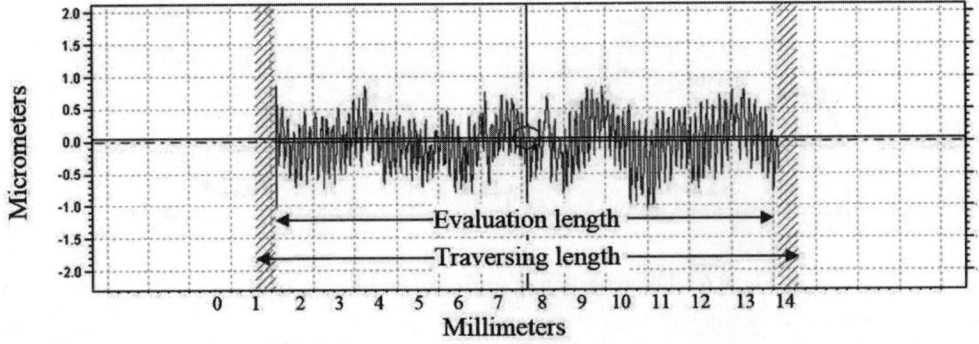

Fig. 8.7 Surface roughness profile of bearing

The stylus method is, however, not the accurate method of measurement as several measurement errors are associated with it. These errors are introduced due to the size of the stylus, stylus load, stylus speed and lateral deflection of the stylus by the asperities. These errors become more significant with the decrease in the curvature of the peaks and valleys and with the increase in the slope of peaks and valleys [Thomas, 1999]. Therefore, non-contact type of 3-D optical surface profile measurement method [Wyant et al., 1984] is becoming more common.

The other parameters like the maximum profile peak height Z_p, maximum profile valley depth Z_v, and maximum height of profile R_z are also obtained during the stylus method of measurement. These parameters do not completely characterize the surface and thus additional information is required for complete characterization of the tribo–surface. The information regarding the shape of the asperities, their slope, frequency, etc., are essential to predict the surface behaviour which affects the tribological performance. The shape and spacing parameters are used to quantify the shape of asperities and spacing among peaks and valleys of the surface. These parameters characterize the process used to generate the surface; for example, surface generated using a particular grit size of grinding wheel or using a particular value of feed in turning, etc.

The mean peak spacing S_m, as per International Organization for Standardization (ISO), is defined as the average spacing between two successive negative crossings of the profile through the mean line as shown in Fig. 8.8.

$$S_m = \frac{1}{n} \sum_{i=1}^{n} S_{mi} \tag{8.6}$$

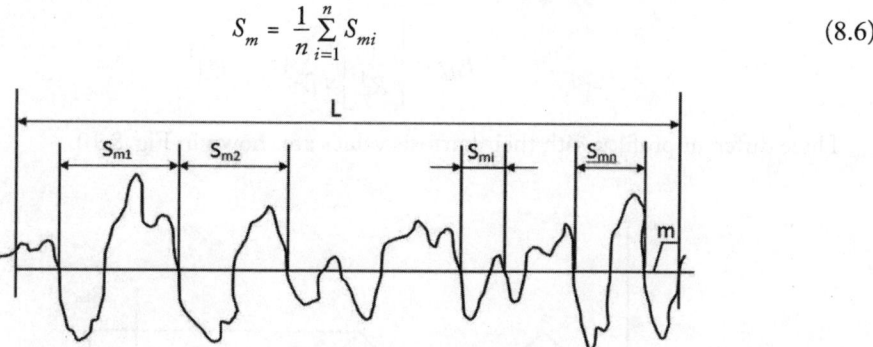

Fig. 8.8 Mean peak spacing parameter Sm [ISO 4287/1]

It is interesting to note that simple mean peak spacing parameter does not find much application in tribology. To illustrate this let us consider the surface profiles shown in Fig. 8.9, where the dash line represents the mean line.

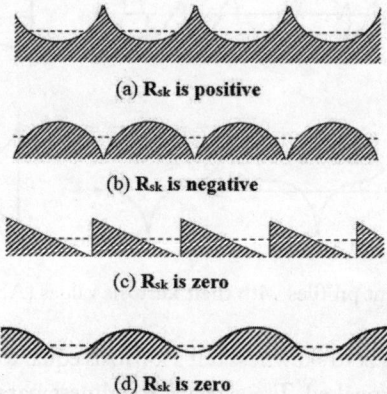

(a) R_{sk} is positive

(b) R_{sk} is negative

(c) R_{sk} is zero

(d) R_{sk} is zero

Fig. 8.9 Surface profiles having same R_a and S_m values

The profiles shown in Fig. 8.9 (b) and (d) represent a good (lesser friction and lesser wear) bearing surface whereas profiles shown in Fig. 8.9 (a) and (c) do not qualify as a good bearing surface. However all these surfaces have same R_a and S_m values. The skewness parameter R_{sk}, helps differentiate between these profiles. It defines the symmetry of the profile about the mean line and describes the asymmetry of the shape of the profile. It is expressed as:

$$R_{sk} = \left[\frac{1}{R_q^3}\right]\frac{1}{n}\sum_{i=1}^{n}(z_i - m)^3 \qquad (8.7)$$

The skewness of the profile (a) shown in Fig. 8.9 is positive, whereas for profile (b) it is negative and for profiles (c) and (d) the skewness is zero. Positive skewness generally indicates a bad bearing surface. To understand this, consider the numeric sequence (39, 40 and 41), which provides profile with zero skewness. This profile can be transformed to negative skewed distribution by replacing lowest number (39) with a value below the lowest value (i.e., 20, 40 and 41). Similarly, we can make the profile positive skewed by adding a value far above the maximum value (39, 40 and 60).

In addition to skewness, another parameter that quantifies the sharpness or peakedness of the surface profile is used. It is known as Kurtosis R_{ku}. It is expressed as:

$$R_{ku} = \left[\frac{1}{R_q^4}\right]\frac{1}{n}\sum_{i=1}^{n}(z_i - m)^4 \qquad (8.8)$$

Three different profiles with their kurtosis values are shown in Fig. 8.10.

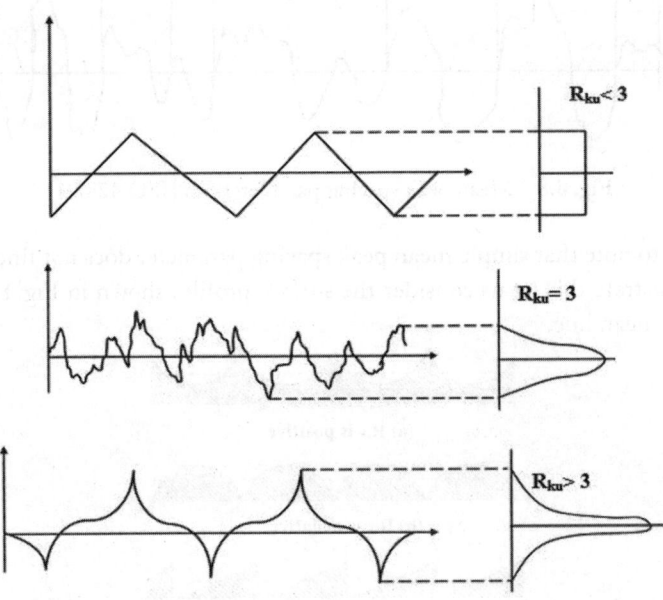

Fig. 8.10 Different profiles with their kurtosis values [ASME B46.1-1995]

The Gaussian distribution has zero skewness but a kurtosis equal to 3. A large kurtosis value reflects that the surface is more sharply peaked. The surface roughness parameters that will be required to qualify a surface to be useful for a given tribo application will be described in section 8.3.

Another surface roughness parameter that combines the height and spacing information for additional description of the surface texture is known as root mean square slope, $R_{\Delta z}$. It requires differentiation of the surface waveform. The root mean square slope, $R_{\Delta z}$, is a general measurement of the slopes which comprise the surface and may be used to differentiate surfaces with similar average roughness. It is related to the degree of surface wetting by various fluids. It is affected both by texture amplitude and spacing. Thus for a given R_a, a wider spaced texture may indicate a lower $R_{\Delta z}$ value than a surface with the same R_a but finer spaced features. The variance of slope depends strongly on the resolution of the roughness-measuring instrument, and is therefore not unique.

In addition to these parameters, statistical description of the surfaces requires the use of several other important spatial parameters. In order to justify/identify the need of such spatial parameters, consider two surface profiles which are idealized by sine wave distributions having same amplitude and different frequencies. The R_a values of these surface profiles will be same, but they will have different spatial arrangements. Under these conditions the slope distribution will not be sufficient to completely describe the spatial distribution as it will reflect only one particular spatial size of feature. Therefore, the spatial functions are employed to represent the properties of all wavelengths (spatial size of feature) of the surface. The most commonly used spatial functions are auto-covariance function (ACVF or $R_{(\tau)}$), autocorrelation function (ACF or $\rho_{(\tau)}$), structure function (SF or $s_{(\tau)}$) and power spectral density function (PSDF or $G_{(\omega)}$). They represent the properties of all wavelengths of the surface feature and are also known as surface texture descriptors [McGillem and Cooper, 1984; Bendat and Piersol, 1986]. The auto-covariance and auto-correlation functions are useful for visualizing the relative degrees of periodicity and randomness in surface profiles. These functions differentiate between the spatial surface characteristics of the surfaces by comparison of their decaying properties.

The ACVF is a measure of how well future values of the function can be predicted based on the past observations and expressed by the following equation:

$$R(\tau) = \lim_{L \to \infty} \frac{1}{L} \int_0^L z(x) z(x + \tau) dx \qquad (8.9)$$

It is an average value of the product of two measurements of the profile separated by distance τ, as shown in Fig. 8.11.

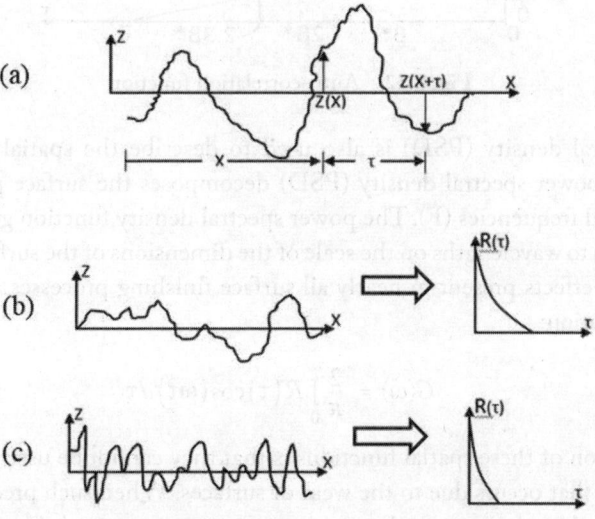

Fig. 8.11 Auto-covariance function

Figure 8.11(a) depicts the function z(x) at two locations, x and (x+ τ). Figures 8.11(b) and (c) depict the auto-variance functions for two different profiles. The profile shown in Figs. 8.11(b) and (c) differ in the horizontal distribution of the roughness and correspondingly, the exponential decay function of the profile shown in Fig. 8.11(b) shows a higher value as compared to the exponential decay function of the profile shown in Fig. 8.11(c). The length of the decay is significantly higher in the case of an open texture surface shown in Fig. 8.11(b) than a closed one as in Fig. 8.11(c). Therefore the form of the decay curve provides information on the horizontal distribution of roughness.

The normalized form of ACVF is called ACF. The normalization is done by the division of auto-covariance expression by its variance (denoted by σ^2). The normalization process will convert the numerical values of the auto-covariance to lie in the range from -1 to +1. It is shown in Fig. 8.12 and expressed by the following equation:

$$c(\tau) = \lim_{L \to \infty} \frac{1}{L\sigma^2} \int_0^L \left[z(x) - m\right]\left[z(x+\tau) - m\right]dx = \left[R(\tau) - m^2\right]/\sigma^2 \qquad (8.10)$$

Since many engineering surfaces have exponential auto-correlation function [Bhushan, 2002], it is also expressed as:

$$\rho(\tau) = e^{-\tau/\beta^*} \qquad (8.11)$$

The correlation length describes the decaying characteristics of the variation in surface profile. This correlation length describes the length over which the auto-correlation function drops to about 10% of its original value. The exponential form of the auto-correlation function has a value of 2.3β, as shown in Fig. 8.12.

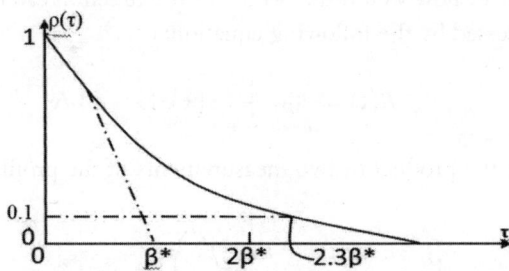

Fig. 8.12 Auto-correlation function

The power spectral density (PSD) is also used to describe the spatial representation of the surface profile. The power spectral density (PSD) decomposes the surface profile into its Fourier components or spatial frequencies (F). The power spectral density function generally increases with spatial wavelength up to wavelengths on the scale of the dimensions of the surface. This phenomenon results from random effects present in nearly all surface finishing processes. The PSD is expressed by the following equation:

$$G(\omega) = \frac{2}{\pi} \int_0^\infty R(\tau)\cos(\omega\tau)d\tau \qquad (8.12)$$

The main limitation of these spatial functions is that they cannot be used to predict the changes in the surface profile that occurs due to the wear of surfaces. When such predictions are done, even though not desirable, then the worn and unworn surfaces may appear quite similar.

8.3 Characterization of Surface

From tribological point of view, a surface is characterized from its roughness. As mentioned in previous heading, the roughness is characterized by asperities of varying amplitude and spacing.

From tribology point of view R_q (root mean square roughness) is preferred over R_a (average roughness). To exemplify it, consider the following two surfaces as shown in Figs. 8.13(a) and (b). In these figures, the value of R_a is the same for both the surfaces, but the value of R_q for the first surface (Fig. 8.13(a)) is higher as compared to R_q for the second surface (Fig. 8.13b), because R_q imposes higher penalty on peaks and valleys. Therefore, for a good bearing surface a lower value of R_q is preferred and the second surface is better than first surface from tribological point of view.

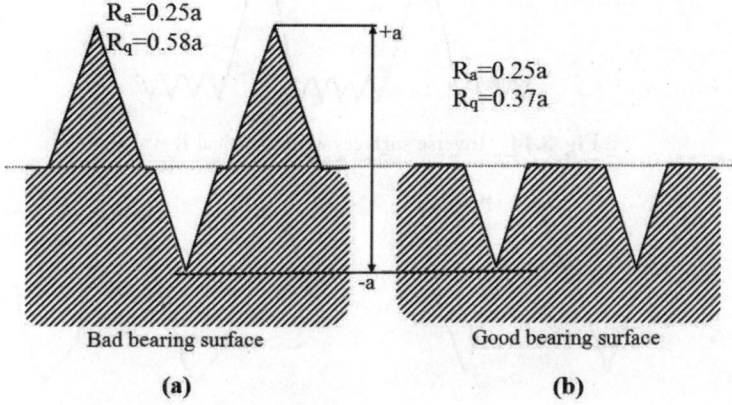

Fig. 8.13 Comparison of bearing surfaces based on R_q values

When average surface roughness is used, the effect of a single highest peak or valley is averaged out and has only a small effect on the final value and this is the main disadvantage of this parameter that it gives similar values for different surfaces. Since RMS roughness is weighted by the square of the heights, it is more sensitive than average roughness to deviations from the mean line and provides a correct surface feature. Based on these considerations, R_q (root mean square roughness) is generally preferred over R_a (average roughness) for the comparison of surfaces.

Due to positive deviations (roughness above the nominal surface), the contact between solids confines to a very small fraction of nominal area (i.e., δA) and as a result estimated contact stresses are much higher in magnitude compared to nominal stresses as expressed below:

Stress on smooth surface = F/A

Stress on rough surface = F/δA

Based on this understanding it can be inferred that the two surfaces which have the same R_a values but inverted, as shown in Fig. 8.14, are least preferred from tribological point of view.

Since the contact between the highest asperities will be more pronounced and because the asperity contacts affect the tribological performance, consideration of only R_a values will not be sufficient to describe the tribological surface.

On the other hand, the R_q value also is not able to describe all the features of the surface. Consider two surface roughness profiles as shown in Figs. 8.15(a) and 8.15(b). Both these surface have the same average surface roughness of 2.848. Surface shown by Fig. 8.15(a) have a higher RMS roughness of 4.758 as compared to that of surface shown in Fig. 8.15(b) with RMS roughness of 4.459. If only numerical values are considered then surface shown in Fig. 8.15(b) would be preferred, but it contains sharp asperity peaks which may result in more wear. The R_q value does not provide any information about the shape of the asperities.

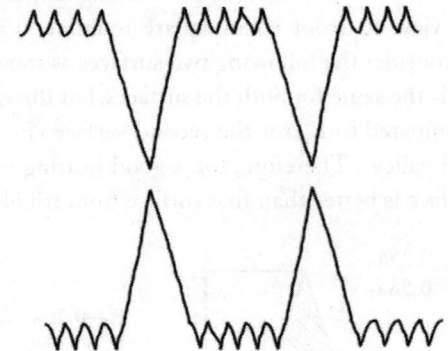

Fig. 8.14 Inverse surfaces with identical R_a value

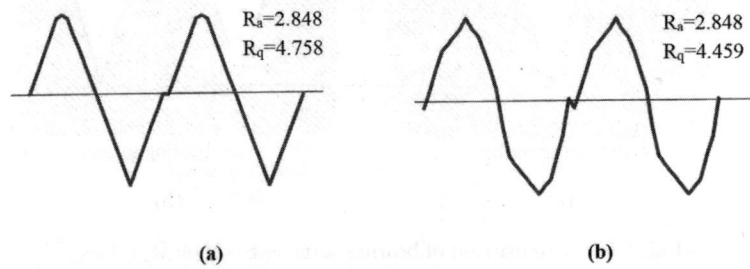

(a)   (b)

Fig. 8.15 Comparison of bearing surfaces

This means, the standard roughness parameters R_a and R_q, which are normally used in practice for identifying and classifying contact surfaces, are not sufficient to determine the tribological properties of contact surfaces. The tribological properties of contact surfaces, either under dry or boundary lubricated conditions cannot be anticipated solely on the basis of these parameters. Surface roughness parameters R_{sk} and R_{ku} are found to show a good correlation to the tribological properties of contact surfaces and could be used for planning surfaces and surface topographies with desired tribological behaviour in boundary lubrication regime [Marco et al., 2012].

8.4 Boundary Lubrication

The existence of the boundary mode of lubrication was discovered by English Scientist W. Hardy in 1922. Boundary lubrication is defined as the type of lubrication when friction and wear between relatively moving surfaces are determined by the behaviour of the surfaces and the lubricant material, differing from the tri-dimensional behaviour [ISO 4378, part 3]. Generally, all tribo–systems subjected to heavy loads operate in the boundary lubrication (BL) mode at some point in time (during starting and stopping, when the speeds are low) or permanently (in systems operating at heavy loads, low velocities and/or elevated temperatures). In several other systems which operate at low speeds (at low loads and low temperatures), the hydrodynamic lubrication is never achieved and they operate in boundary lubrication regime. The characteristic feature of boundary lubrication is the formation of a low shear strength protective layer (few nanometres thick) on the tribo–surfaces and complete absence of hydrodynamic lubrication. In boundary lubrication, the lubricant viscosity does not contribute significantly as compared to its physical and chemical interactions with the tribo–surfaces.

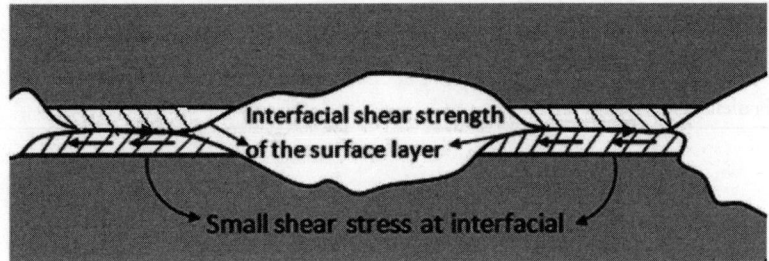

Fig. 8.16 Formation of low shear strength layer at the asperity interface

The reduction in friction and wear does take place in the boundary lubrication but it is very meagre reduction as compared to that obtained by hydrodynamic lubrication. Therefore, it is not a true lubrication as such but due to the unavoidable operating conditions, and cost and space constraints of the system, it is permitted to exist. In this mode, direct contact of the friction members is minimized by the formation of boundary layers on the contact asperities as shown in Fig. 8.16. The low shear strength protective layer in boundary lubrication is formed by the following mechanisms:

1. Physical adsorption (also called 'physisorption')
2. Chemical adsorption (also called 'chemisorption')

In physisorption, physical attraction or physical attachment of lubricant layer on the surface takes place, as shown in Fig. 8.17.

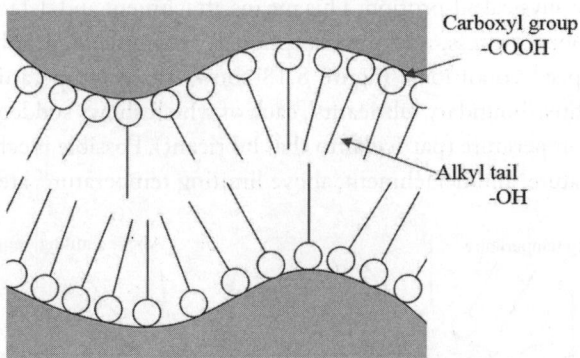

Fig. 8.17 Physisorption in boundary lubrication

In this figure, linear molecules of lubricant hydrocarbon align themselves normally to the contacting surfaces producing a thin protective layer thereby preventing asperity interaction. The free ends of lubricant molecules, due to bond polarity, get attached to the asperity–surfaces and lower the energy. One typical physisorption example is used of oleic acid in mineral oil as listed in the Table 8.2. From this table it can be inferred that on mixing boundary additive in lubricant oil reduces friction coefficient, but up to a certain limit.

Table 8.2 Effect of additive concentration on friction coefficient in boundary lubrication

Lubricant	Friction coefficient
Pure mineral oil	0.360
2% oleic acid in mineral oil	0.249
10% oleic acid in mineral oil	0.198
50% oleic acid in mineral oil	0.198

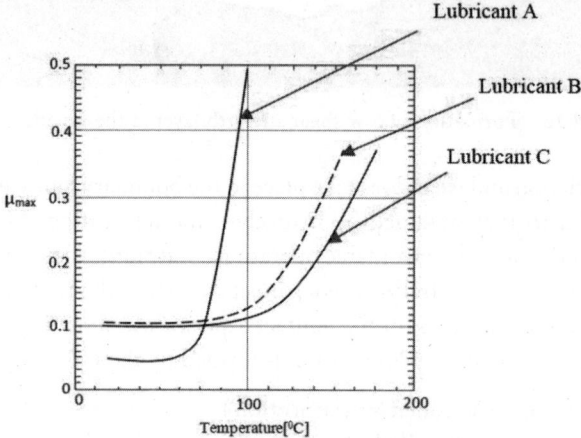

Fig. 8.18 Effect of temperature on friction coefficient in boundary lubrication [Bowden and Tabor, 1950]

Increase in operating temperature and sliding between contacting surfaces causes the detachment of lubricant layer made by physical adsorption. This means attachment and detachment of the lubricant layer remain a continuous process; and physisorption is recommended only for low temperature and relatively lesser speed conditions. Figure 8.18 shows three temperature–friction coefficient corresponding to the three 'boundary lubricants', each of which shows sudden increase in coefficient of friction at limiting temperature (particular to that lubricant). Possible mechanisms of attachment, below limiting temperature, and detachment, above limiting temperature, are depicted in Fig. 8.19.

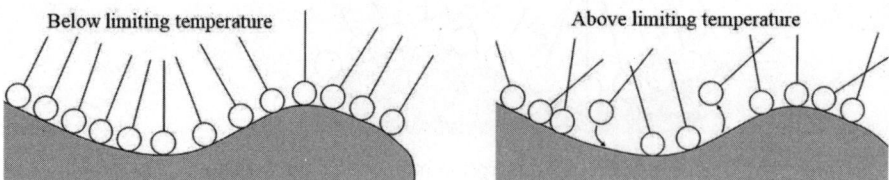

Fig. 8.19 Effect of temperature on boundary layer

The chemisorption occurs due to chemical action. The strength of chemical bonding between the lubricant and the tribo–surface depends on the reactivity of the tribo–surface material, as shown in Fig. 8.20. For chemisorption to occur, a chemically active group like chlorine, sulphide, phosphorus is required to react with tribo–surface and form a boundary layer. The chemical additives will be useful only for the metals as chemical reaction does not occur in case of polymers. The additives react and makes oxide layer on the metal surface that gets removed due to sliding of surfaces, so it acts as sacrificial layer. In order to form a new layer, the additives again react with surface. In other words, the chemisorption is an irreversible process due to occurrence of chemical bonding between lubricant and tribo–surface, so it causes damage to the surface due to detachment of the boundary layer. In addition, with time decrease in the concentration of additives happens and requirement to replenish the additives emerges.

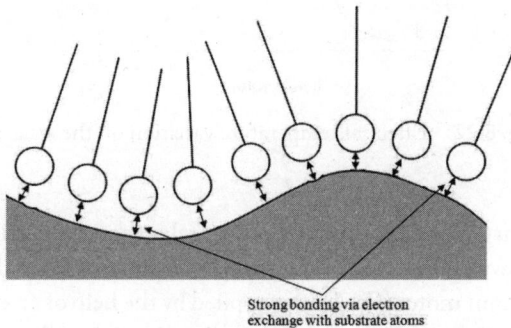

Fig. 8.20 Chemisorption

It is interesting to note that even hydrodynamic and mixed lubrication during operation may progress to boundary lubrication regime. To illustrate this let us consider Fig. 8.21, in which the variation of wear rate with load and temperature for three lubrication regimes has been depicted. It may be noted that the operative lubrication regime is dependent upon the instantaneous operating conditions. Increase in operating load and operating temperature changes lubricant regime from hydrodynamic to mixed and mixed to boundary lubrication as shown in Figs. 8.21 and 8.22.

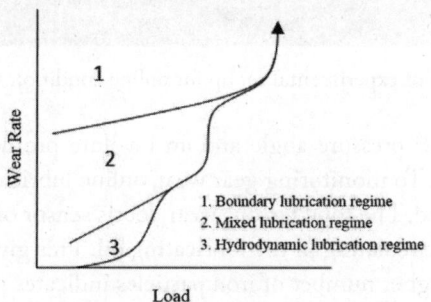

Fig. 8.21 Effect of load variations on the wear rate

With the continuous increase in the load, the lubricant is not sustained on the interface and this causes increase in the wear rate, as shown in Fig. 8.21. In this situation, lubricant with EP additives, which chemisorbed on the surface, will be preferred, and requires more energy for detachment. A similar variation of the wear rate with the increase in temperature is observed and it is shown in Fig. 8.22.

Based on the nature of wear behaviour as shown in Figs. 8.21 and 8.22, it is concluded that an appropriate lubricant is required for a particular operating condition. It is also required to consider the useful life of the tribo–pair for deciding the type of lubricant that will be suitable for the given application. Therefore, the study of boundary lubrication is important.

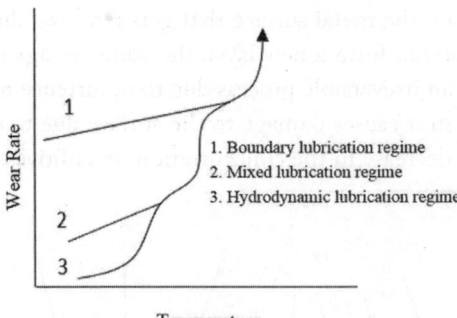

Fig. 8.22 Effect of temperature variation on the wear rate

Case Study 1:

To investigate the boundary lubrication, an experimental study was carried out on spur gears. The experimental set up is shown in Fig. 8.23. It consists of a spur gear box with a provision to apply an input torque using a DC shunt motor. The load is applied by the help of an eddy current dynamometer. The torque is measured by the torque sensor and displayed on an indicator.

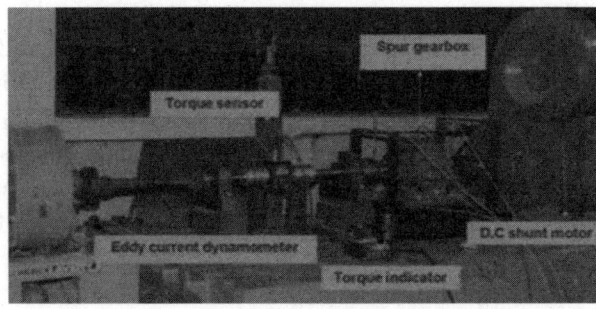

Fig. 8.23 Photograph of experimental set up for online condition monitoring of gear wear.

A gear pair, with gear (20° pressure angle and an involute profile with a module of 2.5 mm) reduction ratio of 2 was used. To monitoring gear wear, online lubricant condition monitoring unit, as shown in Fig. 8.24, was used. The total ferrous wear debris sensor of the monitoring unit provides the number of iron particles circulating in the lubricating oil. This gives us an indication of the rate at which wear takes place. Higher number of iron particles indicates more wear.

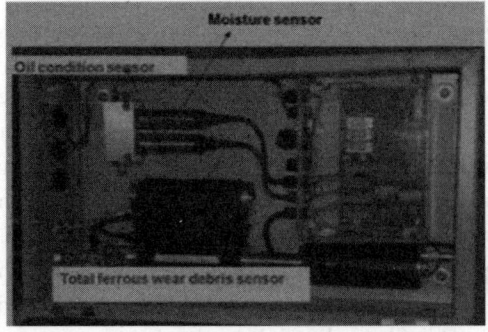

Fig. 8.24 Photograph of online condition monitoring unit

During the initial experiments, (since the new gears require initial running-in), the system was operated at 500 rpm under no load conditions. The lubricant used was 80W90 having a good EP additive package. The experimental results indicated wear of the gears reflected in terms of presence of Fe particle concentration (in ppm) in the lubricant. The concentration of Fe particles (in ppm) initially increased from 74 to 84 after 2 hours of gear box operation and then it remained steady until about 10 hours showing 'bath tub curve' type behaviour. The motor rpm was increased from 500 rpm to 1000 after tenth hour of operation. Experiments were performed for eight hour duration and intermediate results (as listed in Table 8.3) were recorded. No significant variation in Fe particle concentration was observed. Subsequent to this, the test was conducted at increased rpm of 1500 and it was observed that the Fe particle concentration increased significantly indicating the existence of boundary lubrication. This could not have happened in the mixed lubrication operative regime. The high initial wear was attributed to misalignment and running-in of the gears. But more importantly it was attributed to the lack of boundary layer formation on the surface. Since the gears were operated at no load conditions, the boundary lubricant additives did not function under these conditions as the surfaces were lightly loaded and thus the formation of boundary layer was delayed. It may be mentioned here that in case of rolling contact bearings the maximum wear occurs when the bearing is operated at a load less than 3% of the rated load. Therefore operation of gears at no load is not recommended.

Table 8.3 Experimental results showing concentration of wear particles (no load condition)

Serial No.	Operating Time (Hour)	Fe Concentration (ppm)		
		$N = 500$ rpm, $T = 0.35$ Nm	$N = 1000$ rpm, $T = 0.37$ Nm	$N = 1500$ rpm, $T = 0.40$ Nm
1	0	74	66	76
2	1			93
3	2	84	68	95
4	3			111
5	4	77	71	121
6	6	73	73	
7	8	69	74	
8	10	65		

Lubricating oil 80W-90   R.H. 45% to 55%

In the second set of experiments the lubricant was replaced by fresh one and the results obtained from experiments are listed in Table 8.4.

The wear results at 500 rpm indicate a decrease in the Fe particle concentration with passage of time, which means that running-in of gears at 500 rpm is complete. As per results of Table 8.4, on operating gears at 2000 rpm and at 2500 rpm a significant increase in Fe concentration occurs. It can be inferred that the escalation in the wear rate occurs with the increase in the speed. This increase in the wear rate is due to the non-existence of boundary lubrication, with more surface wear/damage taking place. Therefore change in the operating conditions affects the boundary lubrication layer.

Table 8.4 Experimental results of Fe concentration after oil replacement

Serial No.	Operating Time (Hour)	Fe Concentration (ppm)		
		N = 500 rpm, T = 0.35 Nm	N = 2000 rpm, T = 0.42 Nm	N = 2500 rpm, T = 0.45 Nm
1	1	43	89	219
2	2	47	136	283
3	3	30	141	344
4	4	32	143	615
5	5	34	130	540
6	6	35	124	658
7	7	36	120	

R.H. 38% to 52%

The tests were further conducted at load and speed combinations and the results are given in Table 8.5.

Table 8.5 Experimental results under loaded condition

S. No.	Operating Time (Hour)	Fe Concentration (ppm)							
		N=200 rpm T=5 Nm	N=200 rpm T=8.5 Nm	N=600 rpm T=20 Nm	N=600 rpm T=28 Nm	N=200 rpm T=5 Nm	N=800 rpm T=28 Nm	N=1000 rpm T=35 Nm	N=1000 rpm T=40 Nm
1	1	152	100	67	45				
2	1.5								33
3	2	154	94	68	42				
4	2.5	138	90	66	41	30	30	33	
5	3	135	90	63	40				30
6	4	130	86	62	40				
7	4.5								30
8	5	127	86	61	39	30	32	30	
9	6	100	67	56	36				29
10	7	104	60	54	37				
11	7.5	107	60	56	36	32	32	33	32
12	8			53	34				
13	9			53	33				
14	10			52	32				
15	11			47					

It is observed from these results that with the increase in the load and speed, the Fe particle concentration (in ppm) is decreasing with the passage of time, indicating that the formation of the boundary layer is taking place and the boundary additives are active. Under these conditions attachment and detachment of the boundary layer occurs and the system reaches to a steady state condition. With further passage of time the Fe particle concentration (in ppm) is stabilized indicating that the bedding-in time is over. The Fe particle concentration (in ppm) finally stabilizes to about 30 ppm and it was difficult to obtain further reduction.

This experimental study on boundary lubrication indicates that the wear rate during the no load operation is high as compared to the wear that occurs when the gears are operated under load. This indicates that under no load conditions the formation of boundary layer does not take place. When the gears are operated under load, the boundary additive becomes active, contributing to the boundary lubricant layer formation and thereby causing reduced wear. This indicates that the boundary layer phenomenon is very significantly dependent upon the operating conditions.

Case Study 2:
In another case study, tribological performance of heavily loaded journal bearing operating at low speed was carried out. These bearings operate under mixed or boundary lubrication regimes. One such application is the bearing used in sugar mills, where the journal rotates at about 4–5 rpm while being subjected to a load of about 3 MN. The bearing material is solid tin bronze and highly viscous lubricants, namely SM175 or SM85 (with EP additives, and having dynamic viscosity of 0.0664 Pa.s) are used. The bearings are cooled by circulation of water through the water chambers in the bearing casings. The bearing diameter is 530 mm, length 675 mm and radial clearance is taken to be about 5 mm. It has been reported that these bearings fail prematurely under operation. Figure 8.25 depicts a photograph of the failed sugar mill bearing.

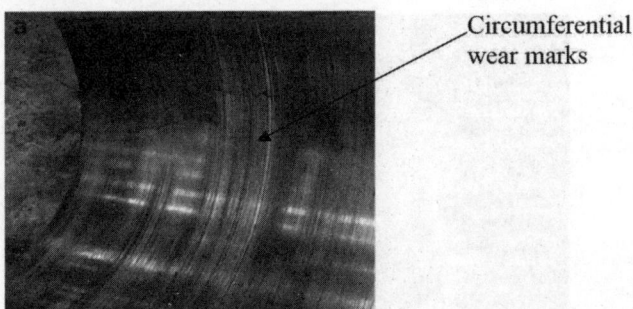

Fig. 8.25 Photograph of the failed sugar mill bearing [Muzakkir et al., 2011]

The visual inspection of the failed bearings shows scratches in circumferential direction leading to conclude that the failure of bearings is caused by lubrication breakdown. Long cracks in bearings often create confusion of 'fatigue cracking', which occurs due to repetitive stress cycle. But the bearings in the present case experience relative speed of only 5 rpm (bearing life < 30,000 relative-rotations) and load does not fluctuate in rotation (fatigue cycles are much lesser than 30,000). In other words, very high load is responsible for failure of the bearings. A theoretical study of these journal bearings is first carried out. The theoretical load carrying capacity was determined using the Hirani's method [Hirani et al., 1997]. The piezo–viscous effect of lubricant assuming lubricant viscosity as function of pressure [Eq. 3.1.2.10, chapter 3] was considered. The theoretical load carrying capacity was thus determined and is given in Table 8.6.

Table 8.6 Theoretical load carrying capacity of the sugar mill bearings

S. No.	Eccentricity ratio (ε)	Theoretical load capacity (W) (N)	Maximum pressure (MPa)
1	0.95	2.187×10^5	2.8
2	0.98	5.718×10^5	11.4
3	0.99	12.527×10^5	36.8

On comparison of the theoretical load capacity, listed in Table 8.6, with the actual applied load (3 MN), it can be inferred that bearing subjected to very heavy load and low rpm condition, operates under boundary/mixed lubrication condition. It is also observed that even eccentricity ratio equal to 0.99 is unable to generate required load capacity. This means contact among asperities cannot be avoided. Secondly, the value of maximum pressure increases very rapidly with increase in eccentricity ratio and may exceed 50 MPa, which is the limit of maximum pressure for tin bronze material used in sugar mill bearings.

The specific film thickness, Λ, determined for this case ($\Lambda = 0.5883$) comes out to be less than 1, which means the operative lubrication regime is boundary lubrication. This establishes that the hydrodynamic lubrication does not exist in sugar mill bearings and metal to metal contact occurs. This results in high friction, and consequently, rise in contact temperature. This increased temperature breaks the protective lubricant film resulting in significant wear, thereby causing a premature failure.

In order to validate these theoretical results and findings, experiments were conducted on a fully automated Journal Bearing Test Rig (JBTR), shown in Fig. 8.26, using three different lubricating oils with varying boundary additives. The operating conditions of the sugar mill roller bearings were simulated in the test rig by taking a load and speed combination of 5000 N and 10 rpm, respectively.

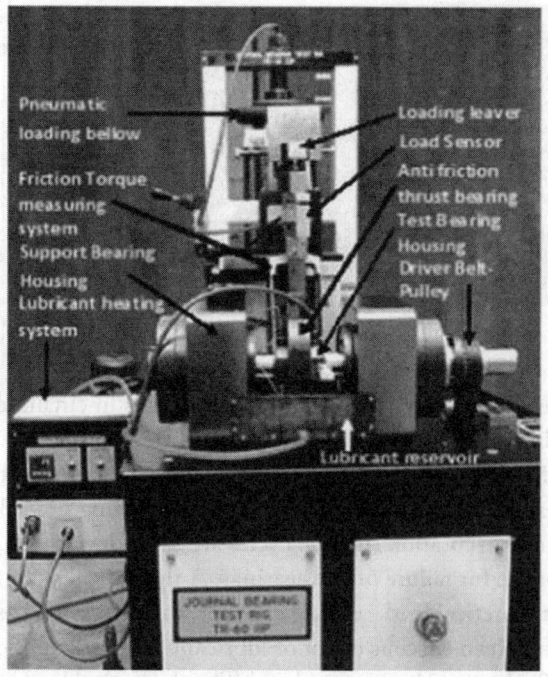

Fig. 8.26 Journal bearing test rig [Muzakkir et al, 2011]

There is a provision to apply the load by the help of bellows provided with pressurized air supply. The load arm is attached to ball bearing to ensure torque free motion of journal. The lubrication system comprises pump for continuous supply of lubricant and temperature controller to regulate the temperature. The belt drive is used to rotate journal of 50 mm diameter. The frictional torque is measured with the help of friction force load cell.

Several bearings, of the same material as that of the sugar mill bearing, were fabricated for conducting the experiments with radial clearance ranging from 25 µm to 40 µm. Figure 8.27 depicts the test bearing drawing indicating the important dimensions.

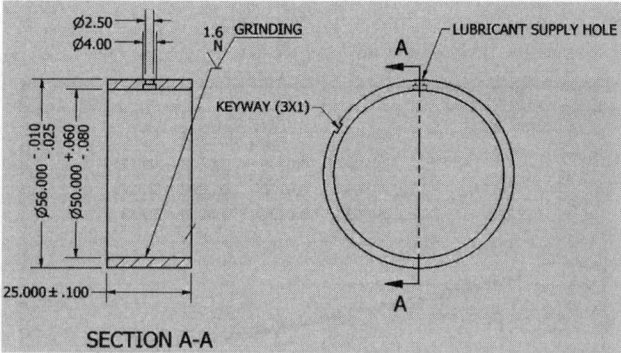

Fig. 8.27 Drawing of the test bearing [Muzakkir et al., 2013]

The diameter and radial clearance of the fabricated test bearings are listed in the Table 8.7.

Table 8.7 Diameter and radial clearance of fabricated bearings

Bearing no.	Diameter	Radial clearance (µm)
1	50.4511	225.550
2	50.1344	67.200
3	50.0862	43.100
4	50.0857	42.850
5	50.0808	40.400
6	50.0783	39.150
7	50.0615	30.750
8	50.0613	30.650
9	50.0472	23.600
10	50.0420	21.000
11	50.0381	19.050
12	50.0191	9.550

Two bearings (bearing numbers 1 and 2) were rejected due to excessive radial clearance and one bearing (bearing number 12) was rejected due to very low radial clearance. The experiments were conducted on the remaining bearings. The lubricant temperature was regulated at 35 ± 2 °C. After attaining the desired lubricant temperature the journal was rotated at the desired speed of 10 rpm, and then the test bearing was loaded using the pneumatic loading lever. A load of 5000 N was applied and resistive friction torque was recorded. Every experiment was carried out at least for 3 h duration. After the completion of the tests, the test bearing were thoroughly cleaned with the help of the solvent, allowed it to dry and weighed. The weight of the bearing before and after the experiments was measured to estimate the amount of wear. First three bearings were experimented using SM 175, SM 120 and SM 85 oil respectively. Experimental results of coefficient of friction for all three bearings shown in Fig. 8.28, indicates SM 85 is the least preferred oil compared to other two (SM 120 and SM 175) oils.

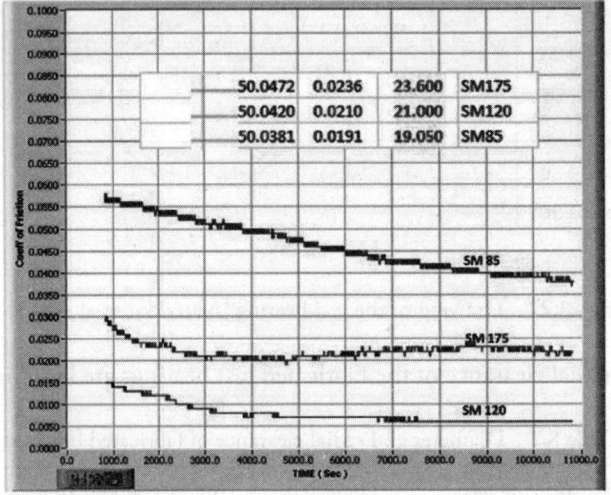

Fig. 8.28 Coefficient of friction of test bearings 1, 2, 3 [Muzakkir et al, 2011]

Therefore in further experiments only two oils SM 120 and SM 175 were used. Results of experiments performed on bearing numbers 7 and 8 are shown in Fig. 8.29, which indicate that performance of oil SM 175 improves with increase in radial clearance.

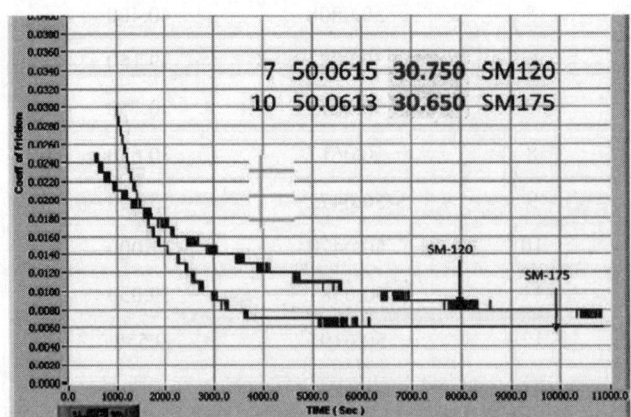

Fig. 8.29 Coefficient of friction of test bearings 7 and 8 [Muzakkir et al, 2011]

To ensure this, the experiments on bearings 5 and 6 were performed and results of which are illustrated in Fig. 8.30.

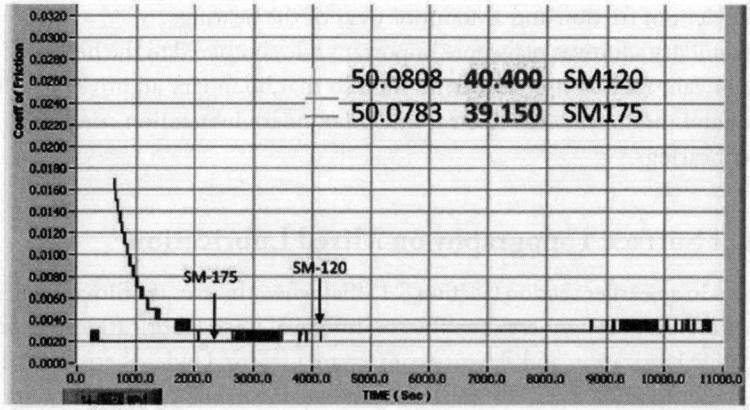

Fig. 8.30 Coefficient of friction of test bearings 5 and 6 [Muzakkir et al, 2011]

These results indicate that coefficient of friction reduces with increase in clearance from 30 μm to 40 μm. Performance of SM 175 and SM 120 oils were found to be comparable. The results of four of the test bearings are listed in Table 8.8.

Table 8.8 Wear of test bearing measured in terms of weight loss

Bearing no.	Weight before experiment (gm)	Weight after experiment (gm)	Reduction in weight (gm)	Reduction in weight (%)	Duration of the test (h)
1	105.620	105.609	0.011	10	3.10
2	109.477	109.471	0.006	5.4	3.10
3	106.018	106.014	0.004	4	3.18
4	110.866	110.861	0.005	4.5	3.10

The results given in Table 8.8 indicate severe wear for bearing number 1. This bearing was lubricated with SM 85 oil. Wear in the case of bearings 2, 3 and 4 is lesser. In other words the bearing number 1 showed excessive wear compared to other bearings. The photograph of bearing 1 after the test is shown in Fig. 8.31.

Fig. 8.31 Photograph of bearing 1 showing excessive wear after the test [Muzakkir et al, 2011]

Based on the theoretical study it was established that due to very heavy load and slow speed conditions the bearing operates in boundary/mixed lubrication regime. The simulated experiments with different lubricating oils indicate that the use of lubricating oil with EP additives significantly reduce the coefficient of friction and amount of wear of the bearings.

Therefore, boundary additives play more important role compared to the high-viscous oil. In other words, liquid lubricant may be used as carrier fluid so that boundary additive completely covers the bearings surface and minimize bearing wear under boundary lubrication conditions experienced by sugar mill roller bearings.

8.5 Effect of Surface Topography on Mixed Lubrication

It has been shown in an earlier section (section 8.1) that when the average film thickness is of the same order as that of the surface roughness then the conditions of mixed lubrications are predominant and both hydrodynamic lubrication and asperity contact are present. Under these conditions, the surface roughness assumes importance and it significantly affects the tribological performance of the system. Since the surfaces produced by the manufacturing processes have non-Gaussian height distributions, the skewness and kurtosis, which defines the non-Gaussian height distribution, will in general be the most influencing parameters in respect of the tribological performance. It has been observed that the surfaces produced by single point cutting tools like shaping and turning processes generate positively skewed surfaces. While the surfaces produced by the multi-point cutting tools like grinding, honing, etc., generates negatively skewed surfaces and in addition have high kurtosis values. The typical value of kurtosis for the grinding is 3, which means surfaces obtained after grindings have Gaussian height distribution (even distribution of peaks and valleys about the mean).

The surfaces that have large kurtosis value have smaller real contact area. In other words, surface having kurtosis >3 will be subjected to more contact stresses.

In a theoretical study conducted by Wang et al [Wang et al., 2006], several rough surfaces were generated using computer simulation. The contact and lubrication performance of those surfaces were determined using the model of Zhu and Hu [Zhu and Hu, 2001]. As per Wang et al [Wang et al., 2006], the skewness and kurtosis both reduce after the sliding of the surfaces.

Therefore skewness and kurtosis have a significant influence on the real contact area, load shared by asperities and maximum pressure generated at the interface. For very rough surfaces (having large R_{rms}) skewness and kurtosis parameters may not be considered important, but for relatively smoother surfaces (having R_{rms} lesser than one micrometre) these parameters are essential. Therefore, a surface with proper values of R_{rms}, skewness and kurtosis must be chosen for operation under mixed lubrication regime.

8.6 Asperity Temperatures in Mixed Film Lubrication

The contact of asperities cannot be avoided in boundary and mixed lubrication regimes. The sliding of contacting asperities causes an increase in the localized temperature. The rise in localized temperature affects the lubricant viscosity [section 3.1.2.1, Eq. 3.1.2.2, chapter 3] and friction coefficient. The flow of the lubricant and the side leakage may reduce the temperature rise at the interface. To estimate the localized asperity temperature, a fully coupled thermal mixed-film model [Zhai and Chang, 2000], which imitates solid/fluid interactions may be used. To reduce the complexity of considering solid/

fluid interactions and predicting conservative (higher) temperature rise, elastic spherical Hertzian dry contact–model [Tian and Kennedy, 1994] can be used. The comparison of these two methods, at two different sliding speeds, is given in Fig. 8.32. In this figure the temperature distribution on the surfaces of the global Hertzian junction (a region of high contact stresses located at the asperity contact) around the asperity contact has been shown. From Fig. 8.32, it can be observed that the solid/fluid interaction reduces the asperity temperature as heat is carried by the lubricant.

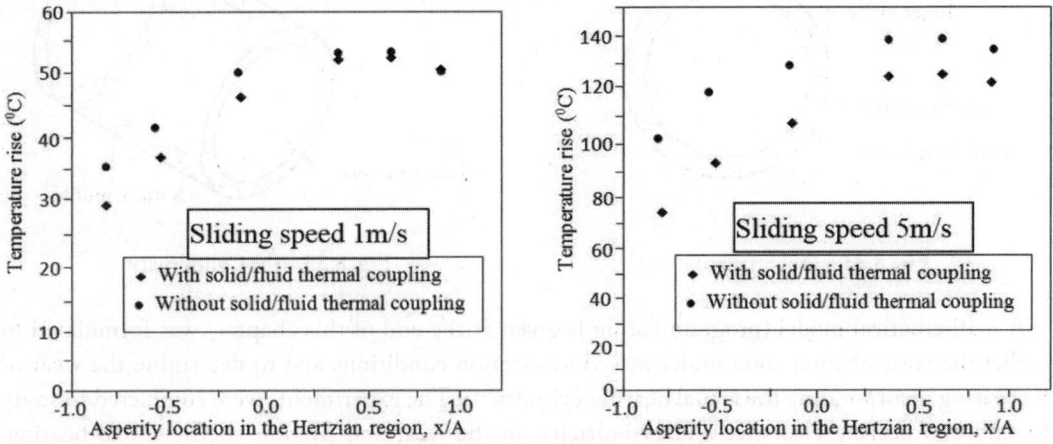

Fig. 8.32   Effects of the solid/fluid interaction on asperity thermal [Zhai and Chang, 2001]

8.7 Tribological Performance of Bearing Operating in Mixed Lubrication Regime

In this article, a case study involving the investigation of the tribological performance of a bearing operating in mixed lubrication regime is presented. This heading provides significance of the overall geometry of the contacting surfaces; and highlights the role of geometric dimensioning and tolerances in determining the tribological performance of the bearing.

A journal bearing that is subjected to heavy load and slow speed operates in the mixed lubrication regime. The operative lubrication regime is identified by the determination of the specific film thickness parameter and lifting-off speed. It has already been mentioned in sections 8.1, 8.2 and 8.3 that the surface roughness (quantified by surface roughness parameters) of the tribological element is dependent upon the method of manufacturing [refer to Fig. 8.3]. However the manufacturing variability introduces another important and very significant geometric errors: the out-of-roundness (or circularity) and cylindricity. These geometric errors are particularly induced in the manufacturing of cylindrical tribological elements like journal and bearing. The Fig. 8.33 [Muzakkir et al., 2014] depicts the circularity and cylindricity of circular tribological elements.

It is observed from Fig. 8.33 that the circularity describes the dimensional variation of the geometric element at any particular section along the axis, whereas cylindricity combines this effect over the complete bearing length. This suggests that the cylindricity is a comprehensive estimate of manufacturing variability as compared to circularity. In research conducted by Vijayaraghavan et al. [Vijayaraghavan et al., 1991] and Iwamoto and Tanaka [Iwamoto and Tanaka, 2001], the out-of-roundness was idealized in the form of geometrical shapes which did not correspond to the actual profile of the bearing. Therefore, in the present heading performance of bearing, analysis of which was carried out by incorporating actual bearing profile in the theoretical model, has been presented.

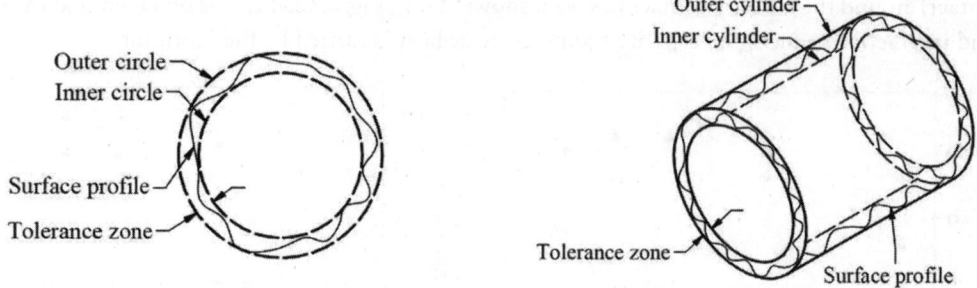

Fig. 8.33 (a) Circularity   **Fig. 8.33** (b) Cylindricity

A mathematical model (program listing is given at the end of this chapter) was formulated to predict the wear phenomenon under mixed lubrication conditions and to determine the wear of the bearing incorporating the actual bearing cylindricity. The experiments were conducted to study the effect of bearing clearance and cylindricity on the wear and friction-coefficient of bearing. Phosphorus bronze bearings having 50 mm nominal diameter and 25 mm length were fabricated for testing. The detail of the bearing with geometric and dimensional tolerance is shown in Fig. 8.34(a) and (b).

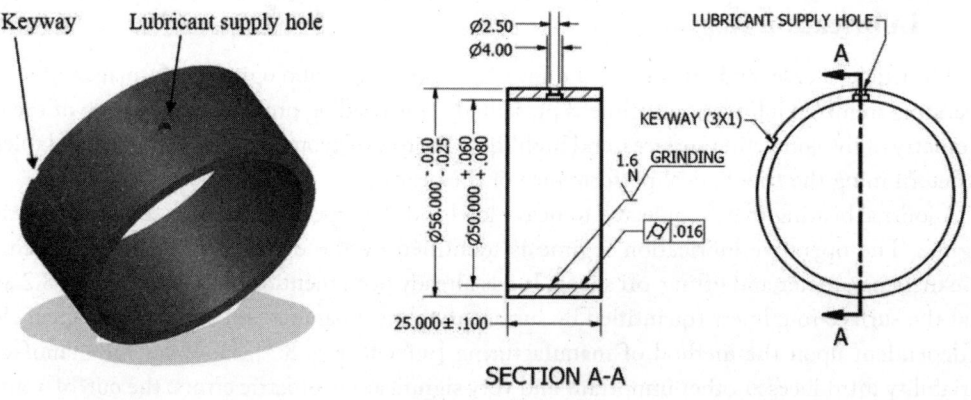

Fig. 8.34 (a) Test bearing   **Fig. 8.34** (b) Test bearing drawing

The bearing radial clearances were decided to be between 35–45 μm. The cylindricity tolerance was proposed to be within 16 μm. The measurement of the circularity and cylindricity was done on Coordinate Measuring Machine (CMM). Table 8.9 provides the measurement data of the fabricated bearings arranged in the increasing order of radial clearance.

Figure 8.35 depicts a representative actual bearing surface profile of bearing No.3.

Table 8.9 Measurement data of the test bearings

Bearing No.	Bearing Diameter (mm)	Radial clearance (μm)	Circularity (μm)	Cylindricity (μm)
1	50.0626	35.30	0.1	5.6
2	50.0761	42.05	10.0	15.3
3	50.0771	42.55	5.9	5.9
4	50.0796	43.80	9.1	13.8
5	50.0800	44.00	1.4	4.6

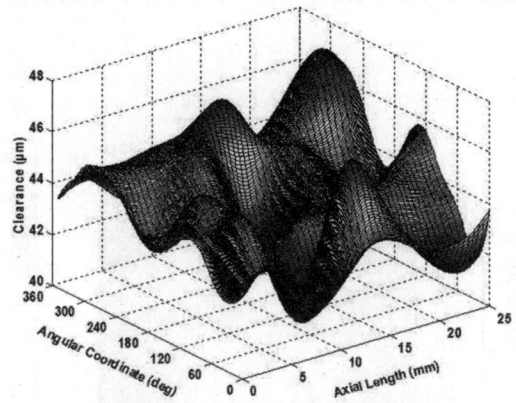

Fig. 8.35 Actual surface profile of bearing 3 with respect to journal surface [Muzakkir et al., 2014]

The experiments were conducted on a journal bearing test rig (Fig. 8.26). The bearings were operated at two loads (5000 N and 7500 N) and a speed of 10 rpm. The test duration was decided to be 6 hours so as to obtain measureable wear of the bearing. The wear was measured as a weight loss of the bearing. The bearings were cleaned using a solvent and bearing weight was measured before and after the test using a precision weighing balance having an accuracy of 0.1 mg. The lubricant inlet temperature was kept at 75 °C. The experiments were conducted keeping running-in distance of 850 m for each bearing so that a distinct performance is achieved for the test bearings. This value of running-in distance was determined experimentally using the lubricity tester. The results of the tests are given in Table 8.10.

Table 8.10 Experimental results (load 7500 N, speed 10 rpm)

Experiment number (Bearing No.)	Experimental results (running-in test of one and half hour duration)		Experimental results (normal test of six hours duration)	
	Wear (mg)	Coefficient of friction (10^{-2})	Wear (mg)	Coefficient of friction (10^{-2})
1	No running –in test conducted		4.9*	0.2580
2	2.9	0.0742	11.3	0.5630
3	0.0	0.1674	33.4	1.1673
4	41.3	1.3054	15.4	7.79754
5	-4.4	0.6165	54.5	1.21546

*load 5000N

It is observed from the experimental results listed in Table 8.10 that the bearing operates in the mixed lubrication regime and wear occurs. The rate of wear in the initial duration of 1.5 hours and 6 hours after running-in are different. During the initial sliding, more asperities get plastically deformed and result in wear. With continued sliding the bearing surface profile gets modified thereby changing the cylindricity. Ideally, initial running-in shall modify the bearing surface profile, increase the contact area and reduce the wear rate. After observing the results listed in Table 8.10, it is difficult to state that the initial running-in improves the bearing performance. To understand this variation, the surfaces of the bearings were analyzed using Taylor Hobson 3-D optical profiler. An area of 325 × 300 μm² of the bearings near the minimum film thickness was selected for surface profile measurement. The profile of the worn out surface of the bearings was examined. The surface profile of bearing No. 3 is shown in Fig. 8.36.

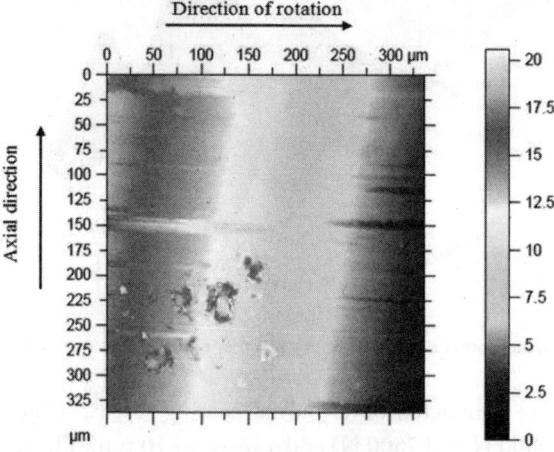

Fig. 8.36 Surface profile of bearing No. 3 after test [Muzakkir et al., 2014]

The observation of the surface profile of the bearing 3 shows material deposits (material transfer). Probably this is the reason for getting zero (negligible) wear during the running-in test. The surface indicates significant wear in next six hours running of the shaft. The maximum profile height in the vertical direction is 20 μm. This suggests that the wear is a localized phenomenon. The location of wear depends on the local cylindricity value in the contact zone. This cylindricity value relocates the zone of wear away from the theoretically determined position of minimum film thickness. Thus the wear may occur at any location where local cylindricity is high. The figures 8.37(a) and 8.37(b) depicts the actual zone of wear with respect to the vertical axis that suggests the dependence of wear upon the local cylindricity.

To understand the effect of cylindricity on the amount and location of wear, a mathematical model, considering elastic–plastic deformation, can be developed and the wear (considering the Archard's wear model) of the bearing can be estimated. It is worthy to note that the bearing surface profile gets modified due to the wear that takes place in each rotation of the journal therefore the subsequent wear of the bearing in the next rotation of the journal is dependent on this modified profile. This effect must be incorporated in the computer program. The variation (peaks and valleys) in the bearing surface profile significantly changes the local bearing clearance. The bearing clearance at a particular location is thus dependent on the local surface profile of the bearing. To incorporate these, the actual surface profile of the bearing must be provided as input to the computer program. Muzakkir et al. [Muzakkir et al., 2014] have provided the details of the mathematical model and its implementation to estimate the wear and modified surface profile of the bearing for each rotation of the journal. The parameters considered by Muzakkir et al. [Muzakkir et al., 2014] for computation in the model are given in Table 8.11.

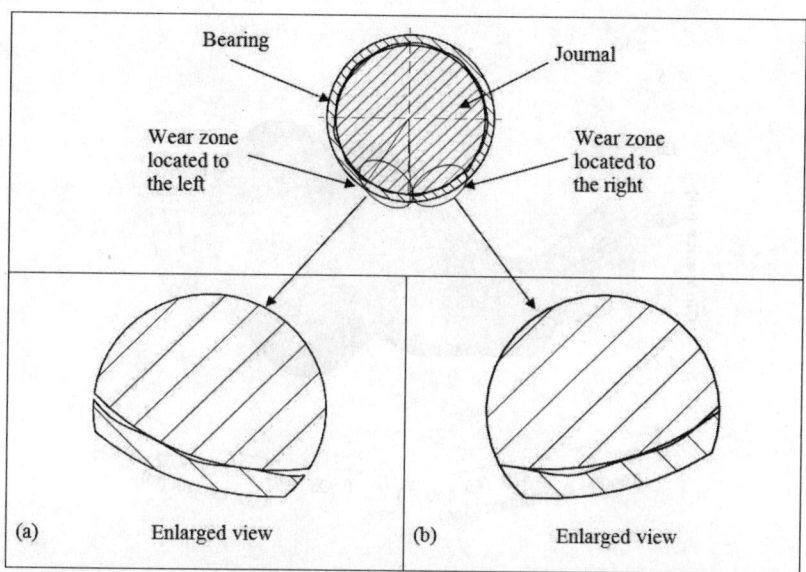

Fig. 8.37 (a) Actual location of maximum wear zone located towards the left of minimum film thickness zone [Muzakkir et al., 2014], (b) Actual location of maximum wear zone located towards the right of minimum film thickness zone [Muzakkir et al., 2014]

Table 8.11 Parameters used in mathematical model

S. No.	Parameter	Values
1.	Dynamic viscosity	0.099 Pa.s
2.	Nominal Radial Clearance	40 μm
3.	Young's Modulus (journal)	209 GPa
4.	Poisson's Ratio (journal)	0.29
5.	Young's Modulus (bearing)	91 GPa
6.	Poisson's Ratio (bearing)	0.343

Figure 8.38 depicts the worn-out surface profile after the completion of 4500 rotations of the journal. Theoretical results of wear, obtained by Muzakkir et al. [Muzakkir et al., 2014], for different surface profiles of the bearing having different cylindricy, are shown in Fig. 8.39.

As may be observed from Fig. 8.39, the running-in wear and steady state wear for each of the bearing surface profiles are different. The initial high running-in (transient) wear for all the bearings is observed during the initial 2400 rotations of the journal after which the steady-state wear is reached, that continues uptil 4500 rotations of the journal.

It is inferred from this study that for the journal bearing operating in mixed lubrication regime, the variation in the geometrical profile of the bearing at the contact zone significantly affects the amount of wear and the use of idealized geometric shapes that are used to represent the bearing is not justified. The effect of the surface roughness is dwarfed by the effect of geometrical errors that are an order higher than the scale of surface roughness.

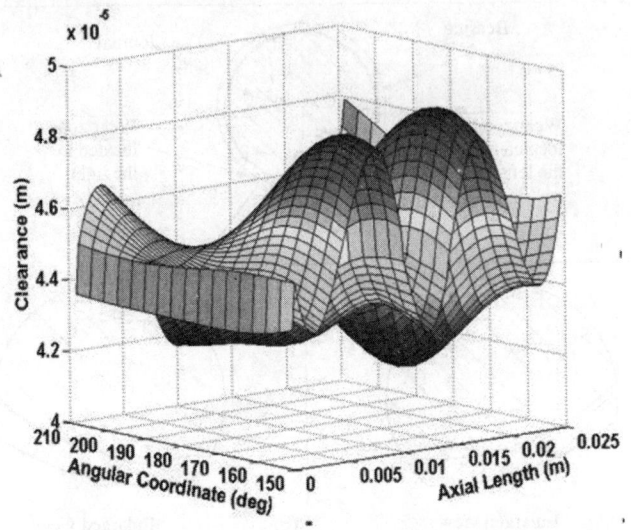

Fig. 8.38 Profile of the worn out bearing 1 [Muzakkir et al., 2014]

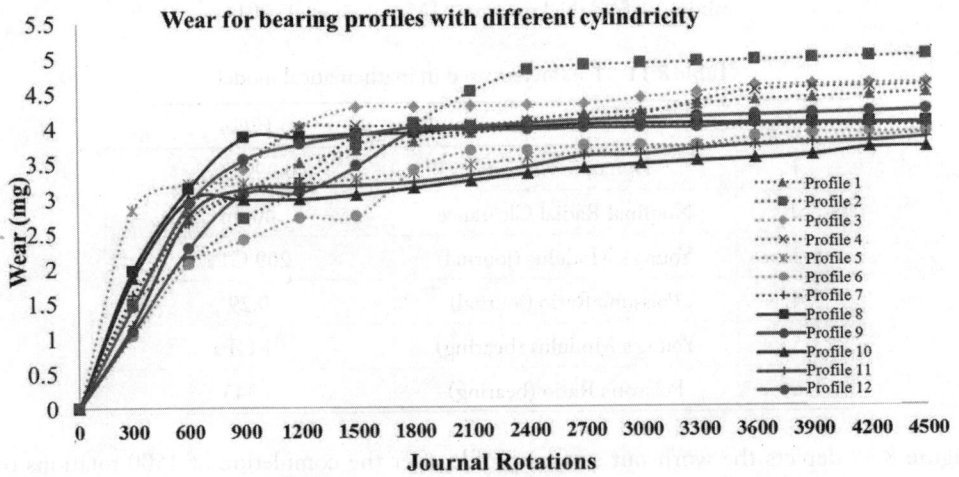

Fig. 8.39 Cumulative wear with respect to journal rotations for different bearing profiles [Muzakkir et al., 2014]

Therefore, the tribological elements operating in mixed lubrication regime require specific consideration of the errors in the geometric form in addition to the surface roughness considerations.

Frequently Asked Questions

Q.1. Define mixed lubrication.

Ans. Mixed lubrication is an operating state (regime) of a lubricated contact in which surface roughness significantly affects the performance of the contact. It may occur with conformal contact lubrication, such as journal bearing lubrication. Traditionally, as suggested by the term mixed, it is thought that both hydrodynamic lubrication and asperity contact have to be present for a contact to be considered as functioning in the mixed regime. However between the theoretical limit of the mixed lubrication regime and the full hydrodynamic regime, there is a region where surface roughness still has a significant influence on contact performance, even if asperity contact is not present.

Q.2. Which factor affects the tribological performance in mixed lubrication regime?

Ans. The tribological performance in mixed lubrication regime is affected by boundary lubricants, skewness and kurtosis of surface roughness.

Q.3. How does mixed lubrication occur?

Ans. When the applied load is shared between full-film lubrication and asperity contact regions, mixed lubrication occurs. This lubrication mechanism is common in many machine components like gears, cams and heavily loaded sliding bearings.

Q.4. Name the most common method to identify the operative lubricating regime.

Ans. The most common method of identifying the operative lubricating regime is by specifying the specific film thickness (Λ).

Q.5. What are the parameters necessary to describe surface topography?

Ans. Various parameters necessary to describe surface topography are:
- Arithmetic average (AA) or centre line average (CLA) R_a
- Root mean square (RMS) R_q
- Maximum profile peak height (Z_p)
- Maximum profile valley height (Z_v)
- Maximum height of profile (R_z)
- Mean peak spacing (R_{sm})
- Root mean square slope ($R_{\Delta z}$)
- Skewness (R_{sk})
- Kurtosis (R_{ku})

Q.6. What is the average surface roughness?

Ans. The average surface roughness is the arithmetic mean deviation of the surface height from mean line (equal areas of profile above and below it) through the profile.

Q.7. What do you understand by mean peak spacing S_m?

Ans. The mean peak spacing S_m as per ISO, is defined as average spacing between two successive negative going crossings of the mean line.

Q.8. Why R_q (root mean square roughness) is preferred to the R_a (average roughness) in evaluating the tribological performance?

Ans. Average surface roughness (R_a) is a measure to find the variation of surface from perfect plane. It is difficult to differentiate the bad and good surfaces on the basis of R_a values.

R_q is weighted by the square of the heights; therefore, it is more sensitive than R_a to deviation from the mean line and provides a better measure of surface roughness.

Q.9. Why is a low shear strength film required at the asperity contact and how is it formed when a tribo–pair is operating in boundary lubrication regime?

Ans. The low shear strength film is required at the asperity contact so that it can provide easy sliding of the contacting asperities. This reduces friction and heat generation.

The low shear strength films are formed either by physisorption or chemisorption process.

Q.10. What are the factors that prevent the chemisorption in boundary lubrication?

Ans. Since the chemisorptions occur due to formation of covalent bonds between the additive and metal surface, the presence of moisture on the surface prevents this bonding. Low temperature/pressure conditions also prevent the formation of chemisorped layer on the surface.

Q.11. What happens when a tribo–pair is operated at no load conditions?

Ans. The "no load conditions" in a tribo-pair generally represent the absence of the external load. However in real situations the surfaces will interact with each other due to several unintended excitation forces/moments and thus the chances of wear exists. In fact the tribo pairs must not be operated at a load lesser than 3% of its rated capacity as it will drastically increase the wear rate.

Q.12. How does the boundary lubricant reduce wear rate of a tribo-pair?

Ans. The boundary lubricants help to form a layer on the surfaces of contact bodies. Due to that layer, direct contact between asperities does not takes place and so wear reduces.

Q.13. Why is the determination of specific film parameter not an absolute method to identify the operative lubrication regime?

Ans. The specific film parameter uses only the R_q (root mean square roughness) and the minimum film thickness values of the composite surface. It does not incorporate all the influencing factors, like grain size, hardness of surface, clearance, etc. As most of these parameters are statistical, defining any absolute method to characterize the operative lubrication regime is not possible.

Q.14. Comment on 'lubricant viscosity does not play a major role in boundary lubrication'.

Ans. Since the asperity contact occurs during the boundary lubrication, the lubricant additives that can be easily adsorbed (physically or chemically) on the surface to form the boundary layer, perform better. The lubricant viscosity does not have much effect on the boundary layer formation and therefore it does not affect the tribological performance.

Q.15. Is chemisorption an irreversible process?

Ans. Yes, the chemisorption is an irreversible process due to occurrence of chemical bonding between lubricant and tribo surface. It causes damage to the surface due to detachment of the boundary layer. The additive gets consumed in the process and require replenishment.

Multiple Choice Questions

Q.1. In mixed lubrication, which lubrication mechanisms are operative?
   (a) Hydrodynamic and hydrostatic   (b) Hydrostatic and EHL
   (c) Hydrodynamic and boundary   (d) Hydrodynamic and EHL

Q.2. The most common method for determination of operative lubrication regime is:
   (a) Sommerfeld's number   (b) Reynold's number
   (c) specific film thickness parameter   (d) minimum film thickness

Q.3. In which operative lubrication regime a comparatively higher wear of surface is generally observed?
(a) Hydrodynamic lubrication regime (b) Hydrostatic lubrication regime
(c) Boundary lubrication regime (d) Mixed lubrication regime

Q.4. The conditions that shift the hydrodynamic lubrication regime to mixed lubrication regime are:
(a) high load, low speed, high operating temperature
(b) high load, high speed, high operating temperature
(c) low load, high speed, high operating temperature
(d) low load, high speed, low operating temperature

Q.5. The most commonly used parameters to define the surface roughness are:
(a) $R_{average}$ and R_{rms}
(b) $R_{average}$ and R_p
(c) R_z and R_p
(d) Z_p and Z_w

Q.6. If the specific film thickness is lesser than 1, then the tribo pair operates under:
(a) boundary lubrication regime
(b) mixed lubrication regime
(c) EHL regime
(d) hydrodynamic lubrication regime

Q.7. The mean peak spacing S_m, as per ISO is:
(a) the average spacing between two successive peaks of the surface profile
(b) the average spacing between any two peaks of the surface profile
(c) the average spacing between two successive negative crossings of the surface profile
(d) the average spacing between the two successive positive crossings of the surface profile

Q.8. The skewness parameter R_{sk} is used to:
(a) quantify the asymmetry of the profile about the mean line
(b) quantify the asperity heights of the profile about the mean line
(c) quantify the asperity slope of the profile
(d) quantify the asperity curvature of the profile

Q.9. The positive skewness generally indicates:
(a) a good bearing surface
(b) a bad bearing surface
(c) an average bearing surface
(d) a non-usable bearing surface

Q.10. Boundary lubricant additives will be the most effective when:
(a) the material of the contacting surfaces are metallic
(b) the material of the contacting surfaces are different
(c) one surface is metallic and other ceramic
(d) the material of the contacting surfaces are ceramic

Q.11. The boundary layer formed during boundary lubrication:
(a) has very low shear strength than that of parent material
(b) has very high shear strength than that of parent material
(c) has same shear strength as that of parent material
(d) increases the shear strength of the parent material

Q.12. The physisorption phenomenon is limited due to:
(a) interface pressure
(b) interface temperature
(c) interface pressure as well as temperature
(d) lubricant viscosity

Q.13. The chemisorption phenomenon:
(a) occurs mostly in metallic tribo pairs
(b) occurs mostly in non-metallic tribo pairs
(c) is limited due to lubricant viscosity
(d) is not dependent upon interface temperature

Q.14. The coefficient of friction in physisorption phenomenon:
(a) increases with increase in temperature
(b) increases with decrease in temperature
(c) decreases with increase in temperature
(d) does not get affected with the change in temperature

Q.15. In boundary lubrication:
(a) asperity temperature reduces with the reduction in lubricant side leakage
(b) asperity temperature increases with the reduction in lubricant side leakage
(c) asperity temperature increases with the increase in lubricant side leakage
(d) asperity temperature increases with the increase in lubricant flow rate

Answers

Q.1. (c) Q.2. (c) Q.3. (c) Q.4. (a) Q.5. (a) Q.6. (a)
Q.7. (c) Q.8. (a) Q.9. (b) Q.10. (a) Q.11. (a) Q.12. (b)
Q.13. (a) Q.14. (a) Q.15. (b)

References

Lu, X. and Khonsari, M. M. 2005. 'On the Lift-off Speed in Journal Bearings'. *Tribol Lett.* 20:299–305.

Moshkovich, A. Perfilyev, V. Gorni, D. Lapsker, I. and Rapoport, L. 2011. 'The effect of Cu grain size on transition from EHL to BL regime (Stribeck curve)'. *Wear.* 271:1726–32.

Khonsari, M. M. Booser, E.R. 2006. 'Proper Film Thickness key to Bearing Survival'. *MaSch Des.* 4–7.

Buyanovskii, I. A. 2010. 'Boundary lubrication by an adsorption layer'. *J Frict Wear.* 31: 33–47.

Thomas, T.R. 1999. *Rough Surfaces.* Second ed., Imperial College Press, London.

Bowden, F.P. and Tabor, D. 1950. *The Friction and Lubrication of Solids.* Part 1. Clarendon Press. Oxford.

Muzakkir, S.M. Hirani, H. Thakre, G. D. and Tyagi, M. R. 2011. 'Tribological Failure Analysis of Journal Bearings used in Sugar Mills'. *Engineering Failure Analysis.* 18: 2093–2103.

Wang, S. Cusano, C. Conry, T. F. 1991. 'Thermal analysis of elastohydrodynamic lubrication of line contact using the Ree–Eyring model'. *ASME J Tribol.* 103(2):232–44.

Zhai, X. and Chang, L. 2000. 'A deterministic thermal model for mixed–film contact'. *Tribol Trans.*43(3): 427–34.

Tian, X. Kennedy, F. E. 1994. 'Maximum and average flash temperatures in sliding contacts'. *ASME J Tribol.*118:33–42.

Zhai, X. Chang, L. 2001. 'Some insights into asperity temperatures in mixed-film lubrication'. *Tribology International.* 34 :381–387.

Wang, W.Z., Chen, H., Hu, Y.Z., and Wang, H., 2006. 'Effect of surface roughness parameters on mixed lubrication characteristics'. *Tribology International*. 39: 522–527.

Zhu, D. Hu, Y. Z. 2001. 'A computer program package for the prediction of EHL and mixed lubrication characteristics, friction, subsurface stresses and flash temperatures based on measured 3-D surface roughness'. *Tribology Transactions*. 44(3): 383–90.

Moshkovich, A. Perfilyev, V. Gorni, D. Lapsker, I. and Rapoport, L. 2011. 'The effect of Cu grain size on transition from EHL to BL regime (Stribeck curve)'. *Wear*. Vol. 271: pp. 1726– 1732.

Lu, X. and Khonsari, M. M. 2005. 'On the Lift-off Speed in Journal Bearings'. *Tribology Letters*. Vol. 20, Nos. 3–4: , pp. 299–305.

Majumdar, A. and Bhushan, B. 1990. 'Role of fractal geometry in roughness characterization and contact mechanics of solids'. *ASME J. Trib.* 112: 205–216.

Khonsari, M. M. and Booser, E. R. 2001. *Applied Tribology: Bearing Design and Lubrication*. John Wiley and Sons.

Wyant, J. C. Koliopoulos, C. L. Bhushan, B. and George, O. E., 1984. 'An optical profilometer for surface characterization of optical media'. *ASLE Trans*. 27: 101–113.

Iwamoto, K. Tanaka, K. 2005. 'Influence of manufacturing error of roundness for characteristics of cylindrical journal bearing'. *Tribol Interface Eng Ser*. 48:751–754.

ISO 4378-3:2009 Plain Bearings-Terms, Definitions, Classification and Symbols-Part 3.

Vijayaraghavan, D. Brewe, D. E. Keith, T. G. 1991. 'Effect of out-of-roundness on the performance of a diesel engine connecting-rod bearing'. *STLE/ASME Tribol. Conf.* Missouri, USA.

Muzakkir, S. M. Lijesh, K. P. Hirani, H. Thakre, G. D., 2014. 'Effect of Cylindricity on the Tribological Performance of Heavily Loaded Slow-Speed Journal Bearing'. *Proceedings of the Institution of Mechanical Engineers. Part J: Journal of Engineering Tribology*. Vol. 229, Issue 2, pp 178-195. doi:10.1177/1350650114548053.

Program Listing in MATLAB for Figure 8.39

```
% Estimation of wear of journal bearing
% L-Length of bearing (m)
% R= Radius of the bearing (m)
% Clc= Clearance (m)
% w = (2*pi*N/60)- angular velocity (N is the rotational speed)
% eta- Viscosity of lubricant (Pas)
% N = Number of division in angular direction
% M = Number of Division in Axial Direction
% W = Applied load (N)
% V= Velocity of Journal
% E1= Young's Modulus of Journal
% E2= Young's Modulus of Bearing
% mu1= Poisons ratio of journal
% mu2= poisons ratio of Bearing
% Hard- hardness of the bearing
% M= Number of data points in circumferential direction
% N= Number of data points in axial direction

clear all
%data for true circular profile make AA=zeros(m,n)
```

```
% for general profile, enter path of the file containing the surface profile
% below

%%Input Block
AA=xlsread('C:\Users\DELL\Documents\MATLAB\journal data.xlsx','sheet1');
M=(12*m+4)*1;
N=(n*16+4)*1;
shift=7;
K=zeros(N,M);
h11=zeros(N,M);
x=AA(14,(1:7));
y=AA((1:13),8);
xxi=0;
mu1=0.343;
E1=91*10^9;
mu2=0.29;
E2=205*10^9;
EE=((1-mu1^2)/E1+(1-mu2^2)/E2)^-1;
Hard=2.8*152e6;
eta=0.1;
Z=0.025;
R = 0.05/2;
deltet = 2*pi/N;
delz = Z/M;
WW=7000;
w=0.52359*2;
s0=1;
Tin=40+273;
Teff= 41.0021+273;
Pres=0;
B1=((Teff-138)/(Tin-138))^-s0;
a1=1e-8;
a12=5.1*10^-9;
A1=log(eta+9.67);
W=7000;
ITER = 10000;

%%Initialization Block
for i=1:M-1
    xxi(i+1)=xxi(i)+0.25;
end
WWap=0;
WWa=0;
yyi=0;
for i=1:N+1
    yyi(i)=22/7*2/N*(i-1);
end

ff=0;
```

```
gg=0;
shift=0;
for i =1:N
    for j=1:M

        yy=i+round(100*shift/12);
         if yy<=100
        KKK(yy,j)=interp2(x,y,AA((1:13),(1:7)),xxi(j),yyi(i),'spline');
         else
          KKK(i-(N-round(100*shift/12)),j)=interp2(x,y,AA((1:13),(1:7)),xxi(j)
,yyi(i),'spline');
        end
        end
end
k1 = (R*deltet/(Z*delz))^2;
zz=al2/(5.1*10^-9*(log(eta)+9.67));
T=10;
sum1= sum(KKK);
avg=sum(sum(KKK))/(N*M);
lamda=0;
BB=(KKK-avg)*8e-6;
 deltet=2*pi/100;
delz=Z/100;

for i=1:N
    tet(i)=deltet*i;
end

for i=1:M
    z(i)=delz*i;
end


for I=1:N
    tet(I)=deltet*(I);
    zz(I)=delz*(I);
    p(N+1,I)=-000;
end


CC=43.8e-6;
c = CC-BB/8;
WWa(1)=0;
rms1=0.5e-6;
rms2=rms(rms(BB/8));
rmss=sqrt(rms1^2+rms2^2);

for I=1:N
   for J=1:M
```

```
h11(I,J)=0;
Del(I,J)=0;
WWap(I,J)=0;
WWa(I,J)=0;
Rp1(I,J)=0;
ddel(I,J)=0;
        h11(I,J)=0;
        dell3(I,J)=0;
        cc(I,J)=c(I,J);
cc1(I,J)=c(I,J);
elips(I,J)=0;
h11(I,J)=0;
        dell(I,J)=0;
        BB1(I,J)=0;
    end
end
sum(1)=0.0;

%% Calculation of radii using three point method
for I=4:N-1
for J=4:M-1
r1=0;   r2=0;  r3=0;  tet1=0;  tet2=0;   tet3=0;
r1=c(I,J);
r2=c(I,J+1);
r3=c(I,J-1);
tet1=2*deltet*R; %2*delz;
tet2=3*deltet*R;
tet3=deltet*R;
R1=0; R2=0; R3=0;
R1(I,J)=sqrt(r1^2+tet1^2);
R2(I,J)=sqrt(r2^2+tet2^2);
R3(I,J)=sqrt(r3^2+tet3^2);
s = 0.5*(R1(I,J) + R2(I,J) + R3(I,J));
Km(I,J) = sqrt(s.*(s-R1(I,J)).*(s-R2(I,J)).*(s-R3(I,J)));
Rp21(I,J) = (R1(I,J)*R2(I,J)*R3(I,J)/(4*Km(I,J)));
end
end

for J=4:N-1
for I=4:M-1
r1=0;   r2=0;  r3=0;  tet1=0;  tet2=0;   tet3=0;
r1=c(I,J);
r2=c(I+1,J);
r3=c(I-1,J);

tet1=2*delz; %2*delz;
tet2=3*delz;
tet3=delz;
R11=0; R21=0; R31=0;
```

```
R11(I,J)=sqrt(r1^2+tet1^2);
R21(I,J)=sqrt(r2^2+tet2^2);
R31(I,J)=sqrt(r3^2+tet3^2);
s = 0.5*(R11(I,J) + R21(I,J) + R31(I,J));
Km(I,J) = sqrt(s.*(s-R11(I,J)).*(s-R21(I,J)).*(s-R31(I,J)));
Rp11(I,J) = (R11(I,J)*R21(I,J)*R31(I,J)/(4*Km(I,J)));
end
end
Rp1=1./(1./Rp21+1./Rp11);

for I=1:N+1
p(I,1)=0;
end
for I=1:N+1
p(1,I)=-000;
end

for I=1:N+1
p(I,N+1)=-000;
end

e(1)=0;
eps=1-20*rmss./(CC++h11+Del);
eps2=1-1*rmss./(CC++h11+Del);
e(2)=0.43e-4; %eps(1,1)*CC(1,1);%min(min(cc))*0.99;% 0.48
e2=c;
FFFehl=0;
WWapehl=0;
WWaehl=0;
KKKK=zeros(100,100,100);
WWWap=zeros(100,100,100);
WWap2=zeros(N,M);

for KK=1:2
eps=abs((e(KK+1)./(cc)));
for I=1:N
for J=1:M
p(I,J)=0;
V(I,J)=0;
end
end
    for I=1:N
for J=1:M
                if eps(I,J)>1
            eps(I,J)=0.99 ;
                else
                eps(I,J)=eps(I,J);
                end
            end
    end
```

```
%%Iteration Block
for K = 1:ITER
 K=K+1;
sumij=0;
for I=2:N
for J=2:M
rms(I,J)=rmss./cc(I,J);
tet(I)=deltet*(I-1);
h0(I,J)=(1+eps(I,J).*cos(tet(I)));
h(I,J)=h0(I,J).*(1+7/6*(rms(I,J)./h0(I,J)).^2)+h11(I,J)./cc(I,J)+Del(I,J)./
cc(I,J)+++dell(I,J)/cc(I,J);
A=((-3/2.*eps(I,J)*sin(tet(I)))./(h(I,J)*deltet))+(1/deltet^2);
B=((3/2.*eps(I,J)*sin(tet(I)))./(h(I,J)*deltet))+(1/deltet^2);
C=(R^2/(delz)^2);
D=2*(1/deltet^2 +R^2/(delz^2));
E=eps(I,J)*sin(tet(I))/h(I,J).^3;
p(I,J)=C/D*p(I,J+1)+C/D*p(I,J-1)+E/D+A/D*p(I+1,J)+B/D*p(I-1,J);
if p(I,J)<0
                p(I,J)=0;
else
                p(I,J)=p(I,J);
end
sumij=sumij+p(I,J);
end
end
sum(K+1)=sumij;
percentage = abs(sum(K+1)-sum(K))/abs(sum(K+1));
if percentage < 1e-6
    break
    end
end
for I=1:N
for J=1:M
Pr(I,J)=p(I,J).*(6*eta*w*(R./cc(I,J))^2*exp(A1*(B1-1)));
end
end
for I=1:N
    for J=1:M
h11(I,J)=0;
Del(I,J)=0;
dell(I,J)=0;
dell2(I,J)=0;
    end
end
max(max(Del));
PP=max(Pr);
Pmax=max(PP);
```

```
for i=1:N+1
    tett(i)=deltet*(i-1);
end
lamda=h0./rms;
hh=h0./h;
for i=2:N
    for j=2:M
if lamda(i,j)<3
Prehl(i,j)=1./(1-hh(i,j)^(6.3)).*Pr(i,j);
else
    Prehl(i,j)=0;
end
end
    end

Pehl=Pr+Prehl;
for i=2:N
      for j=2:M
      Fehl(i,j)=dblquad(@(t,z)Pehl(i,j)*R*sin(t),tett(i),tett(i+1),0,delz);
      end
end

for i=2:N
    for j=2:M
        FFehl(i,j)=dblquad(@(t,z)Pehl(i,j)*R*cos(t),tett(i),tett(i+1),0,delz);
    end
end

Fxehl=0;
Fyehl=0;
for i=2:N
    for j=2:M
        Fxehl=Fxehl+(Fehl(i,j));
Fyehl=Fyehl+(FFehl(i,j));
    end
end
fi1=atand(Fxehl/Fyehl);
FFFehl=abs((Fyehl*cosd(fi1)+Fxehl*sind(fi1)));

for I= 2:N
    for J=2:M
Ffehl(I,J)=sqrt((Fehl(I,J).^2+FFehl(I,J).^2));
    end
end

  for I=2:N
    for J=2:M
        if lamda(I,J)<5
```

```
            del=0;
for i=2:N-1
            AA=(1/(pi*EE));
            for j=2:M-1
                    if I==i && J==j
                    j=j+1;
                        else
del=del+Pehl(i,j)*R*deltet*delz/sqrt((I-i)^2*(R*deltet)^2+ (J-j)^2*(delz)^2);
                    end
                end
end
Del(I,J)=Del(I,J)+AA*del;
        else
            J=J+1;
        end
      end
    end
  end
  for I=1:N
    for J=1:N
    e3(I,J)=e(KK);
     e3(I,J)=e3(I,J)-e3(I,J).*(1+eps(I,J)*cos(tet(I)));
    end
  end


for I=1:N-1
for J=M/2-round(M/12):M/2+round(M/12)
        dell(I,J)=e3(I,J)-e2(I,J);
            if dell(I,J)>0
            dell(I,J)=dell(I,J)-dell(I,J).*(1+eps(I,J)*cos(tet(J)));
            else
            dell(I,J)=0;
            end
            delc(I,J)=(1/R+1/Rp1(I,J))^-1.*(Hard/(EE))^2;
            if dell(I,J)<delc(I,J);
            WWa(I,J)=4/3*EE*dell(I,J).^(3/2)*((1/R+1/Rp1(I,J))^-1).^0.5;
            else
        dell2(I,J)=(dell(I,J));
WWap(I,J)= 3*pi*(1/R+1/Rp1(I,J))^-1.*(2*dell(I,J)-delc(I,J))*Hard/2.8;
        end
    end
end
WWapehl=0;
WWaehl=0;
for i=2:N
    for j=2:M
      WWapehl=WWapehl+(WWap(i,j));
       WWaehl=WWaehl+(WWa(i,j));
    end
```

```
end

for k=1:2
    if k==1
        WWap1=WWap2;
        dell3=dell2;
    else
        WWap2=WWap;
        dell2=dell;
    end
end

WWWap(KK+1,:,:)=WWap;
VV=0;
if WW>FFFehl+WWaehl+WWapehl
for I= 2:M-1
    for J=2:N-1
            V=w*1.2*60*60*9e-13*(WWap2(I,J)-WWap1(I,J)).*R;
h11(I,J)=9e-13*((WWap2(I,J)-WWap1(I,J))*2*pi*R*10*60*1.2/(delz*deltet))^(2/3);
            VV=VV+V;
    end
end

else
    for I= 2:M-1
    for J=2:N-1
            V=w*1.2*60*60*9e-13*(WWap2(I,J)-WWap1(I,J)).*R;
            h11(I,J)=0.66*R^(-1/3)*Z^(-2/3)*V^(2/3);
            VV=VV+V;
    end
    end
end


ww(1)=0;
Wa=FFFehl+WWaehl+WWapehl;
ww(KK+1)=WW-Wa;
ww(1)=0;

  if WW>FFFehl+WWaehl+WWapehl
e(KK+2)=e(KK+1)+ww(KK+1)./(abs(ww(KK+1)-ww(KK))/abs(e(KK+1)-e(KK)))/(57);
  else
      e(KK+2)=e(KK+1)+e(KK+1)/100000000;
  end

  for J=M/2-round(M/12):M/2+round(M/12)
for I=1:N-1
   Del(I,J)=Del(I,J)-Del(I,J).*(1+eps(I,J)*cos(tet(J)));
   h11(I,J)=h11(I,J)-h11(I,J).*(1+eps(I,J)*cos(tet(J)));
```

```

Vv(KK)=VV
max(max(h11))
max(max(eps))
max(max(Del))
max(max(dell))
mass(KK)=Vv(KK)*8.6e6
max(max(WWapehl))
cc=cc-h11-Del-(dell2-dell3);
cc1=cc1-h11-(dell2-dell3);
KKKK(KK,:,:)=cc;
bb1=(BB-dell);
end
```

Chapter 9

Tribological Aspects of Rolling Motion

It is known that rolling motion is efficient and preferable to the sliding motion. The concept of rolling motion is applied in all types of wheels, rolling element bearings and various transmission (belt–drive, cam–follower, gears) drives. Rolling motions are often associated with tangential tractions in contact zone (i.e., wheel and its track) or with some degree of slip (i.e., rolling element bearings). In the present chapter, tribo–pairs related to rolling motion, such as rolling element bearings, tyre–road interaction and wheel on rail track have been described. Finally, an analysis of failure of four row cylindrical roller bearing is detailed.

9.1 Rolling Element Bearings

The rolling element bearings find applications in auto industry, electric motors, power tools, machine tools, conveyors, escalators, fans, pumps, gyroscope, aircraft, ships, land vehicles, skates, dental drills, etc. There are more than 20000 varieties of rolling bearings (with variations in materials: steel, polymers, ceramics; with and without cage; various types of cages; with or without seal/shield; with and without circlip; single and multi-row; weight: few grams to 50 tonnes). These bearings are often referred to as anti-friction bearings and require relatively less lubricant than that of sliding bearings. In normal lubrication conditions, the friction and wear in rolling element bearings are very low and the bearings have long operational life. However, in the absence of lubrication, tangential and contact stresses increase significantly, both of which shorten the bearing life.

9.1.1 Bearing terminology

There are four main components of a rolling bearing: inner ring, rolling elements, cage and outer ring, as shown in Fig. 9.1. The rolling elements can be balls (Fig. 9.1a), cylindrical rollers (Fig. 9.1b), needle rollers (Fig. 9.1c) tapered rollers (Fig. 9.1d) or spherical rollers (Fig. 9.1e). Generally, needle diameter is much smaller than its length and bearings made of needle rollers are suited for the conditions where radial space is limited. To avoid sliding and collision between rolling elements, rolling elements are arranged in the cage. Cages are generally thin to form a shape to guide the rolling elements. Depending upon the requirement, the cage material can be polymer if the operating temperature is below 60 °C, or can be made of steel coated with molybdenum disulphide, graphite, etc. The rolling element bearings can be made without any cage. A typical example of cage-less bearing is shown in Fig. 9.1(c). In such bearings, there are rolling elements, outer ring and inner ring. The cage-less bearings are often referred as 'full complement' bearings. The rolling elements in 'full complement' bearing collide with each other and increase relative sliding/rubbing. Such bearings are recommended for low speed high rigidity applications. Increase in the length of bearing (shown in Fig. 9.1) increases the load carrying capacity of the bearings. Apart from cylindrical and needle rollers, there are taper roller (Fig. 9.1d) and spherical roller (Fig. 9.1e) used in rolling element bearings.

In most of the assemblies, the inner ring is mounted on the shaft and outer ring is engaged in the housing. The type of fitting between inner ring and shaft, and outer ring and housing affects the performance of rolling element bearing. Improper bearing fits, too loose or too tight, causing excessive wear and shortens the bearing life. Typically, the rotating ring requires an interference fit, while the non-rotating ring needs a transition fit. To rotate the inner ring with the angular speed of shaft, interference fit between the shaft and inner ring of the bearing is required. Similarly to fix the outer ring in housing, transition fit is required.

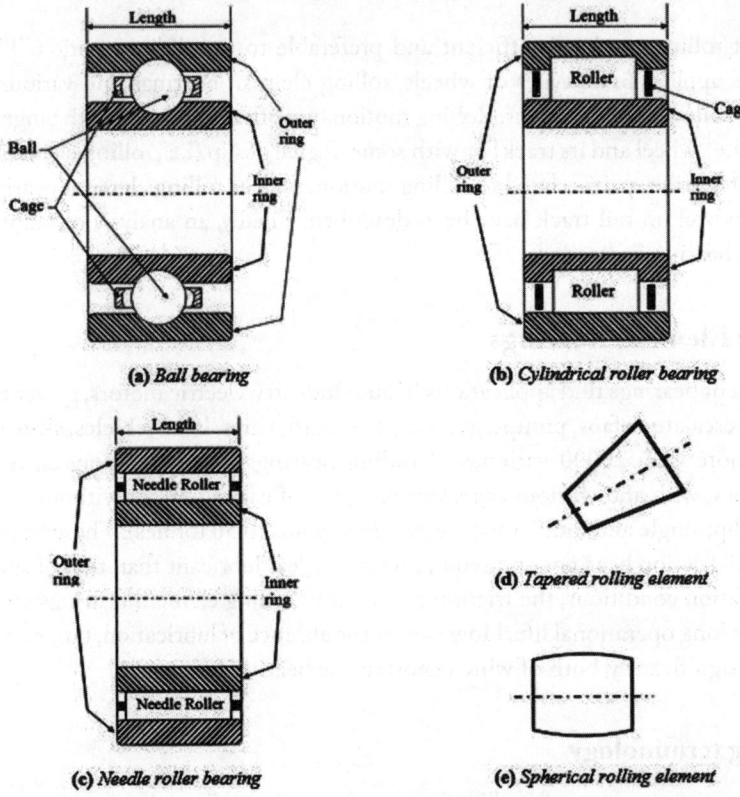

Fig. 9.1 Four main components of rolling element bearings

Three arrangements of ball bearings with different curvatures of the inner and the outer raceways of the contacting surfaces are shown in Fig. 9.2. In Fig. 9.2, rolling curvatures of outer surface of inner ring and inner surface of outer ring have been modified. Figure 9.2(a) depicts a bearing where the radius of the inner raceway is taken to be 0.52 times the ball diameter (i.e., 'r' > 0.52d, where d is the diameter of ball) so as to reduce the coefficient of friction. In this case bearing can sustain radial load and roughly points of contact between ball and rings can be maintained. Even though such open race curvature reduces the friction, it increases maximum contact stresses and consequently reduces bearing fatigue life.

In the second configuration, shown in Fig. 9.2(b), the ball and outer ring are conformal. This configuration allows radial and one-side axial load. In the third configuration, shown in Fig. 9.2(c), conformity between ball and inner ring as well as ball and outer ring is maintained. This kind of bearing will be able to sustain axial load in both directions in addition to radial load.

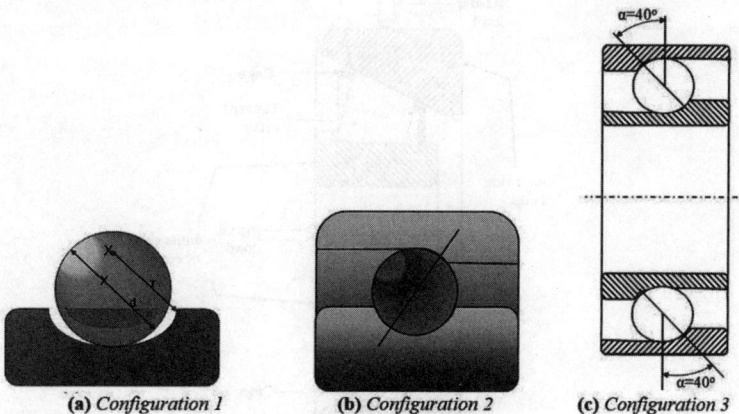

Fig. 9.2 Ball bearings

Figure 9.3 shows the two arrangements of cylindrical roller bearings. In the first arrangement (Fig. 9.3a) the inner ring is flat and there is no groove arrangement. In the second arrangement (Fig. 9.3b) the inner ring has groove for the placement of rollers. This groove is located on the outer surface of the inner ring. The roller bearing, shown in Fig. 9.3(b), can sustain the axial load (load that is parallel to the roller axis). The bearing shown in Fig. 9.3(a) may not be able sustain the axial load applied on the inner ring. In that case, the ring may slide in one direction; while in the arrangement shown in Fig. 9.3(b), the shoulder restricts the motion of the slider. There is a possibility of plastic deformation or rupture of the shoulder depending on the strength of the shoulder and the applied load.

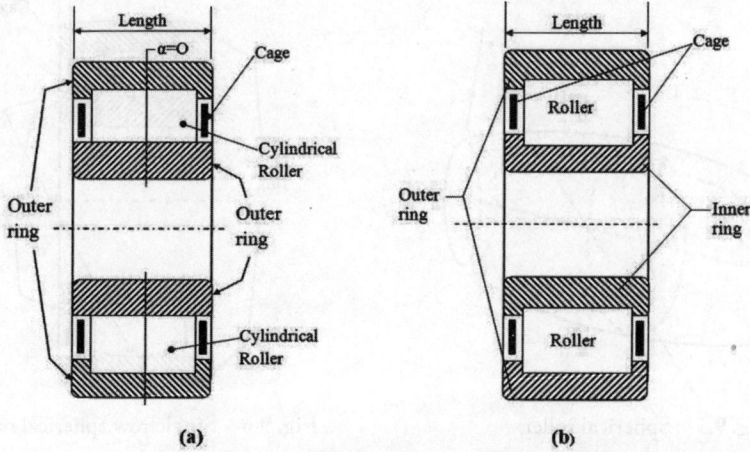

Fig. 9.3 Cylindrical roller bearings

Let us consider the taper roller bearing as shown in Fig. 9.4. In such bearings, rollers are tapered (frustum of a cone), and inner and outer rings are also inclined. In other words axes of the roller, the outer surface of the inner ring and the inner surface of the outer ring are inclined to the shaft axis. In taper roller bearing terminology differs. The inner ring is known as cone and outer ring is termed as cup. To separate the rolling elements a cage, shown in Fig. 9.4, is used.

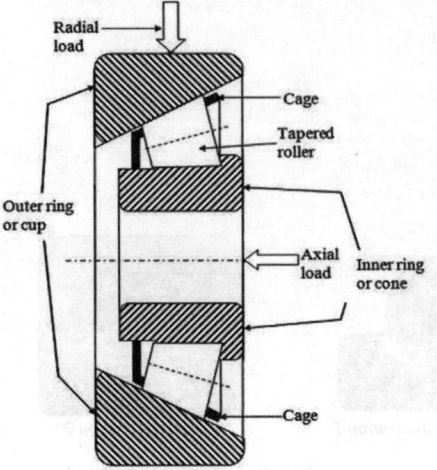

Fig. 9.4 Taper roller bearing

The taper roller bearing can sustain the radial load, which is perpendicular to the axis of the direction and as well as one-direction axial load, which is parallel to the axis of rotation. To deal with misalignment, the roller may be made spherical as shown in Fig. 9.5. In this figure R_y represents the curvature of the roller.

Let us consider the case of single row spherical roller bearing as shown in Fig. 9.6. In this bearing, curvature of spherical roller permits some degree of misalignment.

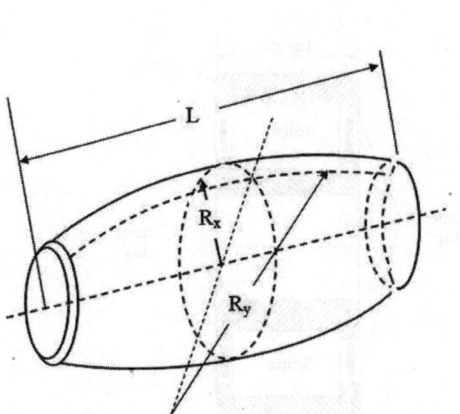

Fig. 9.5 Spherical roller

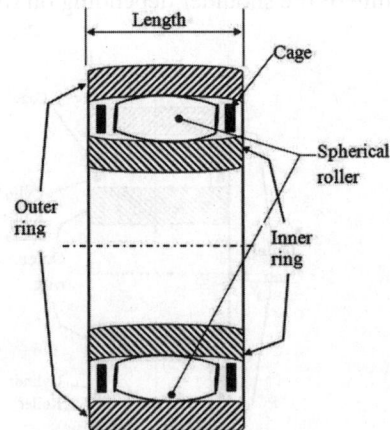

Fig. 9.6 Single row spherical roller bearing

9.1.2 Classification of rolling bearings

The bearings can be classified based on the capability of bearing the load and geometric construction of the bearing. Based on the load, bearings can be classified as:

1. Radial (load) bearings i.e., journal bearing
2. Thrust (load) bearings i.e., axial bearing
3. Combined (load) bearings i.e., taper roller bearing

Based on the requirements, bearings can be made on some angle i.e., the angle of race and the axis of rotation. Often sustainability of load depends on the geometric configuration of rolling elements. Few arrangements of rolling bearings are sketched in Fig. 9.6. In this figure, angle α represents the angle between the shaft axis and the plane of curvature of the rings.

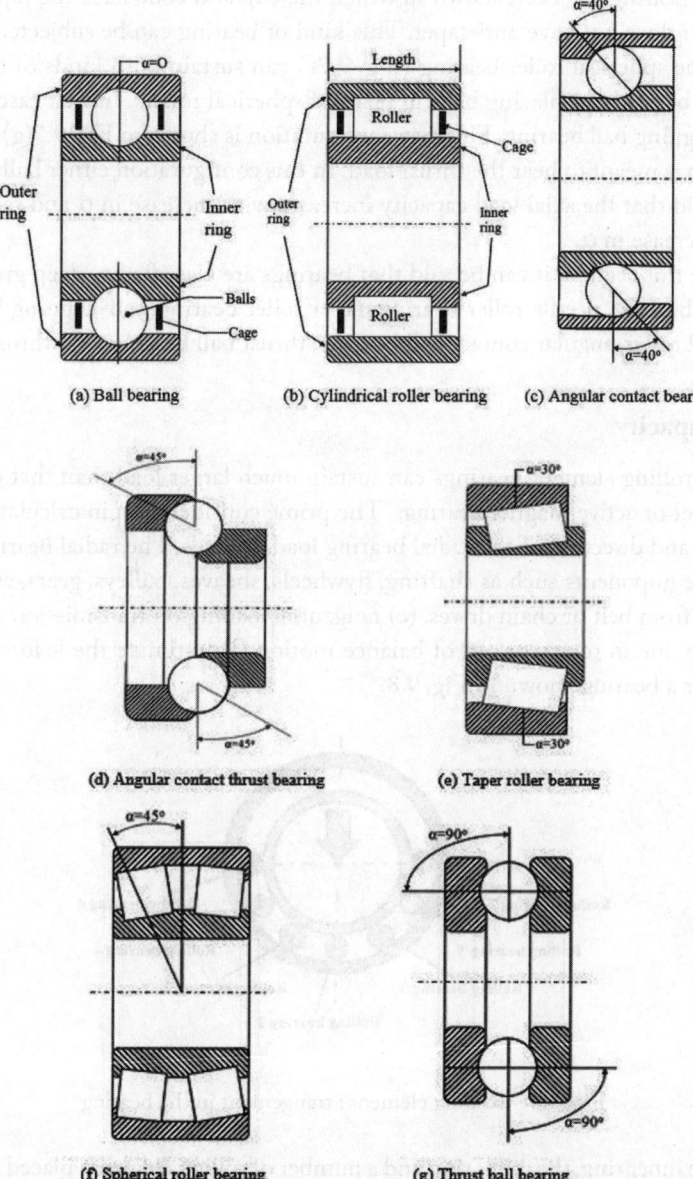

Fig. 9.7 Arrangements of rolling element bearings

In Figs. 9.7(a) and 9.7(b), the angle between the axis and the plane of curvature of the rolling element is zero (i.e., $\alpha = 0°$); therefore, both of these bearings shall be used mainly for radial load applications. The bearing shown in Fig. 9.7 (a) is termed as ball bearing or sometime deep groove (due to deeper groove in outer ring to accommodate the rolling balls) ball bearing. If roller diameter is much lesser than its length, then the bearing shown in Fig. 9.7(b) will be called as needle roller bearing (Fig. 9.1c). Otherwise it is known as roller bearing. In the third and the fourth arrangements (Figs. 9.7c and 9.7d), the angle of contact between the axis and the plane of curvature is 45° (i.e., $\alpha = 45°$), which means these kinds of bearings can sustain the thrust load as well as the radial load. In Fig. 9.7e, a roller bearing has been shown in which the cup and cone have the taper ($\alpha = 30°$) side, whereas, the roller does not have anti-taper. This kind of bearing can be subjected to radial as well as thrust load. The spherical roller bearing (Fig. 9.7f) can sustain both kinds of the loads. Similar arrangement can be made by placing balls in place of spherical rollers. In that case, bearings will be termed as self-aligning ball bearing. Finally a configuration is shown in Fig. 9.7(g) with $\alpha = 90°$ and this configuration is meant to bear the thrust load. In this configuration either balls or rollers can be used. It can be said that the axial load capacity increases with increase in α and radial load capacity decreases with increase in α.

To summarize this section, it can be said that bearings are classified as deep groove ball bearing, cylindrical roller bearing, needle roller bearing, taper roller bearing, self-aligning ball bearing, self-aligning spherical roller, angular contact ball bearing, thrust ball bearing, and thrust roller bearing.

9.1.3 Load capacity

For a given area, rolling element bearings can sustain much larger load than that of hydrodynamic, permanent magnet or active magnet bearings. The prime consideration in calculating bearing loads is the magnitude and direction of the radial bearing load or loads. The radial bearing load is (a) due to the weight of components such as shafting, flywheels, sheaves, pulleys, gears, etc., (b) due to the tension resulting from belt or chain drives, (c) tangential loading on transmission devices (i.e., gear) or inertia loading due to rotary or out of balance motion. To estimate the load capacity of rolling bearings, consider a bearing shown in Fig. 9.8.

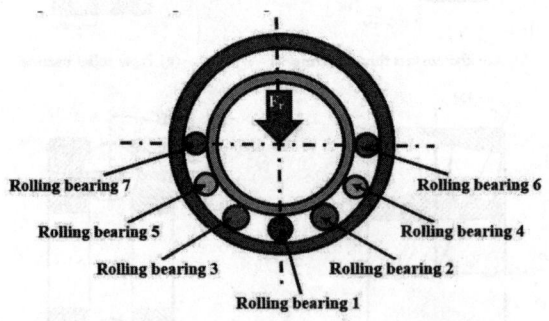

Fig. 9.8 Rolling element arrangement in the bearing

In Fig. 9.8, the inner ring, the outer ring and a number of rolling elements placed between the inner and outer rings have been shown. In such an arrangement either the inner or the outer ring rotates, and as a result, rolling elements rotate. Due to rotation of rolling elements, each rolling element is subjected to variable load (zero to maximum load). In other words, even if there is steady radial load, rolling elements are subjected to fatigue loading.

There is a possibility of some clearance between the inner ring, the outer ring and the rolling element. When either the inner ring or the outer ring is fixed and the other ring is free to move, displacement can take place in either an axial or radial direction. This amount of displacement (radially or axially) is termed the internal clearance and, depending on the direction, is called the radial internal clearance or the axial internal clearance. However, in some applications few types of (i.e., angular contact ball and tapered roller) bearings are fitted to create negative internal clearance before operation. This is called 'preload' and is commonly applied to make the bearing extremely rigid so that even when load is applied to the bearing, radial or axial shaft displacement does not occur. Thus, the natural frequency of the shaft is increased, which is suitable for high speeds.

On exerting radial load (F_r) on the inner ring, whole load gets transferred to the rolling elements and finally the load gets assigned to the outer ring. The load on any rolling element depends on its angular position with respect to load line. For example, the rolling element 1 shown in Fig. 9.8 is subjected to the maximum load. The rolling element 2 has lesser load intensity as compared to the rolling element 1. The load on the rolling element 3 is even lesser compared to that of the rolling element 1 and rolling element 2. If the clearance between the rolling element and the rings is almost zero then only 50% of the outer ring is subject to the load, while remaining 50% of the outer ring remains unloaded. In the presence of clearance, load on each rolling element increases as the load portion of the outer ring decreases. To quantify this, the load angle (ψ_1) that represents 50% of angular extent of the load zone, is used.

$$\psi_1 = \cos^{-1}(C_d/2\delta_r) \tag{9.1}$$

where, C_d is the diametric clearance and δ_r is radial shift of outer ring. From this expression it can be inferred that when C_d is zero the load angle (ψ_1) is 90°. It is desirable to keep C_d as small as possible.

Under load, each rolling element becomes elastically deformed and the deflection of rolling element (δ) under load can be derived as

$$W = K.\delta^n \tag{9.2}$$

where, W is the applied load, K is the proportionality constant, n = 1.5 for balls and n = 1.11 for rollers. The load on any rolling element can be estimated using Eq. 9.3 [Brandlein et al., 1999].

$$W_\psi = W_{max}(\cos\psi)^n \tag{9.3}$$

If ψ is zero, then load will be the maximum as shown on the rolling element 1 in Fig. 9.7. The total load (F_r) transmitted to the outer ring can be expressed as follows:

$$F_r = \sum_{\psi=-\psi_1}^{\psi=\psi_1} W_\psi \cos\psi \tag{9.4}$$

The total load (F_r) on rolling bearing is given as [Brandlein et al., 1999]:

$$F_r\big|_{ball} = (Z/4.37).W_{max}$$

$$F_r\big|_{roller} = (Z/4.06).W_{max} \tag{9.5}$$

where, Z is the number of rollers in roller bearing or number of balls in ball bearing; W_{max} is the maximum load on any rolling element. From Eq. 9.5, it can be inferred that for an applied load (F_r), increasing numbers of rolling elements reduces the value of maximum load on each rolling element (W_{max}). The static load capacity, expressed in Eq. 9.5, is generally given in terms of permanent deflection of rolling elements. In the catalogues bearing load capacity is defined in terms of static load rating and dynamic load rating. The dynamic load rating (C) is the radial load (thrust load for thrust bearings) which a group of identical bearings with stationary outer rings can theoretically endure for one million revolutions of inner ring. Similarly the static load rating (C_0) is defined as radial load (thrust load for thrust bearings) which causes permanent deflection equal to 0.01% of diameter of rolling elements.

The dynamic load capacity (C) can be used to estimate fatigue life of the bearing. This can be found using the Lundberg Palmgren approach [Brandlein et al., 1999], given as

$$(C)^a 10^6 = P_1^a L_1 = P_2^a L_2 = P_3^a L_3 \qquad (9.6)$$

The exponent 'a' is equal to 3 (a = 3) for ball bearings, while it is equal to 10/3 ($a = \frac{10}{3}$) for roller bearings. If $P_1 > C$, then the bearing life L_1 will be less than 10^6 rotations. Equation 9.6 is valid only for reliability (survivability) of 90%. Sometimes the term L10 life is used to indicate probability of failure is equal to 10%. For bearings operating at constant speed it may be more convenient to deal with a basic rating life in operating hours, by rearranging the Eq. 9.6:

$$\text{Bearing life in hours} = \left(\frac{C}{P}\right)^a \frac{1000,000}{60 \text{ Speed}} \qquad (9.7)$$

Some of the important observations from Eq. 9.6 are:

- Bearing failure is very sensitive to the applied load.
- If P = 0.5C ... Life is increased by 10 times.
- If P = 0.25C ... Life is increased by 100 times.

The bearing life has 90% reliability as expressed by the previous relation. In order to get more reliability, life adjustment factor (a_1), must be multiplied to right hand side of Eq. 9.7. This life adjustment factor is expressed in Table 9.1.

Table 9.1 Failure probability vs. factor a_1

Failure probability (%)	Factor a_1
10	1
5	0.62
4	0.53
3	0.44
2	0.33
1	0.21

In the presence of adequate lubricant and absence of contamination, bearings may attain a much longer life than predicted by standardized life calculation (Eq. 9.7) method.

Example 9.1: A rolling bearing is required to support a shaft of 25 mm diameter. The bearing will be subjected to radial load of 2224 N at 1500 rpm shaft-speed. There is a possibility of some shock loading or some transient loading in between and to account that loading factor equal to 1.5 may be considered. The minimum operating life of bearing must be 8 (hours in a day)*5 (days in a week)*52 (weeks in a year)*5 (years). Find a suitable rolling bearing.

Answer: Ball or roller bearing may be selected for the specified application. To choose a bearing for a desirable service life, it is necessary to estimate the required dynamic load capacity expressed in Eq. (9.6).

$$(C)^a 10^6 = P_1^a L_1 \tag{9.6}$$

In the present case,

$$P_1 = \text{Load*load factor}$$
$$\Rightarrow P_1 = 2224 * 1.5$$
$$L_1 = 8*5*52*5*1500*60$$

On substituting values of P_1 and L_1,

C ≥ 2224*1.5*(10400*1500*60/10⁶)^(1/a)
C ≥ 32,633 N for Ball Bearings
C ≥ 25,978 N for Roller Bearings

This means if roller bearing is selected then dynamic load capacity needs to be greater than 25978 N. Similarly, if ball bearing is selected then dynamic load capacity must be more than 32633 N. The roller/ball bearing can be chosen from bearing catalogue [SKF catalogue]. Roller bearing N 205ECP can be picked as its dynamic load capacity is 28600 N and permissible operating speed is 11000 rpm. Similarly, ball bearing 6405 can be selected as its dynamic load capacity is 35800 N and permissible speed is 9000 rpm.

In the example 9.1, bearings (roller and ball) have been selected based on the applied radial load. The selection of bearing depends on the loading conditions.

For example, deep groove ball bearings can sustain radial load with slight axial load. These bearings are advantageous due to their lower friction and quiet running operation. Spherical roller bearings are suitable for heavy load application requiring self-aligning characteristics.

Table 9.2 provides superficial guidance in selecting rolling bearing based on the applied load conditions.

Table 9.2 Rolling element bearing vs. load

Application of load	Rolling element bearing
Pure radial load	Cylindrical, Needle roller
Pure axial load	Thrust (Cylindrical roller, ball), four point angular contact ball bearings
Combined load	Taper roller, spherical roller, angular contact ball bearings

In addition to load, operating speed, running accuracy and misalignment capacities affect the bearing selection. For high running accuracy, deep groove, angular contact and cylindrical roller bearings are preferable. Table 9.3 provides better guidance, compared to Table 9.2, to select the bearing.

Table 9.3 Misalignment capabilities of rolling element bearing

Rolling element bearings	Load direction			Misalignment capacity		
	Radial	Axial	Both	High	Med	Low
Deep groove ball	Y		Y	Y		
Cylindrical Roller	Y		Some types			Y
Needle	Y					Y
Taper Roller	Y	Y	Y			Y
Self-Aligning Ball	Y		Y	Y		
Self-Aligning Spherical Roller	Y		Y	Y		
Angular contact ball		Y	Y			Y
Thrust ball/roller		Y				Y

Figure 9.9 depicts a bearing subjected to both radial and axial loads.

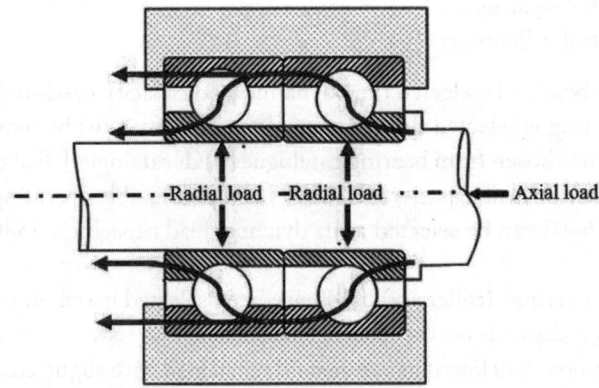

Fig. 9.9 Load transfer under combined loading

The axial and radial loads are always in phase of 90°. Therefore, the resultant load should be calculated using $P = \sqrt{F_r^2 + F_a^2}$. But the damage done by the thrust load is different than damage caused by radial load, so it is important to find equivalent radial load that causes the same damage as combination of thrust and radial loads. It needs extensive experimentations. To simplify the situation, the factors X and Y are used to determine the equivalent dynamic load as expressed in the following equation.

$$P = VXF_r + VF_a \qquad (9.8)$$

where, V is the rotation factor (1 if the inner ring rotates, 1.2 if the outer ring rotates), X is the radial factor, F_r is the applied radial load, Y is thrust factor and F_a is the applied thrust load.

The values of X and Y factors depend upon the bearing geometry, number of balls and size of balls. Therefore the values of X and Y factors shall be determined from the bearing catalogue. These factors for few bearings are listed in Table 9.4. In this table 'e' is a dimensionless ratio. As illustrated in Table 9.4, if $F_a/F_r \leq e$ then the axial load does not affect at all or the equivalent load will be radial load itself.

Table 9.4 Rolling element bearing catalogue

Bearing type			Inner ring		Single row		Double row				e
			Rotating	Stationary	$F_d/VF_r > e$		$F_d/VF_r \leq e$		$F_d/VF_r > e$		
			V	V	X	Y	X	Y	X	Y	
Deep groove ball bearing	F_d/C_o		1	1.2	0.56	2.30	1	0	0.56	2.30	0.19
	0.014					1.99				1.99	0.22
	0.028					1.71				1.71	0.26
	0.056					1.55				1.55	0.28
	0.084					1.45				1.45	0.3
	0.11					1.31				1.31	0.34
	0.17					1.15				1.15	0.38
	0.28					1.04				1.04	0.42
	0.42					1.00				1.00	0.44
	0.56										
Angular Contact ball bearing	20		1	1.2	0.43	1.0	1	1.09	0.70	1.63	0.57
	25				0.41	0.87		0.92	0.67	1.44	0.68
	30				0.39	0.76		0.78	0.63	1.24	0.80
	35				0.37	0.66		0.66	0.60	1.07	0.95
	40				0.35	0.57		0.57	0.57	0.93	1.14
Self-aligning ball bearing			1	1	0.4	0.4 cotα	1	0.42 cotα	0.65	0.65 cotα	1.5 tanα

$$P = Fr$$

If a bearing is clean, properly lubricated, mounted and sealed, and operated at reasonable temperature then metal fatigue will be the only cause of failure, which is manifested in the form of pit formation. The sub-surface cracks are initiated due to shear stresses and propagate to the surface due to cyclic loading and unloading on the outer ring. A chunk of the material is removed and it results in pitting of the surface. This is depicted in Fig. 9.10.

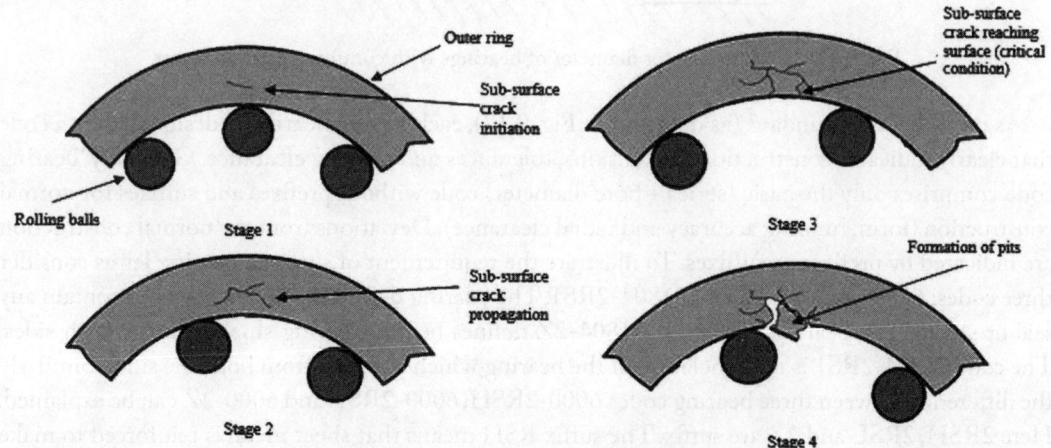

Fig. 9.10 Stages in pitting failure in rolling motion

9.1.4 Standardization

The rolling bearing industry with the help of ASTM committee has standardized the rolling bearings to reduce the manufacturing cost. As per ASTM, the world market for rolling bearing is about $20.5 billion annually. The dimensions, materials, types and fitting of rolling element bearings are standardized to promote exchangeability. As the standard rolling bearings are produced in mass as per the specified standards, the bearings are economic, and the assembly and disassembly of the bearing is relatively easier. Now due to standardization appropriate bearing can be selected based on the operational requirements (space, load, speed, etc.). For example, bearings with different outside diameters OD_1, OD_2, OD_3, OD_4 etc., $(OD_4>OD_3>OD_2>OD_1)$, as shown in Fig. 9.11, are available with the same inner diameter (ID). This variation in OD is often represented as 'diameter series' in bearing code. The diameter series is represented with a number such as 8, 9, 0, 1, 2, 3, or 4. Number 8 represents least dimension of outer diameter for any radial rolling bearing and number 4 represents the maximum dimension of outer diameter of that bearing. The default number remains 0 (zero) to indicate nominal dimension of outer diameter. Generally increase in the dimension of outer diameter increases the load capacity of the bearing. This means the '8 number' diameter series will have the least load carrying capacity and '4 number' diameter series will have the maximum load capacity for that bearing.

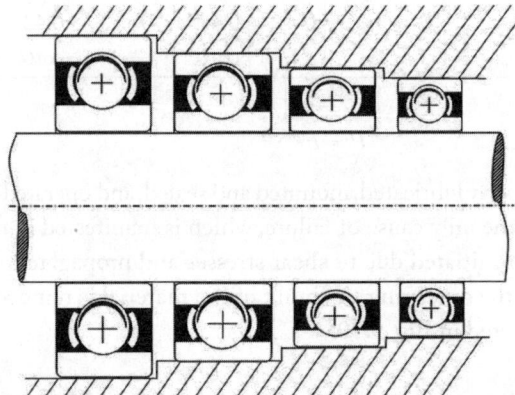

Fig. 9.11 Different outer diameter of bearings with common shaft diameter

As per DIN-623 standard (as depicted in Fig. 9.12), each rolling bearing is designated by a code that clearly indicates construction, dimensions, tolerances and bearing clearance. Generally 'bearing code' comprises only the basic (series + bore diameter) code without prefixes and suffixes for normal construction (form, running accuracy and radial clearance). Deviations from the normal construction are indicated by prefixes or suffixes. To illustrate the requirement of suffix in bearing let us consider three codes: 61804, 61804-2Z and 61804-2RS1. The bearing defined as 61804 does not contain any seal or shield. The bearing with code 61804-2Z defines bearing having shields on the both sides. The code 61804-2RS1 is nomenclature of the bearing which is sealed from both the sides. Similarly the difference between three bearing codes 6000-2RSH, 6000-2RSL and 6000-2Z can be explained. Here 2RSH, 2RSL and 2Z are suffix. The suffix RSH means that sheet metal is reinforced to make it as contact seal. The word H stands for high coefficient of friction and L stands for low coefficient of friction. It is important to know that sealed (2RS1) bearings can retain the grease and keep dust away from the bearings, but these bearings cause higher power loss.

The 'series code' generally contains three digits. First digit represents the type of rolling bearing (i.e., '0' for double row angular contact ball bearing; '1' for self-aligning ball bearing; '3' for taper roller bearing; '4' for double row deep groove ball bearing; '5' for thrust ball bearing; '6' for single row deep groove ball bearing). Sometimes instead of first digit, alphabets (i.e., 'K' is used for needle roller bearing; 'N' is used for cylindrical bearing; 'QJ' is used for four-point contact ball bearing) are used. The second digit of 'series code' signifies its width series, and last digit denotes the bearing outer diameter.

More information about standard dimensions of bearings is available in catalogues of bearing manufacturers (SKF, Timken, NTN, FAG, and Nachi)

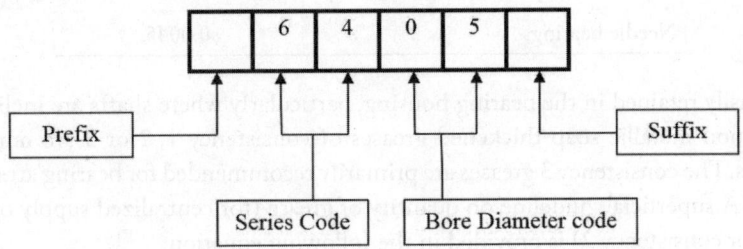

Fig. 9.12 Bearing standard series

Let us consider 6405 bearing code. Here, series code is 64 and bore diameter code is 05. There is no prefix and there is no suffix. To find the diameter of bearing bore, the bore diameter code given in the standard is multiplied by 5. This means bore diameter of 6405 bearing code is 5×5 = 25 mm. However there are some exceptions. Bore diameter code '00', '01', '02', and '03' means 10 mm, 12 mm, 15 mm and 17 mm as diameters of bearing bore, respectively. The 'bore diameter code' from 04 to 99 is multiplied by 5 to get the bore diameter of the bearing. It is interesting to note that bore diameter code is restricted from '00' to '99'. If diameter is greater than 95 mm, then actual dimension of bearing bore (511/530) is represented. For example 511/530 means series code is 511, while 530 is the diameter (530 mm) of bearing bore. Similarly unconventional diameter of bearing bore such as 8 mm is represented using slash symbol (i.e., 618/8).

9.1.5 Tribology of rolling bearings

Standardized bearings have very good surface finish and if the applied load is within 3 to 10% of dynamic load capacity (with appropriate lubrication conditions), the bearing operates in elasto-hydrodynamic lubrication regime. Appropriate lubricant minimizes sliding between rolling elements and rings, thereby minimizing friction and wear. In addition, the lubricant also protects the bearing from corrosion.

The coefficients of friction, among well lubricated rolling bearings, are listed shown in Table 9.5. Without lubricant, coefficient of friction will increase. However, excessive lubricant is harmful as it increases the coefficient of friction. As a thumb rule, at least 50% of the free space should be without any lubricant. The lubricant with the different viscosity and boundary additives may provide different outcomes. The selection of lubricant often depends on operating conditions and environment. For example, if operating speed is lesser than 1500 mm/min, solid lubricants are more suitable compared to grease or liquid lubricant. However, if speed is greater than 1500 mm/min but less than 1500 m/min, grease lubrication provides better results. Oil lubrication is more suitable to carry away the heat from bearings.

Table 9.5 Comparative study of coefficient of friction

Bearings	Coefficient of friction (f)
Self-aligning ball bearings	0.0010
Cylindrical roller bearing	0.0011
Thrust ball bearings	0.0013
Single-row deep-groove ball bearings	0.0015
Tapered and spherical roller bearings	0.0018
Needle bearings	0.0045

Grease is easily retained in the bearing housing, particularly where shafts are inclined or vertical. To reduce friction metallic soap thickened greases of consistency 1, 2 or 3 are normally used for rolling bearings. The consistency 3 greases are primarily recommended for bearing arrangements with vertical shafts. A superficial guideline on quantity of grease (for centralized supply of grease having consistency 1 or consistency 2) is provided in the following equation:

$$Q = 0.00004 \cdot D \cdot B \text{ gm/hour}$$

where D is the bearing outside diameter in mm
B is the bearing width/height in mm

It is important to note that the friction coefficient is proportional to the size of contact area between rolling elements and raceways. For example, ball bearing with four or three points of contact exhibits greater friction than bearing with two points of contacts.

Calculation of friction in a loaded bearing between rolling elements and raceways is a complex phenomenon. The elastic hysteresis, sliding, lubricant shearing and mechanical contact of seals are the main contribution of friction in anti-friction bearing. The use of chrome steel in making rolling element and rings is preferred as it has minimum coefficient ($\mu \cong 0.0001$) of elastic hysteresis in rolling. Appropriate selection and mounting of bearing maintains the low sliding speed between rolling elements and guiding surfaces of the cage. The lubricant shearing can be calculated using the following equation:

$$\tau = \eta \, du/dz$$

An empirical relation can be provided to determine the seal-friction.

Load dependent friction moment can be given as [Brandlein et al., 1999]:

$$M_P = fP(d/2) \quad \text{N.mm} \quad (9.9)$$

where

f = Coefficient of friction as given in Table 9.5.
P = Resultant load = $\sqrt{F_r^2 + F_a^2}$, N

d = Bore diameter, mm

Lubricant and speed dependent friction moment (M_L in N.mm) may be expressed as [Brandlein et al., 1999]:

$$M_L = 10^{-7} f_L \left(vN\right)^{2/3} d_m^3 \quad \text{if } vN \geq 2000 \quad (9.10)$$

$$M_L = (1.6e-5) f_L d_m^3 \quad \text{if } vN < 2000 \quad (9.11)$$

where

v = Operating viscosity of oil, mm²/s
N = Rotaional speed, rpm
f_L = Lubrication factor as given in Table 9.6.

Table 9.6 Lubrication factor f_L

Bearing type	Grease	Oil spot	Oil bath	Vertical shaft in oil bath
Deep groove ball	0.75-2	1	2	4
Self-aligning ball	1.5-2	0.7-1	1.5-2	3-4
Angular contact ball	2	1.7	3.3	6.6
Cylindrical roller	0.6-1	1.5-2.8	2.2	4
Needle roller	12	6	12	24

Seal dependent friction moment is given as [Brandlein et al., 1999]:

$$M_s = f_2 + \left(\frac{\text{Min dia} + \text{Max dia}}{f_1}\right)^2 \text{ N.mm} \qquad (9.12)$$

where f_1 and f_2 are friction factors as listed in Table 9.7

Table 9.7 Friction factors for seals [Brandlein et al., 1999]

Bearing type	f_1	f_2
Deep groove ball	20	10
Self-aligning ball	20	10
Angular contact ball	20	10
Cylindrical roller	10	25
Needle roller	20	50

where the bearing is only sealed at one side, $M_s/2$ should be used instead of M_s.

Total friction moment, $M = M_P + M_L + M_S$

Example 9.2: Estimate friction moment of 6214-2RS1 bearing running at 1,200 rpm under 5000 N radial load when jet lubricated by synthetic ester jet engine oil having a viscosity of 6 mm²/s (cSt) at operating temperature.

Answer: The bore diameter of 6214-2RS1 bearing is 70 mm and its OD is 125 mm. Using Eq. 9.9 and Table 9.5,

$$M_P = fP(d/2) \text{ N.mm}$$

$$M_P = 0.0015 * 5000 * (70/2) \text{ N.mm}$$

$\Rightarrow \qquad M_P = 262.5 \text{ mm}$

To use Eq. 9.10 or Eq. 9.11, evaluate

$$vN = 6 \times 1200$$

$\Rightarrow \qquad vN = 7200$

$$M_L = 10^{-7} f_L (vN)^{2/3} d_m^3 \quad \text{as } vN \geq 2000$$

$$\Rightarrow \qquad M_L = 10^{-7} \times 1 \times (7200)^{2/3} \left(\frac{125 + 70}{2}\right)^3 \qquad (9.10)$$

$$\Rightarrow \qquad M_L = 34.56 \text{ N.mm}$$

Finally seal friction can be calculated using Eq. 9.12

$$M_s = f_2 + \left(\frac{\text{Min dia} + \text{Max dia}}{f_1}\right)^2$$

$$M_s = 104 \text{ N.mm}$$

Total friction moment, $M = M_P + M_L + M_S = 401$ N.mm

A well installed and well lubricated rolling bearing does not fail due to wear, but it may fail due to fatigue and attains its rated life. Unfortunately, sometimes a bearing does not survive its calculated rating life. There may be many reasons for this: heavier loading than has been anticipated, inadequate or unsuitable lubrication, careless handling, or too tight fits are among those causes. There is possibility of development of surface cracks on the bearing inner and outer rings and rolling element, but heavy load and/or contaminations in lubricant play major role in the development of those cracks.

9.1.6 Case study: Analysis of failure of four row cylindrical roller bearing

In the present section, analysis of failure of four–row cylindrical roller bearings (Inner ring ID = 865 mm and Inner ring OD = 945 mm; Outer ring ID = 1073.0 mm, Outer ring OD = 1180 mm) used in back-up roll assembly of cold rolling mills has been described. An analytical approach has been utilized to determine the maximum load on roller and outer ring raceway interface. The maximum principal normal and shear stresses have been reported considering coefficient of friction equal to 0.001, 0.1, 0.2 and 0.3. Three dimensional static finite element analysis performed to explain the cracking of outer ring of bearing, has been briefed.

Four–row cylindrical roller bearings are commonly used in cold rolling mills (CRM) to bear considerable radial load (3500 kN) at relatively low rotational speed (~ 250 rpm). These large size bearings are available in 'pin-type steel' and 'pronged machined solid brass' cages. Solid brass cage roller bearings are relatively economic ($ 80,000) compared to pin-type-cage roller bearings ($ 95,000). Due to economics of brass caged bearings a company replaced pin-type-four-row-roller-bearings of company X with brass-cage-four-row-roller bearings of company Y in their cold-rolling-mills (CRM). The solid cage occupies more space and thus reduces the space available for rollers. In a typical roller bearing having ID = 865 mm and OD = 1180 mm, 34 rollers could be placed in every row of solid-brass-cage bearing, while 38 rollers were arranged in pin-type-cage bearing. Reduction in the number of rolling elements reduces the static and dynamic capacity of brass-cage four–row roller bearings compared to pin-type-bearing, as depicted in Table 9.7.

Table 9.7 Load carrying capacity of pin type and brass type cage

Load carrying capacity	Pin type cage	Brass type cage
Dynamic Load Capacity	23,300 kN	21,600 kN
Static Load Capacity	67,000 kN	60,000 kN

To determine the effect of reduced load capacity on bearing life, the following equation can be used

Roller bearing life $\propto (\text{Dynamic capacity})^{10/3}$

or, $\dfrac{L_{Solid-brass-cage}}{L_{pin-type-cage}} = \left(\dfrac{21600}{23300}\right)^{10/3} = 0.8$

This equation clearly suggests that replacing 'pin-type cage bearing' with 'solid-brass-cage bearing' may reduce the bearing life by 20%. Under normal conditions (oil viscosity at 40°C= 470 cSt, oil viscosity at 100°C = 31 cSt, operating temperature < 55°C, and radial load = 3400 kN ± 5%) pin-type bearing provides an estimated life of 40,000 hours, therefore under the same conditions solid brass cage bearing should survive for 32,000 hours. Unfortunately, after installation of solid-cage-four-row roller bearings, these bearings were uninstalled within 300 operating hours due to excessive vibration and noise from CRM. Figure 9.13 shows the failure of outer rings of two gigantic roller bearings. Outer ring of one bearing was splintered to pieces (Fig. 9.13 (a)), while 25% of outer ring of second bearing (Fig. 9.13 (b)) disintegrated to 10–12 pieces. Such pre-mature failure of two bearings cannot be just a chance; therefore author was motivated to investigate the root cause of bearing failure. It was necessary to pinpoint the root cause of bearing failure so that it can be prevented from happening again.

(a) Outer ring after 300 operating-hours (b) Outer ring after 105 operating hours

Fig. 9.13 Failure of four–row roller bearing

The visual observation of bearing outer ring, shown in Fig. 9.14, indicates the most common rolling contact fatigue failure, which is also termed as 'pitting failure'. There is no point to analyze such failures, as pitting (spalling) of rolling surfaces is a natural fatigue phenomenon. On comparing Fig. 9.13 with Fig. 9.14, it is clear that failure of outer ring, shown in Fig. 9.13, does not fall in the category of 'pitting failure'. According to Brandlein et al. [Brandlein et al., 1999], if a bearing fails earlier than its predicted fatigue life (like in the present study predicted life = 32,000 hours, while bearing failed within 300 hours), then overloading by external forces, poor lubrication, contaminants in the lubricant or faulty installation are possible causes of bearing failure. All these failures leave their traces on bearing surfaces, which can qualitatively be identified by visual observation.

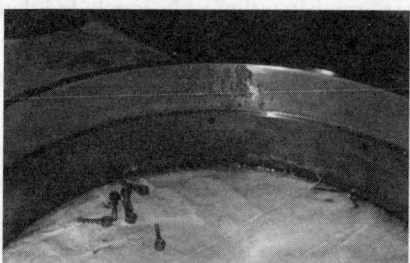 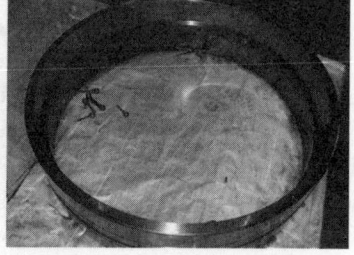

Fig. 9.14 Spalling failure of outer ring of pin-type roller bearing

Ai and Nixon [Ai and Nixon, 2000] performed experimental study on reduction of roller bearing fatigue life due to debris denting, and concluded that ductile particles cause severe life reduction than brittle particles. Visual observation of failed bearing, carried out in the present study, revealed mild abrasion (Fig. 9.15) on the roller surfaces. Therefore, particle abrasion cannot be the root cause of outer ring failure.

Fig. 9.15 Mild abrasive wear of rollers

Figure 9.16 illustrates the possibility of corrosion or fretting–corrosion of outer ring of failed bearing. Cantley [Cantley, 1977] studied the effect of water on bearing fatigue life and concluded detrimental effects of water concentration in lubricating oil on the fatigue life of rolling bearings. Bearings that operate in an environment where water is absorbed in lubricant may be subjected to pitting corrosion by hydrolysis. Sulphur and chlorine of lubricant could react with the water to attack on the steel on a microscopic scale. However in the present study, outer ring was disintegrated in a number of pieces, which just cannot happen due to water–ingress. Therefore water contamination cannot be the root cause failure of outer ring in question.

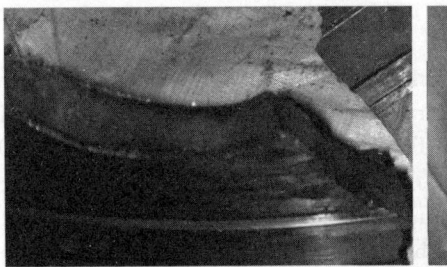 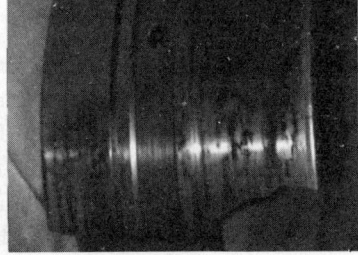

Fig. 9.16 Fretting corrosion of outer ring

Pearson [Pearson, 1990] after analysis of 14 bearing failures indicated that tensile stresses as low as 140 to 200 MPa were sufficient to cause cracks to grow in fatigue. He concluded that when tensile stresses drive the cracks in radial direction, fracture rather than spalling would occur. Figure 9.13 of the present study, depicts the cracking of outer ring. Prerequisite for such cracking is initiation of cracks and determining the source of tensile stress which leads crack to grow in fatigue.

Figure 9.17 shows an interesting failure of the outer ring. It shows beach marks which are indicative of fatigue failure. In addition, this figure shows ruptured surface, which is indicative of excessive loading on the outer ring raceway. In short, the visual examination of Figs. 9.13–9.17 clearly indicates 'CRACKING' as the main source of bearing damage. Mainly excessive load cracks rolling surfaces. Therefore identifying the reasons for excessive load should be the first target of the study.

 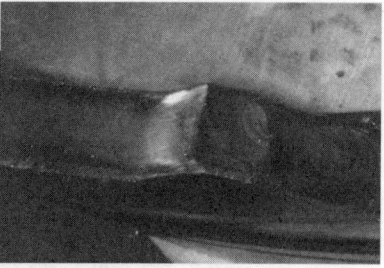

Fig. 9.17 Beach marks indicating fatigue failure of the outer ring

Excessive load occurs due to: (a) wrong tolerances between housing (choke) and outer ring, (b) excessive operational load (c) improper load distribution, or (d) misalignment. After analyzing the detail dimensions, tolerances and operational load data, it was concluded that the only possibility of bearing failure was improper load distribution.

On analyzing logbook load data it was observed that the nearly constant direction 3000–3400 kN load was supported by four-row-roller-bearing. Assuming uniform distribution of load among four rows of roller bearing, leads to conclude that 850 kN (= 3400/4 kN) was applied on each row of rollers. Subsequently this 850 kN load from rollers is transferred to a certain portion of cylindrical outer ring. The load distribution on outer ring depends on the number of rollers and radial internal clearance. The radial clearance for multi-row gigantic bearings was in the range of 0.59–0.73 mm. Such large clearance generally reduces the extent of load zone and increases the maximum value of load. As per available design of the bearing, only a quarter of the outer ring supports complete load. Therefore the side face of the outer ring is divided into four zones indicated by I–IV. When a four-row-roller-bearing is mounted first time, zone I is positioned in the direction of the applied load and after a period of 1000 operating hours the outer ring is supposed to be turned through 90°. Outer ring of roller bearing, studied in the present paper, contains maximum of nine (~34/4) rollers, therefore 850 kN load was shared by nine roller at a time. In solid brass cage, each roller was placed at angular separation (ψ) of 10.59° (= 360/34) from nearby two rollers. For static equilibrium, the applied radial load (850 kN) must equal the sum of the vertical components of the rolling element loads [Harris, 2001],

$$850,000 = Q_{max}\left[1 + 2\left(\sum_{\psi=10.59, 21.18}^{31.77, 42.36}\left[1 - \frac{\delta_r}{\delta_r - radial\ clearance}(1 - \cos\psi)\right]^{1.11}\cos\psi\right)\right] \quad (9.11)$$

where ring radial shift, δr, is evaluated using following equation [Harris, 2001],

$$\delta_r = \frac{radial\ clearance}{\cos\left(angular\ extent\ of\ load\ zone\ /\ 2\right)}$$

Or,

$$\delta_r = \frac{0.66\ mm}{\cos(\pi / 4)} = 0.93\ mm \quad (9.12)$$

The maximum load, Q_{max}, on any roller can be determined by employing Eq. (9.11), which comes out to be 166,550 N. The load distribution among nine rollers in load zone at a moment of time is shown in Fig. 9.18.

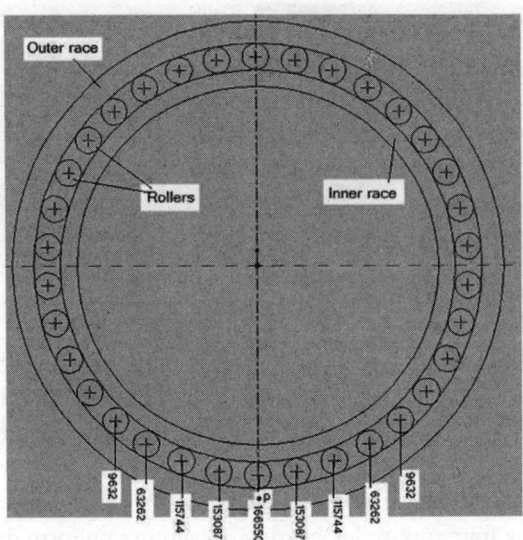

Fig. 9.18 Load (N) distribution in roller bearing

This load distribution indicates that every roller is subjected to variable load which varies from zero to maximum 166,550 N. Therefore, rollers are subjected to fatigue loading even though applied load is unidirectional and nearly constant magnitude. Due to dynamic loading on rollers, the outer ring also experiences dynamic load. The point p on outer ring (being stationary) as shown in Fig. 9.18, experiences $\frac{34}{2} \times (1 - 64/1009) \approx 16$ times load variation in one rotation of inner ring.

Therefore it can be said that outer ring is subjected to dynamic load up to 166,550 N and frequency of this variation is 240000 (= 15.92*250*60) per hour. Such high radial load with high frequency (due to rolling) along with tangential load (due to sliding) induces contact stress (which causes pitting failure), shear stress (which causes crack initiation and may cause crack propagation) and tensile stress (which causes rapid failure) in the outer ring. To determine all three stresses analytically, it is necessary to model cylinder-on-cylinder geometry having roller radius equal to 32 mm and outer-ring radius equal to − 536.5 mm (negative sign indicates concave surface). The plane strain condition with x axis aligned to the direction of rotation, and the y-axis radial to the rollers can be considered. The total stress can be determined by superimposing the component due to the normal and tangential loads [Norton, 2001]:

Stress = Stress due to normal + Stress due to frictional loading

$$\sigma_x = \sigma_{xn} + \sigma_{xt}$$
$$\sigma_y = \sigma_{yn} + \sigma_{yt}$$
$$\sigma_z = \nu(\sigma_{xn} + \sigma_{xt})$$
$$\tau_{xy} = \tau_{xy_n} + \tau_{xy_t} \qquad (9.13)$$

In Eq. 9.13 subscript n denotes the normal load while subscript t denotes the tangential friction force. To find the largest tensile and compressive stresses, it is necessary to find principal normal stresses. The expression relating the applied stresses to the principal stresses is [Norton, 2001]

$$\begin{bmatrix} \sigma_x - \sigma & \tau_{xy} & 0 \\ \tau_{yx} & \sigma_y - \sigma & 0 \\ 0 & 0 & \sigma_z - \sigma \end{bmatrix} \begin{bmatrix} n_x \\ n_y \\ n_z \end{bmatrix} = 0 \qquad (9.14)$$

where σ is the principal stress magnitude and n_x, n_y, and n_z are the direction cosines of the unit vector n. The maximum shear stress can be determined from the values of the principal normal stresses using [Norton, 2001]:

$$\tau_{max} = \left| \frac{\sigma_{max} - \sigma_{min}}{2} \right| \qquad (9.15)$$

To evaluate the stress levels, Young modulus equal to 205.8 GPa and Poisson ratio equal to 0.27 have been employed. Using Eqs. 9.14–9.15 and Viete's cubic root finding method, the maximum principal normal stress, σ_{max} equal to 1.205 GPa (compressive), the minimum principal normal stress, σ_{min} equal to 546.75 MPa (compressive) and the maximum shear stress, τ_{max} equal to 329 MPa have been calculated for coefficient of friction, μ, equal to 0.1. However increasing the value of μ = 0.2, increases σ_{max} to 1.349 GPa, increases σ_{min} to 563.42 MPa and τ_{max} to 393 MPa. Increasing μ to 0.3, increases σ_{max} to 1.494 GPa, increases σ_{min} to 580 MPa and τ_{max} to 457 MPa. For frictionless (μ = 0.001) interface σ_{max} = 1.062 GPa, σ_{min} = 530 MPa and τ_{max} = 266 MPa has been calculated. These results indicate that increasing coefficient of friction increases shear stresses significantly, which may cause the shear fracture of roller bearings. However, this stress analysis clearly indicates that tensile stress is completely absent. As per Pearson [Pearson, 1990], fracture of large-size-ring mostly happens due to tensile stresses. Therefore it was necessary to revisit the failed bearing.

Fig. 9.19 Material handling holes in bearing outer ring

Figure 9.19 shows the outer ring fastened with chains through bolts. This figure indicates that heavy weight outer ring was made with four tapped hole to facilitate its handling. This figure also indicates that fourth hole, which is not shown in figure, was placed in load zone and was subjected to 166,550 N of load. Surprisingly these four holes were missing in the original drawing of outer ring supplied by manufacturer. On interaction with shop–floor worker, it was revealed that four holes of 3/8' 10 UNC 3B of 45 mm depth were drilled and tapped to facilitate the handling of outer ring. As per IS 2473 [Indian Standards, 2002], centre hole in any work piece reduces the load area. If the hole at the centre is to be removed from the finished product then parting off dimension needs to be greater than the length of hole. In other words, the effective (load bearing) length will be less than the length of work piece having a hole at the centre. Further, due to hole at the centre of ring thickness, effective thickness of outer ring decreases to half, thus making outer ring very flexible. Increase in flexibility of outer ring reduces the extent of load zone and increases the value of the maximum load on roller–raceway interface. Figure 9.20 shows the fracture near hole, which confirms the guess that weakest portion of the outer ring was subjected to maximum load and resulted fracture of outer ring. The detailed study using the finite element modelling of this bearing has been provided by Hirani [Hirani, 2009].

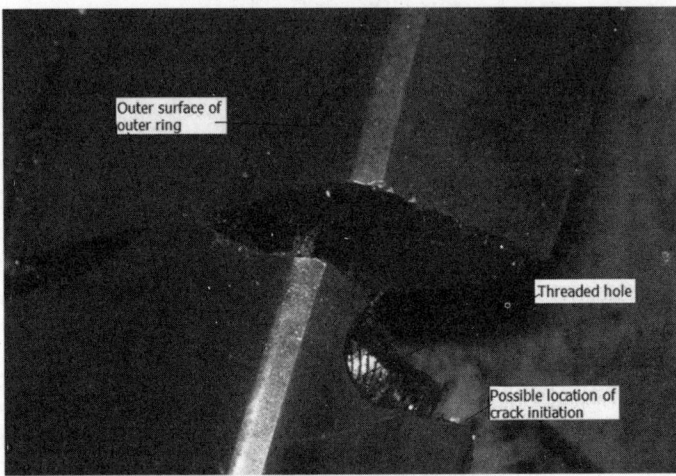

Fig. 9.20 Fracture near hole

9.2 The Mechanics of Tyre–Road Interactions

Tyre plays an important part of a vehicle as it stays in direct contact with the road surface and transfers all forces from road to vehicle and vehicle to the road. The interaction of tyre and road is subjected to diversified conditions due to variation in air pressure in the tyres, longitudinal or side forces during driving, and road-surface irregularities. The performance of tyre on the road is expected in terms of 'low rolling resistance', 'high comfort', 'low wear', 'very safe even on wet roads' and 'minimum emission of particles'. The expected requirements from tyre are often contradictory. For example, rolling resistance is related to fuel consumption and control on this resistance is important. Wear of the tyre on road surfaces is of concern as it is related to emission of particles into the environment and service–life of the tyre. To reduce rolling resistance and tyre–wear a hard rubber is required, but this rubber has less traction (i.e., reduction in dry traction). Under wet road conditions, channels in the tyre are required to drain off water, and this reduces contact area and as a result dry traction is further reduced. To control the vehicle, there is a need of reasonable traction and this means softer tyre is preferable from traction point of view. There is a need to understand the physics of tyre–road interaction to improve the performance of tyre.

Ideally a tyre must experience pure rolling on the road, but due to the applied normal load and the requirement of motion–control (braked/driven wheel), normal and tangential tractions in the contact zone between the tyre and the road are induced. Since the tyre is made from deformable (i.e., rubber) materials, it deforms under load and loses pure rolling motion. In fact, the rubber tyre under load experiences much larger deformation compared to any metallic (i.e., steel) tyre. To deal with large deformation problem, it is necessary to revise concepts of small deflection. For small strains the material properties can be defined by the elastic properties (Young's modulus E and Poisson's ratio ν).

$$E = 2(1 + \upsilon)G \tag{9.2.1}$$

$$\upsilon = \frac{1}{2}\frac{3K - 2G}{3K + G} \tag{9.2.2}$$

where G is shear modulus and K is the modulus of bulk compression. Equations 9.2.1 and 9.2.2 are valid for material subject to:

- small elastic strain conditions and no possibility of permanent deformation; and
- immediate recovery of deformation on release of stress.

In bodies made of tyre, the size of deformation is several orders of magnitude greater than that happen in rolling elements, made of steel in rolling element bearings. One such deformation of tyre is shown in Fig. 9.21. The Poisson's ratio of rubber is approximately 0.495, which means $E \cong 3G$ (as per Eq. 9.2.1) and $K \gg G$ (as per Eq. 9.2.2). If rubber (having ν~0.495) is compressed, it shears outward and this behaviour cannot be modelled using equations (9.2.1–9.2.2) of simple elastic material. To model this behaviour, instead of modulus a strain energy density function, as a function of stretch ratios $\lambda_1, \lambda_2,$ and λ_3 in the principle directions, is used. Such models are termed as hyperelastic material models [i.e., Neo–Hookean, Mooney–Rivlin, Ogden] and the models are of empirical nature and cannot be used without performing experiments (i.e., uniaxial, biaxial and planar) on the specified material. Detailing such behaviour of tyre is beyond scope of the present book.

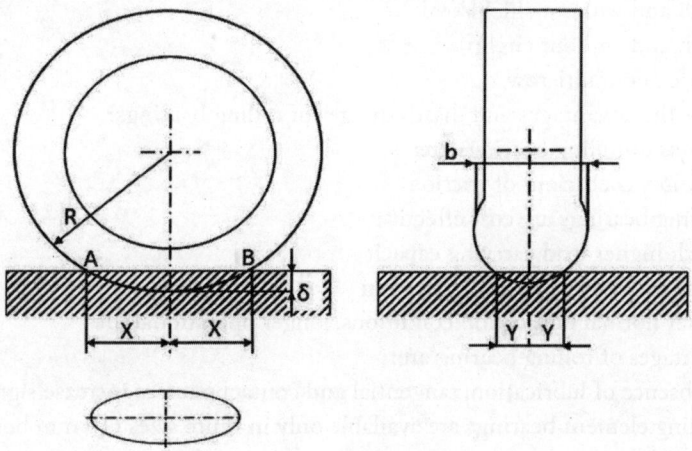

Fig. 9.21 Excessive deformation of tyre

It is interesting to note that due to deformation pure rolling cannot be achieved and it is always accompanied with sliding. Thus the actual sliding velocities occurring in the contact zone during rolling of steel bodies are very small. But rubber, which is the main material to make tyres, is a very stretchable material and does not follow reversible stress–strain relations. When a rubber is stretched and released, the returned energy is lesser than the infused energy. This can be termed as hysteresis loss. As tyre material is subject to excessive deformation, the ratio of sliding to rolling velocity is much higher compared to tyre made of steel material. This means, the force required to produce motion must have minimum of two components. One component must overcome hysteresis loss, and other components needs to accommodate the sliding friction losses. To control the friction and wear, nowadays most of the passenger car tyres are radial tyres. The detailed design of tyre is beyond the scope of the present chapter.

Frequently Asked Questions

Q.1. Name few applications of rolling bearings.

Ans. The rolling bearings find applications in auto industry, electric motors, power tools, machine tools, conveyors, escalators, fans, pump, gyroscope, aircrafts, ships, heavy vehicles, skates, dental drills, etc.

Q.2. Which materials are widely used for rolling elements?

Ans. Steels are widely used to make rolling elements. However, with advances in material technology, polymers and ceramics can also be used.

Q.3. What are the criteria to classify rolling bearings?

Ans. Rolling bearings can be classified based on:
- Materials used (steel, polymers, ceramics)
- With and without cage
- Various types of cages
- With and without shield/seal
- With and without circlip
- Single and multi-row

Q.4. What are the advantages and disadvantages of rolling bearings?

Ans. Advantages of rolling bearings are:
(a) Very low coefficient of friction
(b) Rolling bearings are cost effective
(c) Much higher load carrying capacity
(d) Require relatively lesser lubrication
(e) Under normal lubrication conditions, longer operational life

Disadvantages of rolling bearing are:
(a) In absence of lubrication, tangential and contact stresses increase significantly.
(b) Rolling element bearings are available only in finite sizes (10 mm bore, 15 mm bore, 17 mm bore, etc.).
(c) They are highly sensitive towards misalignment.
(d) They require very careful mounting.

Q.5. Name the major components of rolling bearings.

Ans. There are four main components of rolling bearings: inner ring, rolling element, cage and outer ring.

Rolling elements can be balls, cylindrical rollers, needle rollers, tapered rollers or spherical rollers.

Q.6. What is the purpose of cage in a rolling bearing?

Ans. Cages are provided to avoid sliding and collision between rolling elements in bearings. Cages are made of polymer (if operating temperature is lesser than 60°C) or steel sheet coated with MoS_2.

Q.7. What is the meaning of 'Full Complement' bearing?

Ans. The cage-less bearings are often referred as 'Full Complement' bearings.

Advantages
- These bearings have the largest possible number of rolling elements.
- Full complement bearings have extremely high load carrying capacity and stiffness.

Disadvantages
- High rate of rubbing or sliding due to cage less.
- Used only if very high load capacity within the specified space is required.

Q.8. What is the meaning of improper fit in assembly of rolling element bearings?

Ans. Improper bearing installation fit means too loose or too tight.

Improper bearing fit causes excessive wear and reduces the bearing life.

Q.9. What kind of fit is widely used between inner ring and shaft?

Ans. The interference fit is widely used. However if inner ring is stationary, then transition fit may be used.

Q.10. What is the meaning of internal clearance in rolling element bearings? Define radial internal clearance and axial internal clearance?

Ans. The amount of play between inner, rolling element and outer ring is termed as internal clearance.

Radial internal clearance is the maximum relative radial movement of the outer ring with respect to the inner ring.

Axial internal clearance is the maximum relative axial movement of the inner ring with respect to the outer ring.

Q.11. What is the meaning of 'Pre-load' in rolling bearings?

Ans. 'Pre-load' is defined as load required in creating negative internal clearance during installation of rolling bearing. It is commonly used to increase the rigidity of shaft bearing assembly.

Q.12. Define 'Bearing load capacity'.

Ans. 'Bearing load capacity' is defined in terms of static load rating (C_0) and dynamic load rating (C).
- Dynamic load rating (C) is the radial load (thrust load for the thrust bearing), which is a group of identical bearings with stationary outer rings can theoretically endure for one million revolutions of inner ring. Dynamic load capacity (C) is required to estimate fatigue life of the bearing.
- Static load rating (C_0) is defined as the radial load which causes permanent deflection of 0.01% of rolling element diameter.

Q.13. How 'Series Code' is used in standardization of rolling bearings?

Ans. The 'Series Code' generally contains three digits.
- First digit represents the type of rolling bearing.
- Second digit signifies its width series.
- Last digit denotes the bearing outer diameter.

Q.14. What are the main factors which affect friction in rolling bearings?

Ans. The hysteresis, sliding, lubricant shearing and mechanical contact of seals are main contributor of friction in rolling bearing. To reduce hysteresis, chrome steel is used to manufacture rolling elements and rings.

Appropriate selection and mounting of bearing maintains the low sliding speed between rolling elements and guiding surface of the cage.

Q.15. How does excessive load occur in rolling bearing?

Ans. Excessive load occurs due to
- Wrong tolerances between housing and outer ring.
- Excessive operating load.
- Improper load distribution.
- Misalignment.

Q.16. How coefficient of friction affects the bearing life?

Ans. Increasing coefficient of friction increases shear stresses, which increases the chances of the shear fracture of rolling bearing.

Multiple Choice Questions

Q.1. The rolling contact bearings are designed to operate under which lubrication regime?
(a) Hydrodynamic (b) Hydrostatic
(c) Mixed (d) EHL

Q.2. In respect of rolling contact bearings, there is:
(a) more starting frictional resistance compared to running frictional resistance
(b) less starting frictional resistance compared to running frictional resistance
(c) almost similar starting and running frictional resistance
(d) starting frictional resistance always remain very high compared to the running frictional resistance.

Q.3. The lubricant requirement for rolling contact bearings is?
(a) Very less as compared to hydrodynamic bearings
(b) More as compared to hydrodynamic bearings
(c) Almost similar to hydrodynamic bearings
(d) No lubricant is required

Q.4. The most common lubricant used in rolling contact bearings operating under normal conditions:
(a) Lubricant without any additives
(b) Semi-solid lubricant like grease
(c) Lubricant with low viscosity but with high thermal conductivity
(d) Lubricant with high viscosity but with low thermal conductivity

Q.5. The operation of rolling contact bearings under normal operating conditions (without misalignment) is:
(a) More noisy as compared to hydrodynamic bearings
(b) Less noisy as compared to hydrodynamic bearings
(c) Not much different from hydrodynamic bearings
(d) Noise decreases with increase in operating speed

Q.6. For a comparable service life, the:
(a) Wear is more for a rolling contact bearing as compared to sliding contact bearing
(b) Wear is less for a rolling contact bearing as compared to sliding contact bearing

(c) Wear is similar for rolling contact bearing and hydrodynamic bearing
(d) No wear in rolling contact bearing

Q.7. The main function of a cage in rolling contact bearings is to:
(a) Separate the rolling elements from each other
(b) Keep the rolling elements from spilling out of bearing
(c) To provide continuous lubrication to rolling elements
(d) To provide support to the rolling elements

Q.8. A full complement rolling element bearing:
(a) Does not have a cage
(b) Does not have rolling elements
(c) Does not have inner ring
(d) Does not have outer ring

Q.9. In a full complement rolling element bearing:
(a) Rolling elements collide with each other
(b) Rolling elements penetrate each other
(c) Rolling elements are separated from each other
(d) Rolling elements remains far apart from each other

Q.10. When the inner ring is not stationary, then generally it requires:
(a) Sliding fit with the shaft
(b) Transition fit with the shaft
(c) Interference fit with the shaft
(d) Running fit with the shaft

Q.11. When the inner ring is not stationary, then generally the outer ring requires:
(a) Sliding fit with the stationary housing
(b) Transition fit with the stationary housing
(c) Interference fit with the stationary housing
(d) Running fit with the stationary housing

Q.12. The diameter of the ball is:
(a) Kept smaller than the inner ring diameter
(b) Kept larger than the inner ring diameter
(c) Both have same diameter
(d) Kept larger than the outer ring diameter

Q.13. When the radius of ball is smaller than the radius of curvature of outer surface of inner ring, then:
(a) Contact stresses are increased
(b) Contact stresses are reduced
(c) No change in the contact stresses
(d) Contact stresses first increases and then decreases

Q.14. A deep groove ball bearing is able to support:
(a) Pure radial load
(b) Pure axial load
(c) Mainly radial load but also some axial load
(d) Both radial and axial loads

Q.15. The cylindrical radial roller bearings with grooves on both inner and outer rings:
(a) Supports pure radial load
(b) Supports axial load
(c) Supports radial load and also some axial load
(d) Both radial and axial loads in any proportion

Q.16. A single row tapered roller bearing supports:
  (a) Pure radial load
  (b) Pure axial load
  (c) Radial load and axial load in only one direction
  (d) Radial load and axial load in only both directions

Q.17. In tapered roller bearings:
  (a) The rollers are spherical
  (b) The rollers are hollow
  (c) The cage may or may not be used
  (d) The cage is not required

Q.18. The spherical roller bearings:
  (a) tolerate shaft misalignments
  (b) are meant to handle thrust load
  (c) The inner ring has to be spherical
  (d) The outer ring has to be spherical.

Q.19. The pitting in rolling bearings may occur on:
  (a) Inner ring
  (b) Outer ring
  (c) Rolling elements
  (d) All of the above

Q.20. In a radial ball bearing:
  (a) All the balls are equally loaded
  (b) The balls are subjected to variable loading even when the load is steady
  (c) The balls are subjected to stationary loading even when the load is variable
  (d) The balls are subjected to constant loading

Q.21. The load carrying capacity of the rolling element bearing:
  (a) Directly proportional to the number of rolling elements
  (b) Inversely proportional to the number of rolling elements
  (c) Directly proportional to the square root of number of rolling elements
  (d) Inversely proportional to the square root of number of rolling elements

Q.22. The equivalent radial load:
  (a) Causes same damage as the radial load only
  (b) Causes same damage as the axial load only
  (c) Causes equivalent damage as the radial and axial load
  (d) Do not causes any damage

Answers

Q.1. (d)	Q.2. (c)	Q.3. (a)	Q.4. (b)	Q.5. (a)	Q.6. (b)
Q.7. (a)	Q.8. (a)	Q.9. (a)	Q.10. (c)	Q.11. (b)	Q.12. (a)
Q.13. (a)	Q.14. (c)	Q.15. (c)	Q.16. (c)	Q.17. (c)	Q.18. (a)
Q.19. (d)	Q.20. (b)	Q.21. (a)	Q.22. (c)		

References

Ai, X. and Nixon, H. P. 2000. 'Fatigue life reduction of roller bearings due to debris denting: Part II – Experimental validation'. *Tribology Transactions*. 43:2, 311–317.

ASTM: http://www.astm.org/SNEWS/JULY_2000/july_bearings.html.

Brandlein, J. Eschmann, P. Hasbargen, L. and Waigand, K. 1999. *Ball and roller bearings*. John Wiley & sons. Third edition, p.426, John Wiley & Sons.

Cantley, R. E. 1977. 'The Effect of Water in Lubricating Oil on Bearing Fatigue Life'. *Tribology Transactions*. 20:3. 244–248.

FAG: 'http://rolling.hu/webshop/pdf/FAG_alapismeretek.pdf'.

Harris, T. A. 2001. *Rolling Bearing Analysis*. John Wiley & Sons. Fourth edition. Printed in USA.

Hirani, H. 2009. 'Root cause failure analysis of outer ring fracture of four row cylindrical roller bearing'. *Tribology Transactions*. 52 (2): 180–190.

Indian Standards (2002) Dimensions for Center Holes, IS: 2473.

Nachi: 'http://nachi-tool.jp/bearing/pdf/Tech.pdf'

Norton, R. L. 2001. *Machine Design: An Integrated Approach*. Pearson Education Asia. Second Edition. Delhi (India). 462–498.

NTN:'http://www.ntn-snr.com/portal/de/de-de/file.cfm/Bearing-HBen.pdf?contentID=8897'

Pearson, P. K. 1990. 'Fracture and fatigue of high hardness bearing steels under low tensile stresses'. *SAE 901628*. 510–515.

SKF: 'http://www.skf.com/group/products'.

Timken: 'http://www.timken.com/en-in/products/Documents/Engineering%20Section.pdf'.

Chapter 10

Tribological Aspects of Gears

Gears are machine elements having projections, known as teeth. Gears operate in pairs (needs to have geometric compatibility: same type and same module), and gear teeth maintain positive engagement between both the gears. Gears are rugged, durable and compact torque transmitting tribo–pairs. Properly designed and maintained gear-systems can run over decades. For successful operation of gears bending and contact stresses must be less than allowable stresses. In practice, the strength of gears deteriorates over time due to continuous wearing out of gear material. Therefore the chemically inert gear material and appropriate operating conditions (ultra-low wear rate 10^{-10} mm/min) are required to maintain good service life.

It is noteworthy that although gears are designed to provide positive drive without slip, but there is always sliding at surfaces of gear teeth that contributes as one major source of power loss and causes tribological failure (i.e., wear, scuffing, pitting) of gears (Fig. 10.0). The mixed lubrication mechanism, discussed in chapter 8 can be used to diagnose tribological failures in these gears. In the present chapter tribology (friction, wear and lubrication) of gears, particular spur gears, has been described. In addition, condition monitoring of spur gear has been detailed.

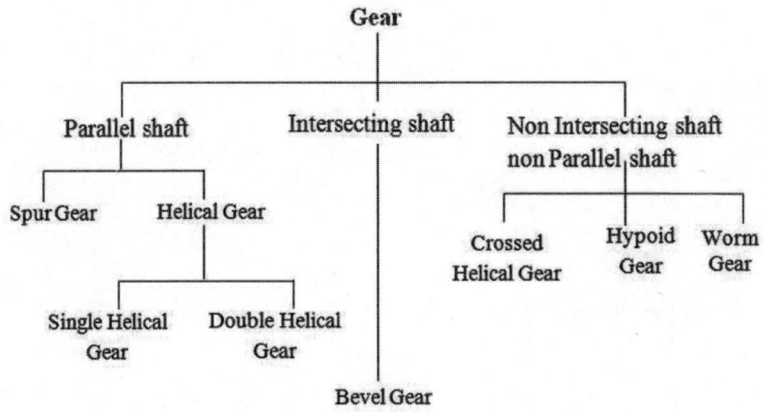

Fig. 10.0 Various types of gears

10.1 Spur Gears

Spur gears are designed to transmit power between parallel shafts and the usual reduction ratio is limited to 1:8. For higher speed reduction 'two to three stage gear train' is required. At each stage, an efficiency of 96–99% can be expected.

The geometry of spur gears is relatively simple. The gear teeth are straight along the length and are parallel to the axis. The details of conjugate action and design of gear teeth are available in the mechanical engineering books [Buckingham E., 1988].

A brief description of the geometry and operation of spur gears was provided in chapter 6 (Figs. 6.7.1–6.7.4). In the present chapter, tribology performance of spur gears, which depends on surface roughness, base oil and its additives, operating temperature, lubricant flow rate, etc., is detailed. Figure 10.1.1 shows a pair of gears (pinion as driving gear) with surface roughness. To improve the performance of these gears and to reduce the extent of boundary lubrication, it is necessary to minimize the surface roughness by precise machining, surface treatments, or specialized hard surface coatings. Engineering the surfaces of gear teeth by reducing roughness, enhancing surface strength and increasing hardness are key factors to increase durability and efficiency of gears.

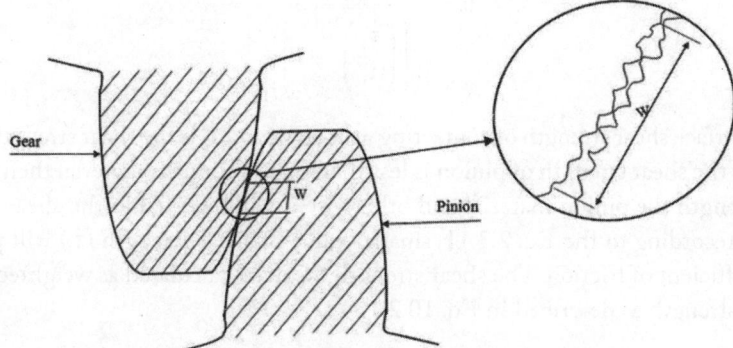

Fig. 10.1.1 Contact between asperities of gear and pinion

10.2 Friction and Wear of Spur Gears

Friction at gear tooth contact interface consists of contact friction at local contact spots, and of hydrodynamic friction due to lubricant shearing. Contact friction can be readily estimated experimentally, while hydrodynamic friction can be predicted if the frictional behaviour of the lubricant is known. A liquid lubricant having low shear stress and high thermal conductivity are preferred than high–viscosity liquid lubricant, provided friction modifiers to reduce boundary friction are added.

It is very difficult to model the friction characteristics of the gear–pair. The friction depends on the lubrication mechanism, which can be decided based on the specific film thickness (Λ),

$$\Lambda = \frac{h_{\min}}{\sqrt{R_{q1}^2 + R_{q2}^2}} \tag{10.1}$$

Here $h_{\min}$ is the minimum separation between pinion and gear teeth (expressed in Eq. 6.7.4 of chapter 6); and R_{q1} & R_{q2} are surface roughness of pinion and gear teeth respectively. In Eq. 10.1 $h_{\min}$, R_{q1} & R_{q2} need to be expressed in the same units (i.e., micrometer, micro-inch) so that Λ is dimensionless.

If Λ is less than 1, gears operate under boundary lubrication and metallic gears experience significant friction. On increasing film thickness or reducing roughness, the value of Λ increases. If the specific film thickness is greater than 1 but less than 3, then gears function under mixed lubrication (Fig. 10.2.1). In such a mechanism, liquid lubricant is used with lubricant additives and the load is supported by lubricant additives, metal to metal contact and full film (EHL) lubricants.

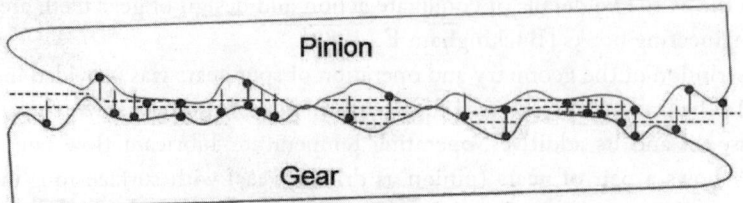

Fig. 10.2.1 Mixed lubrication between gear and pinion surface

To estimate the coefficient of friction, the relation (Eq. 2.2.11) derived in the chapter 2, can be used.

$$\mu = \frac{0.5}{\sqrt{\left(\left(\frac{\tau_y}{\tau_i}\right)^2 - 1\right)}} \qquad (2.2.11)$$

Here τ_i is interface shear strength of contacting gear teeth and τ_y is the shear strength of the pinion/gear material. If the shear strength of pinion is less than that of the gear material then τ_y will be equal to the shear strength the pinion material and otherwise it will be equal to the shear strength of the gear material. According to the Eq. 2.2.11, smaller value of shear strength (τ_i) will provide smaller value of the coefficient of friction. The shear strength τ_i can be calculated as weighted summation of different shear strength as described in Eq. 10.2.1.

$$\tau_i = w_1 \tau_{mi} + w_2 \tau_{bli} + w_3 \tau_{EHLi} \qquad (10.2.1)$$

Here τ_{mi} is the shear strength of metal interface, τ_{bli} is the shear strength of bond lubricant interface, τ_{EHLi} is the shear strength caused by elasto–hydrodynamic mechanism. The symbols w_1, w_2, and w_3 represent the weighting factors. The summation of weighting factors is equal to one, as expressed in Eq. 10.2.2

$$w_1 + w_2 + w_3 = 1 \qquad (10.2.2)$$

To avoid wear of gear tooth, it is essential to minimize metal-to-metal contact (w_1 approaches 0) and maintain a relatively thin oil film between mating gear teeth. However, in practice, low speed, heavy load, extreme temperatures, relatively rough and irregular surfaces or inadequate oil supply is sufficient to create partial lubrication. Under such conditions, there will be metal to metal contact between the mating tooth surfaces and a number of actions such as shearing of surface films, rubbing, deformation of metal, and penetration of hard asperities in the softer material occur, which all result in detachment of wear particles (shown in Fig. 10.2.2) and creation of new asperities. New asperities are formed, some of which are ploughed off to form wear particles. As wear cannot be avoided, attempt should be made to keep wear in ultra-mild or mild regime. Gear operation with very slow loss of material takes millions of cycles to show any noticeable wear, which will have little effect on the satisfactory performance within life of gears.

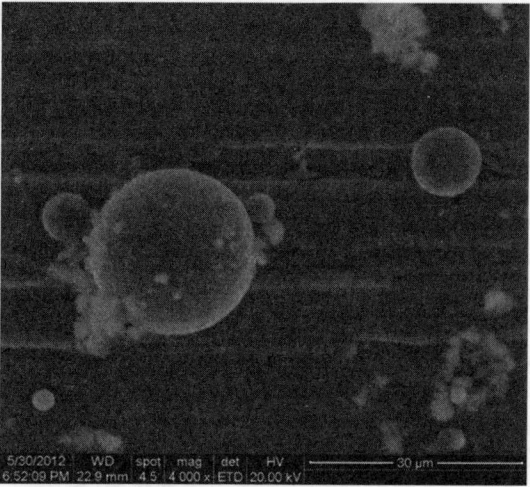

Fig. 10.2.2 SEM analysis at gear tooth

As far as possible the metal to metal contact between the gear pair is avoided ($w_1=0$). If w_2 is assumed to be equal to α then Eq. 10.2.1 will be:

$$\tau_i = \alpha\tau_{bli} + (1-\alpha)\tau_{EHLi} \tag{10.2.3}$$

Generally τ_{EHLi} is much less than τ_{bli} and by reducing the value of α, the coefficient of friction can be reduced. There is a close relation between values of α and Λ (specific film thickness, expression in Eq. 10.1). If $\Lambda = 3$ there is high probability of $\alpha = 0$. Similarly if $\Lambda < 1$, there is very high probability of $\alpha = 1$. This means friction generated at the interface of gear teeth is highly dependent on value of Λ. As per the available literature [Muraro, 2012] wear life also depends on Λ, as shown in Fig. 10.2.3.

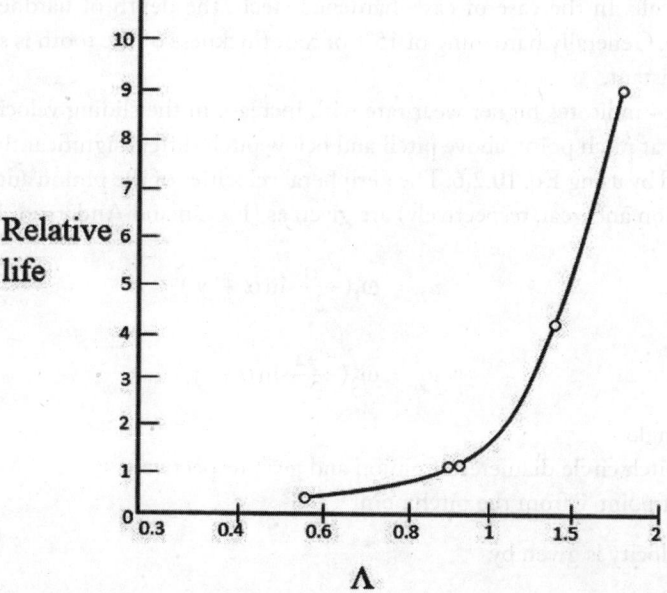

Fig. 10.2.3 Relative life with various film thickness / surface roughness ratio

Figure 10.2.3 shows that as Λ increases towards 1, there is a sudden change in the relative life. When the Λ increases from 1 to 1.5, there is a substantial jump in the relative life. This can be achieved by decreasing the surface roughness (R_{q1} & R_{q2}) and increasing film thickness by using extreme–pressure/anti-wear additives. However, it is necessary to ensure positive drive and maintain high efficiency of the gear drive. For that purpose, mechanical contact cannot be completely avoided. In other words, mechanical wear is inherent in gear drive. The continuous loss of material from the tooth–surface and deviation from the original involute profile is inevitable. Gears cannot be used for life long. It will have only finite life such as 20000 hours, 30000 hours, 40000 hours, etc. If there is a need to increase the service life, then wear rate should be reduced. The wear of gears can be reduced by suitable selection of materials and lubricants. Based on the amount of material loss from contacting teeth surfaces wear can be sub grouped as mild wear, moderate wear, destructive wear, scratching (a severe form of abrasive wear), and corrosive wear.

Flodin and Anderson [Flodin and Anderson, 1997] developed a mild wear model for spur gears. They found the wear over the involute profile using single point (assuming point 'i') observation method and employing following Archard's linear wear equation,

$$\text{Rate of Wear Volume} = \frac{k_1 \times \text{Load} \times v_i}{3 \times \text{Hardness}} \qquad (10.2.4)$$

According to Archard's equation, the wear rate can be reduced by reducing the sliding velocity (v_i), increasing the hardness, reducing the load and reducing k_1. As pinion runs faster than the gear, it wears out faster losing more material than that of gear and it is important to make the pinion harder than the gear. The hardness can be increased by surface–hardening and through–hardening treatment. Surface–hardening produces a hard case on the tooth surface, leaving the core comparatively soft. Surface–hardening can be achieved by the common case–hardening processes, such as, case Carburizing, Nitriding, Induction–hardening, etc. For that purpose, gears are heat treated to the desired hardness. For example, automobiles gears are hardened by case carburizing process up to 60 Rockwell C-scale. In the case of case–hardened steels, the depth of hardness of the case is of prime importance. Generally hardening of 15% of root thickness of the tooth is sufficient to provide desirable wear resistant.

Equation 10.2.4 indicates higher wear rate with increase in the sliding velocity. It is noteworthy that sliding speed at pitch point, above pitch and below pitch differs significantly. The sliding speed can be determined by using Eq. 10.2.6. The peripheral velocities of the pinion and gear are (subscript 1 and 2 is for pinion and gear, respectively) are given as [Flodin and Anderson, 1997]:

$$u_{1i} = \omega_1 \left(\frac{d_{p1}}{2} \sin \alpha + y_i \right)$$

$$u_{2i} = \omega_2 \left(\frac{d_{p2}}{2} \sin \alpha - y_i \right) \qquad (10.2.5)$$

α = pressure angle
d_{p1} and d_{p2} = pitch circle diameter of pinion and gear, respectively
y_i = distance of point 'i' from the pitch point

The sliding velocity is given by,

$$v_i = u_{1i} - u_{2i} \qquad (10.2.6)$$

Figure 10.2.4 shows the zero sliding zones between the flank and face of the tooth. The face of tooth is the surface between the pitch cylinder and the top land. The surface between the pitch cylinder and bottom land is termed as flank of the tooth. The wear shall be zero near pitch line (due to zero sliding velocity, $y_i \sim 0$). At every other contact point, there is a relative sliding between the mating gear teeth, which results in an increased friction and wear. But at the flank surface due to more sliding, wear rate will be higher than that of at the face.

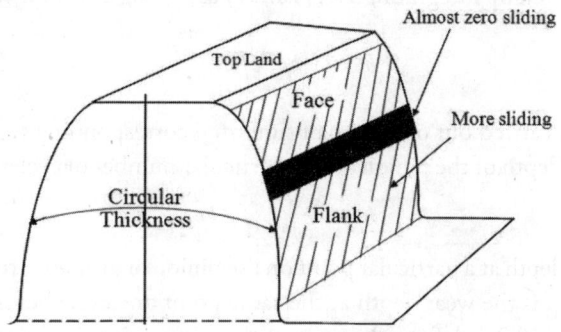

Fig. 10.2.4 Gear tooth profile

When material is worn off from the surface, deviation from the original involute profile is inevitable. It is therefore necessary to model the surface profiles to predict how past wear affects the future wear. This is of great importance particularly when wear is not uniform throughout the surface of the teeth.

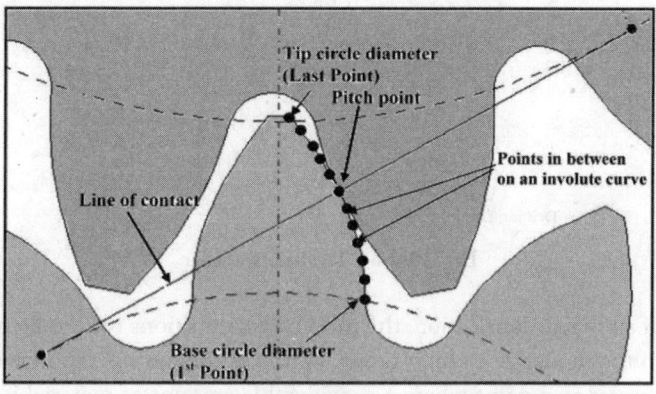

Fig. 10.2.5 Meshing points on the gear

To study the wear of spur gears the involute tooth surface can be divided into number of points (as shown in Fig. 10.2.5) and predict the variation in contact pressure, relative sliding and local surface temperature. Sliding wear at any point 'i' on dry, mixed or boundary lubricated surface can be described by Eq. 10.2.4. On diving both sides of Eq. 10.2.4 by apparent area A, and replacing $k_1/3H$ with a dimensional wear coefficient, following wear equation can be obtained.

$$\frac{dh}{dt} = kpv \qquad (10.2.7)$$

where k is the dimensional wear coefficient, which is a function of material properties, surface roughness, lubrication properties and operating conditions. Due to the complexity of the gear contact problem and dependency on so many parameters, value of 'k' will be different for dry, mixed or boundary lubricated surface. Flodin and Andersson [Flodin and Andersson, 1997] considered k equals to 5e-10 mm²/N. Flodin [Flodin, 2000] mentioned the range of 'k' for different lubrication conditions. For partial EHL regime, k, is less than 1.e-9 mm³/Nm, for boundary lubrication regime k is between 1.e-6 to 1.e-8 mm³/Nm and for unlubricated regime the value of k is greater than 1.e-5 mm³/Nm.

For any point 'i' on the contact surface on pinion (described by superscript p for pinion and g for gear) wear depth is given by integrating Eq. [10.2.7] assuming constant value of wear coefficient,

$$h_i^p = \int_0^t k p_i^p v_i^p \, dt \tag{10.2.8}$$

The integration is carried out over a small time step corresponding to small angle increments of the pinion. The wear depth of the pinion after a particular number of cycles 'n' can be computed using,

$$h_{i,n} = h_{i(n-1)} + k p_i v_i t_i \tag{10.2.9}$$

where $h_{i,n}$ is the wear depth at a particular point on the pinion or gear after running the defined number of cycles 'n' and $h_{i(n-1)}$ is the wear depth at the same point one cycle before. The second term in the right hand side of Eq. 10.2.9 defines the wear of the same point 'i' in n^{th} cycle under consideration. This equation indicates that wear depth at any point (other than pitch point) increases with increase in the number of cycles. Ding et al. [Ding et al., 2007] studied spur gear dynamics to show the effect of mild wear as well as the excessive wear on involute profile. They used Archard's wear model and Hertzian contact model to find the wear of tooth.

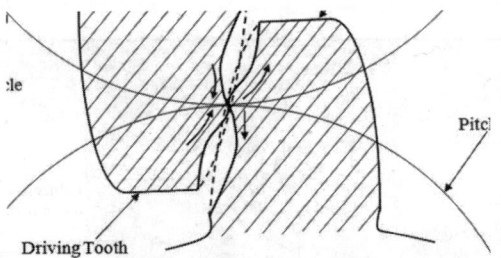

Fig. 10.2.6 Destructive wear

In the absence of sufficient lubrication, the mild wear conditions cannot be maintained and the wear rate remains in the moderate to high range. With continuous operation even under moderate wear, the wear progresses to a point where a considerable amount of material is removed from the surfaces. Such wear occurs over most of the gear tooth except at the pitch line. Figure 10.2.6 shows that the wear pattern on the tooth surface under moderate wear rate which finally leads to destructive wear. Dotted and full lines show the tooth surface before wear and after wear respectively.

In absence of lubrication, wear between steel (the most commonly used material) gears becomes significant (k is greater than 1.e-5 mm³/Nm) and usage of the non-ferrous materials is preferable. Bronze is the most common alloy used as gear material, mainly because of its easy manufacturability and ability to withstand sliding conditions without significant wear. Phosphorus bronze, tin bronze, silicon bronze, lead bronze, etc., find wide application as gear materials.

If there is any restriction on the usage of liquid lubricant, plastic materials (i.e., nylon, duracon, **UHMW-PE,** polyamide) can be selected as gear materials because of their quietness, low maintenance, light weight, self-lubricating properties, corrosion resistance, and low manufacturing cost. It is interesting to note that even though the tensile strength of these polymeric materials may be less than metallic materials, but their resilience and compressive strength are high. Similarly, the values of the moduli of elasticity of these materials are low; the resulting elastic deformation under load takes care of the usual detrimental effects of errors, such as tooth shape error, spacing error, etc. These errors, therefore, have little effect on the overall strength and performance of the gears, which is not so in case of metallic gears. There is a need to be familiar with the dimensional instabilities of plastic gears which is due to their high shrinkage rate, larger coefficient of thermal expansion and tendency to absorb moisture. For example, if a plastic gear with a shrinkage rate of 0.025 mm/mm (2.5% shrinkage rate) has a pitch diameter of 25 mm while in the mould, the pitch diameter after cooling will be reduced by (25)*(0.025) or 0.625 mm, and becomes 24.375 mm. Sometimes, the form of the gear itself changes (i.e., pressure angle increases) as a result of shrinkage and disproportionate gear results, therefore there is need to consider shrinkage aspects at the time of manufacturing. As per some unpublished results, the best tooth form for plastic gears is the 20° stub-tooth system.

In addition to shrinkage, there are dimensional changes as the result of moisture absorption as some plastics are hygroscopic. Dimensional changes on the order of 0.1% or more can develop in the course of time, if the humidity is sufficient. Therefore, a good design of plastic gears requires allowance for a greater amount of backlash (> 0.3*module) than for metal gears. It will be more advisable to use one of the gears of a mated pair made of metal to sink the heat to combat the temperature rise. Under certain conditions, solid lubricants such as molybdenum disulphide, can be used to reduce tooth friction and temperature rise.

10.3 Contact Stresses

Contact stress plays important role to estimate the pitting failure of the gear tooth. A pitting is a surface failure (to be discussed in section 10.5), which occurs due to fatigue load conditions. For most gears, the geometry of the tooth contact at any point along the contacting path during the entire engagement can be determined from the Hertzian theory for two cylinders. The primary concern is to find contact areas that can initiate surface distress and lead to contact fatigue.

Contact stress calculation is based on Hertz's theory. In chapter 6, Hertz's theory was applied on non-conforming contact (Fig. 6.7.4) between involute gear pair. The interaction between pinion and gear teeth was treated as contact between cylinders. The radii of the cylinders were expressed by following equation

$$r_{p,i} = r_p \cos\varphi \tan\varphi_{pi}$$
$$r_{g,i} = r_g \cos\varphi \tan\varphi_{gi}$$
(6.7.2)

The effective radius of curvature (R_i^*) at the interface was expressed as following expression:

$$\frac{1}{R_i^*} = \frac{1}{r_{p,i}} + \frac{1}{r_{g,i}}$$
(6.7.3)

The maximum value of pressure (contact stress) at the interface of cylindrical contacts was expressed by following expression

$$P_i = \frac{2}{\pi}\frac{F}{b_i L}$$
(6.2.10)

where F is the normal contact force, L is the gear/pinion length and b_i is half width of the elliptical contact zone at contact point i. The b_i can be represented as [Budynas and Nisbett, 2010]

$$b_i = \sqrt{\frac{2FR_i^*}{\pi L E^*}} \qquad (10.3.1)$$

where E^* is the equivalent modulus of elasticity, as expressed in chapter 6 using the following equation

$$\frac{1}{E^*} = \frac{1 - v_1^2}{E_1} + \frac{1 - v_2^2}{E_2} \qquad (6.2.2)$$

On substituting expression of b_i from Eq. 10.3.1 into Eq. 6.4.10,

$$P_i = \frac{2}{\pi} \frac{F}{\sqrt{\frac{2FR_i^*}{\pi L E^*}} L}$$

On rearranging,

$$P_i = \sqrt{\frac{2E^* F}{\pi L R_i^*}} \qquad (10.3.2)$$

The contact stress on each gear tooth varies from zero to P_i (Eq. 10.3.2). The number of pinion-teeth is less than that of the gear. Therefore, the pinion has to complete more cycles than that of the gear cycles. In other words, pinion teeth are more frequently stressed than the teeth of the mating gear. This means endurance strength of pinion material needs to be more than that of gear-material. In addition, endurance strength needs to be greater than the contact stress (expressed in Eq. 10.3.2) to avoid 'pitting' caused by contact stress. In other words, the surface–fatigue failure can occur even with proper lubrication and results primarily from repeated stressing beyond endurance limit of the material. In such failures, the material of gear-tooth in the fatigue region gets removed and a pit is formed. The pit itself will cause stress concentration and soon the pitting spreads to adjacent region till the whole surface is covered with pits. The initial pitting with associated enlarged view has been shown in Fig. 10.3.1.

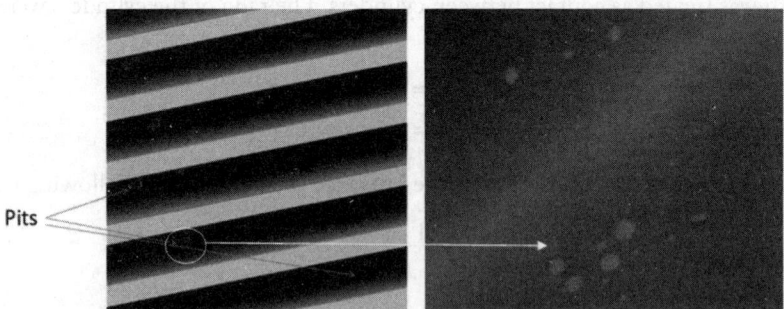

Fig. 10.3.1 Initial pitting

In the stage of initial pitting, there is a possibility of corrective actions, i.e., better lubricant, liquid lubricant with solid lubricant, reduction in the applied load, etc. However, if pits are allowed to grow or excessive load is applied even after the start of initial pitting, the pits break into each other and the size of pits increases. Eventually, tooth shape gets destroyed, and gear becomes noisy. Figure 10.3.2 shows the destructive pitting predominantly in the dedendum area.

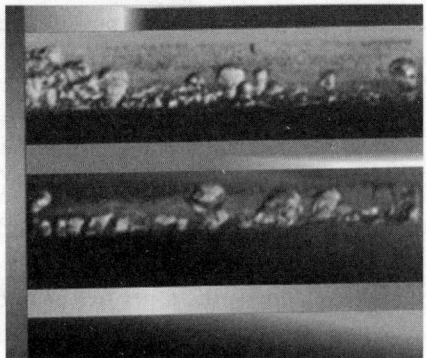

Fig. 10.3.2   Destructive pitting

Very similar to destructive pitting is spalling failure. In this case, the pits are usually larger in size and quite shallow compared to destructive pitting failure. Figure 10.3.3 shows gear with spalling failure, in which the material progressively gets removed away from the surface until a large irregular surface patch is formed.

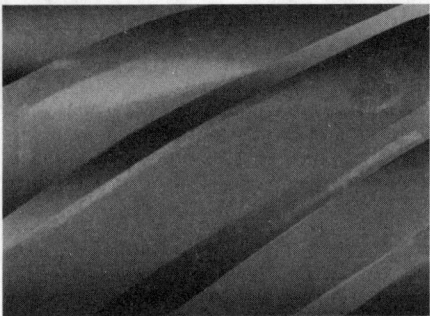

Fig. 10.3.3   Spalling failure

Dudley [Dudley, 1980] provided extensive coverage on how the relation between the pitting initiation life and contact stresses improves as the regime of lubrication or the EHL film changes from boundary to mixed and finally to full film lubrication. In each regime, Dudley proposed a different empirical exponent to relate the contact stress and fatigue life. This relation was also suggested by Kubo et al. [Kubo et al., 1991] in their review of gear lubrication. They also contributed a chart for determining the regime of lubrication when the EHL film thickness and the gear teeth surface roughness are known.

The contact stress developed between two contacting tooth surfaces is proportional to the square root of the tooth-load. This means to increase the load capacity, there is a need to increase endurance strength of the gear tooth. As a thumb rule, the following minimum safety factor (ratio of endurance strength to contact stress) needs to be 1.2 (for normal conditions), 1.4 (for few reversals in direction of rotation) and 2 (for continuous change in rotation direction). It is interesting to note that endurance strength of gear tooth is proportional to hardness, which means hardness plays an important role in reducing wear (discussed in previous heading of this chapter) and increasing endurance strength.

However there is an upper limit on the surface hardness of gear tooth. At high hardness material becomes brittle and its notch sensitivity increases, thereby rendering the teeth weaker in bending.

In general, case–hardened gears can withstand higher loads than through–hardened ones. Due to its comparative softness at the core, the case–hardened gears possess interior toughness and can sustain impact loading. Sometime as a post-treatment, hardened gears are tempered (somewhat reduction in hardness) to permit final finishing the teeth.

10.4 Lubrication of Spur Gears

In the absence of any lubrication (dry surfaces), there is a possibility of strong adhesion in contacting teeth of metallic gears. In such a case, the coefficient of friction is majorly decided by adhesion phenomenon. In case of bad surface roughness, the asperities penetrate in contacting surface and possibility of cold welding between pinion and gear increases. Fortunately, in the normal atmospheric conditions, there always exists some sort of oxide layer (nanometre level thick layer) on the metallic surfaces. Figure 10.4.1 shows pinion and gear surfaces having thin oxide layer. The shear strength of these layers is generally less than the shear strength of the parent materials, due to which the adhesion between contacting teeth decreases.

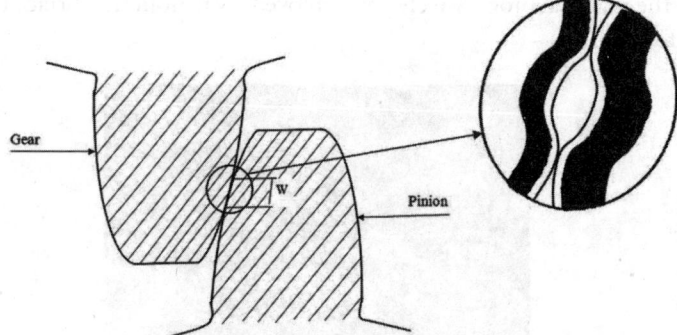

Fig. 10.4.1 White colour contamination layer on gear and pinion surfaces

The oxide layer on the gear teeth depends on operating temperature, moisture and other environmental conditions. To avoid these kinds of uncertainties, the gear teeth must be coated with some lubricant (liquid or solid lubricant) or must be made of non-metallic materials.

One of the suitable solid lubricants for the gears is phosphate. The phosphate layer(s) on the surface of gear teeth helps to reduce coefficient of friction. Similar to phosphate, graphite, molybdenum di-sulphide and tungsten disulphide materials can be used as solid lubricants. One of the drawbacks of solid lubricant coating on spur gear is its continuous wearing away from the gear tooth. To counter this problem, solid lubricants are used as additives in liquid lubricants. Due to circulation of the liquid lubricant, solid lubricants at the interface get replenished. Generally a 10–15 µm thick layer of solid lubricant is sufficient for gearing operation if liquid lubricant with solid lubricant as additive is used. In absence of liquid lubricant, the thickness of solid lubricant layer needs to be thicker to provide service for certain number of operating hours before any refurbishment.

Apart from solid lubricants as additives in liquid lubricants, there are a number of other additives required for the satisfactory performance of a liquid lubricant. Anti-wear (AW) additives, anti-scuffing extreme–pressure (EP) additives, rust inhibitors, pour–point depressant, viscosity index improvers, and foam suppression additives are among the useful lubricant–additives. The higher percentage of

EP additives (consisting of sulphur, phosphorus, chlorine, nitrogen and boron) is recommended for gear lubricants. For example, a multi-grade mineral oil (75W90) contains 7% of EP additives. In severe operating conditions, liquid lubricants based on synthetic base oils, due to their thermal and oxidative stability, are preferred over mineral based oils. The EP additives (3.5–10% by weight) are also used in synthetic base gear oils.

The quantity of liquid lubricant matters, more the lubricant quantity, higher the churning losses. Higher friction may lead higher temperature that may lead more volume of the wear.

The viscosity of liquid lubricant must be decided based on the optimum thermal balance or thermal equilibrium. The viscosity is a strong function of operating temperature; it reduces significantly with increase in the temperature. The rise in the temperature occurs due to heat generated by rubbing sliding surfaces of pinion and gear teeth. This means temperature rise will be a function of relative sliding speed (Eq. 10.2.6), normal contact pressure (P_i), expressed in Eq. 10.3.2), and friction coefficient (f) between contacting gear teeth, such as:

$$\Delta T_i \propto f |u_{1i} - u_{2i}| P_i \tag{10.4.1}$$

The expression (10.4.1) comes from the rate of heat generation per unit area, $f|u_{1i} - u_{2i}|P_i$. In addition to depending on rate of heat generation, the temperature rise depends on the coefficient of thermal conductance (expressed in Eq. 10.4.2) and tangential speed (V) of the individual gear tooth.

$$\beta = \text{thermal conduction} \times \text{specific heat} \times \text{density}$$

$$\beta = \lambda\, C\, \rho \tag{10.4.2}$$

Here, λ is the thermal conductivity, C is the specific heat and ρ is the density.

Larger the β value, lower will be the temperature. If the thermal conductivity is high, it can dissipate the heat faster which leads to faster thermal equilibrium. So, there will be overall lower temperature. Similarly larger tangential velocity of gears will reduce the value of the maximum temperature. Here there is need to understand the difference between sliding speed (Eq. 10.2.6) and the tangential speed. According to Eq. 10.4.1 temperature rise varies with relative sliding speed, which itself changes at various contacting points (Fig. 10.2.5). In other words, heat source travels along the gear tooth surface and non-uniform distribution of temperature occurs. At high enough speed of travel, the temperatures tend to concentrate themselves [Block, 1963] in a surface area which is very narrow compared to gear tooth face/flank area. It is interesting to note that the maximum temperature occurs somewhere between the middle and the trailing edge [as shown in Fig. 10.4.2].

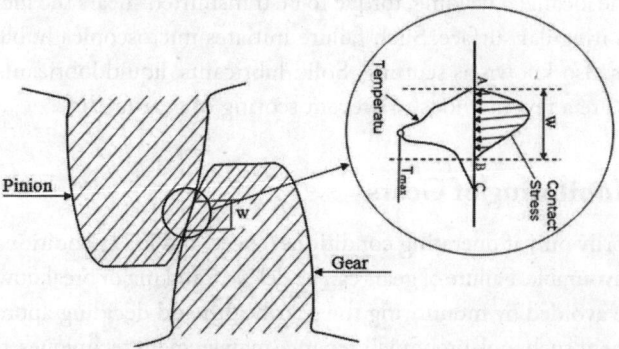

Fig. 10.4.2 Non-uniform distribution of operating temperature

On combining Eq. 10.4.1, concept of coefficient of thermal conductance and tangential speed [Block, 1963], the temperature rise can be expressed as

$$\Delta T_i \propto \frac{f|u_{1i} - u_{2i}|P_i}{\sqrt{V(\beta_1 + \beta_2)}} \qquad (10.4.3)$$

Equation 10.4.3 provides an estimation of temperature rise, which is useful in deciding the type of lubricant and requirements of cooling the gearbox (i.e., external cooling to the gearbox casing, continuous circulation of cooled lubricant, etc.).

10.5 Surface Failures

The gear failures can be grouped into 'plastic flow failure', 'breakage of gear tooth' and 'surface failures'. Failure due to plastic flow of materials occurs due to excessive yielding of gear tooth under heavy loads. The breakage of gear tooth is defined as fracture of a substantial portion of a tooth. It is very rare to find gear failure due to either 'plastic flow' or 'breakage'. Therefore, in the present chapter only 'surface failures' of gear tooth have been described.

The wear and fatigue pertaining to surface failures of gear teeth has been dealt in sections 10.2 and 10.3 of the present chapter. In the present section, attempts have been made to explore the effects of surface contact temperature and foreign particles. Ingress of abrasive particles in the lubricant causes the surface failure of the gear tooth. Scratching of gear teeth occurs when hard (i.e., dirt, sand, wear debris) particles slide under pressure, across the tooth surface.

In absence of appropriate lubricant, there is a possibility of overheating of gear tooth in mesh. The temperature at gear interface depends on the rate of friction heat generation, the rate of heat dissipation from the gear, and from the gear housing to the outside. From heat dissipation point of view steel/plastic mating pair is preferred over plastic/plastic mating pair. In case of steel/plastic mating pair, case–hardened steel gear having surface quality of $R_t < 8$ µm shall be used. The driving pinion is always subject to greater wear. Therefore the pinion should always be of the more wear resistant material (e.g., steel) in steel/plastic mating gear pair. In addition, the plastic gear face should be slightly narrower than that of the steel gear to avoid any grooving into the face of plastic gear. To reduce the temperature rise, completely open gear assembly (if appropriate lubrication is ensured) is preferred over partially closed gear assembly.

One of the surface failures, which occurs due to overheating of the tooth mesh zone, is scoring. In scoring, failure occurs due to tearing out of small particles that weld together as a result of excessive local heating. After the localized welding, torque to be transmitted shears the metal from pinion/gear surface and produces irregular surface. Such failure initiates microscopically, but progresses rapidly. Scoring is sometimes also known as scuffing. Solid lubricants, liquid lubricants with EP additives, profile modifications are a few methods to prevent scoring of gear teeth.

10.6 Offline Monitoring of Gears

Gear works satisfactorily only if operating conditions (speed and load conditions, materials strength, lubrication, etc.) are favourable. Failure of gears can be delayed, and major breakdown of machine caused by gear failure can be avoided by monitoring the gear health and deciding appropriate maintenance scheme(s). To implement such fool-proof maintenance management techniques, the first and foremost step is the right selection of monitoring technique(s) to detect the condition of gears. Mostly gears are operated under lubrication and the used lubricant coming out from the gearbox contains lots of information regarding the envelope in which it circulates. Early detection of changes in lubricant condition and consistent monitoring of wear metal debris coming along with lubricant provide greater insight of the actual condition of gearbox. In the present chapter only monitoring the condition of lubricants has been described.

The condition of lubricant can be monitored online (during machine operation) and/or offline (observing a sample picked from working site). Few oil analysis techniques are depicted in Fig. 10.6.1. In the present section offline condition monitoring of spur gears has been discussed.

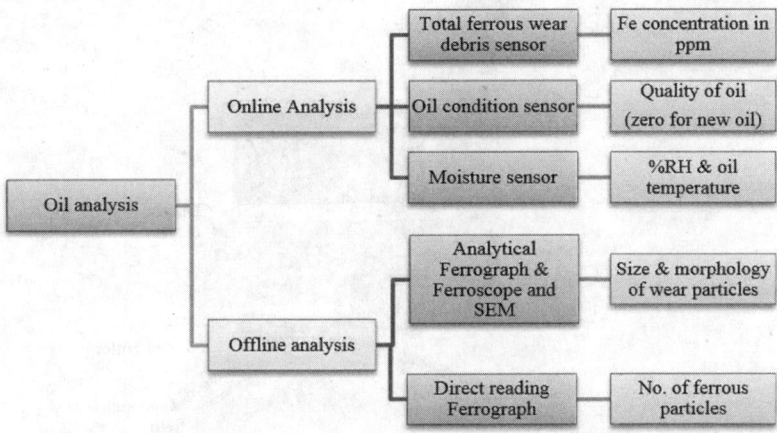

Fig. 10.6.1 Flowchart of oil analysis techniques included in this study

In an offline oil analysis techniques, greater emphasis is given to ferrography (direct and analytical), which separates wear debris and contaminant particles from a lubricant and arranges them according to the size on a transparent substrate (glass slides) for examination. The particles can be examined using a standard optical microscope or laser scanning microscope. Particle type, overall surface characteristics and colour can be studied using these microscopes. To observe the root cause of failure related to gears, scanning electron microscope (SEM) will be used.

Direct reading ferrography is a quantitative analysis of ferrous particles present in the used oil. It is used to separate larger (DL > 5µm) and smaller (DS < 5µm) size of the ferrous particles present in the oil sample. The numbers given do NOT relate to any other numbers such as ppm. The direct reading ferrograph is shown in Fig. 10.6.2. In this equipment magnetic field is used to separate the ferrous particles. At the starting point of magnetic field, particles of 5 micron or greater are collected, and towards the end of magnetic field, 1–2 micron particles are collected. The densities of particles are found by shinning a monochromatic light through the bottom of the glass part of the precipitator tube and the amount of light cut off by the particle build up in the glass tube is read above the glass tube by sensors which are converted into DL and DS readings. The performance of lubricant is compared based on one the indices: Wear rate index (WR = DL+DS), Wear intensity index (WII = DL-Ds), and Wear severity index (SI = DL*(DL-DS)/DS).

In normal rubbing wear, the majority of the particles are small, so DS will be comparable with DL, and SI will be small. If the wear regime is more severe, DL will be large compared to DS; i.e., SI increases with increasing severity of wear. If SI readings are beyond the normal trend a ferrogram sample slide is made with the oil for examination by optical microscopy. A ferrogram on a fixed slide containing wear/ferrous particles is made for microscopic examination and photographic documentation. Such analytical ferrography analysis is based on two aspects—the qualitative and quantitative methods. The first method describes the characteristics of the morphology of wear particles whereas other method is based on the size and concentration of the wear debris.

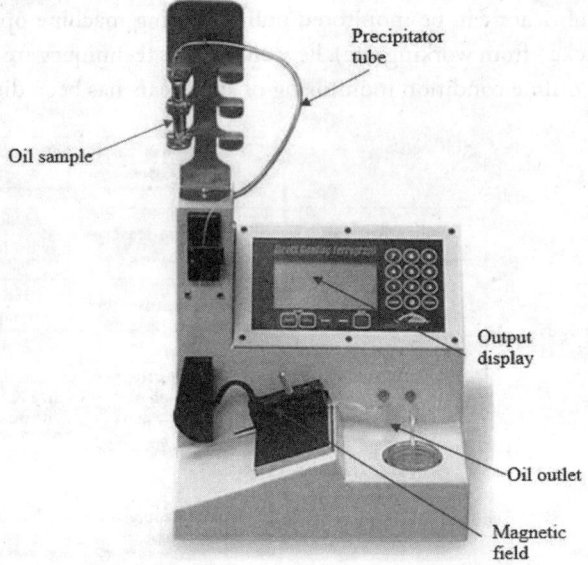

Fig. 10.6.2 Direct Reading Ferrograph (http://www.tricocorp.com/product/direct-reading-ferrograph)

The ferrogram analyser shown in Fig. 10.6.3, consists of a pump to deliver an oil sample at a low rate (approximately 0.2 ml/min), a magnet to provide a high gradient magnetic field near its poles and a ferrogram on which the particles are deposited. The oil sample, diluted with a special solvent to promote the precipitation of the wear particles, is pumped across a ferrogram which is mounted at a slight inclination; the magnetic particles adhere to the substrate, distributed approximately according to size. Larger ferrous particles are deposited near the entry zone where the field is weakest, and smaller particles, are deposited further downstream as shown in Fig. 10.6.4.

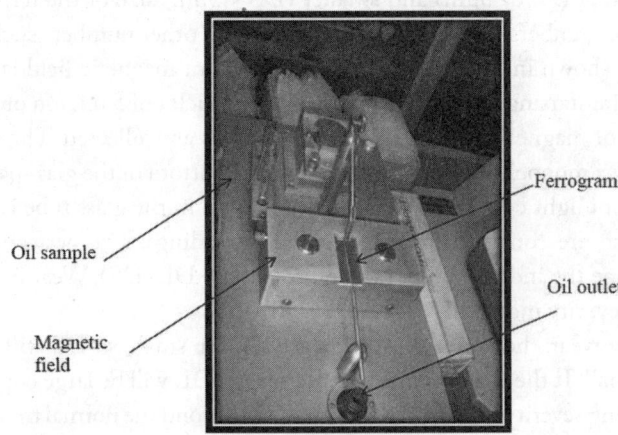

Fig. 10.6.3 Ferrogram analyser

After the particles have deposited on the ferrogram, a solvent is used to flush away the oil residue or water-based lubricant. After the solvent evaporates, wear particles remain permanently attached to the glass substrate and are ready for microscopic examination. Scanning Electron Microscope (SEM) and Energy Dispersive X-ray analysis (EDX) are used to detect the size, shape and material of collected wear particles.

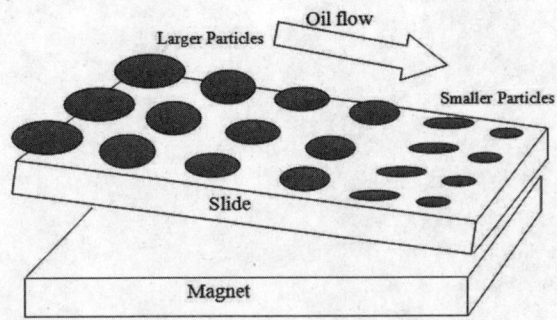

Fig. 10.6.4 Separation of ferrous particles from oil on ferrogram/slide

10.6.1 Offline condition monitoring – a case study

To understand the offline condition monitoring, oil samples from a gearbox were collected. Direct reading ferrograph was calibrated with fixer oil by passing it through precipitation tube. Test samples were made by mixing 50 units of fixer liquid in one unit of the lubricant collected from gearbox. After passing the test sample through precipitator tube, recorded the number of larger sized (> 5 µm) particles DL and number of smaller sized (< 5µm) particles DS. Based on the recorded values, estimated the wear rate index (WR= DL+DS), wear intensity index (WII= DL-DS), and wear severity index (SI = DL*(DL-DS)/DS).

Using the specified procedure, four samples were analyzed.

Sample 1: GL-4 EP-90, 1 ml oil + 50 ml fixer (1:50 dilute) at 420 rpm, 60 Nm, operated for 30 min. The SEM image of the collected wear particles is shown in Fig. 10.6.1.1.

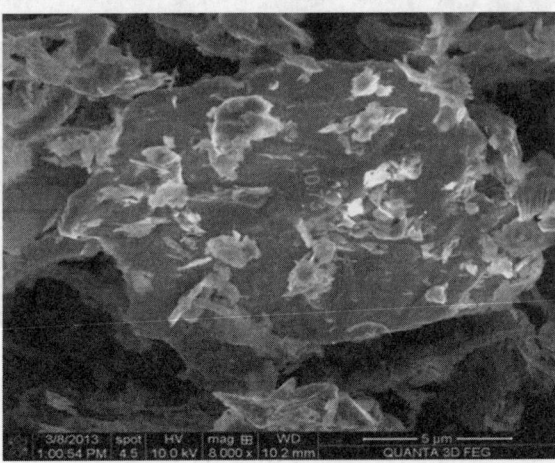

Fig. 10.6.1.1 SEM result of the wear debris extracted from sample 1

Sample 2: GL-5 80w-90, 1 ml oil + 50 ml fixer (1:50 dilute) at 657 rpm, 50 Nm, operated for 25 min. The SEM image of the particles collected from sample 2 is shown in Fig. 10.6.1.2.

Fig. 10.6.1.2 SEM result sample 2 (largest particle size is 6.61 μm)

Sample 3: GL-5 80w-90, 1 ml oil + 50 ml fixer (1:50 dilute) at 824 rpm, 50 Nm, operated for 45 min. The largest particle size of this sample, as shown in Fig. 10.6.1.3, is 8.86 μm.

Fig. 10.6.1.3 SEM result sample 3

Sample 4: GL-4 EP-90, 1 ml oil + 50 ml fixer (1:50 dilute), operated for 10 hr. As per the Fig. 10.6.1.4, the maximum size of wear debris is 28.82 μm.

Fig. 10.6.1.4 SEM result sample 4 (largest particle size is 28.82 μm)

The results of various indices (WRI, WII, WI)) is provided in Table 10.6.1.1. The results indicate an increase in all indices with increasing the operating cycles. This means gears were operated in mixed lubrication conditions. The 28.82 μm size of wear debris, obtained from sample 4, intimated the need of careful investigation.

Table 10.6.1.1 Direct reading ferrography Results

S. No.	Samples	Operating cycles	Torque (Nm)	DL	DS	WRI	WII	WI
1.	GL- 4 EP-90	12600	60	33.7	15.4	49.1	18.3	40.04
2.	GL-5 80w-90	16425	50	40.0	19.2	59.2	20.8	54.02
3.	GL-5 80w-90	37080	50	67.0	32.6	99.6	34.4	70.69
4.	GL- 4 EP-90	300000	(20 to 60)	91.9	44.8	136.7	47.1	96.61

On performing EDX analysis (Fig. 10.6.1.5) of the wear particles of sample 4, three main materials Fe (~80.78%), Ni (~18.07%) and Cr (~1.16%) were diagnosed.

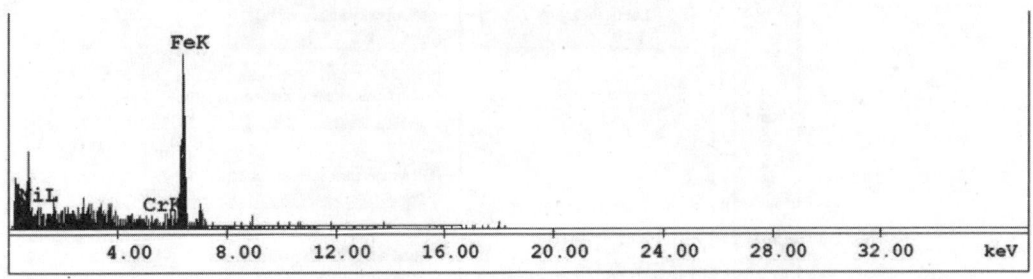

Fig. 10.6.1.5 EDX of debris from sample 4

10.7 Online Monitoring of Gears

The most common online monitoring techniques, relevant to the detection of gear faults are: vibration monitoring [Randall, 2011] and oil analysis [Loutas et.al., 2011]. In the present section, online oil analysis has been described. The oil analysis includes both lubricant analysis (very useful for low speed lubricated machinery) and wear–debris analysis (useful for detecting pitting fatigue failure/wear of spur gears). When gear teeth are moved against one another with a sufficient normal force, wear occurs. In the presence of liquid lubricant, the wear particles (debris) get mixed in the lubricant oil. Analyzing those wear debris help diagnosing the wear related gear faults. Moreover material removed from contacting surfaces contaminates lubricant and deteriorate the oil quality, hence lubricant analysis is also a good indicator of gear faults.

In online oil analysis, the oil coming out from the gearbox is pumped by tubes to online sensors suite and then from the sensors the oil enters the gearbox. Online analysis sensors generally include oil condition/quality sensor, moisture sensor, and total ferrous wear debris sensor. Total ferrous wear debris sensor is used for measuring ferrous concentration (ppm) present in the oil samples. Ferrous debris is expressed in parts per million (ppm) by weight. As wear increases the ferrous concentration (ppm level) also increases which gives the indication of fault. Oil condition sensor and moisture sensor are used for analyzing the change occurring in oil properties. Oil condition sensor is used for monitoring the physical and chemical properties of lubricant (i.e., TAN, TBN, glycol, contaminants, and additives). The oil condition sensor indicates the condition of the oil using a 0–100 index scale, generally zero is set as reference for new oil. As the oil degrade the oil quality number increases from zero level. The moisture sensor is used for detecting the moisture (water) content of oil up to it saturation point and the oil temperature. The moisture sensor expresses the moisture content as % relative humidity (i.e., 100 % is the saturation point).

Figure 10.7.0 summarize the oil analysis as study of total acid number (TAN), total base number (TBN), viscosity measurement, oil temperature, moisture content, number of particles in oil sample, and wear debris concentration in lubricating.

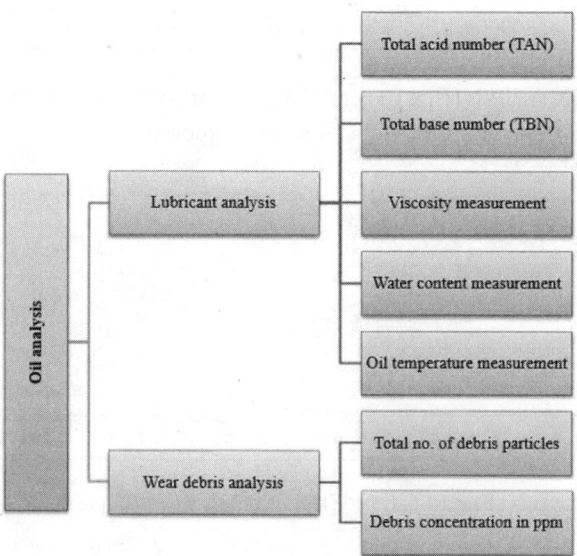

Fig. 10.7.0 Flowchart stating oil analysis techniques

10.7.1 Online condition monitoring – a case study

The gear test rig used for condition monitoring of spur gear has been shown in Fig. 10.7.1. The test rig assembly used for experimentation consists of spur gearbox, driven by shunt electric motor (30 kW DC), and a speed controller to regulate motor speed in the range of 0–3000 rpm. The torque (0–75 Nm) is applied on the gears with the help of eddy current dynamometer coupled with the output shaft. The test gearbox consists of pair of standard involute profile spur gear and bearing (URB 32306 bearing for the driving shaft and URB 30307 bearing for the driven shaft). The motor is connected to input shaft of gearbox through coupling. Similarly output shaft of the gearbox is connected to shaft of torque sensor through coupling. To take care of the angular and linear misalignments, the universal coupling is used between dynamometer and load shafts.

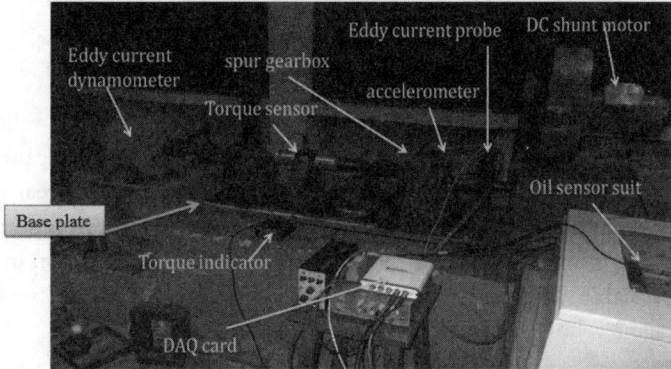

Fig. 10.7.1 Spur gear experimental test rig

The specifications of gearbox are listed in Table 10.7.1. The sampling point, from where the oil sample enters into online sensors (Fig. 10.7.2), was arranged below the gear mesh where a high circulating flow predominates. The oil from the gearbox passed through the sensors and pumped back to gearbox. The equipment (oil analysis sensor suit) reports metallic ferrous wear debris, oil condition, and the moisture content of the oil.

Table 10.7.1 Specification of gearbox

S. No.	Parameters	Pinion	Gear
1.	Module	2	2
2.	No. of teeth	27	53
3.	Pitch diameter	54	106
4.	Outer diameter	58	110
5.	Base diameter	50.7434	99.6074
6.	Face width	33	33
7.	Pressure angle	20°	20°
8.	Contact ratio	1.697	1.697
9.	Circular tooth thickness	3.1415	3.1415
10.	Material	EN19	EN19

Fig. 10.7.2 Online oil analysis sensor suite

Figure 10.7.3 shows the image of liquid lubricant. It shows fresh oil is yellow in colour. This fresh oil was used in the gear box. After 100 operating hours, the used oil became completely black in colour containing a number of wear particles. This clearly indicates that gear was working under mixed lubrication mechanism. Figure 10.7.4 shows the concentration of iron particles circulating in the gear oil. The results of oil sensor suit at relatively higher torque are shown in Fig. 10.7.5. These do not show any noticeable variation in oil temperature and moisture content, but there is continuous increase in percentage level of iron particles.

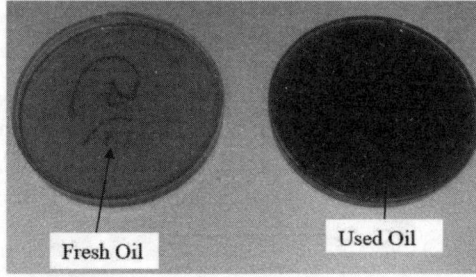

Fig. 10.7.3 Fresh oil and used oil

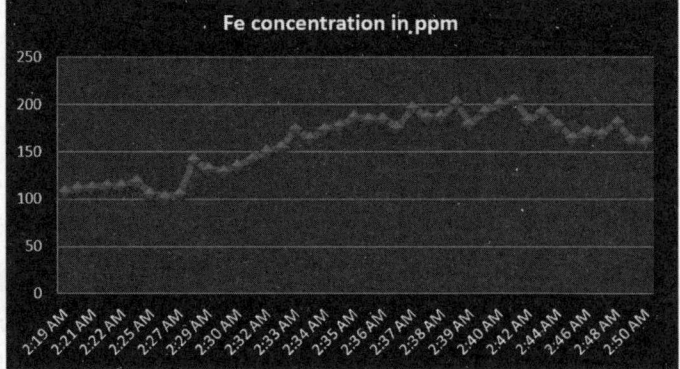

Fig. 10.7.4 Fe concentration in ppm for 9000 wear cycles (speed = 300 rpm, Torque = 15 N-m)

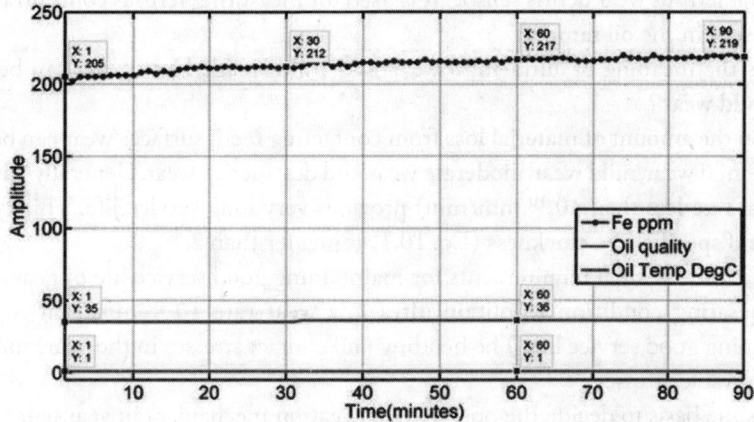

Fig. 10.7.5 Results oil sensor suit at 500 rpm, 50 N-m torque after 90 minutes of operation

Frequently Asked Questions

Q.1. Why is two or more stage gear trains required for obtaining higher reduction ratio?

Ans. The maximum reduction ratio for transmitting power between parallel shafts is generally taken as 1:8, so as to limit the size of the gear. A higher reduction will mean a larger size of the gear which increases the overall space requirement of a gear box. Therefore, higher reduction ratio is obtained using two or more stages.

Q.2. What is the meaning of 'engineering the surface of gear teeth'?

Ans. 'Engineering the surface of gear teeth' means reducing surface roughness, enhancing surface strength and increasing hardness of the gear surface to increase durability and efficiency of gears. This could be achieved by precision machining, surface treatments or providing specialized hard surface coatings on gears.

Q.3. What is the meaning of 'interface shear strength of contacting gear teeth'?

Ans. When two surfaces of gear teeth contact each other, some of their surface asperities interact and forms welded junctions. The shearing of those welded junctions occurs during the gear rotation. The shear strength of these junctions is known as interface shear strength of contacting gear teeth.

Q.4. What is the meaning of R_{q1} & R_{q2} surface roughness of pinion and gear'? What is their significance?

Ans. R_{q1} & R_{q2} are surface roughnesses of pinion and gear teeth respectively. They are used in the determination of the specific film thickness for identification of operative lubrication regime.

Q.5. Name the type of sensors and their use in online analysis.

Ans. The following sensors are used:
- Oil condition\Quality sensor: Oil condition sensor indicates the condition of the oil using a 0-100 index scale, generally zero is set as a reference for new oil.
- Moisture sensor: It expresses the moisture content as % relative humidity (100% is the saturation point).

- Total ferrous wear debris sensor: It is used for measuring ferrous concentration (ppm) present in the oil sample

Q.6. What is the meaning of 'ultra-mild wear' and 'mild wear'? How gears can be designed for 'ultra-mild wear'?

Ans. Based on the amount of material loss from contacting teeth surfaces wear can be sub grouped as ultra-mild wear, mild wear, moderate wear and destructive wear. Generally 'ultra-mild' wear (removal rate less than 10^{-10} mm/min) provides very long service life. This type of wear is possible if specific film thickness (Eq. 10.1) is greater than 2.

Q.7. What are the essential requirements for maintaining good service life of gears?

Ans. The operating condition to obtain ultra-low wear rate 10^{-10} mm/min is necessary for maintaining good service life. The bending and contact stresses in the gears must also be less than allowable stresses.

Q.8. What is the basis to decide the operative lubrication mechanism in gear pair?

Ans. The operative Lubrication mechanism is decided by the Specific film thickness (Λ) value.

$$\Lambda = \frac{h_{min}}{\sqrt{R_{q1}^2 + R_{q2}^2}}$$

where h_{min} is the minimum separation between pinion and gear teeth and R_{q1} & R_{q2} are surface roughness of pinion and gear teeth respectively.

If $\Lambda < 1$, gears operate under boundary lubrication and their service life is severely affected. If $\Lambda > 1 < 3$, then gears function under mixed lubrication and have relatively longer service life.

Q.9. How coefficient of friction is estimated?

Ans. There are two parts of coefficient of friction: the major part of friction is caused due to adhesion, while minor part is caused due to the surface roughness. The coefficient of friction caused due to adhesion is expressed by the following relation:

$$\mu = \frac{0.5}{\sqrt{\left(\frac{\tau_y}{\tau_i}\right)^2 - 1}}$$

where τ_i is interface shear strength of contacting gear teeth and τ_y is the shear strength of the pinion or gear material. If the shear strength of pinion is less than that of the gear material then τ_y will be equal to the shear strength the pinion material otherwise it will be equal to the shear strength the gear material. On reducing shear strength of interface, increasing the hardness and reducing the surface roughness reduces the overall coefficient of friction.

Q.10. Name the type of wear that occurs in contacting teeth surfaces.

Ans. The following wear types may occur in contacting gear teeth surfaces: mild wear, moderate wear, destructive wear, scratching (a severe form of abrasive wear) and corrosive wear

Q.11. What is Archard's linear wear equation?

Ans. According to Archard's equation, wear rate is proportional to the sliding velocity (v_i), hardness and load. It is given by following expression:

$$\text{Rate of Wear volume} = \frac{k_1 \times \text{Load} \times v_i}{3 \times \text{hardness}}$$

Where, Sliding velocity $(v_i) = u_{1i} - u_{2i}$,

$$u_{1i} = w_1 \left(\frac{d_{p1}}{2} \sin \alpha + y_i \right)$$

$$u_{2i} = w_2 \left(\frac{d_{p2}}{2} \sin \alpha - y_i \right)$$

α = pressure angle
d_{p1} and d_{p2} = pitch circle diameter of pinion and gear respectively
y_i = distance of point 'i'

where k_1 is a dimensionless constant which expresses the probability of removing a wear particle. If $k_1 = 1$, then it means that every contact junction involved in the friction process produces a wear fragment. If $k_1 = 0.1$, then 1/10th of the friction junctions produce wear fragments.

Q.12. Why it is necessary to model the surface profile of the gear?

Ans. The wear of the material from the gear surface results in the deviation from the original involute profile. It is therefore necessary to model the surface profiles to predict how past wear affects the future wear. This is of great importance particularly when wear is not uniform throughout the surface of teeth.

Q.13. Name the most common non-ferrous alloy used as gear material.

Ans. Bronze is the most common non-ferrous alloy used as a gear material, mainly because of its easy manufacturability and ability to withstand sliding conditions without significant wear. Phosphorus bronze, tin bronze, silicon bronze, lead bronze etc., find wide applications as a gear material.

Q.14. What is the best tooth form for plastic gear?

Ans. A 20° stub-tooth system is the best tooth form for plastic gears. The 20° pressure angle is usually preferred due to its stronger tooth shape and reduced undercutting compared to the 14.5° pressure angle system. A stub tooth experiences less bending stresses than a full depth tooth.

Q.15. Name the common plastic material used for gears.

Ans. Nylon, duracon, UHMW-PE, polyamide are the most commonly used plastic materials for gears.

Q.16. Mention the advantages and disadvantages of the plastic material used for gears.

Ans. Plastic material is used for gears because of their quietness, low maintenance, light weight, self lubricating properties, corrosion resistance and low manufacturing cost. Even though the tensile strength of these polymeric material is less than the metallic material, but their resilience and compressive strengths are high. The main disadvantages of plastic material are high shrinkage rate, large coefficient of thermal expansion and tendency to absorb moisture.

Q.17. What is pitting failure? Which stresses cause pitting failure?

Ans. The pitting is a surface–fatigue failure. In pitting, material of gear–tooth in the fatigue region gets removed and a pit is formed. It occurs due to fatigue load conditions (repeated stressing beyond endurance limit of the material). The endurance strength must be more than the contact strength to avoid 'pitting' caused by contact stress.

Q.18. Are the strength requirements same for the pinion and gear materials?

Ans. The number of teeth on pinion is less than that on gear, therefore pinion complete more number of cycles than gear in a given time duration. So the pinion teeth are more frequently stressed than that of gears. This requires that the endurance strength of pinion material to be more than that of gear-material.

Q.19. What is the relation between contact stress and tooth–load?

Ans. The contact stress developed between two contacting tooth surfaces is proportional to the square root of the tooth–load. Therefore, to increase the load capacity, the endurance strength of the gear tooth must be increased.

Q.20. How can the oxidation of the gear tooth be prevented?

Ans. The gear teeth must be coated with some lubricant (liquid or solid) to avoid oxidation of the surface layer, or non metallic materials may be used.

Q.21. Name the solid lubricants used in lubrication of gears. How are they used for lubrication? Which method of lubrication is preferable?

Ans. Phosphate, Graphite, MoS_2 and WS_2 are the common solid lubricants used to lubricate the gears. They may be used either as a coating on the gear teeth surface or as additive in liquid lubricant. Since the solid lubricant coating on spur gear continuously wears away from the gear tooth, the preferred method is the use of solid lubricants as additives in oil.

Q.22. Name different lubricant additive used for gears.

Ans. The following lubricant additives are used for gears: Anti-wear additive, Anti-scuffing additive, extreme-pressure (EP) additive, Rust inhibitors, Pour-point depressant, Viscosity index improvers, and Foam suspension additives. The usage of EP additive is almost essential for gear lubrication.

Q.23. Name the different types of gear failure.

Ans. The different ways in which the gears may fail are:

Plastic flow failure: It occurs due to excessive yielding of gear tooth under heavy load.

Breakage of gear tooth: It is defined as a fracture of a substantial portion of a gear tooth.

Surface failure: Wear and fatigue pertaining to surface failure of gear tooth. It is also occurs due to overheating of tooth mesh zone (scoring). However the most common failure type is surface failure and it is very rare to find gear failure due to either 'plastic flow' or 'breakage'.

Q.24. Name the methods to prevent scoring or scuffing of gear teeth.

Ans. The scoring or scuffing of gear teeth is prevented by the use of solid lubricant as coating in tooth surface, use of liquid lubricant with EP additives and profile modification.

Q.25. What is the difference between online and offline oil analysis techniques?

Ans. In online analysis, sensors are used to find the Fe concentration (ppm), quality of oil, % RH and oil temperature of the lubricant while the gear box is operating. In offline analysis a sample of oil is taken and analytical methods are used to find size, morphology of wear particles and number of ferrous particles.

Q.26. What is ferrography and how it is done?

Ans. It is a method to analyze the ferrous particles present in the oil as wear debris. It is done by two methods, one is direct and other is analytical method.

In direct ferrography, magnetic field is used to separate the ferrous particles. Particles of different sizes are collected in the precipitator tube and monochromatic light is passed through the tube to find the density of particles. It gives sizes of ferrous particles (DL and DS) present in the lubricant.

In analytical method pump, magnet and ferrograph are used. Oil sample is pumped across a ferrogram, particles are deposited on the ferrogram, and then a solvent is used to flush away the oil residue. *SEM* and *EDX* are used to detect the size, shape and material of collected wear particles.

Q.27. What objectives are achieved by conducting the oil analysis in online monitoring of gear?
Ans. The analysis of oil quantifies the the total acid number (TAN), Total base number (TBN), viscosity, oil temperature, moisture content, number of particles in oil sample and wear debris concentration in lubricating which are the indicators based on which decision regarding continuance of the oil is made.

Multiple Choice Questions

Q.1. Which lubrication mechanism is the most commonly observed in the operating gear pair?
  (a) Hydrodynamic   (b) Hydrostatic
  (c) Mixed   (d) EHL

Q.2. The tribological performance of a gear pair does not depend upon:
  (a) Surface roughness of gear   (b) Lubricant properties
  (c) Operating conditions   (d) Endurance strength

Q.3. Which method does not reduce the extent of the boundary lubricant and improve the service life of a gear pair?
  (a) Minimization of surface roughness (b) Increasing surface hardness
  (c) Providing surface coating   (d) Reducing the operating speed

Q.4. Which type of lubricant is preferred for gears?
  (a) Lubricant without any additives
  (b) Lubricant without any friction modifiers
  (c) Lubricant with low viscosity but with high thermal conductivity
  (d) Lubricant with high viscosity but with low thermal conductivity

Q.5. If the surface roughness is reduced, the specific film thickness
  (a) Decreases   (b) Increases
  (c) Remains same   (d) None of the above

Q.6. If the specific film thickness is less than 1, then the gears operates under
  (a) Boundary lubrication regime   (b) Mixed lubrication regime
  (c) EHL regime   (d) None of the above

Q.7. If the specific film thickness is more, then the service life of a gear will:
  (a) Be more   (b) Be less
  (c) Not be affected   (d) Be infinite

Q.8. Which method is not used for the case hardening of a gear?
  (a) Carburizing   (b) Nitriding
  (c) Induction hardening   (d) Tempering

Q.9. The wear rate is higher at:
  (a) Face of the gear   (b) Flank of the gear
  (c) Top land of the gear   (d) Bottom land of the gear

Q.10. The wear depth at any point (other than the pitch point)
  (a) Increases with increase in number of cycles
  (b) Decreases with increase in number of cycles
  (c) Remains same with increase in number of cycles
  (d) None of the above

Q.11. Use of non-ferrous alloys is preferable when
   (a) There is sufficient lubrication
   (b) There is insufficient lubrication
   (c) There is very viscous lubricant
   (d) None of the above

Q.12. The tensile strength of the polymeric materials is:
   (a) More than metallic materials
   (b) Less than metallic materials
   (c) Same as that of metallic materials
   (d) None of the above

Q.13. Which amongst the following is not the drawback of plastic gears?
   (a) Large tensile strength
   (b) High shrinkage rate
   (c) Large coefficient of thermal expansion
   (d) Tendency to absorb moisture

Q.14. What is the best tooth form for plastic gears?
   (a) 20° stub-tooth system
   (b) 20° full depth tooth system
   (c) 14½° stub-tooth system
   (d) None of the above

Q.15. The endurance strength of the pinion material must be:
   (a) More than gear material
   (b) Less than gear material
   (c) Same as that of gear material
   (d) None of the above

Q.16. Pitting occurs in:
   (a) Addendum area
   (b) Dedendum area
   (c) Both addendum and dedendum area
   (d) None of the above

Q.17. In spalling failure:
   (a) Pits are larger in diameter and shallower than in pitting
   (b) Pits are smaller in diameter and shallower than in pitting
   (c) Pits are larger in diameter and deeper than in pitting
   (d) Pits are smaller in diameter and deeper than in pitting

Q.18. The contact stresses developed between two contacting gear tooth surfaces is:
   (a) Directly proportional to cube root of tooth load
   (b) Directly proportional to square root of tooth load
   (c) Inversely proportional to tooth load
   (d) Directly proportional to tooth load

Q.19. Endurance strength of the gear is:
   (a) Directly proportional to hardness
   (b) Inversely proportional to hardness
   (c) Directly proportional to surface roughness
   (d) Inversely proportional to surface roughness

Answers

Q.1. (c)	Q.2. (d)	Q.3. (d)	Q.4. (c)	Q.5. (b)	Q.6. (a)
Q.7. (a)	Q.8. (d)	Q.9. (b)	Q.10. (a)	Q.11. (b)	Q.12. (b)
Q.13. (a)	Q.14. (a)	Q.15. (a)	Q.16. (c)	Q.17. (a)	Q.18. (b)
Q.19. (a)					

References

Block, H. 1963. 'The Flash Temperature Concept'. *Wear*. 6: 483–494.

Buckingham, E. 1988. *Analytical Mechanics of Gears*. Dover Publications, Inc. New York.

Budynas, R. G. and Nisbett, K. J. 2010. Shigley's Mechanical Engineering Design. Mcgraw-Hill.

Dudley, D. W. 1980. Gear wear, in *Wear Control Handbook*, Peterson, M.B. and Winer, W.O. (Eds.). The American Society of Mechanical Engineers. New York. 755–830.

Flodin, A. and Andersson, S. 1997. 'Simulation of mild wear in spur gears'. *Wear*. 207: 16–23.

Flodin, A. 2000. 'Wear of spur and helical gears'. Doctoral Thesis, Department of Machine Design. Royal Institute of Technology, Stockholm.

Kubo, A. and Townsend, D. P. 1991. 'Gear lubrication', in *Dudley's Gear Handbook*, Townsend, D. P. (Ed.). McGraw-Hill, New York, 15:1–31.

Loutas, T. H. Roulias, D. Pauly. E. and Kostopoulos, V. 2011. 'The combined use of vibration, acoustic emission and oil debris on-line monitoring towards a more effective condition monitoring of rotating machinery'. *Mechanical Systems and Signal Processing*. V 25: pp 1339–1352.

Muraro, M. A. Koda, F. Reisdorfer, U. and Silva, C. H. 2012. 'The influence of contact stress distribution and specific film thickness on the wear of spur gears during pitting tests'. *Journal of the Brazilian Society of Mechanical Sciences and Engineering*. Vol. 34(2). Rio de Janeiro.

Randall, R. B. 2011. 'Vibration-based Condition Monitoring: Industrial, Automotive and Aerospace Applications'. John Wiley and sons. First edition.

Index

3-D Optical surface profile measurement, 310
Abrasion, 13, 57, 74–77, 79, 81, 388
Abrasive wear, 12, 31, 74–77, 101, 388, 404
Additives
    interface, 38
    need, 34–35
    types, 35–37
Adhesion, 13, 21, 26, 52, 56–59, 61, 63–65, 79, 82, 90, 410
Adhesive wear, 10–11, 74, 77–86, 89
Aerodynamic, 2, 120–122, 210, 298–299, 301, 315
Aerostatic, 115, 120–122, 209–210, 298–299, 309–318
Animal fats, 30
Anti-foaming agents, 36
Anti-oxidant additive, 35, 37–38
Anti-wear additive, 35–36, 404
Archard's model, 80–82, 100–101, 354, 404, 406
Arithmetic average, 330–331
Asperity temperatures, 350–351
Auto-correlation function, 335–336
Auto-covariance function, 335–336
Average surface roughness, 328, 337
Axial ball bearing, 39

Barus relation, 131–132, 270, 274
Bearing, 39–40
Bearing code, 382–383
Bearing life, 345, 371–372, 378, 386–387
Bearing profile, 352, 356
Bearing stiffness, 218, 222, 225–226
Bearing terminology, 39, 371–374
Boundary lubricants, 30, 33, 103–106, 120, 340, 343, 345
Boundary lubrication, 17–18, 34, 52–110, 338–350, 401–402, 406

Bowden-Tabor theory, 57–59, 61–62
Bulk modulus, 155
Burwell and Strang's wear equation, 100

Coulomb theory, 57
Cam, 11, 20, 89, 92–98, 263
Cam wear analysis, 92–98
Capillary, 213–215, 251–252, 312–313, 315
Cavitation, 34, 153–156, 300
Center line average, 330–331
Center of pressure, 201–202
Ceramic and cermet coating, 21, 26
Chemical adsorption, 104, 339
Chemisorption, 104, 106, 339–341
Chemistry, 1–3
Chocked flow, 311–312, 318
Circular flat plates, 229–231
Circular step thrust bearing, 215–219, 226
Circularity, 351–353
Classification, 14, 27–29, 53–54, 74–98
Coefficient of friction, 16, 20–21, 52–55, 59–61, 63–68, 82, 90, 103–104, 123–124, 132, 142–143, 153, 159–161, 163, 165, 168, 177, 262, 340, 348–350, 353, 383
Composite surface roughness, 18, 160
Concave contact, 94–95
Conservation of mass, 135, 317
Constant flow system, 254
Constant pressure system, 214
Contact between two solid cylinders, 263
Contact between two solid spheres, 263
Contact stresses, 1, 12–13, 73, 89–90, 93–94, 96, 98, 337, 350–351, 400, 407–410
Contaminations, 5, 34, 38, 63–65, 85, 210, 300, 386, 388, 410
Continuity equation, 135–136, 300
Convex contact, 94, 263

430 Index

Corrosive wear, 75, 86–87, 404
Cylindrical roller bearing, 40–41, 371, 373, 376, 379, 384, 386
Cylindricity, 351–354

Damped vibration, 70–72
Deformation, 73, 75, 78–79, 81, 85, 90–91, 100, 102, 121, 131, 261–264, 266–267, 278, 280, 326, 354, 373, 393–394, 402, 407
Delamination theory, 102
Design table, 175
Detergents, 34–36, 38
Discretization, 189, 304, 331
Dispersants, 34–36, 38
Dry friction, 55–57
Dry lubrication, 52, 82, 101, 107
Dynamic load capacity, 378–379, 383, 386
Dynamic viscosity, 14, 125–126, 159–161, 163, 168, 327, 345, 355

Effective Young's modulus, 97
Elasto-hydrodynamic lubrication, 2, 17–18, 131, 261–288, 383
Elliptic partial differential equation, 155
Energy losses, 5, 56, 226–227
Engine bearing, 228, 237–240
Engine particle (sand) separator, 89
Erosive wear, 75, 87–89
Ertel-Grubin equation, 274–278
Esters, 27–28, 33, 37, 385
Extreme pressure lubrication, 27, 34, 44–45, 106

Fatigue wear, 88–90, 93, 95
Ferrography, 413–415, 417
Finite difference method, 159, 224, 278–279
Fixed pad thrust bearing, 188, 198–200, 207
Flow rate requirement, 218, 221
Flow requirement, 215, 225
Flow through a capillary, 312–313
Flow through orifice restrictors, 311–312
Flow through restrictors, 311, 317
Flow through slot restrictors, 312
Fluid film lubrication, 2, 107, 120–122, 134, 210
Fluid mechanics, 1–3, 121, 132
Fretting wear, 75, 91–92
Friction, 52–110, 202
Friction angle, 65
Friction control, 102–103
Friction due to deformation, 56–57, 59–61
Friction factors for seals, 385
Friction force, 1, 4, 53–59, 61–62, 65–71, 141–142, 175, 177, 188, 202, 347, 390

Friction instabilities, 54, 68, 70–72
Friction models, 57, 70
Friction moment, 123, 384–386
Friction power loss, 226
Friction torque, 161, 163, 168, 219, 348
Full bearings, 169
Full-Sommerfeld boundary condition, 146–149

Gaseous lubricants, 19, 34, 120, 300
Gasket, 4–10
Gaussian distribution, 334
Gear, 13, 40, 73, 75, 107–108
Graphite, 8, 10, 23–27, 36, 58, 371, 410
Grease, 13–14, 26–30, 34, 38
Gumbel's boundary condition, 148, 154

Half-Sommerfeld boundary condition, 146, 148–150, 152–154, 174, 239
Height of profile, 330, 333
Hertz theory, 263–269
Holm's wear equation, 100
Hybrid bearing, 210
Hybrid hydrostatic bearing, 122, 210
Hydrodynamic lubrication, 10, 17–18, 34, 41–42, 107, 122–124, 156, 188, 209, 238
Hydrostatic, 209–248
Hydrostatic journal bearing, 222–226
Hydrostatic lubrication, 122, 209–222
Hyperbolic partial differential equation, 155
Hysteresis, 56, 384, 393–394

Ideal gas, 298
Idealized bearings, 138–153
Improved method of journal bearing design, 173–177
Inclined plane tribometer, 65
Incompressible fluid, 136, 219
Infinitely long journal bearings, 144–150
Infinitely long plane fixed sliders, 138–143
Infinitely long plane pivoted sliders, 143–144

Jakobsson-Floberg-Olsson equation, 155
Journal bearing, 11, 153–156
Journal bearing design, 145, 175
Journal position, 156, 238
Junction growth theory, 61–65

Kinematic viscosity, 125–126, 128
Knudsen number, 300, 307, 309
Kurtosis, 334, 350

Laws, 54–56, 80–81
Liquid lubricants, 19–20, 29–34, 64, 123, 137, 154, 282, 298, 300, 350, 383, 401–402, 406, 408, 410–412, 418, 420

Load, 201
Load angle, 377
Load carrying capacity. 41, 122, 153, 155, 158–159, 201, 217–218, 221, 225, 230, 232, 236, 298, 305–306, 309, 313, 316, 345–346, 372, 382, 386
Load dependent friction moment, 384
Locating journal position, 156
Lock, 4, 7, 13
Long static bearing, 173
Lubricant, 3, 14–17, 64, 72, 78, 80, 86–87, 101, 103–106, 120, 122–126, 129, 138, 210, 213, 223, 228, 297, 307
Lubricant additives, 3, 34–38, 343, 402, 410
Lubricant and speed dependent friction moment, 384
Lubricant
    classification, 14
    properties, 18–19, 99, 246
    selection, 38–39
    supply, 156–158, 161, 215, 246
    type, 19–34
    usage, 34, 39
Lubricating grooves, 188
Lubrication
    application, 14–15
    basic modes, 17–18
    definition, 13–15
    factor, 38, 385
    mechanism, 2, 10, 17–18

M.R. Fluid, 77
Magnetic bearing, 12
Magnetic recording discs with flying head, 307–309
Material science, 1–3
Maximum profile peak height, 328, 330, 333
Mean free path, 300, 307
Mean peak spacing, 333
Measurable wear, 72–73
Measurement, 4, 17, 65–67, 329, 331–333, 352–354, 418
Metal-solid lubricants, 22–23
Micro fatigue, 75
Micro-cutting, 75
Micro-fracture, 75
Mineral oils, 28, 30–34, 104–105, 127, 131–132, 339–340, 411
Minimum film thickness, 124, 131, 141, 143, 147, 158–161, 163–164, 168, 200, 234, 237, 242, 261, 277–278, 282–284, 287, 300, 327, 354–355

Mixed lubrication, 11, 18, 326–356
Multigrade oils, 131
Multi-pad bearings, 223
Multi-recess annular thrust bearing, 223
Multi-recess bearings, 223
Multi-row roller Bearing, 12

Naphthenic oils, 31
Needle roller bearing, 376, 383
Newton's law of viscosity, 158, 211
Newtonian, 144–15, 129–130, 134–136, 138, 202
Non-dimensional minimum film thickness, 160, 193, 283
Non-Newtonian, 14–15, 130, 228

Offline monitoring of gears, 412–415
Oil flow, 38, 155, 188, 204, 240, 242
Oil hole, 157, 240, 242
Online monitoring of gears, 418
Orifice, 213–215, 251–255, 299, 311–313, 315, 318
Oxide films, 85, 99

Pad aerodynamic bearings, 301–306
Paraffinic oils, 31, 35
Partial bearings, 42
Partial groove, 157, 240, 242
Perfluoro polyalky lether, 33
Perfluoropolyethers, 33
Petroff's equation, 158–159
Physical adsorption, 104, 339–340
Physisorption, 104, 106, 339–340
Pin on disk tribometer, 66
Pitting, 11, 73, 92–93, 95, 381, 387–388, 390, 400, 407–409, 418
Pitting life, 95
Pneumatic transportation, 88
Poiseuille flow, 135, 302, 309
Polyglycols, 33
Polymer, 7, 21–22, 32, 36–37, 72, 127, 340, 371
Polytropic gas, 298
Pour point depressants, 37, 410
Power loss, 37, 74, 121, 155, 161, 163, 168, 173, 177, 188, 215, 219, 222, 226–227, 241–242, 245, 248, 297, 382
Power spectral density, 335–336
Power torque, 161
Pressure distribution, 42, 134, 147–152, 173–174, 188–201, 215, 217, 221, 223, 230–232, 237, 239, 262, 264–265, 270, 277–278, 280, 308, 315–316
Pressure–viscosity index, 131
Profile peak height, 328, 330, 333

Profile valley depth, 330, 333
Pumping power loss, 226–227

Rabinowicz's equation, 75, 77, 99–101
Radial aerostatic bearings, 313–316
Radial ball bearing, 39
Raimondi and Boyd method, 159–173, 177
Rectangular plates, 231–232
Rectangular slot, 210–213
Rectangular thrust bearing, 219–222
Restrictor, 213–214, 223, 251, 309–318
Restrictor pressure, 310, 312–316
Reynolds boundary condition, 150, 154–155
Reynolds' equation, 133–138, 154–156, 159, 189, 197, 209, 219–220, 227, 234, 262, 269–274, 300–302, 307, 315
Roelands equation, 131–132
Rolling bearings, 21, 27, 40–41, 83, 89–90, 282, 371, 374–377, 382–386, 388
Root mean square, 17–18, 330, 335, 337
Root mean square slope, 335
Root mean square surface roughness, 328
Rowe's wear equation, 101

Scoring wear, 78
Scratching, 74–75, 404, 412
Seal dependent friction moment, 385
Seals, 4–5, 9–10, 13, 23, 28, 37, 72, 282, 384–385
Sector pad mesh, 198
Self-adjusting separation, 210
Semi-solid lubricant, 19, 26–29
Shear strain rate, 14
Shear stress, 14, 59, 62–63, 93–94, 96, 125, 129–130, 134, 141, 202, 211, 381, 386, 390–391, 401
Shear thinning, 130, 228
Short static bearing, 173
Silicone, 6, 27–28
Skewness parameter, 334
Sled tribometer, 66
Sliding bearings, 40–41, 120, 326, 371
Solid lubricant, 19–27, 30, 38, 72, 120, 300, 383, 407–408, 410, 412
Solid mechanics, 1–2, 131
Sommerfeld, 146–150, 152–154, 158, 160–161, 163, 168, 174, 236, 239, 327
Specific film thickness, 17, 326–327, 329, 346, 351, 401–403
Spherical rolling element, 41, 371–372, 374, 376, 379–380
Spherical shape, 60
Spur gear, 107, 109, 284, 342, 400–401, 404–406, 413, 418–419

Spur gears lubrication, 410–412
Squeeze film, 121–122, 138
Squeeze film lubrication, 209–249
Stachowiak & Batchelor, 126
Static friction, 1, 53–55, 68–69
Stick-slip motion, 68–72
Stribeck diagram, 127
Stylus method, 327, 330–333
Surface failures, 14, 407, 412
Surface films, 64, 74, 79, 82, 98–99, 101, 402
Surface roughness, 2, 5, 17–18, 21, 160, 326–332, 334–335, 337–338, 350–351, 356, 401, 403–405, 409–410
Surface topography, 4, 330–336, 350
Swift-Stieber boundary condition, 155
Synthetic oils, 33–34, 131–132
System power loss, 226–227

Taper roller bearing, 40–41, 281, 373–376, 383
Tapered rolling element, 371, 373, 377, 384
Theories of dry friction, 57–65
Thermal thickening, 297
Thermal thinning, 298
Thickener, 26–27
Three body abrasion, 75–77
Thrust, 187–206
Thrust aerostatic bearings, 313, 316–318
Tilting pad thrust bearing, 188, 200–201
Tomlinson's theory of molecular attraction, 57
Total acid number, 19, 37, 418
Total base number, 19, 36, 418
Towers experiments, 123–124
Tribology
    definition, 1
    design, 4–9
    history, 3–4
    industry, 10–13
Tribotester, 67
Two body abrasion, 75–77
Tyre-road Interactions, 371, 392–394

Universal tribometer, 67

Variable and alternating loads, 232–233
Vegetable oils, 30
Velocity slip, 300
VI improvers, 32
Vibration monitoring, 418
Viscosity, 10–11, 14, 16, 19, 29–30, 32–34, 39, 77, 105, 121–132, 136, 138, 141, 156, 158–163, 175, 188, 192, 196, 198, 202, 211, 214, 223, 227, 236, 241, 270–271, 274, 277, 280, 283, 297–298, 300, 302, 350, 355, 383, 385, 387, 411, 418

Viscosity Index, 31–33, 126–129, 131, 410
Viscosity-pressure, 131–132, 261
Viscosity-shear rate, 129–130
Viscosity-temperature relation, 126–129, 175
Viscous friction, 1, 227
Vogel's equation, 126, 175
Volume of flow, 212
Volumetric flow rate, 161, 165, 168, 218, 221, 309, 316–317

Walther's equation, 126, 175, 242, 244–245
Wear, 52–110
Wear theory, 99–102

Wear analysis, 92–98
Wear classification, 74–92
Wear coefficient, 405–406
Wear debris, 18, 28, 75, 84, 91, 342, 412–419
Wear prevention, 102–103
Wear rate, 4, 10, 13–14, 21, 25–26, 65, 72, 74–75, 77–78, 82, 85–88, 90, 92, 98–102, 107–110, 326, 341–343, 345, 354, 400, 404–406, 413, 415
Wear-mechanism, 74–75, 77, 86
Weddle's rule, 174
Window lifting mechanism, 14

Zero wear, 72–73, 153, 210